DISTILLATION PRINCIPLES
AND PROCESSES

MACMILLAN AND CO., Limited
LONDON · BOMBAY · CALCUTTA · MADRAS
MELBOURNE

THE MACMILLAN COMPANY
NEW YORK · BOSTON · CHICAGO
DALLAS · SAN FRANCISCO

THE MACMILLAN CO. OF CANADA, Ltd.
TORONTO

DISTILLATION PRINCIPLES AND PROCESSES

BY

SYDNEY YOUNG

M.A., D.Sc., F.R.S.

PROFESSOR OF CHEMISTRY IN DUBLIN UNIVERSITY

WITH THE COLLABORATION OF

Lieut.-Col. E. BRIGGS, D.S.O., B.Sc.

T. HOWARD BUTLER, Ph.D., M.Sc., F.I.C.

THOS. H. DURRANS, M.Sc., F.I.C.

The Hon. F. R. HENLEY, M.A., F.I.C.

JOSEPH REILLY, M.A., D.Sc., F.R.C.Sc.I., F.I.C.

WITH TWO HUNDRED AND TEN ILLUSTRATIONS

MACMILLAN AND CO., LIMITED

ST. MARTIN'S STREET, LONDON

1922

PRINTED IN GREAT BRITAIN

PREFACE

THE volume on " Fractional Distillation " was written in the hope that it would be of assistance to chemists in overcoming the difficulties so frequently met with in the laboratory, not only in the actual carrying out of the fractional distillation of a complex mixture but also in the interpretation of the results obtained.

The last copy was sold shortly after the declaration of war, and the question then arose whether a revised second edition of the book in its original form should be published. It was thought, however, by the Publishers that it would be wiser to extend the scope of the work so as to include distillation on the large scale as carried out in the manufacture of important products.

That this change in the character of the book was really advisable became more and more evident as the war increased in intensity and magnitude, because of the immense importance of such materials as acetone, toluene, petrol, glycerine, and so on, the production of which involved processes of distillation on an enormous scale, and for some of which new sources or methods of formation had to be sought and investigated.

It was obvious that the larger book could only be of real value if the sections on manufacturing processes were written by chemists thoroughly conversant with the various subjects dealt with, and it was eventually decided that I should edit the book if I could secure the co-operation of experts in the different branches of manufacture.

In these preliminary negotiations I received most valuable assistance from my friends, Professor F. E. Francis, D.Sc., Ph.D., and Professor W. E. Adeney, D.Sc., and my sincere thanks are especially due to Professor Francis for the great interest he has taken in the production of the book.

Unfortunately the heavy pressure of work entailed on all chemists by the war caused serious delay in commencing the book, and progress has necessarily been slow.

The new work consists of seven sections, the first of which is practically a revision of "Fractional Distillation." In this section full reference is made to the valuable researches of Wade, Merriman, and Finnemore, of Rosanoff and his co-workers, and of Lecat; there is also an additional chapter on Sublimation.

The remaining sections, dealing with manufacturing processes, are as follows :

2. Distillation of Acetone and n-Butyl Alcohol on the Manufacturing Scale, by Joseph Reilly, M.A., D.Sc., F.R.C.Sc.I., F.I.C., Chemist-in-charge at the Royal Naval Factory, Holton Heath, Dorset, and the Hon. F. R. Henley, M.A., F.I.C.

3. Distillation of Alcohol on the Manufacturing Scale, by the Hon. F. R. Henley and Dr. Reilly.

4. Fractional Distillation as applied in the Petroleum Industry, by James Kewley, M.A., F.I.C., Member of Council of the Institution of Petroleum Technologists.

5. Fractional Distillation in the Coal Tar Industry, by T. Howard Butler, Ph.D., M.Sc., F.I.C., Managing Director of William Butler & Co. (Bristol), Ltd., Tar, Rosin, and Petroleum Distillers.

6. The Distillation of Glycerine, by Lieut.-Col. E. Briggs, D.S.O., B.Sc., Technical Director, Broad Plain Soap Works, Bristol.

7. The Distillation of Essential Oils, by Thos. H. Durrans, M.Sc. (London), F.I.C., of Messrs. A. Boake Roberts & Co., Ltd., London.

S. Y.

DUBLIN, *August* 1921.

PREFACE

TO " FRACTIONAL DISTILLATION "

DURING the past eighteen years I have been engaged in investigations which necessitated the preparation of chemical materials in the purest possible state, and as the great majority of these substances were liquids, the process of fractional distillation had, in most cases, to be resorted to for their purification.

The difficulties I met with in some of the separations led me to make a careful investigation of the whole subject, and I was thus enabled to devise some new methods and forms of apparatus, which have been described from time to time in various scientific journals.

It is in the hope that the solution of the difficulties which so often occur in carrying out a fractional distillation may be rendered easier, and that the value and economy of highly efficient still-heads in laboratory work may come to be more widely recognised than is generally the case at present, that this book has been written.

My sincere thanks are due to Professor J. Campbell Brown for the loan of valuable ancient works by Libavius and Ulstadius, from which Figures 2 and 34 [1] have been taken ; to my colleague, Dr. F. E. Francis, for reading the proofs, and for much valuable assistance in compiling the index ; and to Professor R. A. Gregory and Mr. A. T. Simmons, B.Sc., for many useful suggestions regarding the arrangement of the MS. and for the perusal of the proofs.

In the description and illustration of the stills employed in commerce I have derived much assistance from articles in Thorpe's " Dictionary of Applied Chemistry " and Payen's " Précis de chimie industrielle."

I have made much use of the experimental data of Brown, Konowaloff, Lehfeldt, Zawidski, and other observers, and have, as far as possible, made due acknowledgment, but in some of the tables this has not been practicable.

Several fractional distillations and numerous experiments have been carried out while the book was being written and the results have in many cases not been published elsewhere.

S. Y.

BRISTOL, *August* 1903.

[1] Now Fig. 41.

vii

CONTENTS

DISTILLATION PRINCIPLES AND PROCESSES

ix

CHAPTER XX

DISTILLATION OF ACETONE AND *n*-BUTYL ALCOHOL ON THE MANUFACTURING SCALE

CHAPTER XXI

CHAPTER XXII

CHAPTER XXIII

CHAPTER XXIV

CHAPTER XXV

DISTILLATION OF ALCOHOL ON THE MANUFACTURING SCALE

CHAPTER XXVI

CHAPTER XXVII

CHAPTER XXVIII

FRACTIONAL DISTILLATION AS APPLIED IN THE PETROLEUM INDUSTRY

FRACTIONAL DISTILLATION IN THE COAL TAR INDUSTRY

FRACTIONAL SEPARATION OF THE NAPHTHAS AND LIGHT OILS . . 392

THE DISTILLATION OF GLYCERINE

CHAPTER XXXIX

THE DISTILLATION OF GLYCERINE 425

THE DISTILLATION OF ESSENTIAL OILS

CHAPTER XL

THEORETICAL, STEAM DISTILLATION 443

CHAPTER XLI

TECHNICAL, PREPARATION OF RAW MATERIALS 450

CHAPTER XLII

DISTILLATION 455

CHAPTER XLIII

PURIFICATION OF ESSENTIAL OILS 475

APPENDIX 487

INDEX OF NAMES 491

INDEX OF SUBJECTS 497

DISTILLATION PRINCIPLES AND PROCESSES

CHAPTER I

Object of Distillation.—The object of distillation is the separation of a volatile liquid from a non-volatile substance or, more frequently, the separation of two or more liquids of different volatility.

If only one component of a mixture is volatile, there is no difficulty in obtaining it in a pure state by distillation, and in many cases the constituents of a mixture of two or more volatile liquids may be separated—though frequently at much cost of time and material—by means of the simple apparatus described in this chapter. For the fractional distillation in the laboratory of such complex mixtures as petroleum or fusel oil, the improved still-heads described in Chapters X. to XII. must be employed. Still-heads employed on the large scale are described in the sections of the book relating to alcohol, petroleum, etc.

Quantitative Analysis by Distillation.—The determination, by ordinary analytical methods, of the relative quantities of two or more organic compounds in a mixture is often a matter of great difficulty, but, in many cases, the composition of the mixture may be ascertained approximately and, not seldom, with considerable accuracy from the results of a single distillation, if a very efficient still-head be employed. This method has proved of considerable value.

Difficulties Encountered.—The subject of fractional distillation is full of interest owing to the fact that difficulties so frequently occur, not only in the experimental work, but also in interpreting the results obtained.

In the distillation of petroleum, with the object of separating pure substances, such difficulties are of common occurrence and are due to one or other of three causes :—(a) to the presence of two substances, the boiling points of which are very close together ; (b) to the presence of one or more components in relatively very small quantity ; (c) to the formation of mixtures of constant boiling point.

The separation of two liquids which boil at temperatures even 20° or 30° apart, such as ethyl alcohol and water, or benzene and isobutyl alcohol, may be impossible owing to the formation of a mixture of minimum or, less frequently, of maximum boiling point. It is, indeed, only in the case of substances which are chemically closely related to

3

each other that the statement can be definitely made that the difficulty of separating the components of a mixture diminishes as the difference between their boiling points increases.

In any other case, we must consider the relation between the boiling points, or the vapour pressures, of mixtures of the substances and their composition, and unless something is known of the form of the curve representing one or other of these relations, it is impossible to predict whether the separation will be an easy one or, indeed, whether it will be possible.

The form of these curves depends largely on the chemical relationship of the components, and it is now possible, in a moderate number of cases, to form an estimate, from the chemical constitution of the substances, of the extent to which the curves would deviate from the normal form, and therefore to predict the behaviour of a mixture on distillation.

Fractional distillation is frequently a very tedious process and there is necessarily considerable loss of material by evaporation and by repeated transference from the receivers to the still, but a great amount of both time and material may be saved by the use of a very efficient still-head ; and when the object of the distillation is to ascertain the composition of a mixture, very much greater accuracy is thereby attained.

APPARATUS

Ancient Apparatus.—The process of distillation is evidently a very ancient one, for Aristotle [1] mentions that pure water may be obtained from sea-water by evaporation, but he does not explain how

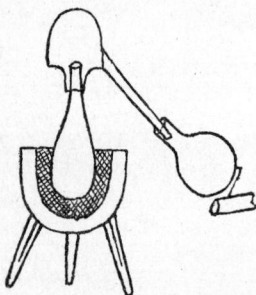

FIG. 1.—Alexandrian still with head, or *alembic*.

FIG. 2.—Ancient still with water condenser.

the condensation of the vapour can be effected. A primitive method of condensation is described by Dioscorides and by Pliny, who state that an oil may be obtained by heating rosin in a vessel, in the upper part of which is placed some wool. The oil condenses in the wool and can be squeezed out of it.

The Alexandrian chemists added a second vessel, the head or cover,

---

[1] Kopp, *Geschichte der Chemie*, Beiträge i, 217.

called by the Arabians the *alembic*, to the boiler or still, and a simple form of apparatus used by them is shown in Fig. 1.

Later on, the side tube was cooled by passing it through a vessel containing water. The diagram, Fig. 2, is taken from Libavius, *Syntagma Alchymiae Arcanorum*, 1611.

Modern Apparatus.—The apparatus employed at the present time is similar in principle, but, in addition, a thermometer is used to register the temperature. In Fig. 3 the ordinary form of apparatus is shown, and we may distinguish the following parts :—The still, A ; the still-head, B ; the Liebig's condenser, C, in which the vapour is deprived of heat by a current of cold water ; the receiver, D ; the thermometer, E. In the laboratory the still is usually heated by means of a Bunsen burner.

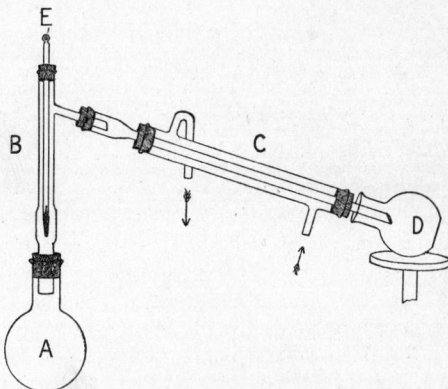

The flask or still is fitted with a cork through which passes the still-head, and the side de-

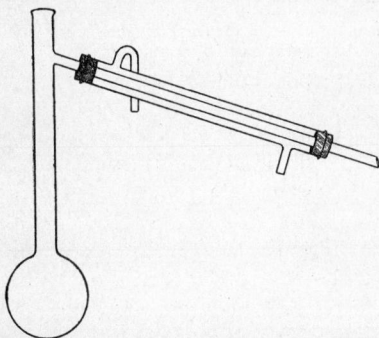

FIG. 3.—Ordinary still, with Liebig's condenser.

livery tube from the still-head passes through a second cork in the condensing tube. For liquids which boil at a high temperature, or which act chemically on cork, it is more convenient to have the still and still-head in one piece and to elongate the delivery tube so that it may pass, if necessary, through the Liebig's condenser (Fig. 4).

The Still.—If a glass flask is used it should be globular, because a flat-bottomed flask is

FIG. 4.—Modified form of still with condenser.

liable to crack when heated with a naked flame. It should not be larger than is necessary for the amount of liquid to be distilled.

The Still-head.—The still-head should not be very narrow, or the thermometer may be cooled slightly below the temperature of the vapour. It is a good plan to seal a short length of wider tubing to the still-head near the bottom, leaving a sufficient length of the narrower tubing below to pass through the cork in the still, as shown in Fig. 3.

The still-head, as supplied by dealers, is often too short. It should, if possible, be long enough for the thermometer to be placed in such a position that not only the mercury in the bulb but also that in the stem is heated by the vapour of the boiling liquid ; otherwise a troublesome and somewhat uncertain correction must be applied (p. 11), and, if the distillation is not proceeding quite steadily, a little air may be carried back from time to time as far as the thermometer bulb and the temperature registered by the thermometer will then fluctuate and will, on the whole, be too low (p. 25).

The longer and wider the still-head and the higher the boiling point of the liquid distilled, the greater will be the amount of con- densed liquid flowing back to the still. The lower end of the still- head should be wide enough to ensure that no priming takes place. With the bottom ground obliquely as in Fig. 6 a much narrower tube may be used than when the end is cut off horizontally (Fig. 3).

The Condenser.—If the boiling point of the liquid to be distilled is higher than about 170°, the condensing tube should not be cooled by running water for fear of fracture. A long tube should be used and the cooling effect of the surrounding air will then be sufficient.

When a Liebig's condenser is used there is no advantage in having either the inner or the outer tube very wide ; an internal diameter of 7 or 8 mm. is sufficient for the inner, and of 15 mm. for the outer tube. If the outer tube is much wider it is unwieldy, and, when filled with water, it is inconveniently heavy. A mistake that is rather frequently made may be referred to here. It is usual to seal a short wide tube to the long, narrow condensing tube for the insertion of the delivery tube from the still-head. The tubes are often sealed together in such a way

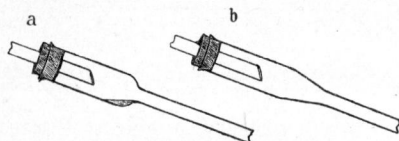

FIG. 5.—Condensing tube of (*a*) faulty, (*b*) correct construction.

that when the distillation is proceeding a little pool of liquid collects at the junction (Fig. 5, *a*), and, in the fractional distillation of a small quantity of liquid, the error thus intro- duced may be serious. The fault is easily remedied by heating the wide tube close to the junction with the narrow one until the glass is soft, and then drawing it out very gently until it has the form shown in Fig. 5, *b*.

When a long still-head is used, it is advisable to bend the narrow tube just below its junction with the wider one, so that the condenser may be vertical in position instead of sloping gently downwards. Much less space is thus taken up on the laboratory bench, and the receivers are somewhat more conveniently manipulated.

The Source of Heat.—For laboratory purposes an ordinary Bunsen burner is usually employed. Wire gauze, asbestos cardboard, sand baths, or water or oil baths are not, as a rule, to be recommended because the supply of heat can be much more easily regulated without them, and a round-bottomed flask, if properly blown, is so thin walled that there is no danger of fracture when the naked flame is applied. The flask should be so placed that the flame actually comes in contact with the bottom of it; this is especially necessary when the liquid to be distilled is liable to "bump." Many substances, such as carbon disulphide, which boil quite regularly under the ordinary atmospheric pressure, bump more or less violently when the pressure is greatly reduced, unless special precautions are taken. Under a pressure of 361 mm. carbon disulphide boils at 25°, and if a quantity of it be distilled under this pressure with the flame placed some distance below the bottom of the flask, it may happen that the whole of the carbon disulphide will pass over without any ebullition whatever taking place. The liquid, however, in these circumstances, becomes considerably superheated, and if a bubble does form there will be a sudden and perhaps violent rush of the extremely inflammable vapour. If, however, the top of the burner be placed only about 2 mm. below the bottom of the flask, so that the minute flame touches the glass, ebullition will take place quietly and regularly.

There are liquids which cannot be prevented from bumping in this way, and the best plan is then to add a few small fragments of porous porcelain [a clay pipe broken in small pieces answers the purpose very well] or pumice-stone, or both, or a number of small tetrahedra of silver or platinum. A method frequently employed is to pass a very slow current of air through the liquid, but a small error in the boiling point is thereby introduced. The explanation of this is given on p. 24. A suitable flask, described by Wade and Merriman,[1] is shown in Fig. 6. A water or oil bath need only be used when a solid substance is present in the flask, as, for instance, when a liquid is distilled over lime or phosphorus pentoxide, or when the liquid is liable to decompose when heated with the naked flame.

It is customary to employ a water-bath for the distillation of ether, but it is doubtful whether this is necessary or even advisable except in the case of the ethereal solution of

FIG. 6.

a solid substance or one that will not bear heating much above 100°. When an accident occurs it is almost invariably because, owing to "bumping," or to the distillation being carried on too rapidly, some of the vapour escapes condensation and comes in contact with a flame in the neighbourhood, generally that below the water-bath. If a naked flame were used the distillation could be much more easily regulated, and there would probably be really less danger than if a water-bath were employed.

Protection of Flame from Draughts.—In order that satisfactory results may be obtained it is necessary that the distillation should proceed with great regularity, and the heat supply must therefore not be subject to fluctuations. The most important point is to guard

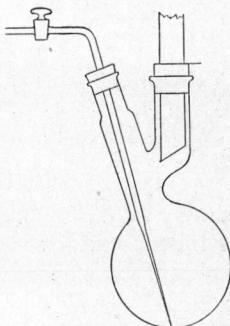

[1] " Apparatus for Fractional Distillation at Pressures other than the Atmospheric Pressure," *Trans. Chem. Soc.*, 1911, **99**, 994.

against draughts, and, to do this, the ordinary conical flame protector may be used, or a simple and efficient guard may be made from a large beaker by cutting off the bottom and taking a piece out of the side (Fig. 7).

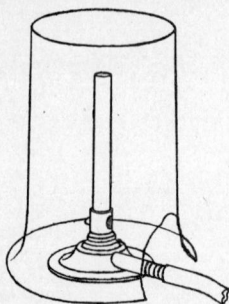

Electrical Heating. — For many purposes, notably for distillation under reduced pressure, it is convenient to employ an electrically heated coil of platinum wire as the source of heat. T. W. Richards and J. H. Matthews [1] strongly recommend this method of heating, and consider that electrical heating gives slightly better separation and far less superheating than ordinary flame heating.

FIG. 7.
Simple flame protector.

In order that the fine platinum wire may still be completely immersed when the quantity of liquid in the flask has become very small, Richards and Matthews recommend a vessel of the form shown in Fig. 8. The coil actually employed consisted of about 40 cm. of platinum wire and had a resistance of 0·7 ohm. A current of ten to fifteen amperes was passed through the coil, the ends of which were sealed into two glass tubes in which were stout copper wires. Connection between the platinum and copper wires was effected by means of a little mercury.

A similar method was recommended by Beckmann.[2]

H. S. Bailey [3] makes use of a flask which is so narrow at the bottom that 90 per cent of the liquid can be distilled with the coil still completely immersed. Bailey recommends a coil of German silver or nichrome wire instead of platinum.

Rosanoff and Easley [4] wind the heating wire on a glass rod bent in the form of a conical spiral, the apex of which extends almost to the bottom of the vessel.

W. R. G. Atkins finds it convenient for many purposes to employ an electrically heated metal plate, placed a little distance

FIG. 8.

below the flask. The amount of heat reaching the flask is regulated by moving a piece of asbestos pasteboard over the heated plate. Superheating of the vapour, when the amount of liquid has become

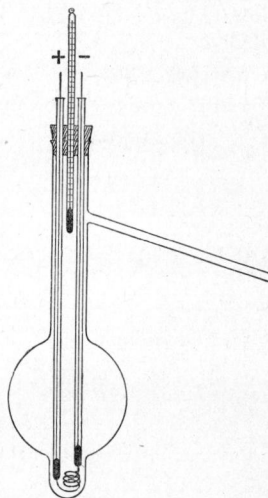

[1] Richards and Matthews, "Electrical Heating in Fractional Distillation," *J. Amer. Chem. Soc.*, 1908, **30**, 1282 ; 1909, **31**, 1200 ; *Zeitschr. physik. Chem.*, 1908, **64**, 120.

[2] Beckmann, "Erfahrungen über elektrisches Heizen bei ebullioskopischen Bestimmungen und bei der fraktionierten Destillation," *Zeitschr. physik. Chem.*, 1908, **64**, 506.

[3] Bailey, "An Electrically Heated Vacuum Fractionation Apparatus," *J. Amer. Chem. Soc.*, 1911, **33**, 447.

[4] Rosanoff and Easley, "Partial Pressures of Binary Mixtures (Apparatus)," *ibid.*, 1909, **31**, 964.

small, can be prevented by allowing the flask to rest on a sheet of asbestos pasteboard with a circular hole cut in the centre, so that the heat only reaches a small area at the bottom of the flask.

Allen and Jacobs [1] encase the distillation flask in the two halves of a pear-shaped mould on the inner side of which is wound a resistance wire which is suitably heated electrically.

Steam as Source of Heat. — On the large scale the still is frequently heated by steam under ordinary or increased pressure (Fig. 9). The steam may be introduced through the pipe A, and the condensed water run off at B (see also pp. 331, 398).

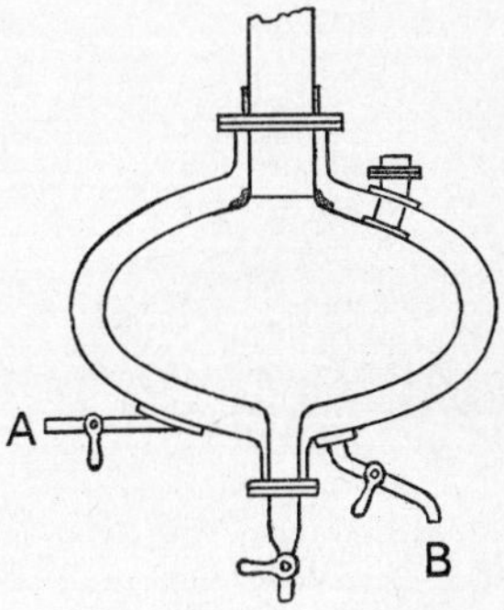

The Thermometer.—In carrying out a fractional distillation one must be able, not only to read a constant or nearly constant temperature with great accuracy, but also to take readings of rapidly rising temperatures. These requirements are best fulfilled by the ordinary mercurial thermometer, which is therefore, notwithstanding its many drawbacks, used in preference to the air or the platinum resistance thermometer. If accurate results are to be obtained the following points must be attended to.

FIG. 9.—Still with steam jacket.

1. **Calibration.**—The thermometer must be carefully calibrated, and it would be a great advantage if all thermometers were compared with an air thermometer, for two mercurial thermometers, constructed of different varieties of glass, even if correct at 0° and 100°, will give different and incorrect readings at other temperatures, more especially at high ones, for various reasons :

(a) In the first place, it is impossible to obtain an absolutely cylindrical capillary tube, and therefore the volume corresponding to a scale division cannot be quite the same in all parts of the tube. Various methods have been devised for calibrating the stem,[2][3] but even when this is done there remain other sources of error.

(b) The position of the mercury in the stem at any temperature depends on the expansion both of the mercury and the glass, and, for both substances, the rate of expansion increases with rise of temperature.

(c) Different kinds of glass have different rates of expansion, so that two thermometers made of different materials—even if the capillary tubes were perfectly cylindrical—would give different readings at the same temperature. It is therefore necessary to compare the readings of a mercurial thermometer with those of an air thermometer, or of another mercurial thermometer which has previously been standardised

[1] Allen and Jacobs, "Electrically Heated Still for Fractional Distillation," *Dept. of Inter., Bur. of Mines, U.S.A., Bull. 19*, 1 ; *J. Soc. Chem. Ind.*, 1912, **31**, 18.

[2] "Methods employed in Calibration of Mercurial Thermometers," *British Association Report for 1882*, 145.

[3] Guillaume, "Traité pratique de la thermométrie de précision," p. 112.

by means of an air thermometer. Or, instead of this, a number of fixed points may be determined by heating the thermometer with the vapours of a series of pure liquids boiling under known pressures.

Table 1 contains a list of suitable substances with their boiling points, and the variation of temperature for a difference of 10 mm. from the normal atmospheric pressure.

TABLE 1

Substance.	Boiling point under normal pressure.	Variation of temperature per 10 mm. pressure.
Carbon disulphide	46·25°	0·40°
Ethyl alcohol	78·3	0·33
Water	100·0	0·37
Chlorobenzene	132·0	0·50
Bromobenzene	156·0	0·51
Aniline	184·4	0·51
Naphthalene	218·05	0·58
Quinoline	237·45	0·59
Bromonaphthalene	280·45	0·64
Benzophenone	305·8	0·63
Mercury	356·75	0·75
Sulphur	444·55	0·87

In this way a table, or curve, of corrections may be constructed, and the error at any scale reading of the thermometer may be easily ascertained.

"Normal" thermometers may now be purchased; they are compared with a standard thermometer before graduation, and true temperatures are said to be registered by them.

2. **Redetermination of Zero Point.**—The zero point of a thermometer should be redetermined from time to time, as it is subject to changes which, in the case of the cheap soda glass thermometers, may be considerable. These changes are of two kinds :—

(a) If a thermometer be graduated shortly after the bulb has been blown, the zero point will be found to rise, at first with comparative rapidity, then more and more slowly, and the elevation of the zero point may go on for many years. If the thermometer be kept at a high temperature—especially, as shown by Marchis, if there are periodical, slight fluctuations of temperature—the rise of the zero point takes place with much greater rapidity, and up to, at any rate, 360°, and probably 450°, the higher the temperature the more rapid is the rise and, apparently, the higher is the final point reached. A rise of more than 20° has several times been observed in the case of soft German glass thermometers on being subjected to prolonged heating at 360°. In all cases the rise, which is rapid at first, becomes slower and slower, and it seems doubtful whether, at any given temperature, actual constancy of zero point has ever yet been attained. If, however, a thermometer has been heated for many hours to a given high temperature and then allowed to cool very slowly, subsequent heating to lower temperatures has very little effect on the zero point. The best thermometers, as first recommended by Crafts, are kept at a high temperature for a long time before being graduated.

(*b*) If a thermometer—even after its zero point has been rendered as constant as possible—be heated and then cooled very rapidly, a slight fall of zero point will be observed ; but after a day or two the greater part of this fall will be recovered, and the remainder after a long period.

3. Volatilisation of Mercury in Stem of Thermometer.—In the cheaper thermometers there is a vacuum above the mercury and, when the mercury in the stem is strongly heated, volatilisation takes place, the vapour condensing in the cold, upper part of the tube ; when, therefore, the temperature is really constant it appears to be gradually falling. The better thermometers, which are graduated up to high temperatures, contain nitrogen over the mercury, a bulb being blown near the top of the capillary tube to prevent too great a rise of pressure by the compression of the gas ; but thermometers which are only required for moderate temperatures, say, not higher than 100° or even 150°, are not usually filled with nitrogen. If, however, such thermometers are used for the distillation of liquids boiling at so low a temperature as 100°, or even 80°, a quite perceptible amount of mercury may volatilise and, after prolonged heating, errors amounting to 0·2° or 0·3° may occur. It would be much better if all thermometers required to register temperatures higher than 60° were filled with nitrogen.

4. Correction for Unheated Column of Mercury.—As already mentioned, the thermometer should, if possible, be so placed in the apparatus that not only the mercury in the bulb but also that in the stem is heated by the vapour of the boiling liquid ; otherwise the following correction, which, at the best, is somewhat uncertain, must be applied :—

To the temperature read, add $0·000143(T - t)N$, where T is the observed boiling point, t the temperature of the stem above the vapour, and N the length of the mercury column not heated by the vapour, expressed in scale divisions.

The coefficient 0·00016—the difference between the cubical expansion of mercury and that of glass—is very frequently employed, but it is found in practice to be too high ; and Thorpe has shown that the value 0·000143 gives better results.

Table 2, on page 12, given by Thorpe [1] may be found useful.

[1] Thorpe, " On the Relation between the Molecular Weights of Substances and their Specific Gravities when in the Liquid State," *Trans. Chem. Soc.*, 1880, **37**, 159.

[TABLE

TABLE 2

$T-t$	$N.$ 10.	20.	30.	40.	50.	60.	70.	80.	90.	100.	110.	120.	130.	140.	150.	160.	170.	180.	190.	200.
10	0·01	0·03	0·04	0·06	0·07	0·09	0·10	0·11	0·13	0·14	0·16	0·17	0·19	0·20	0·21	0·22	0·24	0·26	0·27	0·29
20	0·02	0·06	0·09	0·11	0·14	0·17	0·20	0·22	0·26	0·29	0·31	0·34	0·37	0·40	0·43	0·46	0·49	0·51	0·54	0·57
30	0·04	0·09	0·13	0·17	0·21	0·26	0·30	0·34	0·39	0·43	0·47	0·51	0·56	0·60	0·64	0·68	0·73	0·77	0·82	0·86
40	0·05	0·11	0·17	0·23	0·28	0·34	0·40	0·47	0·52	0·57	0·63	0·69	0·74	0·80	0·86	0·91	0·97	1·03	1·09	1·14
50	0·07	0·14	0·21	0·29	0·36	0·43	0·50	0·60	0·64	0·71	0·79	0·86	0·93	1·00	1·07	1·14	1·22	1·29	1·36	1·43
60	0·08	0·17	0·25	0·35	0·43	0·51	0·60	0·70	0·77	0·86	0·94	1·03	1·12	1·20	1·29	1·37	1·46	1·54	1·63	1·72
70	0·10	0·20	0·30	0·40	0·50	0·60	0·70	0·80	0·90	1·00	1·10	1·20	1·30	1·40	1·50	1·60	1·70	1·80	1·90	2·00
80	0·11	0·23	0·34	0·45	0·57	0·68	0·80	0·91	1·03	1·14	1·26	1·37	1·49	1·60	1·72	1·83	1·94	2·05	2·17	2·29
90	0·13	0·26	0·39	0·51	0·64	0·77	0·90	1·03	1·16	1·30	1·42	1·54	1·66	1·80	1·93	2·05	2·17	2·31	2·45	2·54
100	0·14	0·28	0·43	0·57	0·71	0·85	1·00	1·14	1·29	1·43	1·58	1·71	1·84	2·00	2·15	2·29	2·43	2·57	2·72	2·86
110	0·16	0·31	0·47	0·63	0·79	0·94	1·10	1·26	1·42	1·58	1·73	1·89	2·04	2·20	2·36	2·51	2·67	2·83	2·99	3·15
120	0·17	0·34	0·51	0·69	0·86	1·03	1·20	1·37	1·54	1·71	1·89	2·06	2·23	2·40	2·57	2·74	2·92	3·09	3·26	3·43

5. Superheating of Vapour.—When the amount of liquid in the still is very small, the vapour is liable to be superheated by the flame, and unless the bulb of the thermometer is thoroughly moistened with condensed liquid, too high a temperature will be registered. If a very little cotton wool, or, for temperatures above 230°, a little fibrous asbestos, be wrapped round the bulb of the thermometer, it remains, as a rule, thoroughly moist, and, with a pure liquid, heated by a naked flame, the thermometer registers a perfectly constant temperature until the last trace of liquid in the bulb has disappeared.

With a water or oil bath the danger of superheating is greater, and the cotton wool may become dry at the end of the distillation. In that case the temperature registered may be too high, though, as a rule, the error is not so great as it would be if the bulb were not protected.

If it is of special importance to determine the boiling point of a liquid with great accuracy during the course of a distillation, one of the two forms of still devised by Richards and Barry [1] may be employed with advantage.

6. Correction of Boiling Point for Pressure.—The barometer must always be read and corrected to 0° (p. 229) and, in a long distillation or in unsettled weather, it may be necessary to read it frequently, for the boiling point of a liquid varies greatly with the pressure.

It is impossible to give any accurate and generally applicable formula for correcting the observed boiling point to that under normal pressure (760 mm.), but the following may be taken as approximately correct :—

$$\theta = 0 \cdot 00012 (760 - p)(273 + t),$$

where θ is the correction in centigrade degrees to be added to the observed boiling point, t, and p is the barometric pressure.[2]

This correction is applicable, without much error, to the majority of liquids, but for water and the alcohols a better result is given by the formula

$$\theta = 0 \cdot 00010 (760 - p)(273 + t).$$

Crafts [3] has collected together the data for a number of substances, from which the values of c in the formula $\theta = c\ (760 - p)\ (273 + t)$ are easily obtained.[4]   Table 3 (p. 14) contains the boiling points on the absolute scale, T, the values of c and those of dp/dt for some of the substances referred to by Crafts and also for a considerable number of additional ones.[5]

[1] " An Advantageous Form of Still for the Exact Measurement of Boiling Point during Fractional Distillation," *J. Amer. Chem. Soc.*, 1914, **36**, 1787.

[2] Ramsay and Young, "Some Thermodynamical Relations," *Phil. Mag.*, 1885, [V.], **20**, 515.

[3] Crafts, "On the Correction of the Boiling Point for Barometric Variations," *Berl. Berichte*, 1887, **20**, 709.

[4] There are a few misprints in the table given by Crafts, and since it was published many additional accurate determinations of boiling point and vapour pressure have been made.

[5] Young, "Correction of the Boiling Points of Liquids from Observed to Normal Pressure," *Trans. Chem. Soc.*, 1902, **81**, 777.

The values marked with an asterisk have been determined indirectly, and are not to be regarded as so well established as the others. When a boiling point is to be corrected, the constant c for the substance may usually be found by reference to Table 3. Either the constant for that substance in the table most closely related to the one under examination is to be used, or the constant may be altered in conformity with one or other of the following generalisations.

TABLE 3

Substance.	T.	dp/dt.	c.	Substance.	T.	dp/dt.	c.
Oxygen . .	90·5°	75·9*	0·000146*	Iodobenzene .	461·45°	18·0	0·000120
Nitrogen . .	77·5	89·0*	0·000145*	Bromonaphthal-			
Argon . . .	86·9	83·2*	0·000138*	ene . .	553·45	15·75	0·000115
Chlorine . .	239·4	33·2	0·000126	Methyl ether .	249·4	32·0	0·000125
Bromine . .	331·75	25·2	0·000120	Ethyl ether .	307·6	26·9	0·000121
Iodine . . .	458·3	18·75	0·000116	Acetone . .	330·0	26·4	0·000115
Mercury . .	629·75	13·4	0·000118	Benzophenone .	578·8	15·8	0·000109
Sulphur . .	717·55	12·2	0·000114	Anthraquinone .	650·0	13·6	0·000113
Ammonia . .	240·1	37·7	0·000110	Aniline . .	457·4	19·6	0·000112
Sulphur dioxide	262·9	33·7	0·000113	Quinoline . .	510·45	17·0	0·000115
Carbon disul-				Methyl formate .	304·9	28·8	0·000114
phide . .	319·25	24·7	0·000127	Ethyl formate .	327·3	26·6	0·000115
Boron trichloride	291·25	26·8	0·000128	Propyl formate .	353·9	24·5	0·000115
Phosphorus tri-				Methyl acetate .	330·1	26·8	0·000113
chloride .	346·85	23·45	0·000123	Ethyl acetate .	350·15	25·1	0·000114
Carbon tetra-				Propyl acetate .	374·55	23·5	0·000114
chloride . .	349·75	23·25	0·000123	Isobutyl acetate	389·2	22·5	0·000114
Silicon tetra-				Methyl propion-			
chloride .	329·9	24·0	0·000126	ate . . .	352·7	24·9	0·000114
Stannic chloride	387·1	21·4	0·000121	Ethyl propionate	372·0	23·7	0·000113
Methane . .	108·3	68·2*	0·000135	Propyl propion-			
n-Pentane . .	309·3	25·8	0·000125*	ate . . .	395·15	22·3	0·000114
n-Hexane . .	341·95	23·9	0·000122	Isobutyl propion-			
n-Heptane . .	371·4	22·3	0·000121	ate . . .	409·8	21·4	0·000114
n-Octane . .	398·8	21·1	0·000119	Amyl propionate	433·2	20·4	0·000113
Isopentane . .	300·95	26·2	0·000127	Methyl butyrate	375·75	23·3	0·000114
Di-isobutyl . .	382·1	20·9	0·000125	Ethyl butyrate .	392·9	22·3	0·000114
Hexamethylene .	353·9	22·7	0·000124	Propyl butyrate	415·7	20·9	0·000115
Benzene . .	353·2	23·45	0·000121	Amyl butyrate .	451·6	19·5	0·000113
Toluene . .	383·7	21·75	0·000120	Methyl isobutyr-			
Ethyl benzene .	409·15	20·3	0·000120	ate . . .	365·3	23·8	0·000115
Naphthalene .	491·05	17·1	0·000119	Isobutyl isobutyr-			
Anthracene .	616·0	15·0	0·000108	ate . . .	419·6	20·6	0·000116
m-Xylene . .	412·0	21·1	0·000115	Methyl alcohol .	337·7	29·6	0·000100
Triphenyl meth-				Ethyl alcohol .	351·3	30·35	0·000094
ane . .	626·0	14·8	0·000108	Propyl alcohol .	370·2	28·8	0·000094
Methyl chloride .	249·35	31·9	0·000126	Amyl alcohol .	403·0	25·3	0·000098
Ethylene dibro-				Phenol . .	456·0	20·5	0·000107
mide . .	405·0	20·8	0·000119	Acetic acid .	391·5	23·9	0·000107
Fluorobenzene .	358·2	23·3	0·000120	Phthalic anhy-			
Chlorobenzene .	405·0	20·5	0·000120	dride . .	559·0	16·0	0·000112
Bromobenzene .	429·0	19·3	0·000120	Water . . .	373·0	27·2	0·000099

Relation of Constant c to Molecular Weight and Constitution.—1. In any homologous series, or any series of closely related substances, except the alcohols, acids, phenols, the lower esters and perhaps some others, the higher the molecular weight the lower is the constant. Examples :—The normal paraffins ; methyl and ethyl ethers ; toluene and meta-xylene ; acetone and benzophenone ; benzene, naphthalene and anthracene.

2. Iso-compounds have higher values than their normal isomerides, and if there are two iso-groups the value is still higher. Examples :— Isopentane and normal pentane ; di-isobutyl and n-octane ; methyl isobutyrate and methyl butyrate.

3. When hydrogen is replaced by a halogen, the value is lowered.

Examples :—Benzene and a mono-derivative ; naphthalene and bromo-naphthalene.

4. By replacing one halogen by another no change is usually produced. Examples :—The four halogen derivatives of benzene.

5. All compounds containing a hydroxyl group—alcohols, phenols, water, acids—have very low values. But the influence of the hydroxyl group in lowering the constant diminishes as the complexity of the rest of the molecule increases. Thus, with methyl, ethyl, and propyl alcohols the constants must be as much as 0·000035 lower than those of the corresponding hydrocarbons, but with amyl alcohol it is only 0·000029, and with phenol only 0·000015 lower. On the other hand, the constant tends in general to be lowered as the molecular complexity increases, and these two factors, acting in opposite directions, neutralise each other more or less completely ; thus, in the case of the alcohols at any rate, there is apparently no relation between the values of the constant and the molecular weight.

6. The esters—formed from alcohols and acids—have rather low values, and here again the constant is nearly independent of the molecular weight.

Modifications of the Still.—For ordinary laboratory purposes a round-bottomed glass flask is the most convenient form of still, but if a large quantity of liquid has to be distilled, especially when it is very inflammable, it is safer to employ a metal vessel. Metal vessels are generally made use of on the large scale.

Any alteration in the shape of the still as a rule is merely a matter of convenience and does not call for special mention.

Modifications of the still-head are of great importance and will be considered later in Chapters X. to XIII.

Modifications of the Condenser.—For liquids boiling above the ordinary temperature, but below about 170°, the straight Liebig's condenser is usually employed, but various more compact and efficient forms of condenser have been devised and are advertised by dealers. If a liquid boils at a higher temperature than 170°, there would be danger of fracture if the glass delivery tube were cooled by water. The cooling effect of the surrounding air is, however, sufficient if a long tube be employed. For very volatile liquids, the delivery-tube must be cooled by ice or by a freezing-mixture (pounded ice and salt, or ice and concentrated hydrochloric acid are convenient for moderately low temperatures). In this case a spiral, or " worm," tube

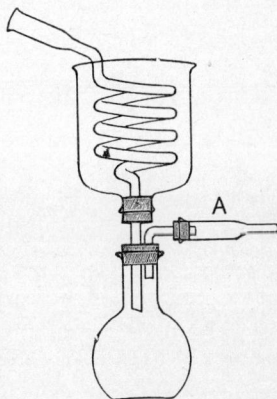

FIG. 10.
Condenser for volatile liquids.

should be used (Fig. 10). Condensation of moisture in the receiver is prevented by the drying tube, A.

Modifications of the Receiver.—If a liquid boils at a very high temperature, or if it suffers decomposition at its ordinary boiling point, it may be necessary to distil it under reduced pressure. For cases of simple distillation the apparatus shown in Fig. 11 may be employed, but if the distillate is to be collected in separate portions, the removal of the receiver would necessitate admission of air into the apparatus and a fresh exhaustion after each change. In order to introduce successive fractions into the still without disturbing the vacuum, Noyes and Skinner[1] fuse a separatory funnel and the still-head to the neck of a Claissen bulb. [The large globe in Fig. 11 serves

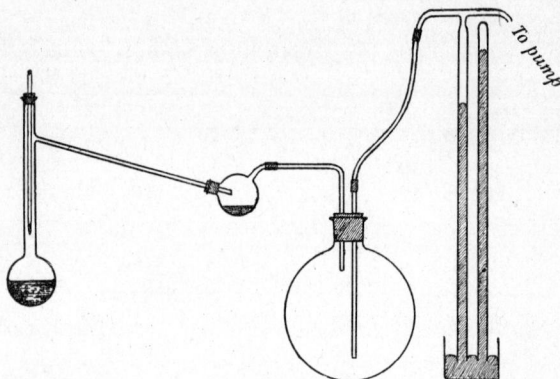

FIG. 11.—Simple apparatus for distillation under reduced pressure.

to keep the pressure steady and to prevent oscillation of the mercury in the gauge.] Various methods have been devised to allow of the receivers being changed without altering the pressure, of which the following may be mentioned.

1. Thorne's Apparatus.—A series of stopcocks may be arranged in such a manner that air may be admitted into the receiver and a fresh one put in its place while the distillation bulb remains exhausted (Fig. 12). The stopcock *b* is closed, and the three-way stopcock *c* is turned so as to admit air into the receiver, which is then disconnected and a fresh one is put in its place. The stopcock *a* is then closed to shut off the still from the pump, and *c* is turned so as to connect the pump with the new receiver, which is then exhausted until the pressure falls to the required amount, when *a* and *b* are again opened.

This method, though comparatively simple, is attended by several disadvantages ; there is some risk of leakage when so many stopcocks are used—even when a three-way stopcock is employed (as in Fig. 12) in place of two simple ones—and this is especially the case because ordinary lubricants cannot as a rule be used for *b*, through which the condensed liquid flows. Moreover, the changing of the receiver, the manipulation of the stopcocks, and the exhaustion of the fresh receiver

[1] Noyes and Skinner, "An Efficient Apparatus for Fractional Distillation under Diminished Pressure," *J. Amer. Chem. Soc.*, 1917, **39**, 2718.

take up some time, during which the progress of the distillation cannot be closely watched.

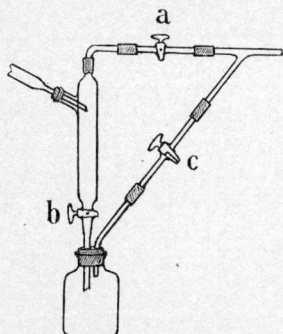

FIG. 12.—Thorne's apparatus for distillation under reduced pressure.

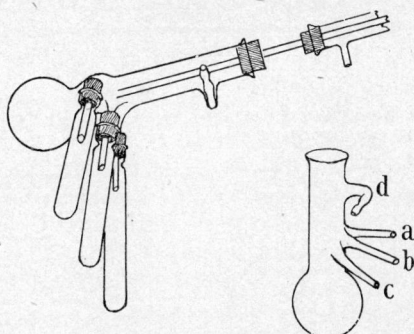

FIG. 13.—Bredt's apparatus for distillation under reduced pressure.

2. Bredt's Apparatus.—To the end of the delivery tube from the still-head is attached a round-bottomed flask with a long neck to which are sealed three narrow tubes *a*, *b*, and *c*, approximately at right angles to it (Fig. 13), and a fourth tube *d*, which serves to admit air when the distillation is completed. The bulb of the flask serves as one receiver, and each of the three narrow tubes is connected with a cylindrical vessel by means of a perforated cork. The long-necked flask is first placed with the three receivers in an inverted position, so that the first fraction collects in the flask; when a change is to be made, the neck of the flask is rotated until the drops of distillate fall into one of the cylindrical receivers and each of these in turn can be brought vertically below the end of the delivery tube.

3. Bruhl's Apparatus.—A number of test-tubes are placed in a circular stand which may be rotated within an exhausted vessel (Fig. 14) so that any one of the tubes may easily be brought under the end of the delivery tube. This arrangement is convenient, as the change of receiver can be effected with the greatest ease and rapidity.

4. Wade and Merriman's Apparatus for Distillation under Constant Low or High Pressure.—To keep the pressure constant a pressure regulator or "manostat" was devised on a principle similar to that of a thermostat. The air inlet passage is auto-

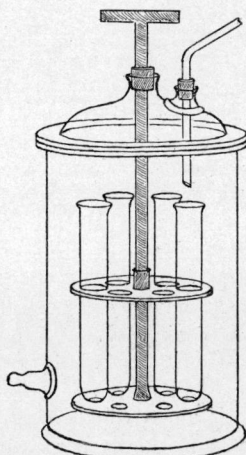

FIG. 14.—Brühl's apparatus for distillation under reduced pressure.

matically left uncovered by the mercury of a manometer when the pressure falls below the limit to which the manostat is set. For the construction of the instrument and the method of using it, the original

paper [1] should be consulted.   In employing the instrument for fractional distillation an air reservoir of about 10 litres capacity is placed
between the manostat and the vacuum pump or air compressor and
a second air reservoir of about twice the capacity between the manostat and the manometer and distillation apparatus. A suitable mercurial air compressor is described in the paper.

In order to collect the fractions without disturbing the distillation
the modification of F. D. Brown's apparatus shown in Fig. 15 was
employed. The separator consists essentially of a tap-funnel B', into
the wide neck of which is sealed a sleeve C, of sufficient calibre to allow
the passage of the condenser tube ; the latter, which should project
well below the sleeve, is made tight by a short length of stout rubber

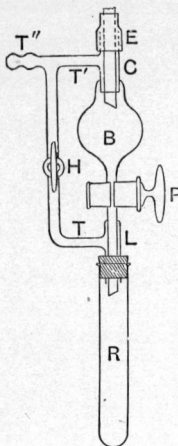

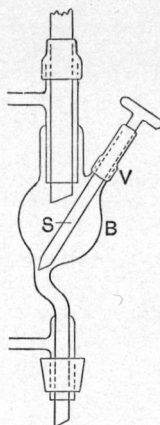

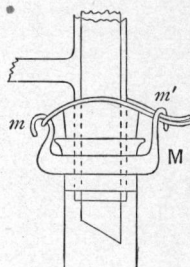

FIG. 15.   FIG. 16 a.   FIG. 16 b.

tubing. The tap P is preferably of large bore, 3 or 4 mm. The stem
of the tap-funnel is sealed to the upper end of a second sleeve L, the
lower end of which carries the rubber stopper of the receiving tube or
flask R. The two sleeves are respectively furnished with T tubes TT',
the ends of which are sealed to opposite ways of a three-way tap H,
the third way of which is open to the air. Connection with the vacuum
pump is made through a branch T'' of the upper side tube T'. The
tap P cannot, as a rule, be lubricated, but as any leakage is inwards no
loss is involved.

Under increased pressure a tap is inadmissible, for the leakage
would be outwards, whilst the pressure would tend to loosen the tap.
For liquids which have no action on rubber a short length of stout
pressure tubing provided with a screw clamp forms an excellent substitute. For other liquids the funnel B is provided with an internal
stopper S (Fig. 16 a) which can be actuated from outside. This stopper
is mounted on a stout glass rod, which passes through a closely-fitting

[1] Wade and Merriman, "Apparatus for the Maintenance of Constant Pressures above and
below the Atmospheric Pressure.   Application to Fractional Distillation," *Trans. Chem. Soc.*,
1911, **99**, 984.

glass sleeve v, to which it is secured by a short length of rubber pressure tubing. This tubing, which is wired to the rod and sleeve, acts as a spring which normally keeps the stopper away from its embouchure in the upper, bent part of the funnel stem. In changing the receiver, the rod is pressed inwards against the elasticity of the rubber tube until it closes the embouchure, and on relieving the pressure by means of the three-way tap, the stopper is held firmly in position by the pressure of the compressed air in the funnel. Any leakage is downwards and internal and the receiver can be changed so quickly that there is no danger of loss. In a distillation under increased pressure all the rubber joints and stoppers must be wired to the respective glass vessels and tubes. A convenient fastening for the receiver is afforded by a stout metal ring M (Fig. 16 b) which fits the test tube fairly closely, and is provided with two stout vertical hooks mm', one of which is radial, the other parallel to the ring ; a loop of fairly stout mild steel wire is passed over the first of these hooks, and the two ends, after traversing the stopper on either side of the sleeve, are sprung under the second, the elasticity of the wire keeping it securely in position.

Convenient separators have been described by Rosanoff and Easley [1] (see also Rosanoff, Bacon and White),[2] and by Hahn.[3]

Prevention of Leakage.—When a liquid is distilled under reduced pressure, it is necessary that all joints in the apparatus should be air-tight. India-rubber stoppers cannot always be used for the still because this substance is attacked or dissolved by many organic liquids, and ordinary corks are seldom quite air-tight. Page,[4] however, finds that all leakage may be effectually prevented by first exhausting the apparatus and then covering the cork with the ordinary liquid gum, sold in bottles (not gum arabic, which is apt to crack when dry). The gum may be conveniently applied with a brush and, if necessary, the application may be repeated several times.

Apparatus for fractional distillation under reduced pressure has also been described by—

Lothar Meyer, *Berichte*, 1887, **20**, 1834.

H. Gautier, *Bull. Soc. Chim.*, 1889, **2**, 675.

H. Wislicenus, *Berichte*, 1890, **23**, 3292.

H. Schulz, *ibid.*, 1890, **23**, 3568.

G. W. A. Kahlbaum, *ibid.*, 1895, **28**, 393.

R. Steinlen, *Chem. Zeit.*, 1898, **22**, 157.

---

[1] *Loc. cit.*

[2] Rosanoff, Bacon and White, "A Rapid Laboratory Method of Measuring the Partial Vapour Pressures of Liquid Mixtures" (Apparatus), *J. Amer. Chem. Soc.*, 1914, **36**, 1806.

[3] Hahn, "Fractionating Device," *Ber.*, 1910, **43**, 1725 ; *J. Soc. Chem. Ind.*, 1910, **29**, 842.

[4] Page, "Cork *versus* Rubber," *Chem. News*, 1902, **86**, 162.

CHAPTER II

THE BOILING POINT OF A PURE LIQUID

THE STATICAL METHOD

THERE are two methods by which the " boiling point " of a liquid under a given pressure may be determined, the *statical* and the *dynamical*. By the first method the pressures exerted by the vapour of the liquid at a series of temperatures are ascertained and plotted against the temperatures, a curve, the vapour pressure curve, being then drawn through the points (Fig. 17).

This curve has a twofold meaning; it represents not only the vapour pressures of the liquid at different temperatures, but also the boiling points of the liquid under different pressures. Thus, the vapour pressure of water at 50° is 91·98 mm., and water boils at 50° under a pressure of 91·98 mm.

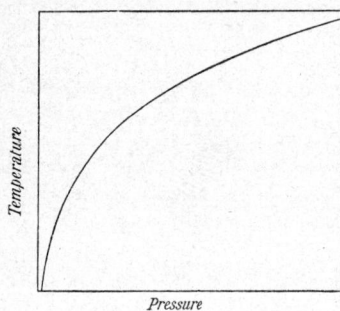

FIG. 17.—Vapour pressure curve.

Evaporation in Absence of Air.—The vapour pressures of a liquid at temperatures lower than its boiling point under atmospheric pressure may be determined by placing some of the liquid over the mercury in a barometer tube and heating the tube to different temperatures. The difference between the height of the barometer and that of the column of mercury in the tube, after correcting for the expansion of the mercury, and, if necessary, for its vapour pressure, gives the vapour pressure of the liquid.

It is necessary to take great care that the liquid introduced is quite free from dissolved air, otherwise, under the reduced pressure and at the higher temperature, some of this air would be expelled and the measured pressure would be the sum of the pressures of the vapour and of the air. If the liquid is pure and free from air, the pressure varies only with the temperature and does not depend on the relative volumes of liquid and vapour so long as both are present. This constancy of vapour pressure is, in fact, a very delicate test of the purity of a liquid.[1]

Evaporation in Presence of Air.—On the other hand, Regnault has shown that almost exactly the same amount of evaporation will take place into a vacuous space as into the same space containing any gas which does not act chemically on the liquid and is only very slightly soluble in it. The only difference is that evaporation into a vacuum usually takes place almost instantaneously, whereas in presence of a gas the process is a slow one, owing to the time required for diffusion between vapour and gas.

If, then, the barometer tube contained air, and we were able to measure its pressure, we could calculate the true pressure of the vapour by subtracting that of the air from the total pressure.

Dalton's law of partial pressures is, in fact, applicable to this case, and in constructing the vapour pressure curve from such results, it is the partial pressures of the vapour, and not the total pressures, that must be plotted against the temperatures.

The Dynamical Method

A liquid is said to boil when it is in a state of ebullition, but the scientific term " boiling point " is not applied to the temperature of a boiling liquid. The temperature of ebullition depends partly on the gaseous, or other, pressure on its surface, partly on the vertical depth of the liquid, and partly on the cohesive force between its molecules and the adhesion of the liquid to the walls of the vessel ; it would be very difficult, if not impossible, to frame a convenient definition to include all these factors.

Ebullition.—If we heat a liquid in an ordinary glass flask by means of a Bunsen burner placed below it, the formation of bubbles of vapour at the lower surface of the liquid, in contact with the glass where the heat is received, is facilitated by the presence of air dissolved in the liquid or adhering as a film to the glass and by sharp points, or roughnesses, on the surface of the glass.

If a minute bubble of air is formed, it will serve as a nucleus for a larger bubble of vapour, but in order that the bubble may increase in size by evaporation from the liquid surrounding it, it is clear that the vapour must overcome the pressure of the column of liquid above it as well as that of the atmosphere. Now the pressure exerted by the vapour of a liquid, in contact with that liquid, depends solely on the temperature, and therefore, under the most favourable conditions, the temperature of the liquid surrounding the bubble must be so high that the vapour pressure is equal to the sum of the pressures of the atmosphere and of the column of liquid.

---

[1] As will be seen later, there are certain liquid mixtures which behave in many respects like pure liquids ; at a given temperature the vapour pressure of such a mixture would not depend on the relative volumes of liquid and vapour, but at a different temperature the vapour pressure would no longer be quite independent of the volumes.

Cause of "Bumping."—But if the liquid is very free from air, and if the walls of the vessel are very smooth and clean, bubbles are formed with much greater difficulty and the temperature of the liquid may rise much higher ; it is then said to be **superheated,** and when a bubble forms, the vapour pressure corresponding to the temperature of the liquid is much greater than the sum of the pressures of the atmosphere and of the column of liquid ; consequently vapour is evolved and the bubble increases in size with great rapidity, and at the same time the temperature of the liquid falls to some extent. Under these conditions the liquid boils irregularly and is said to " bump."

The liability to intermittent ebullition is still greater when the liquid is covered by a layer of another lighter but less volatile liquid, as when water has a layer of oil over it : if a globule of well-boiled water is immersed in a bath of oil of the same specific gravity, so that the water is in contact only with the oil, the formation of bubbles is very difficult indeed, and the temperature may be raised far above 100° ; when a bubble is at last formed, the whole of the water may be suddenly converted into steam with explosive violence.

Definition of the term "Boiling Point."—By the term *boiling point* is to be understood the highest temperature attainable by a liquid, under a given pressure of its own vapour, when evaporating with a perfectly free surface, and when the heat reaches the surface from without. Thus, if we cover the bulb of a thermometer with cotton wool or other porous material, saturate the wool with the liquid under examination and suspend the thermometer in a test tube heated in a bath to a temperature at least 20° higher than the boiling point of the liquid, the temperature will rise, vaporisation will take place, and the air in the test tube will be expelled by the vapour. Under these conditions, so long as the cotton wool remains thoroughly moistened by the liquid, the temperature cannot rise above a maximum which is not influenced by that of the bath, but depends solely on the pressure of the vapour and therefore, since these are equal, on the atmospheric pressure.

This maximum temperature is the true " boiling point " of the liquid under a pressure of its vapour equal to that of the atmosphere. It will be observed that the boiling point of a liquid can only be correctly determined by observing the temperature of the liquid itself under such conditions that ebullition is impossible.

Determination of Boiling Point.—The true boiling point of a liquid is identical with the condensing point of its vapour under the same pressure, provided that some liquid is present and that the vapour is not mixed with an indifferent gas or vapour, and it is usually more convenient to measure the condensing point of the vapour than the boiling point of the liquid. To do this an ordinary distillation bulb is generally employed (Fig. 3). The walls of the vertical tube give up heat to the surrounding air, and some of the vapour condenses, the remainder being therefore in contact with condensed liquid. Condensation also takes place to a slight extent on the thermometer unless the amount of liquid in the still is very small, when the vapour is liable to be superheated owing to the flame, or the heated gases from it, playing against the dry walls of the vessel, and as the thermometer loses heat only very slowly by radiation it may, under these conditions, become too hot for vapour to condense on it. The vapour is especially liable to be superheated when a water or oil bath is employed as the source of heat.

Condensation on the thermometer is promoted by covering it with a little cotton wool,[1] or, for high temperatures, fibrous asbestos, and, in determining a boiling point or carrying out a distillation with the ordinary apparatus, it is always advisable to cover the thermometer bulb in this way. On the other hand, the thermometer should be protected from loss of heat and possible fall in temperature by radiation, and for exact determinations it is advisable to surround the vertical tube with a jacket of the vapour of the boiling liquid. A convenient apparatus is described by A. Edwards [2] and is shown in Fig. 18. The flask A is fitted with a cork through which passes the wider of the two concentric vertical tubes B and C. The inner tube B passes through a cork at the top of C and is itself provided with a cork which holds the thermometer (not shown). The tube B may be enlarged a little near the bottom or the tube C constricted so as to bring the two tubes so close together that the narrow space between them becomes primed with condensed liquid. The vapour from the liquid boiling in A therefore passes up the inner tube, through the hole blown in this near the top, down between the two concentric tubes, and finally up into the narrow side tube D which acts as a reflux condenser. The condensed liquid flows easily through the narrow space between the tubes and returns to the still.

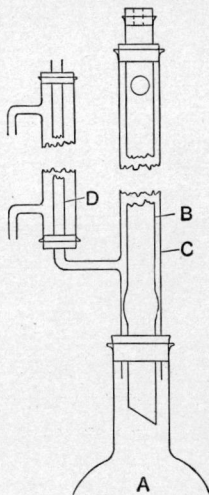

FIG. 18.

If the liquid, the boiling point of which is to be determined, is not known to be quite pure, it is advisable to distil a fair quantity of it ; if it is pure the boiling point should remain quite constant during the whole distillation ; if not, the temperature will rise (unless we are dealing with a mixture of constant boiling point), and some idea of the nature and amount of the impurity will be gained by observing the extent of the rise and whether it takes place in the early or late stages of the distillation, or both, or whether there is a steady rise during the whole period.

Effect of Impurities on Boiling Point.—If there is a rapid rise at first and the temperature afterwards remains steady or nearly so, it

[1] Ramsay and Young, "On a New Method of Determining the Vapour Pressures of Solids and Liquids," *Trans. Chem. Soc.*, 1885, **47**, 42.
[2] *J. Soc. Chem. Ind.*, 1918, **37**, 38 T.

may be assumed that there is a much more volatile liquid present. If the temperature is steady at first but rises rapidly when the distillation is nearly complete, the conclusion may be drawn that a much less volatile liquid is present. In either case the steady temperature will approximate very closely to the true boiling point, but it would be more satisfactory, especially in the second case, to collect the best portion of the distillate separately, and to redistil it and read the boiling point again. If, on the other hand, there is a fairly steady rise of temperature throughout, the presence of one or more substances, not very different in volatility from the pure liquid itself, is probable, and it is impossible to ascertain the boiling point of the pure liquid without carrying out a fractional distillation to remove the impurities.

Reduction of Boiling Point to Normal Pressure.—Assuming that the liquid is pure, that the thermometer has been compared with an air thermometer, and that the precautions mentioned in the last chapter have been attended to, the corrected temperature will give the true boiling point of the liquid under a pressure equal to that of the atmosphere at the time. It is frequently necessary, however, to compare the boiling point of the liquid with that observed by another experimenter, or to compare it with that of some other liquid, and it is therefore convenient to ascertain what the boiling point would be under normal pressure and, in stating the result for future reference, to give this reduced boiling point.

The method of correcting the boiling point of a liquid from observed to normal pressure is given on p. 13.

Prevention of "Bumping."—It frequently happens, especially when the distillation has to be carried out under greatly reduced pressure, that the liquid is liable to boil with bumping (p. 22), and two methods have been proposed to prevent this.

1. As already mentioned, small tetrahedra of silver or platinum, or fragments of porous material such as unglazed porcelain or pumice-stone, may be added to facilitate the formation of bubbles, or

2. A slow current of air may be admitted through a capillary tube which passes nearly to the bottom of the vessel (Fig. 6, p. 7).

The latter method is an excellent one for preventing the bumping, but it must be remembered that an error is caused by the presence of the air which is introduced. The true boiling point of the liquid and the condensing point of the vapour depend on the pressure of the vapour itself and not necessarily on the total gaseous pressure to which the liquid is exposed. If there is air mixed with the vapour, the total pressure remains unaltered but the partial pressure of the vapour is diminished, and the observed temperature is lower than the boiling point under the read pressure. That this is so may be easily proved by altering the rate at which air enters through the capillary tube ; the more rapid the introduction of air the lower will be the observed temperature, and unless the amount of air is exceedingly small the observed boiling point will be sensibly too low.

Law of Partial Pressures.—Dalton's law of partial pressures is, in fact, applicable to the boiling point of a liquid, as indeed is evident from the fact that in determining a boiling point by the statical method, if air is mixed with the vapour, it is the partial pressure of the vapour and not the total pressure that must be taken into account (p. 21). Many experiments might also be described to prove the truth of this statement, but it will be sufficient to mention the following :—

Experimental Proofs.—Water, when distilled in the ordinary manner under a pressure of 15 mm., boils at about 18° ; but Schrötter [1] observed so long ago as 1853 that when some water was placed in a shallow clock glass, supported on a short tripod on a second clock glass containing strong sulphuric acid, the whole being placed under a bell-jar which could be exhausted by an air-pump, the temperature fell to − 3° when the pressure was reduced to 15 mm. Here the aqueous vapour was rapidly absorbed by the strong sulphuric acid, so that, when the total pressure was 15 mm., the partial pressure of the vapour in contact with the water must have been only about 4 mm. The " boiling point " of the water therefore fell below the freezing point ; but in these circumstances bubbles could not be formed (except possibly air bubbles), for the vapour pressure of the water would be far lower than the total gaseous pressure.

More rapid diffusion and removal of vapour, as well as freer evaporation, was effected by suspending in the bell-jar a thermometer, the bulb of which was covered with a piece of sponge soaked in water ; in this case, under a pressure of 40 mm., at which the boiling point of water under ordinary conditions is 34°, the temperature actually fell to − 10°. Here the partial pressure of the vapour must have been only a very small fraction of the total pressure.

The following experiment [2] affords a still more striking proof of the correctness of the statement that the " boiling point " does not necessarily depend on the atmospheric pressure. A copper air bath, through which a current of air could pass freely, was heated to 205°, and a thermometer, the bulb of which was covered with cotton-wool moistened with boiling water, was suspended in the bath through an opening in the top. The pressure of the atmosphere was 748 mm., and the water on the cotton-wool was in a strongly heated chamber, yet its temperature, instead of remaining at nearly 100°, fell rapidly to 66°, and remained nearly constant at this point. When, however, the current of air through the bath was retarded by closing the grating in the side, the temperature of the water rose to about 80°, and when steam was introduced into the bath so as to replace the air as completely as possible by aqueous vapour, the temperature of the water rose to 99°, though that of the bath had fallen slightly. Lastly, on allowing some of the steam to escape, the temperature of the water fell again to 80°.

In this experiment, it is clear that the temperature of the water

---

[1] " On the Freezing of Water in Rarefied Air," *Liebig's Annalen*, 1853, **88**, 188.
[2] Young, " Sublimation," *Thorpe's Dictionary of Applied Chemistry.*

did not depend on that of the air bath or on the atmospheric pressure, both of which remained nearly constant. By limiting the supply of air and by introducing steam, nothing was altered but the relative pressures of the aqueous vapour and of the air surrounding the water. The greater the partial pressure of the vapour, the higher was the temperature reached by the water, and when the air was almost completely replaced by aqueous vapour, the temperature rose very nearly to the ordinary boiling point of water.

Spheroidal State.—It is well known that when a drop of water is allowed to fall on a red-hot sheet of metal, such as platinum, it does not touch the metal, nor does it boil although rapid evaporation takes place, but it assumes the *spheroidal state*, moving about over the hot surface like a globule of mercury on a table. Careful experiments [1] have shown that the temperature of the water under these conditions does not reach 100°, and the explanation of this fact is probably that the aqueous vapour surrounding the spheroidal drop is diluted with air, so that its partial pressure is less than the total atmospheric pressure. There are two reasons, either of which would be sufficient, why ebullition cannot take place :—

1. Heat is received from without towards the surface, from the whole of which evaporation can take place freely, as there is no contact between the water and the heated platinum ;

2. The vapour pressure corresponding to the temperature of the water is lower than the atmospheric pressure, so that a bubble, if formed, would at once collapse.

Wet and Dry Bulb Hygrometer.—Again, the difference in temperature which is usually observed between the wet and dry bulb in the ordinary hygrometer depends on the difference between the partial pressure of the aqueous vapour in the air and the maximum pressure possible at the temperature of the dry bulb, that is to say, the vapour pressure of water at that temperature.

Non-miscible Liquids.—Lastly, as will be shown later on, the fact that when two non-miscible liquids are distilled together, the boiling point is lower than that of either component when distilled alone, may similarly be explained by the law of partial pressures, the vapour of each liquid acting like an indifferent gas towards the other.

[1] Balfour Stewart's *Treatise on Heat* (6th edition), 129.

CHAPTER III

Influence of Molecular Attractions on Miscibility of Liquids and on Heat and Volume Changes during Admixture. — In studying the behaviour of two liquids, A and B, when mixed together, one should consider

1. The attraction of the like molecules—of those of A for each other and of those of B for each other.

2. The mutual attraction of the molecules of A and B.

If the attraction of the unlike molecules is relatively so slight as to be negligible, one may expect the liquids to be non-miscible or very nearly so. With a somewhat greater relative attraction between the unlike molecules there would be miscibility within small limits, and it is reasonable to assume that in the process of mixing there might be slight absorption of heat and slight expansion.

Generally, in the comparison of various pairs of liquids, as the mutual attraction of the unlike increases relatively to that of the like molecules, one would expect increasing and finally infinite miscibility ; absorption of heat at first, diminishing to zero and changing to increasing heat evolution ; and diminishing expansion followed by increasing contraction. These various changes do not, in many cases, run strictly *pari passu*, and, among liquids which are miscible in all proportions, it is not unusual to find a small amount of contraction attended by slight heat absorption, as, for example, when a little water is added to normal propyl alcohol ; but in the case of certain closely related chemical compounds, such as chlorobenzene and bromobenzene, there is neither any appreciable change of volume nor any measurable evolution or absorption of heat when the liquids are mixed together. For such substances it is probable that the different molecular attractions, A for A, B for B, and A for B, are very nearly equal, and that the relation suggested by D. Berthelot[1] and by Galitzine,[2] namely, that $a_{1\cdot2} = \sqrt{a_1 \cdot a_2}$, holds good. [$a_{1\cdot2}$ represents the attraction of the unlike molecules and a_1 and a_2 the respective attractions of the like molecules.]

There would appear, then, to be two simple cases :—

1. That in which the attraction represented by $a_{1\cdot2}$ is relatively so slight that the liquids are practically non-miscible ;

---

[1] D. Berthelot, " On Mixtures of Gases," *Compt. rend.*, 1898, **126**, 1703.

[2] Galitzine, " On Dalton's Law," *Wied. Ann.*, 1890, **41**, 770.

2. That of two closely-related and infinitely miscible liquids which show no heat or volume change when mixed together.

The vapour pressures of some mixed liquids and of non-miscible pairs of liquids have been determined by Regnault, Magnus, Gernez, Konowaloff, and other observers.

Non-miscible Liquids

It was shown by Regnault in 1853 that when two non-miscible liquids are placed together over the mercury in a barometer tube, the observed vapour pressure is equal to the sum of those of the two liquids when heated separately to the same temperature. Each liquid, in fact, behaves quite independently of the other, and so long as both are present in fair quantity and one is not covered by too deep a layer of the other, it does not matter what are their relative amounts or what are the relative volumes of liquid and vapour. If, however, the upper layer is deep, the maximum pressure may not be reached for a considerable time unless the heavier liquid by shaking or stirring is brought to the surface to facilitate its evaporation.

Partially Miscible Liquids

In the case of two partially miscible liquids the vapour pressure was found to be less than the sum of those of the components, but greater than that of either one singly at the same temperature.

Infinitely Miscible Liquids

The vapour pressures of many pairs of infinitely miscible liquids have been determined by several experimenters, and, as with the changes of volume and of temperature on mixing the liquids, so with the vapour pressures of the mixtures, very different results are obtained in different cases. There can be no doubt that the behaviour of mixtures, as regards vapour pressure, depends on the relative attraction of the like and the unlike molecules. When the mutual attraction of the unlike molecules is not much more than sufficient to cause infinite miscibility—for example, with normal propyl alcohol and water—the vapour pressure, like that of a partially miscible pair of liquids, may be greater than that of either component at the same temperature. On the other hand, when that attraction is relatively very great (formic acid and water) the vapour pressure of the mixture may be less than that of either component. It seems reasonable to suppose that, when the attractions of the like and unlike molecules are equal or nearly so, the relation between vapour pressure and composition should be a simple one, and the question what is the normal behaviour of mixtures has been discussed by several investigators.

Normal Behaviour of Mixtures. Guthrie.—Guthrie [1] in 1884 concluded that if we could find two liquids showing no contraction,

---

[1] Guthrie, "On some Thermal and Volume Changes attending Admixture," *Phil. Mag.*, 1884, [V.], **18**, 495.

expansion, or heat change on mixing, the vapour pressures should be expressed by a formula which may be written $P = \dfrac{mP_A + (100 - m)P_B}{100}$

where P, P_A and P_B are the vapour pressures of the mixture, and of the two components A and B, respectively, at the same temperature, and m is the percentage *by weight* of the liquid A. In other words, the relation between the vapour pressure and the percentage composition by weight should be represented by a straight line.

Speyers.—Speyers [1] concludes that the relation between vapour pressure and *molecular* percentage composition is always represented by a straight line when the molecular weight of each substance is normal in both the liquid and gaseous states. The equation, $P = \dfrac{MP_A + (100 - M)P_B}{100}$, where M is the molecular percentage of A, should then hold good.

Van der Waals.—Van der Waals [2] considers that the last-named relation is true when the critical pressures of the two liquids are equal and the molecular attractions agree with the formula proposed by Galitzine and by D. Berthelot, $a_{1.2} = \sqrt{a_1 . a_2}$.

Guthrie was clearly in error in taking percentages by weight, and Speyer has certainly made his statement too general, for there are very many cases known in which the relation does not hold, although both the liquids have normal molecular weights (for example, n-hexane and benzene, or carbon tetrachloride and benzene).

Kohnstamm's Experiments.—In order to test the correctness of the conclusion arrived at by Van der Waals, Kohnstamm [3] has determined the vapour pressures of various mixtures of carbon tetrachloride and chlorobenzene, the critical pressures of which, 34,180 mm. and 33,910 mm., are nearly equal, and he finds that the curvature is not very marked.

At the temperature of experiment, the maximum deviation from the straight line amounted to about 6 mm. on a total calculated pressure of 76 mm., or about 7·9 per cent. It is probable that in this case, the formula $a_{1.2} = \sqrt{a_1 . a_2}$ does not accurately represent the facts.

Closely Related Compounds.—From a study of the alteration of volume produced by mixing various pairs of liquids together, F. D. Brown [4] concluded that this change would probably be smallest in the case of closely related chemical compounds, and he obtained some indirect evidence in favour of this view ; it seems a fair assumption that for such mixtures this, and other physical relations, should be of a simple character.

[1] Speyers, "Some Boiling Point Curves," *Amer. Journ. Sci.*, 1900, IV., 9, 341.

[2] Van der Waals, "Properties of the Pressure Lines for Co-existing Phases of Mixtures," *Proc. Roy. Acad. Amsterdam*, 1900, 3, 170.

[3] Kohnstamm, "Experimental Investigations based on the Theory of Van der Waals," *Inaugural Dissertation, Amsterdam*, 1901.

[4] F. D. Brown, "On the Volume of Mixed Liquids," *Trans. Chem. Soc.*, 1881, 39, 202.

Experimental Determinations by Statical Method.—Direct measurements of vapour pressures by the statical method have, however, in nearly every case been carried out with mixtures of liquids which have no very close chemical relationship, but Guthrie (*loc. cit.*) made such determinations with mixtures of ethyl bromide and ethyl iodide, and his results make it probable, though not certain, that the equation $P = \dfrac{\text{M}P_A + (100 - \text{M})P_B}{100}$ holds good for this pair of substances. It is not known, however, whether the critical pressures of ethyl bromide and ethyl iodide are equal, though it is very probable that they may be.

Linebarger [1] has determined the vapour pressures of a few mixtures of each of the following pairs of liquids by drawing a current of air through them (p. 70) :—benzene and chlorobenzene, benzene and bromo-benzene, toluene and chlorobenzene, toluene and bromobenzene. His results are in fair agreement with the above formula, but unfortunately the method, in one case at least (carbon tetrachloride and benzene), gave inaccurate results, and it is therefore impossible to place complete reliance on the experimental data. The critical pressures of these substances have been determined and those of the components of the mixtures are in no case equal.

Determinations by Dynamical Method.—The vapour pressures of mixtures may be determined by the dynamical method ; the still is kept at a uniform temperature by means of a suitable bath, and the pressures are observed under which ebullition takes place. This method has been adopted by Lehfeldt (p. 64) and by Zawidski (p. 65), and the latter experimenter finds that mixtures of ethylene dibromide with propylene dibromide give results in conformity with the formula $P = \dfrac{\text{M}P_A + (100 - \text{M})P_B}{100}$. The critical pressures of these substances are not, however, known.

Again, the vapour pressures of mixtures may be determined indirectly from their boiling points under a series of pressures. The boiling points of each mixture are mapped against the pressures, or, better, the logarithms of the pressures, and from the curves so obtained the pressure at any required temperature can be read off.

In order to obtain the complete vapour pressure curve for two liquids at a given temperature, the boiling-point determinations for a considerable number of mixtures would have to be carried out through a wide range of pressure, but a less elaborate investigation is sufficient to ascertain whether or not the above formula is applicable.

Suppose that we determine the boiling points of mixtures containing, say, 25, 50 and 75 molecules per cent of the less volatile component, A, at a few pressures above and below 760 mm., in order to ascertain the boiling points under normal pressure with accuracy. If, then, we know the vapour pressures of each component at the three temperatures, we

---

[1] Linebarger, "On the Vapour Tensions of Mixtures of Volatile Liquids," *Journ. Amer. Chem. Soc.*, 1895, **17**, 615 and 690.

may calculate the theoretical vapour pressure of each mixture from the formula.

Chlorobenzene and Bromobenzene.—The boiling points of three mixtures of chlorobenzene and bromobenzene were determined in this way,[1] and the theoretical vapour pressures at these temperatures were then calculated. The results are given in Table 4.

TABLE 4

Molecular percentage of C_6H_5Br.	Observed boiling point.	Vapour pressures at $t°$.			Actual pressure, P'.	$P'-P$.
		P_A. C_6H_5Br.	P_B. C_6H_5Cl.	P. Mixture (calculated).		
25·01	136·75°	452·85	862·95	760·4	760·0	− 0·4
50·00	142·16	526·25	992·30	759·3	760·0	+ 0·7
73·64	148·16	618·40	1153·00	759·3	760·0	+ 0·7
			Mean	**759·7**	Mean	**+ 0·3**

The differences are within the limits of experimental error.

The critical pressures of chlorobenzene and bromobenzene are equal, or nearly so,[2] and it has been found that, when the two liquids are mixed in equimolecular proportions, there is no perceptible alteration of temperature or of volume, and it may therefore be concluded that $a_{1·2} = \sqrt{a_1 \cdot a_2}$. The conditions specified by Van der Waals are therefore fulfilled, and in this case, at any rate, the formula $P = \dfrac{M P_A + (100 - M) P_B}{100}$ gives the vapour pressures accurately.

TABLE 5

Substance.		Critical pressure.	$\triangle$.
Ethyl acetate	. . . .	38·00 atm. ⎫	4·83
Ethyl propionate	. . . .	33·17 ,, ⎭	
Toluene	. . . .	41·6* ,, ⎫	3·5
Ethyl benzene	. . . .	38·1* ,, ⎭	
n-Hexane	. . . .	29·62 ,, ⎫	4·98
n-Octane	. . . .	24·64 ,, ⎭	
Toluene	. . . .	41·6* ,, ⎫	8·5
Benzene	. . . .	50·1* ,, ⎭	
Benzene	. . . .	47·88 ,, ⎫	2·91
Carbon tetrachloride	. . .	44·97 ,, ⎭	
Methyl alcohol	. . . .	78·63 ,, ⎫	15·67
Ethyl alcohol	. . . .	62·96 ,, ⎭	

* As the critical pressures of both toluene and ethyl benzene have been determined by Altschul,[3] it seems best to give his value for benzene in the comparison with toluene.

---

[1] Young, "The Vapour Pressures and Boiling Points of Mixed Liquids, Part I.," *Trans. Chem. Soc.*, 1902, **81**, 768.

[2] Young, "On the Vapour Pressures and Specific Volumes of similar Compounds of Elements in Relation to the Position of those Elements in the Periodic Table," *Trans. Chem. Soc.*, 1889, **55**, 486.

[3] Altschul, "On the Critical Constants of some Organic Compounds," *Zeitschr. physik. Chem.*, 1893, **11**, 577.

Other Mixtures of Closely Related Compounds.—It does not follow, however, that it is only when the critical pressures are equal that the formula is applicable, and it will be seen from Table 6 that the deviations may be exceedingly small when the liquids are closely related but their critical pressures (Table 5) are widely different.[1]

TABLE 6

Mixture.	Molecular percentage of A = M.	Pressures.			Change on mixing in equimolecular proportions.	
		Actual P'.	Calculated P.	$P'-P$.	Volume per cent.	Temperature.
		mm.	mm.	mm.		*
A. Ethyl propionate	25·01	760	756·5	+ 3·5		
	50·00	,,	755·7	+ 4·3	+ 0·015	− 0·02°
B. Ethyl acetate .	74·62	,,	754·6	+ 5·4		
		Mean	**755·6**	**+ 4·4**		
A. Ethyl benzene .	25·02	760	762·8	− 2·8		
	49·97	,,	763·5	− 3·5	− 0·034	+ 0·05
B. Toluene . .	75·00	,,	765·5	− 5·5		
		Mean	**763·9**	**− 3·9**		
A. n-Octane . .	23·31	760	752·0	+ 8·0		
	50·00	,,	760·2	− 0·2	− 0·053	+ 0·06
B. n-Hexane . .	74·99	,,	781·6	− 21·6		
		Mean	**764·6**	**− 4·6**		
A. Toluene . .	24·94	760	760·6	− 0·6		
	50·00	,,	762·9	− 2·9	+ 0·161	− 0·45
B. Benzene . .	72·46	,,	764·5	− 4·5		
		Mean	**762·7**	**− 2·7**		
A. Ethyl alcohol .						
	50·13	760	**759·4**	**+ 0·6**	+ 0·004	− 0·10
B. Methyl alcohol .						

* The temperature changes given here are merely comparative ; they were observed by mixing together, in a round-bottomed flask, equimolecular quantities (22 c.c. in all) of the two substances. The temperature given is the difference between that of the mixture and the mean of those of the components, which never differed by more than 0·2°. The alterations of temperature show the direction and give a rough indication of the magnitude of the corresponding heat changes.

Influence of Chemical Relationship.—For the first three pairs of substances and the last in Table 6, the changes of volume and of temperature are exceedingly small, but the expansion and fall of temperature are much more noticeable with benzene and toluene, and it may be remarked that the chemical relationship is not quite so close in this case, the benzene molecule being wholly aromatic, the toluene partly aromatic and partly aliphatic.

This pair of liquids has been very carefully studied by Rosanoff,

[1] Young and Fortey, " The Vapour Pressures and Boiling Points of Mixed Liquids, Part II.," *Trans. Chem. Soc.*, 1903, **83**, 45.

Bacon and Schulze.[1] Their results confirm those given in Table 6 and are fully considered in Chapter VI.

With regard to the mean differences $P' - P$, it may be said that they are very small, the corresponding temperature differences being only 0·17°, 0·18°, 0·20°, 0·12° and 0·02° respectively, and the values of $100(P' - P)/P$ **0·58, 0·57, 0·60, 0·35** and **0·08** [compare Table 7, p. 34]. The conclusion may be drawn from the results that for mixtures of closely related compounds the relation between vapour pressure and molecular composition is represented by a line which is very nearly, if not quite, straight.

Errors of Experiment.—The greater the difference between the boiling points of the two liquids, the more difficult is it to determine the boiling point of a mixture with accuracy, and the greater also to some extent is the probability of error in the value of P. This may partly account for the irregular results with n-hexane and n-octane.

Comparison of Heat Change, Volume Change, and Vapour Pressure. —It does not appear that change of volume, or temperature, when the liquids are mixed can be relied on as a safe guide to the behaviour of a mixture as regards its vapour pressure, for one would expect that, as a general rule, the actual pressure should be lower than the calculated when there is contraction and rise of temperature, as, in fact, is observed with the second and third pairs of liquids ; and that, when there is expansion and fall of temperature, the observed pressure should be higher than the calculated, as is the case with the first pair. With benzene and toluene, however, the rule is not followed, for the actual pressure is the lower, although there is expansion and fall of temperature on mixing.

T. W. Price [2] has recently found that the vapour pressures of mixtures of acetone and methyl-ethyl ketone agree fairly well with those calculated from the formula $P = \text{M}P_A + (100 - \text{M})P_B$. These two liquids show no change in volume on admixture ; their critical pressures differ by at least 4·7 atm.

Components not closely related.—When we come to consider the behaviour of liquids which are not so closely related, we find in many cases much greater changes of volume and temperature, and also much larger values for $100(P' - P)/P$, as will be seen from Table 7 (p. 34).

It will be seen that contraction is frequently accompanied by absorption of heat, and that $P' - P$ may have a positive value when there is contraction, and even when there is both contraction and evolution of heat on mixing the components. In the case of mixtures of the alcohols with benzene, it is remarkable that, although the value of $(P' - P)/P$ diminish as the molecular weights of the alcohols increase, yet the fall in temperature, from methyl to isobutyl alcohol, becomes

---

[1] " A Method of finding the Partial from the Total Vapour Pressures of Binary Mixtures, and a Theory of Fractional Distillation," *J. Amer. Chem. Soc.*, 1914, **36**, 1993.

[2] " Vapour Pressures and Densities of Mixtures of Acetone and Methyl-ethyl Ketone," *Trans. Chem. Soc.*, 1919, **115**, 1116.

greater, and the very small volume-changes pass from negative to positive.

TABLE 7

Mixture.	Molecular percentage of A = M.	Pressures.			Changes on admixture.			
		Actual P'.	Calcu- lated P.	$\dfrac{100(P'-P)}{P}$.	M.	Volume per cent.	M.	Tempe- rature.
A. Ethylene chloride B. Benzene	50·0	252·4	252·1	+ 0·1	50·0	+0·34	50	− 0·35°
A. Benzene B. Carbon tetrachloride	50·0	760·0	739·85	+ 2·7	50·0	−0·13	50	− 0·69
A. Toluene B. Carbon tetrachloride	50·0	196·0	201·6	− 2·8	50·0	−0·07	50	+ 0·25
A. Ethyl acetate B. Carbon tetrachloride	50·0	314·6	293·2	+ 7·3	50·0	+0·03	50	+ 0·55
A. Chlorobenzene B. Carbon tetrachloride	50·0	82·0	76·0	+ 7·9	50·0	−0·12	50	− 0·4
A. Benzene B. n-Hexane	50·0	760·0	684·2	+11·1	50·0	+0·52	50	− 4·7
A. Carbon disulphide B. Methylal	50·0	702·0	551·1	+27·4	49·7	+1·22	50	− 6·5
A. Acetone B. Carbon disulphide	50·0	646·75	428·05	+51·1	49·9	+1·21	50	− 9·85
A. Chloroform B. Acetone	50·0	254·0	318·8	−20·3	50·0	−0·23	50	+12·4
A. Water B. Methyl alcohol	50·0	282·0	247·0	+14·2	40·0	−2·98	40	+ 7·85
A. Water B. Ethyl alcohol	50·0	372·0	285·7	+30·2	40·0	−2·56	40	+ 2·95
A. Water B. Propyl alcohol	50·0	576·0	387·0	+48·8	40·0	−1·42	40	− 1·15
A. Isobutyl alcohol B. Water	50·0	792·0	476·5	+66·2	60·0	−0·90	60	− 3·15
A. Benzene B. Methyl alcohol	50·0	760·0	475·0	+60·0	50·0	−0·01	50	− 3·8
A. Benzene B. Ethyl alcohol	50·1	760·0	507·3	+49·8	55·0	0·00	50	− 4·2
A. Propyl alcohol B. Benzene	49·0	760·0	535·4	+41·95	50·0	+0·05	50	− 4·65
A. Isobutyl alcohol B. Benzene	50·0	760·0	562·2	+35·2	50·0	+0·16	50	− 6·35
A. Isoamyl alcohol B. Benzene	50·0	760·0	579·6	+31·1	50·2	+0·23	50	− 5·25

The pressures are from observations by Zawidski, Konowaloff, Lehfeldt, Kohnstamm, Jackson and Young, Fortey and Young, or have been specially determined.

General Conclusions.

—The following conclusions may be drawn from the foregoing data :—

1. When the two components are chemically very closely related, the changes of volume and of temperature on mixing the liquids are, as a rule at any rate, inconsiderable.

2. When not only is the chemical relationship very close, but the critical pressures are equal and $a_{1·2} = \sqrt{a_1 . a_2}$ (Van der Waals), the vapour pressure of the mixture is accurately represented by the formula $P = \dfrac{\text{M}P_A + (100 - \text{M})P_B}{100}$.

3. When the components are very closely related, but the critical pressures are not equal, the percentage difference between the observed and calculated pressures—$100(P' - P)/P$—is very small, even, in some cases, when there is molecular association in the liquid state, as with methyl and ethyl alcohol.

4. When the two liquids are not closely related, even if there is no molecular association in the case of either of them, the percentage differences are, as a rule, much greater.

5. When the components are not very closely related and the molecules of one or both of them are associated in the liquid state, the values of $100(P' - P)/P$ are usually very large indeed.

Importance of Chemical Relationship and Molecular Association.

— In considering, then, the probable behaviour, as regards vapour pressure, of two liquids, it would appear that chemical relationship is the chief point to be considered, and that, if the two liquids are not closely related, the question whether the molecules of either or both of them are associated in the liquid state is also of great importance.

Alcohols—Water—Paraffins.—The influence of chemical relationship is well seen by comparing the properties of mixtures of the monhydric aliphatic alcohols with water on the one hand and with the corresponding paraffins on the other.[1]

These alcohols may be regarded as being formed from water by the replacement of one atom of hydrogen by an alkyl group, thus (C_nH_{2n+1})—O—H, or as hydroxyl derivatives of the paraffins, thus $C_nH_{2n+1}(OH)$.

The alcohols are thus related to, and their properties are intermediate between those of water and of the paraffins ; and it is found that, in the homologous series, as the magnitude of the alkyl group increases, the properties of the alcohols recede from those of water and approach those of the corresponding paraffins. This is well seen in Table 8.

TABLE 8

Number of Carbon Atoms.	Boiling points.				
	Paraffin.	Δ.	Alcohol.	Δ.	Water.
1	− 164·7°	+ 229·4°	+ 64·7°	− 35·3°	100·0°
2	− 93	171·3	78·3	− 21·7	,,
3	− 44	141·4	97·4	− 2·6	,,
4	− 0·2	117·2	117·0	+ 17·0	,,
5	36·2	101·8	138·0	38·0	,,
6	69·0	88·0	157·0	57·0	,,
7	98·4	77·6	176·0	76·0	,,
8	125·6	70·4	196·0	96·0	,,
16	287·5	56·5	344·0	244·0	,,

Thus, methyl alcohol boils only 35·3° lower than water, but 229·4° higher than methane, while cetyl alcohol boils 244° higher than water, but only 56·5° higher than the corresponding paraffin.

Water and Alcohols—Action of Dehydrating Agents.—Again, most

[1] Young and Fortey, "The Properties of Mixtures of the Lower Alcohols (1) with Water, (2) with Benzene, and with Benzene and Water," *Trans. Chem. Soc.*, 1902, **81**, 717 and 739.

dehydrating agents, which react or combine with water, behave in a somewhat similar manner towards the alcohols, though to a smaller degree, and to a diminishing extent as the molecular weight increases, and this fact accounts for the unsatisfactory results obtained with them. Thus, phosphoric anhydride gives phosphoric acid with water and a mixture of ethyl hydrogen phosphates with ethyl alcohol; with barium oxide, water forms barium hydrate, while ethyl alcohol forms,

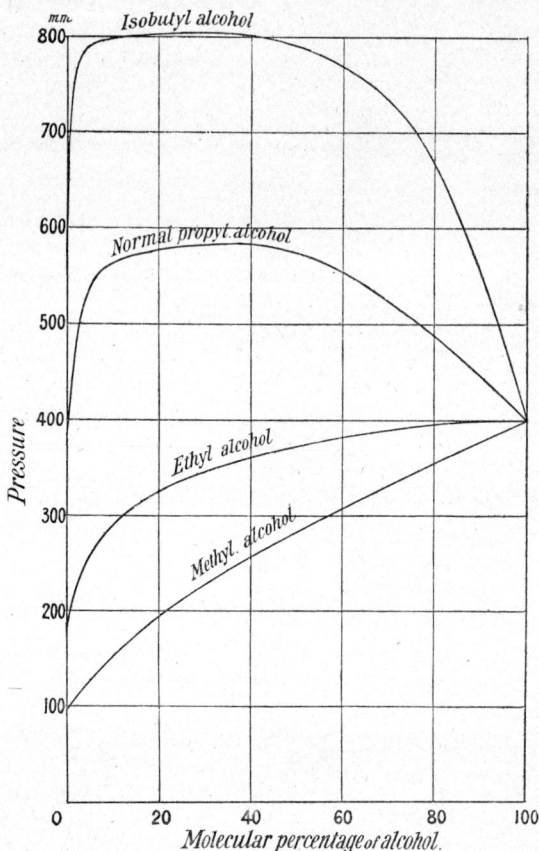

FIG. 19.—Vapour pressures of mixtures of alcohols with water.

according to Forcrand, a compound $3BaO,4C_2H_6O$; sodium acts in precisely the same way on the alcohols as on water, but the intensity of the action diminishes rapidly as the complexity of the alkyl group increases; calcium chloride forms a crystalline hexahydrate with water and a crystalline tetra-alcoholate with methyl or ethyl alcohol; anhydrous copper sulphate dissolves rapidly in water, and, on evaporation, crystals of $CuSO_4,5H_2O$ are deposited; in methyl alcohol the sulphate dissolves slowly, but to a considerable extent, giving a blue solution from which, according to Forcrand, greenish-blue crystals of $CuSO_4.CH_4O$ may be obtained ; anhydrous copper sulphate is, however, quite insoluble in ethyl alcohol, and will extract some water from strong spirit, but it is not a sufficiently powerful dehydrating agent to remove the whole.

Miscibility.—The gradual change in properties of the alcohols is also well shown by the fact that the lower alcohols are miscible with water in all proportions, the intermediate ones within limits, while the higher alcohols are practically insoluble in water.

Volume and Heat Changes.—Lastly, the contraction and heat evolution that take place on mixing the alcohols with water diminish with rise of molecular weight ; there is, in fact, increasing heat absorption in the case of the higher alcohols.

Vapour Pressures.—The lowest member of the series, methyl alcohol, bears the closest resemblance to water, and, as Konowaloff has shown, the vapour pressures of mixtures of these substances are in all cases intermediate between those of the pure components, and the curve representing the relation between vapour pressure and molecular composition does not deviate very greatly from a straight line, as will be seen from Fig. 19, in which the vapour pressures of mixtures of four alcohols with water are given, at the boiling points of the alcohols under a pressure of 400 mm.

With ethyl alcohol and water the deviation is considerable, and more recent and accurate observations have shown that a particular mixture exerts a maximum vapour pressure, but the experiments of Konowaloff are not sufficiently numerous to bring out this point.

When, however, we come to mixtures of *n*-propyl alcohol and water, we find that there is a very well defined maximum pressure, although the liquids are miscible in all proportions, and the curve shows considerable resemblance to that representing the behaviour of the partially miscible liquids, isobutyl alcohol and water.

Mixtures of Maximum Vapour Pressure. — The question whether a mixture of maximum vapour pressure will be formed in any given case depends partly on the deviation of the vapour pressure-molecular composition curve from straightness, partly on the difference between the vapour pressures of the two components. Thus, in the case of ethyl alcohol and water, the difference between the vapour pressures at 63° is 229 mm., and there is a maximum vapour pressure, very slightly higher than that of pure ethyl alcohol, for a mixture containing 89·4 molecules per cent of alcohol (B, Fig. 20). If the difference between the vapour pressures of the pure substances at the same temperature were 350 mm., and the deviation of the curve from straightness were the same, there would be no maximum vapour pressure (A, Fig. 20) ; while, if the vapour pressures of alcohol and water were equal and the deviation the same, the maximum vapour pressure would be far more obvious, and the molecular percentage of alcohol in the mixture that exerted it would be something like 20, instead of 89·4 (c, Fig. 20).

The gradual divergence of the properties of the alcohols from those of water, as the molecular weight increases, is indicated by the increasing curvature of the pressure-molecular composition lines

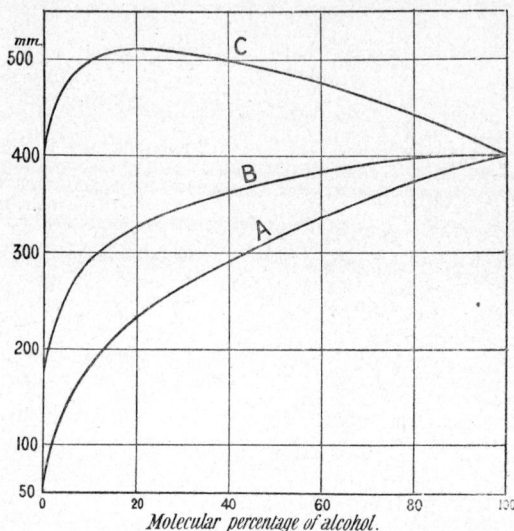

FIG. 20.

(Fig. 19). The maximum differences between the pressures represented by the actual curves and the theoretical straight lines are roughly as follows :—

Methyl alcohol and water	. . .	43 mm.
Ethyl	.	112 ,,
n-Propyl	.	203 ,,
Isobutyl	.	315 ,,

Mixtures of n-Hexane with Alcohols.—We cannot well study the miscibility of the alcohols with the corresponding paraffins, because the first four of these hydrocarbons are gaseous at the ordinary temperature, and most of them are very difficult to prepare in a pure state. Normal hexane, however, may be conveniently prepared by the action of sodium on propyl iodide, and is easily purified. This paraffin is only partially miscible with methyl alcohol at the ordinary temperature, but mixes in all proportions with those of the higher alcohols which have been investigated—ethyl, propyl, isobutyl and isoamyl alcohols. Again, normal hexane forms mixtures of maximum vapour pressure with the lower alcohols, but not with isoamyl alcohol or any other of higher boiling point. Lastly, the fall of temperature, on mixing hexane with the alcohols in equimolecular proportions, diminishes slightly as the molecular weights of the alcohol rise. [Ethyl alcohol −2·55°, n-Propyl alcohol −2·40°, Isobutyl alcohol −2·35°, Isoamyl alcohol $-1.85°$.] In all these respects, therefore, it may be said that the properties of the alcohols approach those of normal hexane, but recede from those of water, as their molecular weights increase.

Mixtures of Benzene with Alcohols.—Benzene is much more easily obtained in quantity than hexane, and behaves in a somewhat similar manner. The lower alcohols are miscible with benzene in all proportions; but while methyl, ethyl, isopropyl, n-propyl, tertiary butyl and isobutyl alcohols form mixtures of maximum vapour pressure with that hydrocarbon, isoamyl alcohol does not, and it is practically certain that no alcohol of higher boiling point would form such a mixture.

It will be seen from Table 7 (p. 34) that while for mixtures of the alcohols with water, the values of $100(P' - P)/P$ show a steady rise as the molecular weights of the alcohols increase, the corresponding values for mixtures with benzene fall regularly. It may also be mentioned that the solubility of the alcohols in benzene, relatively to that in water, becomes greater as the molecular weights of the alcohols increase, for while methyl alcohol can be separated from its benzene solution with the greatest ease, and ethyl alcohol without difficulty by treatment with water, the extraction of propyl alcohol is more troublesome, and that of isobutyl or amyl alcohol by this process is exceedingly tedious. On the other hand, the variations in the temperature and volume changes which occur on admixture are, generally, in the same direction for benzene as for water, though they are much smaller. It should, however, be noted that the relationship of the alcohols to the aromatic hydrocarbon, benzene, is by no means so close as to a paraffin, such as normal hexane; indeed, there is expansion and absorption of heat when benzene and hexane are mixed together and the two substances, in all probability, form a mixture of maximum vapour pressure.

A full account and discussion of all possible types of vapour pressure curves of binary mixtures has been given by A. Marshall.[1]

[1] "The Vapour Pressures of Binary Mixtures," *Trans. Chem. Soc.*, 1906, **89**, 1350.

CHAPTER IV

BOILING POINTS OF MIXED LIQUIDS

Statical and Dynamical Methods of Determination. — It has been stated that the boiling point of a pure liquid under a given pressure may be determined by either the statical or the dynamical method, the curve which shows the relation between temperature and pressure representing not only the vapour pressures of the liquid at different temperatures, but also its boiling points under different pressures ; this statement applies equally to the boiling points of any given mixture under different pressures.

As regards the vapour pressures of mixtures, it has been shown that there are two simple cases :—

(*a*) That of non-miscible liquids, for which the vapour pressure of the two (or more) liquids together is equal to the sum of the vapour pressures of the components at the same temperature ;

(*b*) That of closely related compounds, which are miscible in all proportions, and of a few other pairs of infinitely miscible liquids, for which the formula

$$P = \frac{M P_A + (100 - M) P_B}{100}$$

holds good. It is in these two cases only that the boiling points can be calculated from the vapour pressures of the components.

NON-MISCIBLE LIQUIDS

Liebig had observed in 1832 that when ethylene dichloride and water were distilled together, the boiling point was considerably lower than that of either pure component, and Gay Lussac in the same year pointed out that, since the two liquids are non-miscible, the total pressure must be equal to the sum of the vapour pressures of the pure components at the observed boiling point.

Dalton's law of partial pressures is, in fact, applicable to the case of non-miscible liquids, each vapour behaving as an indifferent gas to the others, and the boiling point of each liquid depends on the partial pressure of its own vapour. The temperature is necessarily the same for all the liquids present, and the total pressure, if the distillation is carried out in the ordinary way, is equal to that of the atmosphere. The boiling point is therefore that temperature at which the sum of

40

the vapour pressures of the components is equal to the atmospheric pressure.

Calculation from Vapour Pressures (Chlorobenzene and Water).

— In Table 9 the vapour pressures of chlorobenzene and of water,[1] two liquids which are practically non-miscible, are given for each degree from 89° to 93°, and also the sum of the vapour pressures at the same temperatures.

TABLE 9

Vapour Pressures in mm.

Temperature.	Chlorobenzene.	Water.	Sum.
89°	201·15	505·75	706·9
90	208·35	525·45	733·8
91	215·8	545·8	761·6
92	223·45	566·75	790·2
93	231·3	588·4	819·7

Thus, when chlorobenzene and water are heated together, say in a barometer tube, the observed vapour pressure at 90° should be 733·8 mm., 761·6 mm. at 91°, and so on. Conversely, when chlorobenzene and water are distilled together in an ordinary distillation bulb, so that the vapours are unmixed with air, the observed boiling point, when the atmospheric pressure is 761·6 mm., should be 91°.

Experimental Verification.

—In an actual experiment 100 c.c. of chlorobenzene and 80 c.c. of water were distilled together when the barometric pressure was 740·2 mm., and it was found that the temperature varied only between 90·25° and 90·35°, until there was scarcely any chlorobenzene visible in the residual liquid, when it rose rapidly to nearly 100°. The theoretical boiling point is calculated as follows :— The increase of pressure for 1° rise of temperature from 90° to 91° is 761·6 − 733·8 = 27·8 mm. It may be assumed without sensible error that the value of dp/dt is constant over this small range of temperature and the boiling point should therefore be $\left(90 + \dfrac{740\cdot2 - 733\cdot8}{761\cdot6 - 733\cdot8}\right)^{\circ} = 90\cdot23°$, which is very close to that actually observed.

LIQUIDS MISCIBLE WITHIN LIMITS

Before considering the behaviour of those infinitely miscible liquids for which the relation between vapour pressure and molecular composition is represented by a straight line, it will be well to take the case of partially miscible liquids.

The boiling point of a pair of such substances is higher than that calculated as above for non-miscible liquids, but if the miscibility is slight, the difference between the observed and calculated temperatures is not serious, and the observed boiling point will be decidedly lower than that of either component.

Aniline and Water.

—Water dissolves only about 3 per cent of aniline, and aniline about 5 per cent of water at 12°, though the

---

[1] Young, "Distillation," *Thorpe's Dictionary of Applied Chemistry.*

solubility in each case is considerably greater at 100°. Fifty c.c. of aniline and 200 c.c. of water were distilled together under a pressure of 746·4 mm., and the boiling point was found to remain nearly constant at 98·75° for some time but afterwards rose slowly to 99·65° while there was still a moderate amount of water visible with the residual aniline. The distillation was then stopped and a large amount of water was added to the residue ; the temperature, when the distillation was recommenced, rose to 98·9°. It would therefore appear that the composition of the liquid in the still may alter considerably without producing any great change in the boiling point, though the temperature does not remain so constant as it does with chlorobenzene and water.

The vapour pressures of aniline and water at 97° to 99° are as follows :—

TABLE 10

Vapour Pressures in mm.

Temperature.	Aniline.	Water.	Sum.
97°	40·5	682·0	722·5
98	42·2	707·3	749·5
99	44·0	733·3	777·3

The calculated boiling point under a pressure of 746·4 mm. would therefore be $\left(97 + \dfrac{746\cdot4 - 722\cdot5}{749\cdot5 - 722\cdot5}\right)^{\circ} = 97\cdot9°$, which is 0·85° lower than the temperature, which remained nearly constant for some time, and 1·75° lower than that which was observed when there was still a moderate amount of water present.

Practical Application.—Advantage is taken of the fact that the boiling point of a pair of non-miscible or slightly miscible liquids is lower than that of either pure component, to distil substances which could not be heated to their own boiling points without decomposition, or which are mixed with solid impurities.

As a rule, water is the liquid with which the substance is distilled and the process is commonly spoken of as "steam distillation."

As an example, the commercial preparation of aniline may be described. Nitrobenzene is reduced by finely divided iron and water with a little hydrochloric acid, the products formed being ferrous chloride, magnetic oxide of iron and aniline. The aniline is distilled over with steam and the greater part of it separates from the distillate on standing. The aqueous layer, which contains a little aniline, is afterwards placed in the boiler which supplies steam to the still, so that in the next distillation the aniline is carried over into the still again.

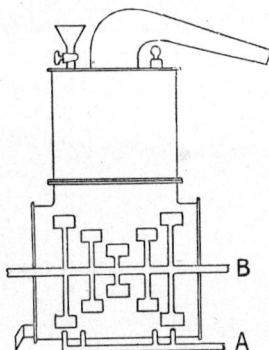

FIG. 21.—Aniline still.

The apparatus is shown diagrammatically in Fig. 21. A is the pipe by which steam is introduced into the still, and B is a rotating stirrer.[1]

LIQUIDS MISCIBLE IN ALL PROPORTIONS

The boiling point of a mixture of two liquids which are miscible in all proportions can be calculated if the vapour pressure of the mixture agrees accurately with that given by the formula $P = \dfrac{\text{M}P_A + (100 - \text{M})P_B}{100}$. The boiling points of such mixtures are, however, not so simply related as their vapour pressures to the molecular composition.

Calculation from Vapour Pressure and Composition.— In order to calculate the boiling points of all mixtures of two such liquids under normal pressure, we should require to know the vapour pressures of each substance at temperatures between their boiling points. Thus chlorobenzene boils at 132·0° and bromobenzene at 156·1°, and we must be able to ascertain the vapour pressures of each substance between 132° and 156°.

The percentage molecular composition of mixtures which would exert a vapour pressure of 760 mm. must then be calculated at a series of temperatures between these limits, say every two degrees, by means of the formula $\text{M} = 100\dfrac{P_B - P}{P_B - P_A}$, where, in this case, $P = 760$.

Lastly, the molecular percentages of A, so calculated, must be mapped against the temperatures, and the curve drawn through the points will give us the required relation between boiling point and molecular composition under normal pressure.

Closely Related Liquids.—For such closely related substances as chlorobenzene and bromobenzene, the ratio of the boiling points on the absolute scale is a constant at all equal pressures, as is also the value of $T . dp/dt$, and it appears, so far as experimental evidence goes, that the ratio of the boiling point (abs. temp.) of any given mixture to that of one of the pure substances at the same pressure is also a constant at all pressures ; it would therefore, strictly speaking, be sufficient to determine the vapour pressures of either substance through a wide range of temperature and the ratios of the boiling points (abs. temp.) of the other pure substance and of a series of mixtures to that of the standard substance at a single pressure, in order to be able to calculate the boiling point of any mixture at any pressure.

The vapour pressures of chlorobenzene and bromobenzene from 132° to 156° are given in the table below, also the values of $P_B - 760$, $P_B - P_A$ and M.

[1] For further information on the use of steam in distillation, reference may be made to Hardy and Richens, "Fractional Distillation by Means of Steam," *Analyst*, 1907, **32**, 197 ; R. Zaloziecki, "Theory of Distillation by Steam," *Petroleum Review*, **3**, 425 ; H. Golodetz, "Fractional Distillation with Steam," *Zeitschr. physik. Chem.*, 1912, **78**, 641.

[TABLE

TABLE 11

Temperature.	Vapour Pressures in mm.				
	P_B. Chloro- benzene.	P_A. Bromo- benzene.	$P_B - 760$.	$P_B - P_A$.	M.
132°	760·25	395·1	0·25	365·15	0·07
134	802·15	418·6	42·15	383·55	10·99
136	845·85	443·2	85·85	402·65	21·32
138	891·4	468·9	131·4	422·5	31·10
140	938·85	495·8	178·85	443·05	40·37
142	988·2	523·9	228·2	464·3	49·15
144	1039·5	553·2	279·5	486·3	57·47
146	1092·9	583·85	332·9	509·05	65·40
148	1148·4	615·75	388·4	532·65	72·92
150	1206·0	649·05	446·0	556·95	80·08
152	1265·8	683·8	505·8	582·0	86·91
154	1327·9	719·95	567·9	607·95	93·41
156	1392·3	757·55	632·3	634·75	99·61

In the diagram (Fig. 22) the values of M have been plotted against the temperatures, and the curve is drawn through the points which are not themselves indicated.

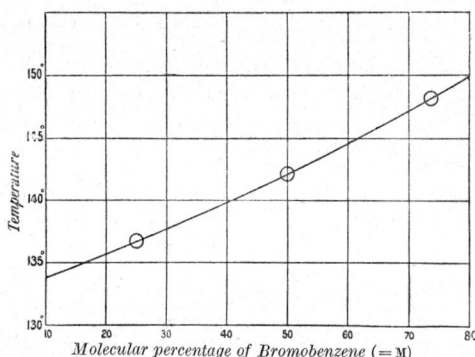

FIG. 22.—Boiling points of mixtures of bromobenzene and chlorobenzene.

Experimental Verification.—The boiling points of three mixtures of chlorobenzene and bromobenzene were determined at a series of pressures between about 690 and 800 mm.[1] The logarithms of the pressures were plotted against the temperatures, and the boiling points under normal pressure were read from the curve with the following results :—

Molecular percentage of bromobenzene .	25·01	50·00	73·64
Boiling point	136·75°	142·16°	148·16°

The points representing these values are indicated in Fig. 22 by circles, and it will be seen that they fall very well indeed on the theoretical curve.

[1] Young, "The Vapour Pressures and Boiling Points of Mixed Liquids, Part I.," *Trans. Chem. Soc.*, 1902, **81**, 768.

Determination by Dynamical Method.—In determining the boiling points of mixed liquids by the dynamical method, it is of great importance that the vapour phase should be as small as possible, because if a relatively large amount of liquid were converted into vapour the composition of the residual liquid would, as a rule, differ sensibly from that of the original mixture. A reflux condenser must be used, and it is advisable that the thermometer should be shielded from any possible cooling effect of the returning liquid. The temperature of the vapour should be read, that of a boiling liquid, even when pure, being higher than the " boiling point " (p. 21); but as the difference between the temperatures of liquid and vapour is practically constant for a given mixture, in a given apparatus with a steady supply of heat, it is a good plan to read both temperatures at each pressure, to subtract the average difference between them from the temperature of the liquid, and to take the mean of the read temperature of the vapour and the reduced temperature of the liquid at each pressure as the true boiling point.

Apparatus.—A suitable apparatus[1] is shown in Fig. 23. It consists of a bulb of about 155 c.c. capacity with a wide vertical tube, to which is sealed a narrow side tube cooled by water to act as a reflux condenser. The upper end of the side tube is connected with an exhaust and compression pump and a differential gauge. The wide vertical tube is provided with a well-fitting cork, through which passes a rather narrower thin-walled tube, which has a hole blown in it just below the cork. This narrower tube is also fitted with a cork, through which passes the thermometer.

The quantity of each mixture placed in the bulb was such that its volume at the boiling point was about 125 c.c.; the volume of vapour in the bulb and vertical tubes was about 75 c.c. The thin-walled tube was pushed down until the bottom of it was about 3 mm. above the surface of the liquid when cold, and the bottom of the thermometer bulb was about level with that of the tube.

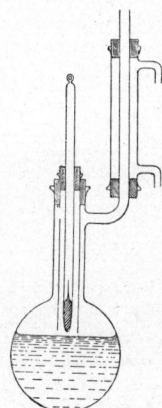

FIG. 23.—Boiling point apparatus for mixtures.

This arrangement possesses the following advantages :—

(a) The liquid that returns from the reflux condenser cannot come near the thermometer, and the amount of liquid that condenses on the thermometer, and on the inner walls of the thin-walled tube, is exceedingly small; on the other hand, with the large quantity of liquid present and the small flame that is required, there is no fear of the vapour being superheated.

(b) It is possible to take readings of the temperature both of the vapour and of the liquid without altering the position of the thermometer, for when the burner is directly below the centre of the bulb, the liquid boils up into the thin-walled tube well above the thermometer

[1] *Loc. cit.*

bulb, but when the burner is moved a little to one side, the surface of the liquid immediately below that tube remains undisturbed and the liquid does not come in contact with the thermometer bulb.

Calculated Boiling Point-Molecular Composition Curves. —The relation between the boiling points and the molecular composition of mixtures of chlorobenzene and bromobenzene is represented by a curve (Fig. 22), the temperatures being lower than if the line were straight, and that is the case for any other pair of liquids for which the formula $P = \dfrac{MP_A + (100 - M)P_B}{100}$ holds good. It is found that the curve approximates more and more closely to a straight line as the difference, Δ, between the boiling points of the components diminishes.

Liquids not closely related.—There is evidence that the actual boiling points of mixtures of liquids which are not closely related are, as a rule, considerably lower than those calculated from the vapour pressures, though occasionally higher, and that the formation of a mixture of constant boiling point is by no means an uncommon occurrence. The convenient term " azeotropic " was proposed by Wade and Merriman [1] and adopted by Lecat [2] for such mixtures, and will be employed in this book in preference to the term " hylotropic " suggested by Ostwald.

AZEOTROPIC MIXTURES

Mixtures of Minimum Boiling Point.—In the great majority of cases where the formation of mixtures of minimum boiling point has been observed, one of the two liquids is a hydroxyl compound—an alcohol, an acid or water—and water also forms such mixtures with all the lower alcohols, except methyl alcohol. It is well known that the molecules of these liquids are more or less associated in the liquid state, and we may therefore conclude that mixtures of minimum boiling point (maximum vapour pressure) are most readily formed when one or both of the liquids exhibit molecular association.

It is probable that the molecules of acetone, and also of the lower aliphatic esters, are associated to a slight extent in the liquid state, and Ryland [3] found in 1899 that the following pairs of liquids form mixtures of minimum boiling point : carbon disulphide and acetone ; carbon disulphide and methyl acetate ; carbon disulphide and ethyl acetate ; acetone and methyl acetate ; acetone and ethyl iodide ; ethyl iodide and ethyl acetate.

In the following year Zawidski [4] showed that the first and last of these pairs of liquids form mixtures of maximum vapour pressure.

Lecat [5] has observed the formation of a considerable number of

[1] *Trans. Chem. Soc.*, 1911, **99**, 1004.

[2] *Le Tension de vapeur des mélanges de liquides, l'Azéotropisme* (Brussels), 1918.

[3] Ryland, " Liquid Mixtures of Constant Boiling Point," *Amer. Chem. Journal*, 1899, **22**, 384.

[4] Zawidski, " On the Vapour Pressures of Binary Mixtures of Liquids," *Zeitschr. physik. Chem.*, 1900, **35**, 129.   [5] *Loc. cit.*

mixtures of minimum boiling point one of the components of which is an ester or one of the lower ketones.

It was thought by Speyers [1] that a mixture of minimum boiling point cannot be formed when both constituents have normal molecular weight at all concentrations, but this conclusion does not appear to be borne out by the facts, and it is clear that when Δ is very small, a comparatively slight deviation from the normal form of the boiling point-molecular composition curve would be sufficient to account for the formation of a mixture of minimum boiling point. It is certain that benzene and carbon tetrachloride ($\Delta = 3 \cdot 44°$) form such a mixture and most probable that benzene and n-hexane ($\Delta = 11 \cdot 3°$) do so. Ryland (*loc. cit.*) states that carbon disulphide and ethyl bromide ($\Delta =$ about $7 \cdot 6°$) form a mixture of minimum boiling point, and Zawidski (*loc. cit.*) finds that methylal and carbon disulphide ($\Delta = 4 \cdot 15°$) form a mixture of maximum vapour pressure. These results have been confirmed and a considerable number of other cases observed by Lecat, who has made an exhaustive study of the literature on this subject and has himself examined more than two thousand mixtures and discovered a great number of azeotropic mixtures.

Mixtures of Maximum Boiling Point.—The occurrence of mixtures of maximum boiling point (or minimum vapour pressure) is comparatively rare ; in most of the known-cases one of the substances is an acid and the other a base or a compound of basic character— formic, acetic and propionic acid with pyridine (Zawidski),[2] hydrochloric acid with methyl ether (Friedel) [3] ; or the liquids are water and an acid—formic, hydrochloric, hydrobromic, hydriodic, hydrofluoric, nitric or perchloric acid (Roscoe) [4] ; but Ryland finds that such mixtures are formed by chloroform and acetone ($\Delta = 4 \cdot 8°$) and by chloroform and methyl acetate ($\Delta =$ about $4 \cdot 3°$) ; the first observation has been confirmed by Zawidski and by Kuenen and Robson,[5] and the second by Miss Fortey (unpublished).

The number of known mixtures of maximum boiling point has been greatly increased by Lecat.

Ternary Mixtures of Minimum Boiling Point.—Benzene and water are practically non-miscible, whilst benzene and ethyl alcohol and ethyl alcohol and water form mixtures of minimum boiling point. It was observed by Young [6] in 1902 that when any mixture of the three liquids is distilled, a ternary azeotropic mixture comes over at a temperature lower than the boiling point of any of the binary mixtures

[1] Speyers, "Some Boiling Point Curves," *Amer. Journ. Sci.*, 1900, IV., **9**, 341.

[2] *Loc. cit.*

[3] Friedel, " On a Combination of Methyl Oxide and Hydrochloric Acid," *Bull. Soc. Chim.*, 1875, **24**, 160.

[4] Roscoe, "On the Composition of Aqueous Acids of Constant Boiling point," *Trans. Chem. Soc.*, 1861, **13**, 146 ; 1862, **15**, 270 ; "On Perchloric Acid and its Hydrates," *Proc. Roy. Soc.*, 1862, **11**, 493.

[5] Kuenen and Robson, "Observations on Mixtures with Maximum or Minimum Vapour Pressure," *Phil. Mag.*, 1902, VI., **4**, 116.

[6] "The Preparation of Absolute Alcohol from Strong Spirit," *Trans. Chem. Soc.*, 1902, **81**, 707.

or of any of the pure substances. It was found that by the distillation of strong spirit with benzene the water came over in the first fraction and anhydrous alcohol was left as the residue.

Ternary azeotropic mixtures of several of the lower alcohols with benzene and water and with n-hexane and water were obtained and examined by Young and Fortey.[1] The formation of other ternary azeotropic mixtures has since been observed by several chemists, Lecat having added a considerable number to the list.

In the following tables data which are not regarded as well established are printed in italics. When any mixture has been examined by more than one observer, the initial letters of the names of those observers whose data are regarded as less well established than the others are printed in italics.

[1] *Trans. Chem. Soc.*, 1902, **81**, 739.

[TABLE

LIST OF KNOWN AZEOTROPIC MIXTURES

TABLE 12

BINARY AZEOTROPIC MIXTURES—MINIMUM BOILING POINT

Substances in mixture.		Boiling points.			Percentage of A. by weight in mixture.	Observer.
A.	B.	A.	B.	Az. mixture.		
Water	Ethyl alcohol	100·0	78·3	78·15	4·43	N. & W., Y. & F., W., Vr., M.
Water	Isopropyl alcohol	100·0	82·44	80·37	12·10	Y. & F.
Water	Tertiary butyl alcohol	100·0	82·55	79·91	11·76	Y. & F.
Water	n. Propyl alcohol	100·0	97·20	87·72	28·31	Ch., K., Y. & F., Vr.
Water	Allyl alcohol	100·0	96·95	88·0	28·0	D., W. & A., L.
Methyl alcohol	Isopentane	64·7	27·95	24·5	4	L.
Methyl alcohol	Trimethylethylene	64·7	37·15	31·75	7	L.
Methyl alcohol	Diallyl	64·7	60·2	47·05	22·5	L.
Methyl alcohol	n. Hexane	64·7	68·95	50·0	26·9	Ry., Y., L.
Methyl alcohol	Benzene	64·7	80·2	58·34	39·55	Ry., Y. & F.
Methyl alcohol	Cyclo-hexane	64·7	80·75	54·2	37·2	L.
Methyl alcohol	Cyclo-hexadiene, 1·3	64·7	80·8	56·38	38·8	L.
Methyl alcohol	Cyclo-hexene	64·7	82·75	55·9	40	L.
Methyl alcohol	n. Heptane	64·7	98·45	60·5	62	L.
Methyl alcohol	Methyl cyclo-hexane	64·7	101·8	60	70	L.
Ethyl alcohol	Diallyl	78·3	60·2	53·5	13	L.
Ethyl alcohol	n. Hexane	78·3	68·95	58·65	21	Y.
Ethyl alcohol	Benzene	78·3	80·2	68·24	32·37	Th., Ry., M., Y. & F.
Ethyl alcohol	Cyclo-hexane	78·3	80·75	64·9	30·5	L.
Ethyl alcohol	Cyclo-hexadiene, 1·3	78·3	80·8	66·7	34	L.
Ethyl alcohol	Cyclo-hexene	78·3	82·75	66·7	35	L.
Ethyl alcohol	Methyl cyclo-hexane	78·3	101·8	73	53	L.
Ethyl alcohol	Toluene	78·3	110·6	76·7	68	Y., L.
Isopropyl alcohol	n. Hexane	82·45	68·95	61	22	L.
Isopropyl alcohol	Benzene	82·45	80·2	71·92	33·3	Ry., Y. & F.
Isopropyl alcohol	Cyclo-hexane	82·45	80·75	68·6	33	L.
Isopropyl alcohol	Cyclo-hexadiene, 1·3	82·45	80·8	70·4	36	L.
Isopropyl alcohol	Cyclo-hexene	82·45	82·75	71	36	L.
Isopropyl alcohol	Toluene	82·45	110·6	80·6	69	Y., L.
Tert. butyl alcohol	Benzene	82·55	80·2	73·95	36·6	Y. & F.
Tert. butyl alcohol	Cyclo-hexane	82·55	80·75	71·8	37	L.
Tert. butyl alcohol	Cyclo-hexadiene, 1·3	82·55	80·8	73·4	38·5	L.
Tert. butyl alcohol	Cyclo-hexene	82·55	82·75	73·7	38	L.
n. Propyl alcohol	n. Hexane	97·2	68·95	65·65	4	Y., L.
n. Propyl alcohol	Benzene	97·2	80·2	77·12	16·9	Ry., Y.
n. Propyl alcohol	Cyclo-hexane	97·2	80·75	74·3	20	L.
n. Propyl alcohol	Cyclo-hexadiene, 1·3	97·2	80·8	76·1	21	L.
n. Propyl alcohol	Cyclo-hexene	97·2	82·75	76·6	21·6	L.
n. Propyl alcohol	Toluene	97·2	110·6	92·6	49	Ry., L.
n. Propyl alcohol	n. Octane	97·2	125·8	95	74	L.
Allyl alcohol	Benzene	97·05	80·2	76·75	17·36	L., W. & A.
Allyl alcohol	Cyclo-hexane	96·95	80·75	74	20	L.
Allyl alcohol	Cyclo-hexadiene, 1·3	96·95	80·8	75·9	21	L.
Allyl alcohol	Cyclo-hexene	96·95	82·75	76·3	21·7	L.
Allyl alcohol	Toluene	96·95	110·6	92·4	50	Ry., L.
Isobutyl alcohol	n. Hexane	108·05	68·95	68·1	2	Y., L.
Isobutyl alcohol	Benzene	108·05	80·2	79·84	9·3	Y.
Isobutyl alcohol	Cyclo-hexane	108·0	80·75	78·1	14	L.
Isobutyl alcohol	Cyclo-hexadiene, 1·3	108·0	80·8	79·35	12	L.
Isobutyl alcohol	Cyclo-hexene	108·0	82·75	80·5	14·2	L.
Isobutyl alcohol	Toluene	108·0	110·6	101·15	44·5	Ry., L.
Isobutyl alcohol	Ethyl benzene	108·0	136·15	107·4	<80	L.
n. Butyl alcohol	Cyclo-hexane	116·9	80·75	79·8	10	L.
n. Butyl alcohol	Toluene	116·9	110·6	105·5	32	L.
Isoamyl alcohol	Ethyl benzene	131·8	136·15	125·9	49	L.
Isoamyl alcohol	Xylene, p.	131·8	138·2	126·8	51	Ry., L.
Isoamyl alcohol	Xylene, m.	131·8	139·0	127·0	53	Ry., L.
Isoamyl alcohol	Xylene, o.	131·8	142·6	128·0	60	Ry., L.
Cyclo-hexanol	Pinene	160·65	155·8	149·9	35·5	L.
Cyclo-hexanol	Mesitylene	160·65	164·0	156·3	50	L.
Cyclo-hexanol	Cymene	160·65	175·3	159	71	L.
Cyclo-hexanol	Carvene	160·65	177·8	159·25	73·5	L.
Sec. Octyl alcohol	Carvene	178·7	177·8	174·4	35	L.
Benzyl alcohol	Carvene	205·5	177·8	176·25	11	L.

E

TABLE 12—*continued*

Substances in mixture.		Boiling points.			Percentage of A. by weight in mixture.	Observer.
A.	B.	A.	B.	Az. mixture.		
Benzyl alcohol	Naphthalene	205·5	218·1	204·3	*60*	L.
Methyl alcohol	Ethyl bromide	64·7	38·4	34·95	4·5	*Ry.* L.
Methyl alcohol	Methyl iodide	64·7	44·5	39·0	7·2	*H.,* L.
Methyl alcohol	Ethylidene chloride	64·7	57·3	49·05	11·5	L.
Methyl alcohol	Chloroform	64·7	61·2	53·5	12·5	*Hy., Ry., P., Ty.,* Ti.
Methyl alcohol	Isobutyl chloride	64·7	68·9	53·05	23	L.
Methyl alcohol	Propyl bromide	64·7	71·0	54·1	20·2	L.
Methyl alcohol	Ethyl iodide	64·7	72·3	54·7	18·5	*Ry.,* L.
Methyl alcohol	Tert. butyl bromide	64·7	73·3	55·6	*24*	L.
Methyl alcohol	Carbon tetrachloride	64·7	76·75	55·7	20·56	*T.,* Y.
Methyl alcohol	Trichlorethylene	64·7	86·95	60·2	36	L.
Methyl alcohol	Dichloro-bromo-methane	64·7	90·2	63·8	40	L.
Methyl alcohol	Isobutyl bromide	64·7	91·6	61·4	42	L.
Methyl alcohol	Propyl iodide	64·7	102·4	63·5	61	L.
Ethyl alcohol	Ethyl bromide	78·3	38·4	37·6	3	*Ry.,* L.
Ethyl alcohol	Propyl chloride	78·3	46·6	44·9	5·5	L.
Ethyl alcohol	Acetylene dichloride, *tr.*	78·3	48·35	46·5	6	Cv.
Ethyl alcohol	Dibromopropylene, *α, cis.*	78·3	57·8	56·4	9	Cv.
Ethyl alcohol	Acetylene dicoloride, *cis.*	78·3	60·25	57·7	9·8	Cv.
Ethyl alcohol	Chloroform	78·3	61·16	59·4	7·0	*Th., Ry., L.,* W.
Ethyl alcohol	Isobutyl chloride	78·3	68·9	61·45	16·3	L.
Ethyl alcohol	Propyl bromide	78·3	71·0	63	17	*H.,* L.
Ethyl alcohol	Ethyl iodide	78·3	72·3	63	13	*Ry., J. & G.,* L.
Ethyl alcohol	Tert. butyl bromide	78·3	73·3	63·8	15	L.
Ethyl alcohol	Carbon tetrachloride	78·3	76·7	64·95	15·85	*Y., Hy., Sch., Ty., Hi.,* L.
Ethyl alcohol	Trichlorethylene	78·3	86·95	70·9	27	L.
Ethyl alcohol	Dichloro-bromo-methane	78·3	90·2	75·5	28	L.
Ethyl alcohol	Isobutyl bromide	78·3	91·6	72·7	*38*	*H.,* L.
Ethyl alcohol	Allyl iodide	78·3	102	75·6	41	L.
Ethyl alcohol	Isobutyl iodide	78·3	120	77·65	70	*Ry.,* L.
Ethyl alcohol	Isoamyl bromide	78·3	120·2	77·6	71·5	*H.,* L.
Ethyl alcohol	Tetrachlorethylene	78·3	120·8	77·95	81	L.
Isopropyl alcohol	Isobutyl chloride	82·45	68·9	63·8	17	L.
Isopropyl alcohol	Ethyl iodide	82·45	72·3	65·8	>15	*Ry.,* L.
Isopropyl alcohol	Isobutyl iodide	82·45	120·0	81·8	76	*Ry.,* L.
Tert. butyl alcohol	Carbon tetrachloride	82·55	76·75	69·5	17	L., *A.*
Tert. butyl alcohol	Trichlorethylene	82·55	86·95	75	16	L.
Allyl alcohol	Carbon tetrachloride	96·95	76·75	72·32	11·5	L.
Allyl alcohol	Trichlorethylene	96·95	86·95	80·95	16	L.
Allyl alcohol	Dichloro-bromo-methane	96·95	90·2	85·85	17·5	L.
Allyl alcohol	Isobutyl bromide	96·95	91·6	83·9	18	L.
Allyl alcohol	Tetrachlorethylene	96·95	120·8	94	*49*	L.
Propyl alcohol	Isobutyl chloride	97·2	68·9	67·2	7	L.
Propyl alcohol	Propyl bromide	97·2	71·0	69·2	9	*H.,* L.
Propyl alcohol	Ethyl iodide	97·2	72·3	70·1	7·8	*Ry.,* L.
Propyl alcohol	Carbon tetrachloride	97·2	76·75	72·8	11·5	*H.,* L.
Propyl alcohol	Trichlorethylene	97·2	86·95	81·75	17	L.
Propyl alcohol	Dichloro-bromo-methane	97·2	90·2	86·4	19·5	L.
Propyl alcohol	Allyl iodide	97·2	102	90·0	29	L.
Propyl alcohol	Propyl iodide	97·2	102·4	90·2	30	L.
Propyl alcohol	Tetrachlorethylene	97·2	120·8	94	54	L.
Isobutyl alcohol	Carbon tetrachloride	108·0	76·75	75·8	5·5	L.
Isobutyl alcohol	Trichlorethylene	108·0	86·95	85·4	9	L.
Isobutyl alcohol	Dichloro-bromo-methane	108·0	90·2	89·3	11	L.
Isobutyl alcohol	Propyl iodide	108·0	102·4	96	*18*	L.
Isobutyl alcohol	Isoamyl bromide	108·0	120·2	103·8	*41*	*H.,* L.
Isobutyl alcohol	Tetrachlorethylene	108·0	120·8	103·05	40	L.
Isobutyl alcohol	Ethylene bromide	108·0	131·5	106·2	60·5	*Ry.,* L.
Isobutyl alcohol	Chlorobenzene	108·0	131·8	107·2	63	L.
Isoamyl alcohol	Isoamyl bromide	131·8	120·2	116·4	*20*	*H.,* L.
Isoamyl alcohol	Tetrachlorethylene	131·8	120·8	116·0	19	L.
Isoamyl alcohol	Ethylene bromide	131·8	131·5	123·2	*30*	*Ry.,* L.
Isoamyl alcohol	Chlorobenzene	131·8	131·8	124·3	36	L.
Cyclo-hexanol	Bromobenzene	160·65	156·15	153·6	33·5	L.
Cyclo-hexanol	Pentachlorethane	160·65	161·95	157·9	36	L.
Methyl alcohol	Ethyl formate	64·7	54·15	50·95	16	L.
Methyl alcohol	Methyl acetate	64·7	57·0	54·0	19	*Ry., H.,* L.
Ethyl alcohol	Ethyl acetate	78·3	77·15	71·8	30·98	*W.,* M.
Ethyl alcohol	Methyl propionate	78·3	79·7	73·2	52	L.
Isopropyl alcohol	Ethyl acetate	82·45	77·05	74·8	23	*Ry.,* L.

TABLE 12—*continued*

Substances in mixture.		Boiling points.			Percentage of A. by weight in mixture.	Observer.
A.	B.	A.	B.	Az. mixture.		
Propyl alcohol	Ethyl propionate	97·2	99·1	93·4	51	L.
n. Butyl alcohol	Methyl isovalerate	116·9	116·3	113	33	L.
Isoamyl alcohol	Isoamyl acetate	131·8	*138·8*	*131·3*	95	*H.*, L.
Ethyl alcohol	Methyl ethyl ketone	78·3	79·6	74·8	40	*Ma.*, L.
Tert. butyl alcohol	Methyl ethyl ketone	82·55	79·6	78·9	—	A., L.
Isopropyl alcohol	Methyl ethyl ketone	82·45	79·6	77·3	*30*	A., L.
Propyl alcohol	Diethyl ketone	97·2	102·2	94·9	*57*	L.
Isobutyl alcohol	Diethyl ketone	108·0	102·2	101·95	*22*	L.
Octyl alcohol	Aniline	178·7	184·35	177·8	65	L.
Allyl alcohol	Epichlorhydrin	96·95	116·4	95·8	78	L.
Propyl alcohol	Epichlorhydrin	97·2	116·4	96·0	77	L.
Isobutyl alcohol	Epichlorhydrin	108·0	116·4	105·0	60·5	L.
Isoamyl alcohol	Epichlorhydrin	131·8	116·4	115·35	19	L.
Allyl alcohol	Nitromethane	96·95	101·2	89	60	L.
Propyl alcohol	Nitromethane	97·2	101·2	89·15	56	L.
Benzyl alcohol	Nitrobenzene	205·5	210·85	204·3	58	L.
Borneol	Nitrobenzene	211·8	210·85	207·75	46	L.
Methyl alcohol	Carbon disulphide	64·7	46·2	37·65	14	*Ry.*, *Go.*, L., A.
Ethyl alcohol	Carbon disulphide	78·3	46·2	42·4	9	*Be.*, *Ry.*, L.
Isopropyl alcohol	Carbon disulphide	82·45	46·2	44·6	8	*Ry.*, L.
Methyl alcohol	Acetonitrile	64·7	81·6	63·45	81	*V. & D.*, L.
Ethyl alcohol	Acetonitrile	78·3	81·6	72·5	56	*V. & D.*, L.
Isopropyl alcohol	Acetonitrile	82·5	81·6	75	*55*	L.
Methyl alcohol	Methylal	64·7	42·2	41·82	18·2	L.
Phenol	Chlorotoluene	181·5	161·3	160·3	8	L.
Phenol	Pentachlorethane	181·5	161·95	160·85	9·5	L.
Phenol	Bromotoluene, *o.*	181·5	181·75	174·35	40	L.
Phenol	Bromotoluene, *p.*	181·5	185	176·2	44	L.
Phenol	Iodobenzene	181·5	188·55	177·7	47	L.
Cresol, *o.*	Bromotoluene, *o.*	190·8	181·75	180·3	16	L.
Cresol, *o.*	Bromotoluene, *p.*	190·8	185	181·7	25	L.
Cresol, *o.*	Iodobenzene	190·8	188·55	185·0	*32*	L.
Propionic acid	Water	140·7	100	99·98	17·7	L.
Isobutyric acid	Water	154·35	100	99·3	21	L.
Butyric acid	Water	163·5	100	99·4	18·4	L.
Acetic acid	Benzene	118·5	80·2	80·05	2	Ne.
Acetic acid	Toluene	118·5	110·7	105·4	28	*Ry.*, L.
Acetic acid	Ethyl benzene	118·5	136·15	114·65	66	L.
Acetic acid	Xylene, *m.*	118·5	139	115·38	72·5	*Ry.*, L.
Propionic acid	Ethyl benzene	140·7	136·15	131·1	28	L.
Propionic acid	Xylene, *p.*	140·7	138·2	132·0	36	L.
Propionic acid	Xylene, *m.*	140·7	139·0	132·65	35·5	L.
Propionic acid	Pinene	140·7	155·8	136·15	58·5	L.
Isobutyric acid	Pinene	154·35	155·8	146·7	35	L.
Isobutyric acid	Carvene	154·35	177·8	151·0	78	L.
Butyric acid	Pinene	163·5	155·8	152·0	25	L.
Butyric acid	Carvene	163·5	177·8	160·75	55	L.
Isovaleric acid	Carvene	176·5	177·8	168·9	41	L.
Chloracetic acid	Carvene	186·5	177·8	167·8	34	L.
Acetic acid	Isobutyl iodide	118·5	120·0	107	34	L.
Acetic acid	Tetrachlorethylene	118·5	120·8	107·35	38·5	L.
Acetic acid	Ethylene bromide	118·5	131·5	114·35	55	B.
Acetic acid	Chlorobenzene	118·5	131·8	114·65	58·5	L.
Propionic acid	Tetrachlorethylene	140·7	120·8	118·95	8·5	L.
Propionic acid	Ethylene bromide	140·7	131·5	127·75	17·5	L.
Propionic acid	Chlorobenzene	140·7	131·8	128·75	18	L.
Propionic acid	Bromobenzene	140·7	156·15	139·85	60	L.
Isobutyric acid	Tetrachlorethane, *s.*	154·35	146·25	144·8	7	L.
Isobutyric acid	Bromobenzene	154·35	156·15	148·6	35	L.
Isobutyric acid	Pentachlorethane	154·35	161·95	152·9	43	L.
Butyric acid	Bromoform	163·5	148·3	146	7	L.
Butyric acid	Bromobenzene	163·5	156·15	152·2	18	L.
Butyric acid	Chlorotoluene, *p.*	163·5	161·3	156·3	27	L.
Butyric acid	Pentachlorethane	163·5	161·95	156·75	26	L.
Isovaleric acid	Bromotoluene, *o.*	176·5	181·75	172·1	39·5	L.
Chloracetic acid	Benzyl chloride	186·5	179·35	172	28	L.
Chloracetic acid	Bromotoluene, *o.*	186·5	181·75	172·95	32	L.
Chloracetic acid	Bromotoluene, *p.*	186·5	185·2	174	47	L.
Chloracetic acid	Iodobenzene	186·5	188·55	175·3	35	L.
Benzene	*n.* Hexane	80·2	68·95	68·87	*19*	J. & Y.
Benzene	Cyclo-hexane	80·2	80·75	77·5	*55*	L.
Benzene	Cyclo-hexene	80·2	82·75	79·45	*85*	L.

TABLE 12—continued

Substances in mixture.		Boiling points.			Percentage of A. by weight in mixture.	Observer.
A.	B.	A.	B.	Az. mixture.		
n. Heptane	Methyl cyclo-hexane	98·45	101·8	<98	>80	L.
Cyclo-hexane	Cyclo-hexadiene, 1·3	80·75	80·8	79·2	52	L.
Pinene	Mesitylene	163·8	164·0	162·7	52	L.
Isopentane	Ethyl bromide	27·95	38·38	23·5	70	L.
Trimethylethylene	Ethyl bromide	37·15	38·38	35·2	40	L.
n. Hexane	Chloroform	68·95	61·2	59·95	28	L.
n. Hexane	Isobutyl chloride	68·95	68·85	66·3	45	L.
n. Hexane	Propyl bromide	68·95	71·0	67·5	67	L.
n. Hexane	Ethyl iodide	68·95	72·3	68	34 (?)	L.
Benzene	Carbon tetrachloride	80·2	76·75	76·75	4·4	Y. & F., Li., Z., Ro.
Ethyl benzene	Propylene dibromide	136·15	141·7	135·95	95	L.
Xylene, p.	Propylene dibromide	138·2	141·7	137·5	78	L.
Xylene, m.	Propylene dibromide	139 0	141·7	138	70	L.
Pinene, α	Bromobenzene	155·8	156·1	153·4	50	L.
Carvene	Benzyl chloride	177·8	179·35	174·8	54	L.
Carvene	Bromotoluene, o.	177·8	181·75	177·3	83	L.
Isopentane	Methyl formate	27·95	31·9	17·05	53	L.
n. Pentane	Methyl formate	36·15	31·9	21·8	47	L.
Trimethylethylene	Methyl formate	37·15	31·9	24·3	46	L.
Xylene, m.	Isoamyl acetate	139·0	138·8	136	50	L.
Xylene, m.	Methyl lactate	139·0	144·8	134	70	L.
Pinene, α	Ethyl lactate	155·8	155	147	49	L.
Pinene, α	Methyl oxalate	155·8	163·3	144·1	61	L.
Mesitylene	Methyl oxalate	164	163·3	154·8	50·2	L.
Carvene	Methyl oxalate	177·8	163·3	156·7	25	L.
Carvene	Propyl lactate	177·8	171·7	166·35	37	L.
Carvene	Isobutyl lactate	177·8	182·15	172·5	60	L.
Carvene	Ethyl oxalate	177·8	185·0	172·2	59	L.
Ethyl bromide	Methyl formate	38·4	31·9	29·85	36	L.
Ethyl iodide	Ethyl acetate	72·3	77·05	70·5	75	Ry., K., Z., Ro., L.
Carbon tetrachloride	Ethyl acetate	76·75	77·05	74·75	57	Y. Z., Ty., L.
Carbon tetrachloride	Methyl propionate	76·75	79·7	75·5	60	L.
Allyl iodide	Ethyl propionate	102	99·1	97·8	40	L.
Allyl iodide	Propyl acetate	102	101·55	99·6	50	L.
Isobutyl iodide	Isoamyl formate	120	123·6	117·5	70	L.
Tetrachlorethylene	Ethyl butyrate	120·8	119·9	118·4	62	L.
Tetrachlorethylene	Ethyl carbonate	120·8	126·0	118·55	74	L.
Chlorobenzene	Isobutyl propionate	131·8	136·9	131·2	76	L.
Bromobenzene	Ethyl lactate	156·1	154·5	153	47	L.
Bromobenzene	Methyl oxalate	156·1	163·3	153·05	72	L.
Chlorotoluene, o.	Ethyl lactate	159	154·5	153·5	35	L.
Chlorotoluene, p.	Methyl oxalate	161·3	163·3	157·0	60	L.
Pentachlorethane	Methyl oxalate	161·95	163·3	157·55	68	L.
Benzyl chloride	Propyl lactate	179·35	171·7	171·2	22	L.
Benzyl chloride	Isobutyl lactate	179·35	182·15	178	70	L.
Bromotoluene, o.	Isobutyl lactate	181·75	182·15	180	56	L.
Bromotoluene, o.	Ethyl oxalate	181·75	185·0	177·35	60	L.
Bromotoluene, p.	Ethyl oxalate	185	185·0	180·4	47·3	L.
Iodobenzene	Ethyl oxalate	188·55	185·0	181·1	40	L.
Ethyl ether	Methyl formate	34·6	31·9	28·25	44	L.
Methyl propyl ether	Methyl formate	38·9	31·9	30·5	20	L.
Acetal	Isobutyl formate	104·5	98·3	97	20	L.
Anisol	Ethyl lactate	153·85	155	150·0	44	L.
Anisol	Isobutyl butyrate	153·85	157	151	67	L.
Phenetol	Methyl oxalate	171·5	163·3	161·25	30	L.
Phenetol	Propyl lactate	171·5	171·7	167·0	50	L.
Amylmethylal	Ethyl benzoate	207·5	213	206·1	85	L.
Anisol	Pinene, α	153·85	155·8	150·45	56	L.
Phenol	Pinene, α	181·5	155·8	152·75	19	L.
Phenol	Carvene	181·5	177·8	169	40·5	L.
Cresol, o.	Carvene	190·8	177·8	175·35	25	L.
Pinene, α	Dichlorhydrin, α	155·8	174·5	152	85	L.
Carvene	Dichlorhydrin, α	177·8	174·5	165·75	57	L.
Carvene	Dichlorhydrin, β	177·8	183	169·3	40	L.
Carvene	Benzaldehyde	177·8	179·2	171·2	57	L.
Pinene, α	Aniline	155·8	184·35	155·25	85	L.
Carvene	Aniline	177·8	184·35	171·35	61·2	L.
Carvene	Methyl aniline	177·8	196·1	174·5	87	L.
Benzene	Methyl ethyl ketone	80·2	79·6	78·35	62·5	Ma., L.
Pinene, α	Methyl acetoacetate	155·8	169·5	150·5	64	L.

TABLE 12—*continued*

Substances in mixture.		Boiling points.			Percentage of A. by weight in mixture.	Observer.
A.	B.	A.	B.	Az. mixture.		
Pinene, α	Ethyl acetoacetate	155·8	180·7	153·35	78	L.
Carvene	Ethyl acetoacetate	177·8	180·7	169·05	57	L.
Terpinene	Ethyl acetoacetate	181	180·7	171	51	L.
Xylene, p.	Ethyl chloracetate	138·2	143·5	137·0	72	L.
Xylene, m.	Ethyl chloracetate	139·0	143·5	137·25	68	L.
Carvene	Ethyl bromisobutyrate, α	177·8	178	174	45	L.
Toluene	Epichlorhydrin, α	110·7	116·45	108·25	74	L.
Pinene, α	Chloracetal	155·8	156·8	151	55	L.
Naphthalene	Chlorophenol, p.	218·1	217	215·9	36·5	L.
n. Pentane	Ethyl hydrogen sulphide	36·15	36·2	32	50	L.
Trimethylethylene	Ethyl hydrogen sulphide	37·15	36·2	32·95	40	L.
n. Pentane	Carbon disulphide	36·15	46·25	35·5	85	L.
Trimethylethylene	Carbon disulphide	37·15	46·25	36·5	83	L.
n. Pentane	Ethyl ether	36·15	34·6	33·4	30	L.
Trimethylethylene	Methylal	37·15	42·25	35·3	27	L.
Toluene	Stannic chloride	110·7	113·85	109·15	48	L.
Tetrachlorethylene	Paraldehyde	120·8	124	118·75	68	L.
Benzylchloride	Benzaldehyde	179·35	179·2	177·9	50	L.
Bromotoluene, o.	Aniline	181·75	184·35	178·65	63	L.
Bromotoluene, p.	Aniline	185	184·35	180·3	40	L.
Iodobenzene	Dimethyl aniline	188·55	194·05	187	85	L.
Isobutyl chloride	Acetone	68·9	56·25	55·8	27	L.
Ethyl iodide	Acetone	72·3	56·25	56·0	20	Ry., L.
Carbon tetrachloride	Acetone	76·75	56·25	55·8	..	Hy., A.
Carbon tetrachloride	Methyl ethyl ketone	76·75	79·6	73·8	71	L., A.
Allyl iodide	Diethyl ketone	102	102·2	100·8	66	L.
Allyl iodide	Methyl propyl ketone	102	102·25	100·9	66	L.
Iodobenzene	Ethyl acetoacetate	188·55	180·7	178·0	52	L.
Isobutyl iodide	Epichlorhydrin, α	120	116·45	110	53	L.
Tetrachlorethylene	Epichlorhydrin, α	120·8	116·45	110·12	48·5	L.
Isopropyl chloride	Carbon disulphide	36·25	46·25	36·0	78	L.
Ethyl bromide	Carbon disulphide	38·4	46·25	37·85	67	Ry., L.
Methyl iodide	Carbon disulphide	42·6	46·25	41·65	59	L.
Propyl chloride	Carbon disulphide	46·6	46·25	45·2	45	L.
Ethylidene chloride	Carbon disulphide	57·3	46·25	46	6	L.
Dichloromethane	Methylal	41·5	42·2	39	62	L.
Methyl iodide	Methylal	42·6	42·2	39·35	57	L.
Benzyl chloride	Cineol	179·35	176·3	175·5	19	L.
Ethylene dibromide	Chlorobenzene	131·5	131·8	129·75	55	L.
Carbon disulphide	Acetone	46·25	56·25	39·25	66	Ry., Y., Ro., M., L.
Carbon disulphide	Methyl ethyl ketone	46·25	79·6	45·85	84·7	L.
Carbon disulphide	Methyl formate	46·25	31·9	24·75	33	L.
Carbon disulphide	Ethyl formate	46·25	54·15	39·35	63	L.
Carbon disulphide	Methyl acetate	46·25	57·0	40·15	70	L.
Carbon disulphide	Ethyl acetate	46·25	77·05	46·1	97	L.
Carbon disulphide	Ethyl ether	46·25	34·6	34·5	1	Gu., Ry., Y., L.
Carbon disulphide	Methyl propyl ether	46·25	38·8	36·2	18	L.
Carbon disulphide	Methylal	46·25	42·25	37·25	46	Z., Y.
Acetone	Diethylamine	56·1	55·5	51·35	38·2	Ma., L.
Diethyl ketone	Nitromethane	102·2	101·2	99·1	45	L.
Methyl propyl ketone	Nitromethane	102·25	101·2	99·15	44	L.
Acetone	Methyl acetate	56·25	57·0	56·1	55	Ry., L.
Methyl ethyl ketone	Methyl propionate	79·6	79·7	79·25	52	L.
Methyl ethyl ketone	Propyl formate	79·6	80·8	79·45	55	L.
Diethyl ketone	Propyl acetate	102·2	101·55	101·35	40	L.
Methyl propyl ketone	Propyl acetate	102·25	101·55	101·35	38	L.
Ethyl chloracetate	Methyl lactate	143·5	144·8	140·4	52	L.
Isoamyl chloracetate	Ethyl oxalate	190·5	185·0	181·5	35	L.
Methyl ethyl ketone	Water	79·57	100·0	73·57	88·6	M., L.
Pyridine	Water	115·5	100·0	92·6	57	Gd., L.

TABLE 13

BINARY AZEOTROPIC MIXTURES—MAXIMUM BOILING POINT

Substances in mixture.		Boiling points.			Percentage of A. by weight in mixture.	Observer.
A.	B.	A.	B.	Az. mixture.		
Water	Nitric acid	100·0	86·0	120·5	32	R.
Water	Hydrochloric acid	100·0	−84	110	79·76	R.
Water	Hydrobromic acid	100·0	−73	126	52·5	R.
Water	Hydriodic acid	100·0	−34	127	43	R.
Water	Hydrofluoric acid	100·0	19·4	120	63	R.
Water	Formic acid	100·0	99·9	107·1	22·5	R.
Water	Formic acid	100·0	100·75	107·3	22·5	L.
Water	Perchloric acid	100·0	110·0	203	28·4	R.
Formic acid	Diethyl ketone	100·75	102·2	105·4	33	L.
Formic acid	Methyl propyl ketone	100·75	102·25	105·3	32	L.
Propionic acid	Acetylacetone	140·7	138	144	70	L.
Hydrochloric acid	Methyl ether	−84	−23·65	−1·5	60	F., L.
Formic acid	Pyridine	100·7	116·7	149	18	Ga., An., Z., L.
Formic acid	Picoline, α	100·7	134	158	25	Ga., L.
Acetic acid	Pyridine	118·5	115·5	139·7	35	Ga., An., L.
Acetic acid	Picoline, α	118·5	134	146	40	Ga., Z.
Propionic acid	Pyridine	140·7	116·7	150–151	31·5	Ga., An., L.
Phenol	Cyclo-hexanol	181·5	160·65	182·45	90·5	L.
Phenol	Sec. octyl alcohol	181·5	178·5	184·65	50	L.
Phenol	Glycol	181·5	197·4	199	22	L.
Cresol, p.	Glycol	201·8	197·4	203	68	L.
Cresol, p.	Benzyl alcohol	201·8	205·5	207·0	38	L.
Phenol	Benzaldehyde	181·5	179·2	185·6	51	L.
Phenol	Aniline	181·5	184·35	186·22	42	L.
Cresol, o.	Dimethyl aniline	190·8	194·05	195·6	<30	L.
Cresol, o.	Methyl aniline	190·8	196·1	196·7	10	L.
Cresol, p.	Toluidine, p.	201·8	200·3	204·35	57	L.
Cresol, p.	Toluidine, o.	201·8	200·7	204·5	53	L.
Cresol, p.	Ethyl aniline	201·8	206·05	207·2	<20	L.
Cresol, o.	Acetophenone	190·8	202	203·7	22	L.
Cresol, p.	Fenchone	201·8	193	205	72	L.
Cresol, p.	Acetophenone	201·8	202	208·45	50	L.
Cresol, p.	Camphor	201·8	208·9	213·15	30	L.
Cresol, m.	Acetophenone	202·8	202	209	52	L.
Cresol, m.	Camphor	202·8	208·9	213·5	35	L.
Phenol	Isoamyl butyrate	181·5	178·6	185·6	61	L.
Phenol	Isobutyl lactate	181·5	182·15	189·05	46	L.
Phenol	Ethyl oxalate	181·5	185·0	189·1	43	L.
Cresol, o.	Ethyl oxalate	190·8	185·0	193·9	67	L.
Cresol, o.	Methyl benzoate	190·8	199·55	200·8	30	L.
Cresol, p.	Methyl benzoate	201·8	199·55	204·7	35	L.
Cresol, p.	Isoamyl lactate	201·8	202·4	207·25	48	L.
Cresol, m.	Methyl benzoate	202·8	199·55	205·6	38	L.
Cresol, m.	Isoamyl lactate	202·8	202·4	207·6	50	L.
Acetone	Chloroform	56·4	61·2	64·7	20	Ry., Th., L., Ty., Y.
Diethyl ketone	Dichloro-bromo-methane	102·2	90·2	103·0	64	L.
Camphor	Benzal chloride	208·9	205·1	209·7	75	L.
Chloroform	Ethyl formate	61·2	54·15	62·7	87	L.
Chloroform	Methyl acetate	61·2	57·05	64·8	77	Ry., L.
Pentachlorethane	Mesitylene	161·95	164·0	166	44	L.
Hydrogen bromide	Hydrogen sulphide	..	..	−70	61	S. & B.

TABLE 14

TERNARY AZEOTROPIC MIXTURES—MINIMUM BOILING POINT

Substances in mixture			Boiling points			Az. mixture	Percentages by weight in Az. mixture			Observer
A.	B.	C.	A.	B.	C.		A.	B.	C.	
Ethyl Alcohol	Water	Benzene	78·3	100·0	80·2	64·85	18·5	7·4	74·1	Y.
Isopropyl alcohol	Water	Benzene	82·45	100·0	80·2	66·5	18·7	7·5	73·8	Y.
Tert. butyl alcohol	Water	Benzene	82·55	100·0	80·2	67·3	21·4	8·1	70·5	Y.
n. Propyl alcohol	Water	Benzene	97·2	100·0	80·2	68·5	9·0	8·6	82·4	Y.
Allyl alcohol	Water	Benzene	97·05	100·0	80·14	68·21	9·2	8·6	82·2	W. & A., L.
Ethyl alcohol	Water	Cyclo-hexane	78·3	100·0	80·75	62·1	17	7	76	L.
Isopropyl alcohol	Water	Cyclo-hexane	82·45	100·0	80·75	64·3	18·5	7·5	74	L.
Tert. butyl alcohol	Water	Cyclo-hexane	82·55	100·0	80·75	65	21	8	77	L.
n. Propyl alcohol	Water	Cyclo-hexane	97·2	100·0	80·75	66·55	10	8·5	81·5	L.
Allyl alcohol	Water	Cyclo-hexane	96·95	100·0	80·75	66·18	11	8	81	L.
Sec. butyl alcohol	Water	Cyclo-hexane	99·6	100·0	80·75	67	...	...	...	L.
Ethyl alcohol	Water	Cyclo-hexadiene, 1·3	78·3	100·0	80·8	63·6	20	7	73	L.
Isopropyl alcohol	Water	Cyclo-hexadiene, 1·3	82·45	100·0	80·8	65·7	...	...	...	L.
Tert. butyl alcohol	Water	Cyclo-hexadiene, 1·3	82·55	100·0	80·8	66·7	...	...	...	L.
n. Propyl alcohol	Water	Cyclo-hexadiene, 1·3	97·2	100·0	80·8	67·75	12	9	80·8	L.
Allyl alcohol	Water	Cyclo-hexadiene, 1·3	96·95	100·0	80·8	67·5	...	...	...	L.
Ethyl alcohol	Water	Cyclo-hexene	78·3	100·0	82·75	64·05	20	7	73	L.
Isopropyl alcohol	Water	Cyclo-hexene	82·45	100·0	82·75	66·1	21·5	7·5	71	L.
Tert. butyl alcohol	Water	Cyclo-hexene	82·55	100·0	82·75	67	...	...	...	L.
n. Propyl alcohol	Water	Cyclo-hexene	97·2	100·0	82·75	68·2	11·5	9	79·5	L.
Allyl alcohol	Water	Cyclo-hexene	96·95	100·0	82·75	67·95	11	8·5	80·5	L.
Isobutyl alcohol	Water	Cyclo-hexene	108	100·0	82·75	69·5	...	...	...	L.
n. Butyl alcohol	Water	Cyclo-hexene	116·9	100·0	82·75	70·2	...	...	...	L.
Ethyl alcohol	Water	Methyl cyclo-hexane	78·3	100·0	101·8	70·5	...	...	...	L.
Ethyl alcohol	Water	Toluene	78·3	100·0	110·7	74·55	...	...	...	L.
Isopropyl alcohol	Water	Toluene	82·45	100·0	110·7	76·2	...	...	...	Y., L.
n. Propyl alcohol	Water	Toluene	97·2	100·0	110·7	80·05	...	...	...	L.
Allyl alcohol	Water	Toluene	96·95	100·0	110·7	80·2	...	...	...	L.

TABLE 14—continued

Substances in mixture.			Boiling points.			Az. mixture.	Percentages by weight in Az. mixture.			Observer.
A.	B.	C.	A.	B.	C.		A.	B.	C.	
Dimethyl ethyl Carbinol	Water	Toluene	102°	100·0°	110·7°	82°	…	…	…	L.
Isobutyl alcohol	Water	Toluene	108	100·0	110·7	83	…	…	…	L.
Isobutyl alcohol	Water	Ethyl benzene	108	100·0	136·15	89·5	…	…	…	L.
Ethyl alcohol	Water	Diallyl	78·3	100·0	60·2	52	…	…	…	L.
Ethyl alcohol	Water	n. Hexane	78·3	100·0	68·95	56·6	…	…	…	Y.
Isopropyl alcohol	Water	n. Hexane	82·45	100·0	68·95	58·2	…	…	…	Y., L.
Tert. butyl alcohol	Water	n. Hexane	82·55	100·0	68·95	58·9	…	…	…	L.
n. Propyl alcohol	Water	n. Hexane	97·2	100·0	68·95	59·95	…	…	…	Y.
Allyl alcohol	Water	n. Hexane	96·95	100·0	68·95	59·7	5	5	90	L.
Sec. butyl alcohol	Water	n. Hexane	99·6	100·0	68·95	61·1	…	…	…	L.
Ethyl alcohol	Water	n. Heptane	78·3	100·0	98·45	69·5	…	…	…	L.
Ethyl alcohol	Water	Acetylene dichloride, tr.	78·3	100·0	48·35	44·4	4·4	1·1	94·5	Cv.
Ethyl alcohol	Water	Acetylene dichloride, cis.	78·3	100·0	60·25	53·8	6·65	2·85	90·5	Cv.
Ethyl alcohol	Water	Chloroform	78·3	100·0	61·15	55·5	4·0	3·5	92·5	W. & F.
Ethyl alcohol	Water	Isobutyl chloride	78·3	100·0	68·85	58·6	13	4·5	82·5	L.
Ethyl alcohol	Water	Propyl bromide	78·3	100·0	71·0	60	12	5	83	L.
Ethyl alcohol	Water	Ethyl iodide	78·3	100·0	72·3	61	9	5	86	L.
Ethyl alcohol	Water	Carbon tetrachloride	78·3	100·0	76·75	61·8	9·7	4·3	86·0	Hi., L.
Ethyl alcohol	Water	Ethylene dichloride	78·3	100·0	83·7	66·7	17	5	78	L.
Ethyl alcohol	Water	Trichlorethylene	78·3	100·0	86·95	67·25	25	5	69	L.
Ethyl alcohol	Water	Isobutyl bromide	78·3	100·0	91·6	69·5	25	8	65	L.
Tert. butyl alcohol	Water	Carbon tetrachloride	82·55	100·0	76·75	64·7*	11·9	3·1	85·0	A. (*768 mm.)
Allyl alcohol	Water	Carbon tetrachloride	96·95	100·0	76·75	65·15	11	5	84	L.
Allyl alcohol	Water	Trichlorethylene	96·95	100·0	86·95	71·4	12·5	7·5	80	L.
Allyl alcohol	Water	Dichloro-bromo-methane	96·95	100·0	90·2	76	…	…	…	L.
n. Propyl alcohol	Water	Carbon tetrachloride	97·2	100·0	76·75	65·4	11	5	84	L.
n. Propyl alcohol	Water	Trichlorethylene	97·2	100·0	86·95	71·55	12	7	81	L.
n. Propyl alcohol	Water	Allyl iodide	97·2	100·0	102	78·15	20	8	72	L.
n. Propyl alcohol	Water	Propyl iodide	97·2	100·0	102·4	78·25	…	…	…	L.

TABLE 14—continued

Substances in mixture			Boiling points			Az. mixture	Percentages by weight in Az. mixture			Observer
A.	B.	C.	A.	B.	C.		A.	B.	C.	
n. Propyl alcohol	Water	Tetrachlorethylene	97·2	100·0	120·8	88				L.
Isobutyl alcohol	Water	Trichlorethylene	108	100·0	86·95	72·7				L.
Isobutyl alcohol	Water	Dichloro-bromo-methane	108	100·0	90·2	77·5	8·4	9·0	82·6	L.
Ethyl alcohol	Water	Ethyl acetate	78·3	100·0	77·15	70·23	20	20	60	W., M.
Propyl alcohol	Water	Diethyl ketone	97·2	100·0	102·2	81·24				L.
Propyl alcohol	Water	Nitromethane	97·6	100·0	101·2	81·85	22·2	3·0	74·8	L.
Methyl ethyl ketone	Water	Carbon tetrachloride	79·6	100·0	76·75	65·7	65	18	17	A.
Diethyl ketone	Water	Nitromethane	102·2	100·0	101·2	82·4				L.
Methyl alcohol	Ethyl bromide	Carbon disulphide	64·7	38·4	46·25	33·92	10	50	40	L.
Methyl alcohol	Methyl iodide	Carbon disulphide	64·7	42·6	46·25	35·95				L.
Methyl alcohol	Ethyl bromide	Trimethylethane	64·7	38·4	37·15	31·4	15	55	30	L.
Methyl alcohol	Methyl iodide	Methylal	64·7	42·6	42·25	38·5				L.
Methyl alcohol	Methyl acetate	Carbon disulphide	64·7	57·0	46·25	37				L.
Methyl alcohol	Methylal	Carbon disulphide	64·7	42·25	46·25	35·55	7	38	55	L.
Ethyl alcohol	Acetone	Isobutyl chloride	78·3	56·25	68·85	52				L.
Isopropyl alcohol	Chloroform	n. Hexane	82·45	61·2	68·95	58·3				L.
Glycol	Ethyl acetate	Cyclo-hexane	197·4	77·05	80·75	68·3				L.
Propionic acid	Aniline	Carvene	140·7	184·35	177·8	162·45				L.
Isobutyric acid	Ethylene bromide	Chlorobenzene	154·35	131·5	131·8	127·5				L.
Isobutyric acid	Bromobenzene	Pinene, α	154·35	156·1	155·8	146·4				L.
Isobutyric acid	Anisol	Pinene, α	154·35	153·85	155·8	143·9				L.
Methyl formate	Ethyl bromide	Isopentane	31·9	38·4	27·95	16·95	52	5	43	L.
Methyl formate	Ethyl bromide	Trimethylethylene	31·9	38·4	37·15	24·1				L.
Methyl formate	Carbon disulphide	Trimethylethylene	31·9	46·25	37·15	24				L.
Methyl formate	Ether	n. Pentane	31·9	34·6	36·15	20·4	40	8	52	L.
Methyl formate	Ether	Trimethylethylene	31·9	34·6	37·15	24				L.
Amyl formate	Tetrachlor-ethylene	Paraldehyde	123·6	120·8	124	117·6	25	45	30	L.
Propyl lactate	Phenetol	Menthene	171·7	171·5	170·8	163·0	31	33	36	L.
Isobutyl lactate	Benzyl chloride	Carvene	182·15	179·35	177·8	172·5				L.

REFERENCES IN TABLES 12 TO 14

A.   =Atkins, *Trans. Chem. Soc.*, 1920, **117**, 218 ; and (with G. Barr) *Reports of Advisory Com. for Aeronautics*, 1916, T, 752.

An.   =André, *Compt. rend.*, 1897, **125**, 1187 ; 1898, **126**, 1105 ; *Bull. Soc. chim.*, Paris, 1899, (3), **21**, 278 and 285.

B.   =Baud, *Bull. Soc. chim.*, Paris, 1909, (4), **5**, 1022.

Be.   =Berthelot, M.P.E., *Compt. rend.*, 1863, **57**, 430 and 985 ; *Ann. Ch. Ph.*, 1864, (4), **1**, 384.

Ch.   =Chancel, *Compt. rend.*, 1869, **68**, 659 ; *Bull. Soc. chim.*, Paris, 1869, (2), **12**, 87.

Cv.   =Chavanne, *Bull. Soc. chim. de Belg.*, 1913, **27**, 205 ; *Compt. rend.*, 1914, **158**, 1698.

D.   =Dittmar, *Chem. News*, 1876, **33**, 53 ; *Proc. Glasgow Philos. Soc.*, 1877, **10**, 63

F.   =Friedel, *Compt. rend.*, 1875, **81**, 152 ; *Bull. Soc. chim.*, Paris, 1875, II., (2), **24**, 160.

Ga.   =Gardner, *Ber.*, 1890, **23**, 1587.

Go.   =Golodetz, Patent 1911, Kl. 12a, No. 286,425.

Gol.   =Goldschmidt and Constam, *Ber.*, 1883, **16**, 2976.

Gu.   =Guthrie, *Phil. Mag.*, 1884., [V.], **18**, 512.

H.   =Holley, *Journ. Amer. Chem. Soc.*, 1902, **24**, 448.

Hi.   =Hill, *Trans. Chem. Soc.*, 1912, **101**, 2467.

Hy.   =Haywood, *Journ. Phys. Chem.*, 1897, **1**, 232 ; 1899, **3**, 317.

J. & Y.   =Jackson and Young, *Trans. Chem. Soc.*, 1898, **73**, 922.

J. & G.   =Jana and Gupta, *Journ. Amer. Chem. Soc.*, 1914, **36**, 115.

K.   =Konowalow, *Wied. Ann.*, 1881, **14**, 34.

L.   =Lecat, *La Tension de vapeur des mélanges de liquides. l'Azéotropisme*, Brussels, 1918.

Li.   =Linebarger, *Journ. Amer. Chem. Soc.*, 1895, **17**, 615 and 690.

M.   =Merriman, *Trans. Chem. Soc.*, 1913, **103**, 1790.

Ma.   =Marshall, *ibid.*, 1906, **89**, 1350.

Ne.   =Nernst, *Zeitschr. physik. Chem.*, 1891, **8**, 110.

N. & W.   =Noyes and Warfel, *Journ. Amer. Chem. Soc.*, 1901, **23**, 463.

P.   =Pettit, *Journ. Phys. Chem.*, 1899, **3**, 349.

R.   =Roscoe, *Quart. Journ. Chem. Soc.*, 1861, **13**, 146 ; 1862, **15**, 270 ; *Proc. Roy. Soc.*, 1862, **11**, 493.

Ro.   =Rosanoff and others, *Journ. Amer. Chem. Soc.*, 1909, **31**, 953 ; 1914, **36**, 1803 and 1993 ; 1915, **37**, 301.

Ry.   =Ryland, *Amer. Chem. Journ.*, 1899, **22**, 384.

S. & B.   =Steele and Baxter, *Trans. Chem. Soc.*, 1910, **97**, 2607.

Sch.   =Schreinemakers, *Zeitschr. physik. Chem.*, 1904, **47**, 445 ; **48**, 257.

Th.   =Thayer, *Journ. Phys. Chem.*, 1898, **2**, 382 ; 1899, **3**, 36.

T.   =Thorpe, *Trans. Chem. Soc.*, 1879, **35**, 544.

Ti.   =Timofeëv, *Bull. de l'Inst. polytech. de Kiev*, 1905, p. 1.

Ty.   =Tyrer, *Trans. Chem. Soc.*, 1912, **101**, 1104.

Vr.   =Vrevskij, *Zeitschr. physik. Chem.*, 1912–13, **81**, 1 ; 1913, **83**, 551.

V. & D.   =Vincent and Delachanel, *Compt. rend.*, 1880, **90**, 747.

W.   =Wade, *Trans. Chem. Soc.*, 1905, **87**, 1656.

W. & A.   =Wallace & Atkins, *ibid.*, 1912, **101**, 1179 and 1958.

W. & F.   =Wade and Finnemore, *ibid.*, 1904, **85**, 938 ; 1909, **95**, 1842.

W. & M.   =Wade and Merriman, *ibid.*, 1911, **99**, 997.

Y.   =Young, *ibid.*, 1902, **81**, 707 and 768 ; 1903, **83**, 68 and 77, and data not published elsewhere.

Y. & F.   =Young and Fortey, *ibid.*, 1902, **81**, 717, 739 and 752 ; 1903, **83**, 45.

Z.   =Zawidski, *Zeitschr. physik. Chem.*, 1900, **35**, 129.

See also Hartman, " On the First Plait in Van der Waals's Free Energy Surface for Mixtures of Two Substances," *Journ. Phys. Chem.*, 1901, **5**, 425.

No ternary mixture of maximum boiling point has yet been discovered, but Lecat has pointed out a probable case of a quaternary mixture of minimum boiling point. Of the four liquids, methylal, methyl iodide, methyl alcohol and carbon disulphide, each of the six possible pairs and each of the four possible triplets form azeotropic mixtures of minimum boiling point, and Lecat considers it probable that a mixture of all four liquids boils at a slightly lower temperature than the most volatile of the ternary mixtures. The data are given below :—

Substance.				Boiling points.		
Methylal .	.	. A 42·25°	AB 39·35°	ABC 38·5°		ABCD 35·5° (?)
Methyl iodide .		. B 42·6	AC 41·82	ABD 37·2 (?)		
Methyl alcohol.		. C 64·7	AD 37·25	ACD 35·55		
Carbon disulphide .		D 46·25	BC 39·0	BCD 35·95		
			BD 41·65			
			CD 37·65			

INFLUENCE OF PRESSURE (TEMPERATURE) ON THE COMPOSITION OF AZEOTROPIC MIXTURES

For many years after the discovery of the formation of mixtures of constant boiling point it was thought that these azeotropic mixtures were definite chemical compounds. Thus Bineau[1] described the azeotropic mixture of hydrochloric acid and water as a hydrate of the acid, and Chancel[2] regarded the azeotropic mixture of propyl alcohol and water as a hydrate of the alcohol.

Roscoe,[3] however, showed that the azeotropic concentration of aqueous mineral acids varied with the pressure, and this would not be the case if definite chemical compounds were formed.

So, also, in the case of mixtures of minimum boiling point (maximum vapour pressure) the composition of the azeotropic mixtures has been found to vary with the pressure (temperature) by Konowalow, Young and Fortey, Homfray, Wade and Merriman, Lecat and others, and in no case has the composition of a mixture of either maximum or minimum boiling point been found to be independent of the pressure.

Perhaps the most interesting case yet investigated is that of ethyl alcohol and water. Very careful determinations of the composition of the azeotropic mixtures formed by distillation of aqueous alcohol under a series of pressures were made by Wade and Merriman,[4] and later by Merriman,[5] who also redetermined the vapour pressures of ethyl alcohol. The results obtained by Merriman are given in the table on the following page.

[1] Bineau, "Sur les combinaisons de l'eau avec les hydracides," *Ann. Ch. Ph.*, 1843, (3), 7, 257.
[2] Chancel, "Nouvelles Recherches sur l'alcool propylique de fermentation," *Compt. rend.*, 1869, 68, 659.
[3] *Loc. cit.*
[4] Wade and Merriman, "Influence of Water on the Boiling Point of Ethyl Alcohol at Pressures above and below the Atmospheric Pressure," *Trans. Chem. Soc.*, 1911, 99, 997.
[5] Merriman, "The Vapour Pressures of the Lower Alcohols and their Azeotropic Mixtures with Water, Part I., Ethyl Alcohol," *ibid.*, 1913, 103, 628.

TABLE 15

Pressure mm.	Boiling points.			Percentage of water in Az. mixture.
	Azeotropic mixture.	Alcohol.	Difference.	
1451·3	95·35°	95·58°	0·23°	4·75
1075·4	87·12	87·34	0·22	4·65
760·0	78·15	78·30	0·15	4·4
404·6	63·04	63·13	0·09	3·75
198·4	47·63	47·66	0·03	2·7
129·7	39·20	39·24	0·04	1·3
94·9	33·35	33·38	0·03	0·5
70·0	...	27·96	...	...

As the pressure falls the percentage of water in the azeotropic mixture diminishes, and at pressures lower than about 75 mm. no such mixture is formed.

Merriman [1] has also very carefully investigated the effect of pressure on the composition of the binary azeotropic mixtures of ethyl acetate and water, ethyl acetate and ethyl alcohol, and of the ternary azeotropic mixtures of ethyl acetate, ethyl alcohol and water. A few of the data are given below :—

TABLE 16

Pressure in mm.	Ethyl acetate and water. Percentage of water.	Ethyl acetate and ethyl alcohol. Percentage of ethyl alcohol.
25	3·60	12·81
50	4·00	14·49
100	4·70	16·97
200	5·79	20·52
400	7·11	25·37
760	8·43	30·98
1000	9·26	33·86
1500	10·04	39·07

TABLE 17

Ethyl Acetate, Ethyl Alcohol and Water

Pressure in mm.	Percentage composition.		
	Ethyal acetate.	Ethyl alcohol.	Water.
25·0	92·0	4·0	4·0
178·5	88·4	5·6	6·0
503·6	84·8	7·2	8·0
760·0	82·6	8·4	9·0
1090·8	79·9	10·6	9·5
1446·2	77·6	12·1	10·3

[1] Merriman, "The Azeotropic Mixtures of Ethyl Acetate, Ethyl Alcohol and Water at Pressures above and below the Atmospheric Pressure," *Trans. Chem. Soc.*, 1913, **103**, 1790.

From these results and others obtained by Young and Fortey, Homfray, Thayer, Lehfeldt, Ryland and Zawidski, Merriman arrives at the conclusion that, in all cases observed except one, the percentage of that component in a binary azeotropic mixture which has the lower value of dp/dt increases as the pressure decreases. Mixtures which follow the rule are ethyl acetate and water, ethyl acetate and ethyl alcohol, propyl alcohol and water, ethyl alcohol and propionitrile, benzene and ethyl alcohol, toluene and acetic acid, carbon disulphide and acetone. The only known exception to the rule is the ethyl alcohol-water mixture.

Vrevsky[1] has made careful determinations of the composition of azeotropic mixtures of alcohols and acids with water at different temperatures and gives the following general rule :—The concentration of that constituent of an azeotropic mixture which has the higher molecular heat of vaporisation increases with rise of temperature if the mixture is one of maximum vapour pressure, and diminishes if it is one of minimum vapour pressure.

The question has been treated analytically by Roozeboom,[2] Kuenen,[3] and by Van der Waals and Kohnstamm.[4]

[1] Vrevsky, "Über Zusammensetzung und Spannung des Dampfes binärer Flüssigkeitsgemische," Zeitschr. physik. Chem., 1912–13, **81**, 1 ; 1913, **83**, 551.

[2] Roozeboom, Die heterogenen Gleichgewichte vom Standpunkte der Phasenlehre, Braunschweig, 1904, vol. ii., Part I., p. 66.

[3] Kuenen, Handbuch der angewandten physikalischen Chemie, Leipzig, 1906, p. 114.

[4] Van der Waals and Kohnstamm, Lehrbuch der Thermodynamik in ihrer Anwendung auf das Gleichgewicht von Systemen mit Gasförmig-Flüssigen Phasen, Zweiter Teil, Leipzig, 1912.

CHAPTER V

COMPOSITION OF LIQUID AND VAPOUR PHASES. EXPERIMENTAL DETERMINATIONS

Evaporation into Vacuous Space.—When two volatile liquids —miscible, partially miscible or non-miscible—are placed together in a vacuous space, such as that over the mercury in a barometer tube, evaporation takes place and, as a rule, the composition of the residual liquid differs from that of the vapour. It is only when the liquids form a mixture of maximum or minimum vapour pressure—and therefore of constant boiling point—and when it is this particular mixture that is introduced into the vacuous space, that the composition of the vapour is the same as that of the liquid. In all other cases the vapour is richer in the more volatile of the two components into which the mixture tends to separate when distilled, these components being either the original substances from which the mixture was formed, or one of these substances and a mixture of the two which has a higher or lower boiling point than that of either of the original constituents.

If the volume of vapour is relatively very small, the composition of the residual liquid will differ only slightly from that of the original mixture, but if it is relatively very large, and if the boiling points of the two components into which the mixture tends to separate are not very close together, the residual liquid will be much richer in the less volatile component than the original mixture.

Methods employed.—The difficulties attending the experimental determination of the composition of liquid and vapour are, in most cases, very considerable, and, unless great care be taken, erroneous and misleading results may be obtained. The chief methods which have been employed are the following :—

1. A mixture of known composition is introduced into a suitable still ; a relatively very small quantity is distilled over, and the composition of the distillate—and, in some cases, of the residue also—is determined either (a) from its specific gravity (Brown),[1] (b) from its refractive power (Lehfeldt),[2] Zawidski,[3] (c) from its boiling point

---

[1] F. D. Brown, "Theory of Fractional Distillation," *Trans. Chem. Soc.*, 1879, **35**, 547 ; "On the Distillation of Mixtures of Carbon Disulphide and Carbon Tetrachloride," *ibid.*, 1881, **39**, 304.

[2] Lehfeldt, "Properties of Liquid Mixtures, Part II.," *Phil. Mag.*, 1898, [V], **46**, 42.

[3] Zawidski, "On the Vapour Pressures of Binary Mixtures of Liquids," *Zeitschr. physik. Chem.*, 1900, **35**, 129.

(Carveth),[1] or (d) by quantitative analysis ; but for organic liquids the last method is not generally suitable. The distillation may be carried out either in the ordinary manner under constant pressure, or at constant temperature.

2. A known volume of air is passed through the mixture at constant temperature [2] ; the total amount of evaporation is ascertained from the loss of weight of the liquid, and the weight of one component in the vapour is determined by quantitative analysis.

3. If a binary liquid whose components are in the ratio $x/(1-x)$ is in equilibrium with a vapour containing the same components in the ratio p_1/p_2, then a saturated vapour of this composition will bubble through the liquid without producing or itself undergoing any change. A method, based on this principle, has been devised and employed by Rosanoff and his co-workers.[3]

4. A distillation is carried out with a still-head kept at a constant temperature.[4] The composition of the distillate is determined in the usual manner ; that of the mixture distilled is ascertained from the temperature of the still-head, the boiling point-composition curve having been previously constructed.

First Method.—Distillations under constant pressure have been carried out by Duclaux [5] and, with great care, by F. D. Brown [6] and Rosanoff, Bacon and White.[7]

Brown's Apparatus.—The apparatus employed by Brown is shown in Fig. 24. " It consists of a copper vessel, s, shaped like an ordinary tin can, but provided with a long neck a. This neck and the upper portion of the vessel are covered with a copper jacket, c c c, which communicates with the inner vessel by means of some small holes round the upper part of a. This outer jacket is terminated below by a strip of copper placed obliquely to the axis of the vessel, and at its lower portion is fitted with a narrow tube, d, which serves to connect the still with the condenser. The vapour rising from the liquid in the vessel s passes through the holes at a, and then descending, passes out at d. The vapour as it rises is thus kept warm, and none of it is condensed until it has entered the outer jacket. Here a slight condensation is of no influence, as both vapour and liquid pass together into the receiver. The inclination of the bottom of the jacket serves to prevent the accumulation of any liquid at that part."

[1] Carveth, "The Composition of Mixed Vapours," *Journ. Phys. Chem.*, 1899, **3**, 193.

[2] Linebarger, "The Vapour Tensions of Mixtures of Volatile Liquids," *Journ. Amer. Chem. Soc.*, 1895, **17**, 615.

[3] Rosanoff, Lamb and Breithut, "Measurement of the Partial Vapour Pressures of Binary Mixtures," *J. Amer. Chem. Soc.*, 1909, **31**, 448 ; Rosanoff and Easley, *ibid.*, p. 953 ; Rosanoff and Bacon, *ibid.*, 1915, **37**, 301.

[4] F. D. Brown, "Fractional Distillation with a Still-head of Uniform Temperature," *Trans. Chem. Soc.*, 1881, **39**, 517.

[5] Duclaux, "Tension of the Vapour given off by a Mixture of Two Liquids," *Ann. Chim. Phys.*, 1878, [5], **14**, 305.

[6] Brown, "The Comparative Value of different Methods of Fractional Distillation," *Trans. Chem. Soc.*, 1880, **37**, 49.

[7] Rosanoff, Bacon and White, "Rapid Laboratory Method of Measuring the Partial Pressures of Liquid Mixtures," *J. Amer. Chem. Soc.*, 1914, **36**, 1803.

After heat has been applied, but before ebullition commences, a good deal of evaporation takes place and the mixture of warm air and vapour passes into the condenser, where most of the vapour is condensed. Brown, therefore, only made use of the data obtained from the first fraction of the distillate to ascertain the composition of the residue in the still at the moment of change from the first to the second fraction.

The form of receiver shown in Fig. 24 was used in order to avoid evaporation and consequent change in composition of the fractions, and also to allow of the distillation being carried on under reduced pressure. The quantity of liquid placed in the still was usually about 900 or 1000 grams, and in each experiment about one-fourth of the total amount was distilled over, and was collected in four fractions. The composition of the residual liquid at the end of the distillation was found in every case to agree satisfactorily with that calculated from the composition of the original mixture and of the fractions collected.

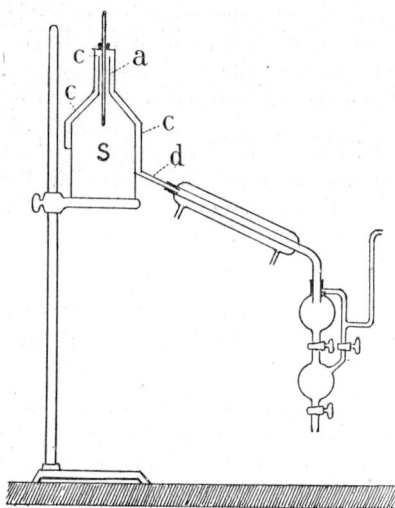

FIG. 24.—Brown's apparatus.

This apparatus was employed for mixtures of carbon disulphide and carbon tetrachloride, a preliminary series of determinations of the specific gravity of mixtures of these substances having been made, in order that the composition of any mixture might subsequently be ascertained from its specific gravity. A similar series of distillations was afterwards carried out under reduced pressure (about 430 mm.), the apparatus being connected with a large air reservoir and with a pump and gauge. In the earlier experiments with mixtures of carbon disulphide and benzene somewhat less satisfactory forms of still were employed.

Lehfeldt's Apparatus.—Lehfeldt's apparatus [1] for distillation at constant temperature is shown in Fig. 25. The still, which is in the form of a large test tube, is provided with a cork perforated with two holes, through one of which passes a thermometer, the bulb of which is covered with a little cotton-wool, which dips just below the surface of the liquid. Through the other hole passes the delivery tube, E, E, connected with the condenser F, which is provided with a tap, G, for drawing off the distillate, and a tube connected with a pressure gauge, air reservoir and air pump. The bell-jar, J, contains either cold water or a freezing mixture. The still is heated by water in

[1] *Loc. cit.*

a large beaker placed on a sand bath ; the water is constantly stirred and its temperature is registered by a thermometer. In order to prevent back condensation in the vertical part of the delivery tube, an incandescent electric lamp, with the ordinary conical shade, is lowered as close as possible to the water bath, and a cloth is hung round the whole ; the top of the apparatus is thus kept at least as hot as the bath.

The quantities of material employed by Lehfeldt were small ; about 30 c.c. of the mixture to be investigated were placed in the still and three fractions of about 1 c.c. each were usually collected and examined separately by means of a Pulfrich refractometer.

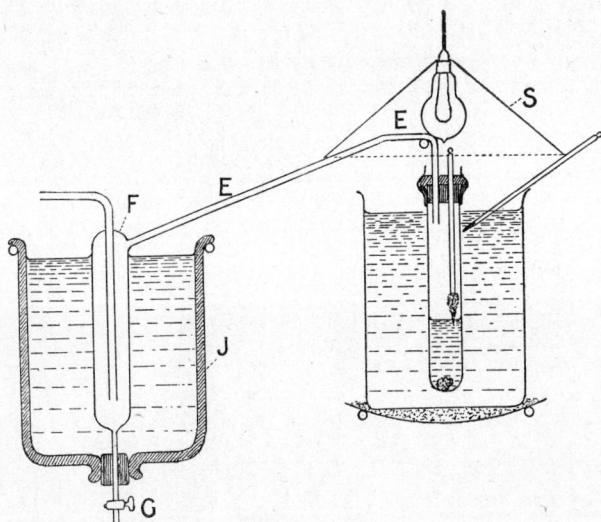

FIG. 25.—Lehfeldt's apparatus.

Steady ebullition was ensured by placing a piece of pumice stone, weighted with copper wire, in the still, and the pressure was adjusted from time to time to keep the boiling point as nearly constant as possible.

Preliminary determinations of the refractive powers of mixtures of the liquids investigated were made.

Zawidski's Apparatus.—Zawidski [1] employed an apparatus which is similar in principle, but more elaborate than Lehfeldt's ; it is shown in Fig. 26. The still, A, of about 200 c.c. capacity, and containing in each experiment from 100 to 120 c.c. of liquid, is heated by a water bath, G, provided with a stirrer and thermostat. Back condensation is prevented by coiling copper wire round the upper part of the delivery tube, H, and heating this with a small flame. Steady ebullition is brought about by means of a fine piece of platinum wire, P, 0·04 mm. diameter, near the bottom of the still, connected to two thicker platinum leads,

[1] *Loc. cit.*

which pass through the side tube opposite the delivery tube and are connected with a battery of three or four accumulators. The fine wire is heated by the current of about 0·4 ampère, and a steady stream of bubbles is thus produced. This method is also recommended by Bigelow.[1] The receiver, B, is of the same form as Lehfeldt's, except that the tube below, instead of being provided with a stopcock, is bent, as shown in the figure, and is connected with a second small receiver, C, into which, by diminishing the pressure in the reservoir, F, the first small portion of distillate, before the temperature and pressure have become constant, is carried over. The distillate required for examination (about 1 c.c.) is then collected in B and, after admission of air, C is removed, and, by slightly raising the pressure, the distillate is forced out of B into a little test tube.

The arrangement of the manometer, the pump and the two air reservoirs, D and F, is shown in Fig. 26. By means of the various

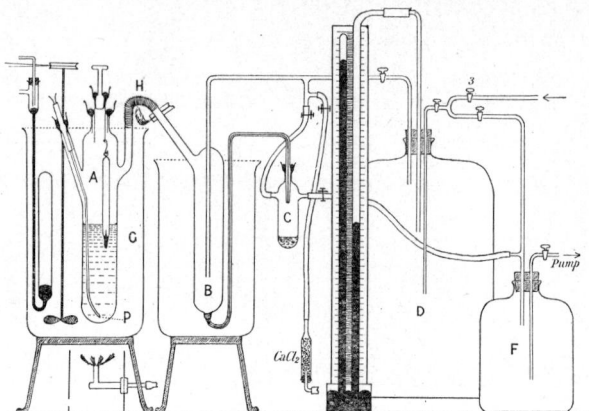

FIG. 26.—Zawidski's apparatus.

stopcocks, the pressure in F can be lowered a little below that in D, and air can be admitted into the apparatus, either through the stopcock 3 into D, or through the calcium chloride tube into C or B. A series of determinations was carried out in the following manner :— About 100 to 120 c.c. of one of the two substances was placed in A, and the pressure under which it boiled at the required temperature, t, was ascertained. A small quantity of the second liquid was then introduced into A, and the first distillation at the same temperature, t, carried out, the pressure being again noted. At the end of the distillation about 1 c.c. of the residual liquid in A was removed and placed in a small test tube for subsequent examination. A further small quantity of the second liquid was then added, and a second distillation was carried out as before, and these operations were repeated until the mixture in A became rich in the second substance.

[1] Bigelow, "A Simplification of Beckmann's Boiling Point Apparatus," *Amer. Chem. Journ.*, 1899, 280.

The series of operations was then repeated, starting with the second liquid in A and adding small quantities of the first.

Rosanoff, Bacon and White's Apparatus.—The method employed by Rosanoff, Bacon and White [1] was devised for rapidity of working and also to get over the difficulty that the composition of the first small fraction is liable to be seriously affected by the presence of even a trace of moisture in the liquid distilled.

The apparatus is shown in Fig. 27. It consists of a pear-shaped vessel with a long neck near the upper end of which are four circular openings for the escape of the vapour. A glass jacket, fused on to the rim of the neck, surrounds the flask and ends below in a tube through which the vapour escapes into a powerful worm condenser, and thence, in liquid form, into a receiver having several compartments for the convenient collection of consecutive fractions. The receiver communicates with the atmosphere through a tube filled with calcium chloride to keep out moisture. The neck of the pear-shaped boiling vessel is permanently stoppered above with a cork, which is made thoroughly vapour- and liquid-tight with shellac and sealing-wax. The cork carries an electric heater of platinum wire and, for the introduction and removal of liquid, an adapter tube reaching nearly to the bottom of the boiling vessel. Liquid is introduced with the aid of a separatory funnel as shown. The jacketed distillation vessel is all but completely immersed in a bath heated somewhat above the highest temperature to be attained by

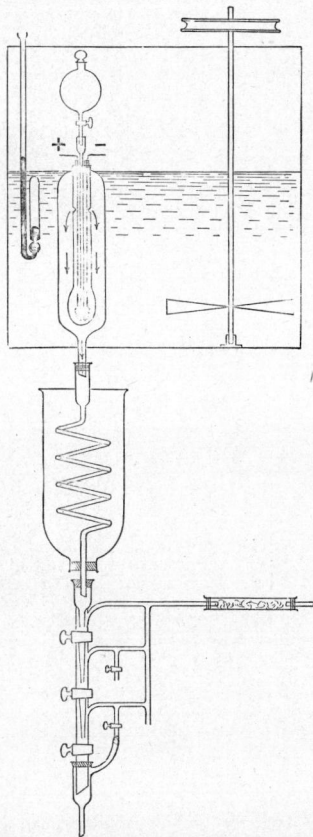

FIG. 27.

the boiling liquid. Reflux condensation in the distillation vessel is thus entirely prevented. In the final determinations a vessel of 125 c.c. capacity was used, and the shape of the platinum heater was such that a residue of barely 25 c.c. was left at the end of the distillation. The composition of the original liquid, the fractions and the residue was ascertained from their refractive indices.

The principle of the method may be best explained by an example. The original mixture of carbon disulphide and carbon tetrachloride contained 36·77 molar per cent of CS_2, the residue after distillation 22·18 molar per cent.

[1] *Loc. cit.*

Table 18

RESULT OF DISTILLATION

No. of fraction.	Weight of distillate in grams.	Molar per cent of CS_2.
1	16·48	59·35
2	19·43	58·96
3	15·86	57·21
4	16·00	55·25
5	23·75	52·73
6	17·80	49·85
7	25·23	46·79

Table 19

TABULATION OF RESULTS

Fractions combined.	Combined weight in grams.	Combined composition. Molar per cent CS_2.
No. 1	16·48	59·35
Nos. 1 + 2	35·91	59·11
„ 1 + 2 + 3	51·77	58·55
„ 1 + 2 + 3 + 4	67·77	57·79
„ 1 + 2 + 3 + 4 + 5	91·52	56·51
„ 1 + 2 + 3 + 4 + 5 + 6	109·32	55·47
„ 1 + 2 + 3 + 4 + 5 + 6 + 7	134·55	53·92
No. 7	25·23	46·79
Nos. 7 + 6	43·03	48·07
„ 7 + 6 + 5	66·78	49·76
„ 7 + 6 + 5 + 4	82·78	50·86
„ 7 + 6 + 5 + 4 + 3	98·64	51·92
„ 7 + 6 + 5 + 4 + 3 + 2	118·07	53·10
„ 7 + 6 + 5 + 4 + 3 + 2 + 1	134·55	53·92

The combined weights of distillate were in each case plotted against the combined composition (Fig. 28) and curves were drawn through the points. Only the point corresponding to the first fraction was badly off the curve which, however, was well defined by the remaining points. Possibly there was a trace of moisture in the mixture which would interfere with the first distillate.

Extrapolating the curves to zero weight of distillate in each case,

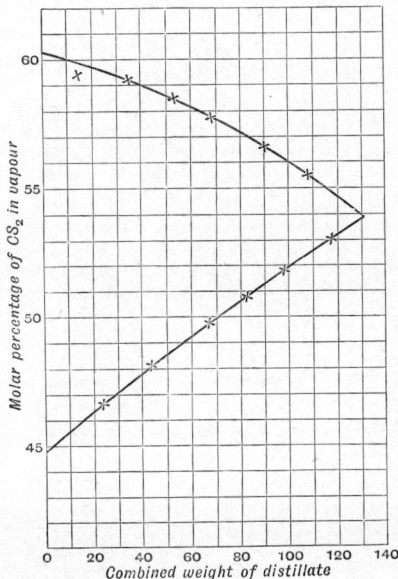

FIG. 28.

the following results are obtained for the composition of the vapour in contact with (a) the original mixture, (b) the residue :—

	Molar percentage of CS_2 in	
	Liquid.	Vapour.
Original mixture . .	36·77	60·35
Residue . . .	22·18	44·85

The results of this and other distillations agreed well with those obtained by the method of Rosanoff and Easley (p. 72).

Sources of Error.—The following sources of error should be noted and guarded against as far as possible :—

1. When a still is heated from below in the ordinary way, the vapour which is evolved before the temperature of the upper portion of the still has reached the boiling point of the liquid will be partially condensed, and the residual vapour will contain an excessive amount of the more volatile component. Partial fractionation will, in fact, go on, and the first portions of distillate will be too rich in the more volatile component.

2. If, while the distillation is progressing, the upper part of the still or delivery tube is exposed to the cooling action of the air, partial condensation of vapour will occur and a similar error to the first will be produced.

3. The air which is in the still before the liquid is heated will become saturated with vapour, and as it passes through the condenser part of this vapour will be condensed. In Brown's case, as already mentioned, this premature condensation of liquid was considerable in amount. The error introduced by the incomplete condensation of the vapour carried over by the air will partially compensate that referred to under No. 1, and, if the top of the apparatus is heated before the liquid in the still is boiled, may possibly more than counterbalance it.

4. If a thoroughly dehydrated liquid is exposed to the air even for a short time, especially if it is poured from one vessel to another so that a large surface of it is so exposed, a certain amount of moisture will almost invariably be absorbed. It is not only liquids which are regarded as hygroscopic which thus absorb moisture, but even substances, like benzene or the paraffins, which are classed as non-miscible with water. As a matter of fact, it is probable that no two liquids are absolutely non-miscible, and certainly all commonly occurring liquid organic compounds can dissolve appreciable quantities of water.

When such a liquid as benzene or carbon tetrachloride, containing a minute amount of dissolved water, is heated, a mixture of minimum boiling point is first formed, and the first small portion of distillate will contain the whole or at least the greater part of the water, and may probably be turbid. Suppose now that a mixture of benzene and carbon tetrachloride is being examined and that a minute amount of water has been absorbed during the preparation of the mixture or in pouring it into the still. The first portion of distillate will then contain the greater part of this water and the specific gravity, refractive index, boiling point and other physical properties of the distillate will be appreciably altered. A considerable error may thus be introduced in the estimation of the composition of the first fraction.

Brown rejected the data derived from the first distillate, and that is probably the safest plan to adopt. Zawidski rejected the first small portion of distillate and avoided the first and second sources of error, partially at any rate, by heating the delivery tube in the manner described, but in spite of this, the first results in some of his series of experiments appear to be less accurate than the later ones. Lehfeldt guarded against the first and second sources of error by heating the whole of the upper part of his apparatus by means of an incandescent lamp, as described.

The fourth source of error is probably of greater influence than is generally recognised. All that can usually be done is to dehydrate the liquids, to dry the apparatus as completely as possible, and to keep the liquids, as far as possible, out of contact with moist air; but Rosanoff, Bacon and White avoided all the errors to which the first fraction is liable by the method just described.

Second Method.—Determinations of the relative composition of liquid and vapour by passing a known volume of air through a mixture at constant temperature have been carried out by Winkelmann,[1] Linebarger,[2] Gahl,[3] and others.

In Linebarger's experiments a known volume of air (from 1 to 4 litres) was passed, at the rate of about 1 litre per hour, through the liquid mixture (40-80 grams) contained in a Mohr's potash apparatus, consisting of five small and two large bulbs. This apparatus was completely immersed in a suitable water bath, the temperature of which was kept constant within 0·05°.

When one of the components of the mixture was an acid, the amount of it in the vapour was estimated by absorption in potash or barium hydrate, and when one component contained sulphur or a halogen, the process of analysis by means of soda-lime was adopted. The total quantity of liquid evaporated, rarely more than 2 grams, was ascertained by weighing the bulbs containing the mixture before and after the experiment.

Preliminary determinations of the vapour pressures of a number of pure liquids were made in order to test the accuracy of the method; the results were more satisfactory with the less volatile liquids—such as chlorobenzene, bromobenzene and acetic acid—than with others, and they appear to be more accurate at low temperatures than at higher in the two cases in which such a comparison was made.

One pair of liquids investigated by Linebarger was benzene and carbon tetrachloride; the results obtained appear improbable and they differ widely from those of Lehfeldt,[2] of Zawidski,[2] and of Young and Fortey,[4] which are themselves in good agreement. It is to be feared, therefore, that for mixtures of volatile liquids the method cannot be regarded as satisfactory.

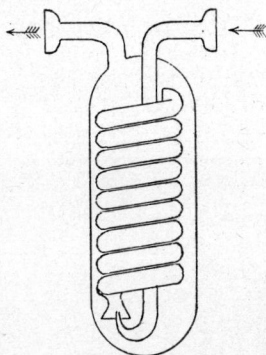

FIG. 29.—Gahl's apparatus.

The tendency—so far as pure liquids are concerned—is apparently for the vapour pressures to be too low, and it may be that the air was not completely saturated with vapour, or that partial condensation occurred before the mixed air and vapour was analysed.

It is possible that the apparatus employed by Gahl (Fig. 29) would give better results.

For liquids of low vapour pressure, on the other hand, the method appears capable of affording accurate results. See, for example, Washburn and Heuse.[5]

[1] Winkelmann, "On the Composition of the Vapour from Mixtures of Liquids," *Wied. Ann.*, 1890, **39**, 1.

[2] *Loc. cit.*

[3] Gahl, "Studies on the Theory of Vapour Pressure," *Zeitschr. physik. Chem.*, 1900, **33**, 179.

[4] Young and Fortey, "The Vapour Pressures and Boiling Points of Mixed Liquids, Part II.," *Trans. Chem. Soc.*, 1903, **83**, 45.

[5] Measurement of Vapour Pressure Lowering by the Air Saturation Method, *J. Amer. Chem. Soc.*, 1915, **37**, 309.

Third Method.—This method was first devised and employed by Rosanoff, Lamb and Breithut; the apparatus was afterwards improved by Rosanoff and Easley.[1] The method depends on the principle that if a binary liquid whose components are in the ratio $x/(1-x)$ is in equilibrium with a vapour containing the same components in the ratio p_1/p_2, then a saturated vapour of this composition will bubble through the liquid without producing or itself undergoing any change.

If the saturated vapour has the composition p_1'/p_2', the composition of the liquid will change until the ratio of its components has become $x'/(1-x')$ corresponding to the composition p_1'/p_2' of the vapour. If a saturated vapour of definite composition at a temperature T be passed through a liquid mixture of the same components, the temperature T' of the liquid will gradually approach and finally reach T as equilibrium is attained.

The vapour bubbling through the liquid after attainment of equilibrium may be condensed in any desired quantity and analysed. Two consecutive samples should have identical composition.

Assuming that the ratio of the partial pressures in the vapour is equal to the molar ratio of the components, which is usually the case, and knowing the total pressure from direct observation, p_1 and p_2 can easily be calculated. Where the assumption is not admissible the relation between the molar ratio and the partial pressures can be determined by separate experiments.

A saturated mixed vapour of constant composition is passed through a liquid mixture of the same substances in a vessel A, Fig. 30, called the equilibrium chamber. It is surrounded by a second vessel B, in which the required vapour of constant composition is produced by boiling a liquid whose composition is kept constant by the introduction of the more volatile component in such quantity as to keep the temperature constant, as indicated by the sensitive thermometer C, the bulb of which is immersed in the liquid. The vapour of constant composition produced in B enters the equilibrium chamber near its bottom, bubbles through the liquid, then passes through a condenser, and the condensed liquid is collected in a receiver, like that shown in Fig. 27, from which consecutive samples of distillate can be withdrawn without interrupting the distillation or disturbing the total pressure in the apparatus. The delicate thermometer D in the equilibrium chamber indicates when true equilibrium has been attained, and this is further demonstrated by the identity in composition of consecutive samples of distillate.

The whole distilling apparatus is placed in a thermostat kept at a temperature slightly higher than that of the boiling mixed liquid.

The additional quantities of the more volatile component are introduced into B by regulated distillation of the liquid by electric heating from a vessel E provided with a condenser. Reservoirs F and G were also provided, from which fresh quantities of the mixtures (or of either component) could be introduced into A or B. There

[1] *Loc. cit.*

was a device to guard against the danger of liquid in A being carried
as spray into the condenser. The form of the electric heater in B
was not as shown in Fig. 30 but was of a conical spiral form so as to
reach almost to the bottom of the tube and to ensure thorough admixture by bubbling. The arrangements for collecting successive portions
of distillate, for regulating and keeping the total pressure constant,
and other devices for avoiding errors, as well as the mode of manipulation, are fully described in the papers.

Experiments were carried out with six pairs of liquids, some of

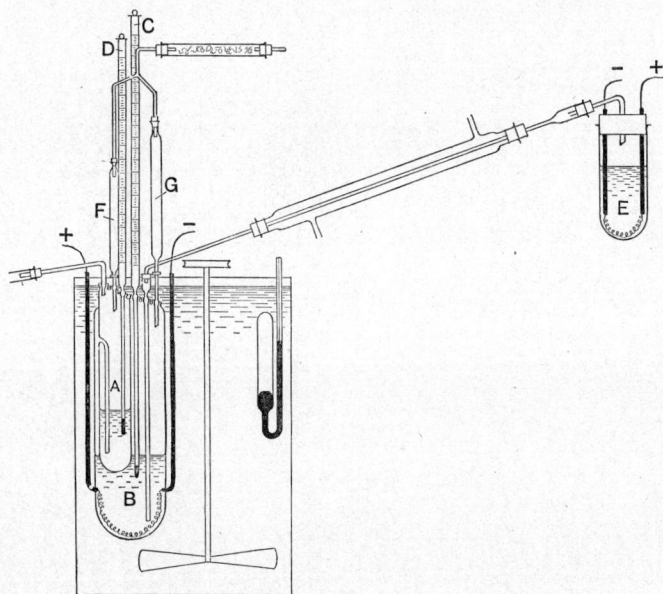

FIG. 30.

which formed mixtures of constant boiling point. There was good
agreement with Zawidski's results, and the authors conclude that
Brown's results are fair but not so accurate as Zawidski's.

Fourth Method.—In the course of his experiments with a still-
head kept at a constant temperature in a bath of liquid (p. 149),
Brown observed that the composition of the distillate was independent
of that of the mixture in the still, and depended only on the temperature
of the still-head. Mixtures of carbon disulphide with carbon tetrachloride and with benzene were specially examined and, for these two
pairs of liquids, Brown had previously determined the relations between
the composition of liquid and vapour, and also the relations between
boiling point and composition of liquid by Method I. From the curves
representing these relations he read :—

1. The composition of the mixtures which would evolve vapour of
the same composition as the distillates obtained with the still-head of
constant temperature ;

2. The boiling points of these mixtures.

He found that these boiling points, read from the curves, agreed closely with the temperatures of the still-head. The actual numbers are given in Table 20.

TABLE 20

I.	II.	III.	IV.	V.
Temperature of still-head $=t_1$.	Percentage of A in distillate.	Percentage of A in mixture evolving vapour of the composition given in Column II.	Boiling point of mixture referred to in Column III. $=t_2$.	$\triangle$ t_1-t_2.
A = Carbon tetrachloride ; B = Carbon disulphide.				
48·1°	6·7	14·1	48·4	− 0·3°
51·5	15·7	32·9	51·8	− 0·3
59·4	35·1	61·3	59·3	+ 0·1
63·7	48·5	73·2	63·6	+ 0·1
66·5	57·6	80·0	66·4	+ 0·1
72·1	78·5	91·6	71·8	+ 0·3
69·6	68·2	86·7	68·9	+ 0·7
70·0	69·6	87·4	69·0	+ 1·0
56·0	26·3	50·7	55·8	+ 0·2
56·5	27·6	52·2	55·9	+ 0·6
A = Benzene ; B = Carbon disulphide.				
48·6°	6·7	16·3	48·9°	− 0·3°
52·1	15·1	36·0	52·5	− 0·4
56·7	24·3	53·1	57·3	− 0·6
63·6	43·0	73·7	63·6	0·0
70·2	62·3	86·7	70·1	+ 0·1

In the last four determinations with carbon disulphide and carbon tetrachloride, the distillation was continued until not more than an occasional drop of distillate fell ; this may probably account for the larger differences between the two temperatures.

The agreement is sufficiently close with both pairs of liquids to allow of the statement being made that the temperature of the still-head is, approximately at any rate, the same as the boiling point of a mixture which, when distilled as in Method I, would give a distillate of the same composition as that actually collected.

Special experiments further to test the truth of this statement were made with mixtures of ethyl alcohol and water, but in this case the comparison was made by determining :—

1. The composition of the distillates when the still-head was kept at 81·8° and 86·5° respectively ;

2. The composition of the distillate from mixtures which boiled at the same two temperatures. The results obtained are shown in Table 21.

TABLE 21

Temperature of still-head and boiling point of mixture $=t°$.	Percentage of alcohol in distillate.	
	With still-head at $t°$.	From mixture boiling at $t°$.
81·8°	76·46	76·32
86·5	65·86	65·88

It will be seen that the agreement is very satisfactory, and it is evident that this method may be used to determine the relation between the composition of liquid and of vapour if the boiling point-composition curve has previously been constructed. Further experiments to test its applicability and the accuracy attainable have been carried out by Rosanoff, Lamb and Breithut,[1] and Rosanoff and Bacon.[2] The authors find that a double-walled metallic cylinder, open at both ends and immersed in liquid kept at constant temperature and vigorously stirred, gives better results than the ordinary spiral form of still-head.

The annular space $a a$, Fig. 31, between the two walls of the cylinder is bounded by a very large condensing surface ; the greatest height of the cylinder is 76 cm., the smallest height 70 cm. ; the mean diameter is 25 cm., and the width of the annular space is 0·95 cm.

A mixture of carbon disulphide and carbon tetrachloride containing

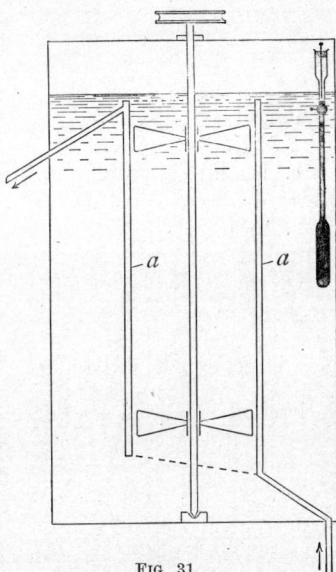

FIG. 31.

at first 26 per cent of the sulphide was distilled through this still-head, the temperature of which was 59·82° throughout the experiment. The distillate was collected in nine fractions,which contained respectively 60·8, 60·7, 60·7, 60·5, 60·7, 60·5, 60·7, 60·7 and 60·7 per cent of carbon disulphide, mean 60·67 per cent. On the other hand the distillate from a mixture which boiled at 59·82° contained 60·8 per cent of the sulphide. The two values are in very good agreement.

Rosanoff, Schulze and Dunphy [3] further investigated the distillation of a ternary mixture through the same still-head. They arrived at the following general conclusions, based on Brown's results and those of Rosanoff and his co-workers. In distillations with a still-head

[1] Loc. cit.
[2] "Fractional Distillation with Regulated Still-heads," J. Amer. Chem. Soc., 1915, 37, 301.
[3] "Fractional Distillation with Regulated Still-heads," J. Amer. Chem. Soc., 1915, 37, 1072.

maintained at a constant temperature, the composition of the distillate is at every instant identical with that of the vapour evolved by a mixture whose boiling point equals the temperature of the still-head. If the mixture is binary, the composition of the distillate is, in the course of a single distillation, constant. In those cases in which the binary boiling point curve passes through a maximum or a minimum, the composition of the distillate, although constant during the distillation, depends on that of the mixture originally placed in the still; it may have either of two values according to whether the mixture in the still is richer or poorer in one component than the mixture of constant boiling point. If the number of substances in the mixture is three or more, the composition of the distillate not only depends on that of the original mixture, but varies in the course of a single distillation. This variation, however, is moderate, and the nearer the constant temperature of the still-head is to the boiling point of the most volatile component, the more nearly constant is the composition of the distillate.

For binary mixtures the method certainly possesses the following advantages :—

1. The first part of the distillate may be altogether rejected, and the errors already referred to thus avoided;

2. It is not necessary to determine the composition of the mixture in the still.

On the other hand, this method would not be suitable for mixtures the boiling points of which vary very slightly with change of composition, such, for example, as mixtures of normal hexane and benzene containing, say, from 1 to 20 per cent of benzene.

Method employed by Carveth.[1]—The relation between the boiling points and the composition of mixtures of the two liquids is first determined under constant pressure, and the curve is constructed. A distillation is then carried out in a special apparatus with the object of determining simultaneously the boiling points of the liquid and of the distillate. If this could be done satisfactorily the composition of the mixture in the still and of the distillate at different stages of the distillation could then be ascertained from the curve.

The apparatus is ingenious, but there are errors involved in the method which, although they partially compensate each other, leave the results somewhat doubtful.

Carveth's actual results differ somewhat widely from Brown's, and the conclusion arrived at by Rosanoff and Easley (*loc. cit.*), who compared their own results with those of Brown, Zawidski and Carveth, is that Carveth's method is not reliable.

In the opinion of the author the methods most strongly to be recommended are those of Rosanoff and his co-workers.

---

[1] Carveth, "The Composition of Mixed Vapours," *Journ. Phys. Chem.*, 1899, **3**, 193.

CHAPTER VI

COMPOSITION OF LIQUID AND VAPOUR PHASES, CONSIDERED THEORETICALLY

Simplest Cases.—It has been seen that when two liquids are placed together in a closed vacuous space, the vapour pressure and the boiling point can only be accurately calculated from the vapour pressures of the components if (*a*) the liquids are non-miscible, or (*b*) they are miscible in all proportions and show no change of temperature or volume when mixed together, this being generally the case when the substances are chemically closely related. A similar statement may be made with regard to the composition of the vapour evolved from the two liquids.

NON-MISCIBLE LIQUIDS

The simplest case is that in which the two substances are non-miscible, for the composition of the vapour—like the vapour pressure and the boiling point—is independent of the relative quantities of the components, provided that they are both present in sufficient quantity and that evaporation can take place freely ; the composition of the vapour can be calculated if the vapour pressures and vapour densities of the non-miscible liquids are known.

Calling the vapour densities D_A and D_B, and the vapour pressures at the temperature t, P_A and P_B, we shall have, in a litre of the mixed vapour, 1 litre of A at $t°$ and P_A mm. pressure and 1 litre of B at $t°$ and P_B mm. The masses of vapour will therefore be $\dfrac{0\cdot0899 \times D_A \times 273 \times P_A}{(273+t) \times 760}$ and $\dfrac{0\cdot0899 \times D_B \times 273 \times P_B}{(273+t) \times 760}$ respectively, and the relative masses will be $\dfrac{D_A P_A}{D_B P_B}$ [Naumann,[1] Brown [2]].

Chlorobenzene and Water.—As an example we may again consider the case of chlorobenzene and water, the vapour pressures of which at 90° to 92° are given below.

---

[1] Naumann, "On the Distillation of Benzene, Toluene, etc., in a Current of Steam," *Berl. Berichte*, 1877, **10**, 1421, 1819, 2015 ; "On a New Method of determining Molecular Weights," *ibid.*, **10**, 2099.

[2] Brown, "Theory of Fractional Distillation," *Trans. Chem. Soc.*, 1879, **35**, 547.

TABLE 22

Temperature.	Vapour Pressures in mm.		
	Chlorobenzene.	Water.	Total.
90°	208·35	525·45	733·8
91	215·8	545·8	761·6
92	223·45	566·75	790·2

The vapour density of chlorobenzene $= 56·2$ and of water $= 9$.

$$\text{At } 90° \text{ the relative masses of vapour } \frac{m'_A}{m'_B} = \frac{56·2 \times 208·35}{9 \times 525·45} = 2·48.$$

$$\text{At } 91° \quad ,, \quad ,, \quad ,, \quad \frac{m'_A}{m'_B} = \frac{56·2 \times 215·8}{9 \times 545·8} = 2·47.$$

$$\text{At } 92° \quad ,, \quad ,, \quad ,, \quad \frac{m'_A}{m'_B} = \frac{56·2 \times 223·45}{9 \times 566·75} = 2·46.$$

The percentages of chlorobenzene by weight will therefore be 71·3, 71·2 and 71·1, respectively, at the three temperatures.

In the actual experiment (p. 41), in which 80 grams of water and 110 grams of chlorobenzene were distilled together under a barometric pressure of 740·2 mm. until about 3 grams of chlorobenzene and 40 of water remained in the flask, the distillate was collected in five fractions, which were found to contain the following percentages of chlorobenzene :—

1	72·5
2	71·5
3	72·0
4	70·4
5	71·7

Mean 71·6   Calculated 71·2

Water is apt to adhere in drops to the walls of a glass tube, but chlorobenzene flows much more freely ; the first fraction is therefore certain to contain too little water, and it would be fairer to reject it and to take the mean from the other four fractions. This would give 71·4 per cent of chlorobenzene, which agrees better with the calculated value.

The vapour pressure, boiling point and vapour composition can be calculated in a similar manner when more than two non-miscible liquids are present.

PARTIALLY MISCIBLE LIQUIDS

If the two liquids are miscible within limits, the observed vapour pressure, boiling point and vapour composition will differ from those calculated in the manner described, but the difference will be small if the miscibility is only slight.

Aniline and Water.—A mixture of 50 c.c. of aniline and 100 c.c. of water was distilled under a pressure of 746·4 mm. (p. 41). The percentage of aniline, calculated on the assumption that the liquids are non-miscible, would be 23·6, while the values observed were 18·7, 20·1, 19·7, 20·4, 20·4, 19·1. After the sixth fraction had been collected the aniline was in great excess ; a large quantity of water was added and the distillation continued, when the distillate contained 19·5 per cent of aniline. The composition of the distillate is clearly independent— within the somewhat wide limits of experimental error—of that of the mixture in the still, but the mean percentage of aniline, 19·9, is 3·7 lower than that calculated.

Isobutyl Alcohol and Water.—When the mutual solubility is greater, the difference between the observed and calculated values, as regards both temperature and vapour composition, is more marked. Thus isobutyl alcohol and water are miscible within fairly wide limits. At 0° a saturated solution of water in the alcohol contains 15·2 per cent of water, and the solubility increases as the temperature rises. At 18° one part of alcohol dissolves in 10·5 of water, but the solubility diminishes with rise of temperature, reaching a minimum at about 52°.

From Konowaloff's data [1] for the vapour pressures of isobutyl alcohol, the boiling point of the alcohol and water, when distilled together under normal pressure, would be 85·7° if the liquids were non-miscible ; the minimum boiling point observed by Konowaloff was actually 90·0°, or 4·3° higher (Young and Fortey, 89·8°). Again, the calculated percentage of isobutyl alcohol in the vapour would be 74·5, while that corresponding to the minimum boiling point is actually 66·8, the difference being 7·7.

It is clear, then, that when two non-miscible liquids are heated together we can calculate the vapour pressure, the boiling point and the vapour composition with accuracy from the vapour pressures of the components, but that, if the liquids are miscible within limits, the calculated values differ from those actually observed, the difference increasing with the mutual solubility of the liquids.

Infinitely Miscible Liquids

In the case of liquids which are miscible in all proportions, it is only, as has already been pointed out, when no appreciable change of volume or temperature occurs on admixture that the vapour pressures and the boiling points can be accurately calculated. The relation between the composition of the mixed liquid and that of the vapour evolved from it appears also to be a simple one only when the vapour pressure of the mixture is accurately expressed by the formula

$$P = \frac{\text{M}P_A + (100 - \text{M})P_B}{100}.$$

Formula of Wanklyn and Berthelot.—Many attempts have been made to find a general formula to represent the relation between the

---

[1] Konowaloff, " On the Vapour Pressures of Mixtures of Liquids," *Wied. Ann.*, 1881, **14**, 34.

composition of liquid and of vapour. In 1863, Wanklyn [1] and Berthelot [2] arrived independently at the conclusion that the composition of the vapour depends (a) on that of the liquid, (b) on the vapour pressures of the pure components at the boiling point of the mixture, (c) on the vapour densities of the components.

Calling m'_A and m'_B the relative masses of the two substances in the vapour, m_A and m_B their relative masses in the liquid, D_A and D_B their vapour densities and P_A and P_B their vapour pressures, the formula would be

$$\frac{m'_A}{m'_B} = \frac{m_A D_A P_A}{m_B D_B P_B}.$$

In 1879, Thorpe [3] observed that carbon tetrachloride and methyl alcohol form a mixture of minimum boiling point, and that, for this particular mixture, when $\dfrac{m'_A}{m'_B} = \dfrac{m_A}{m_B}$ the factor $\dfrac{D_A P_A}{D_B P_B}$ is approximately equal to unity.

Brown's Formula.—The subject was investigated experimentally by F. D. Brown in 1879–1881,[4] and he found that the Wanklyn-Berthelot formula could certainly not be accepted as generally true. A better result was obtained with the formula

$$\frac{m'_A}{m'_B} = \frac{m_A P_A}{m_B P_B},$$

but the agreement between the calculated and observed results was still closer when a constant, c, was substituted for the ratio P_A/P_B.

Applicability of Formula.—Brown's formula,

$$\frac{m'_A}{m'_B} = c \cdot \frac{m_A}{m_B},$$

is not generally applicable to liquids which are miscible in all proportions, and it is obvious that it cannot be true for two liquids which form a mixture of constant boiling point. The experimental evidence, however, which has so far been obtained, points to the conclusion that Brown's law is true for those infinitely miscible liquids for which the relation $P = \dfrac{M P_A + (100 - M) P_B}{100}$ holds good.

The number of such cases so far investigated is, perhaps, too small to allow of the statement being definitely made that the law is true in such cases, but the evidence is certainly strong, for not only does

[1] Wanklyn, " On the Distillation of Mixtures : a Contribution to the Theory of Fractional Distillation," *Proc. Roy. Soc.*, 1863, **12**, 534.

[2] Berthelot, " On the Distillation of Liquid Mixtures," *Compt. rend.*, 1863, **57**, 430.

[3] Thorpe, " A Contribution to the Theory of Fractional Distillation," *Trans. Chem. Soc.*, 1879, **35**, 544.

[4] Brown, " The Comparative Value of Different Methods of Fractional Distillation," *Trans. Chem. Soc.*, 1879, *loc. cit.* ; 1880, **37**, 49 ; " On the Distillation of Mixtures of Carbon Disulphide and Carbon Tetrachloride," 1881, **39**, 304 ; " Fractional Distillation with a Still-head of Uniform Temperature," 1881, **39**, 517.

Brown's formula hold good, within the limits of experimental error, for the two pairs of liquids investigated for which the relation between vapour pressure and molecular composition is represented by a straight line, but also, in other cases, it is found that the closer the approximation of the pressure-molecular composition curve to straightness, the smaller is the variation in the value of the " constant " c.

Zawidski's Results.—Strong evidence is afforded by two pairs of liquids, ethylene and propylene dibromides and benzene and ethylene dichloride, investigated by Zawidski.[1]

In the following tables are given the observed vapour pressures and those calculated from the formula

$$P = \frac{{\rm M}P_{\rm A} + (100 - {\rm M})P_{\rm B}}{100},$$

also the observed molecular percentages of A in the vapour and those calculated from the formula $\dfrac{m'_{\rm A}}{m'_{\rm B}} = c \cdot \dfrac{m_{\rm A}}{m_{\rm B}}$.

This formula may, of course, be written $\dfrac{{\rm M}'_{\rm A}}{{\rm M}'_{\rm B}} = c \cdot \dfrac{{\rm M}_{\rm A}}{{\rm M}_{\rm B}}$, where ${\rm M}'_{\rm A}$, ${\rm M}'_{\rm B}$, ${\rm M}_{\rm A}$ and ${\rm M}_{\rm B}$ represent gram-molecules instead of grams.

TABLE 23

$A = $ Propylene dibromide ; $B = $ ethylene dibromide ; $c = 1\cdot31$;
Temperature $= 85\cdot05°$; $P_{\rm B}/P_{\rm A}$ at $85° = 1\cdot357$.

Molecular percentage of A in liquid.	Vapour pressures.			Molecular percentage of A in vapour.		
	Observed.	Calculated.	△.	Observed.	Calculated.	△.
0·00	172·6	172·6	0·0	...	...	...
2·02	171·0	171·7	+0·7	1·85	1·55	−0·30
7·18	168·8	169·3	+0·5	6·06	5·60	−0·46
14·75	165·0	165·9	+0·9	12·09	11·66	−0·43
22·21	161·6	162·5	+0·9	18·22	17·89	−0·33
29·16	158·7	159·4	+0·7	23·50	23·90	+0·40
30·48	158·9	158·8	−0·1	23·96	25·08	+1·12
40·62	154·6	154·2	−0·4	34·25	34·31	+0·06
41·80	153·4	153·6	+0·2	34·51	35·41	+0·90
52·63	149·6	148·7	−0·9	45·28	45·89	+0·61
62·03	143·3	144·4	+1·1	55·35	55·50	+0·15
72·03	140·5	139·9	−0·6	65·86	66·28	+0·42
80·05	136·8	136·3	−0·5	74·94	75·39	+0·45
85·96	133·9	133·6	−0·3	82·45	82·38	−0·07
91·48	130·9	131·1	+0·2	89·50	89·13	−0·37
93·46	130·2	130·2	0·0	92·31	91·60	−0·71
96·41	128·4	128·8	+0·4	96·41	95·35	−1·06
98·24	127·3	128·0	+0·7	99·39	97·71	−1·68
100·00	127·2	127·2	0·0	...	...	...

The agreement between the observed and calculated percentages of propylene dibromide in the vapour is not altogether satisfactory,

[1] Zawidski, " On the Vapour Pressures of Binary Mixtures of Liquids," *Zeitschr. physik. Chem.*, 1900, **35**, 129.

G

but Zawidski states that the errors of experiment were much greater for this pair of substances than for others, owing to the small quantity of material at his disposal. The last two observed molecular percentages of propylene dibromide in the vapour, 96·41, and 99·39, are obviously too high ; the last should indeed be less than 98·24.

It will be seen later that when there is a real variation in the values of c, it is in the nature of a steady rise or fall from $A = 0$ to $A = 100$ per cent, whereas, in this case, the calculated values of c would be low at each end of the table.

The value of c, 1·31, does not differ greatly from the ratio of the vapour pressures, 1·357.

This series of experiments is of special interest because the two liquids are very closely related. Zawidski, however, found that the relation $P = \dfrac{M P_A + (100 - M) P_B}{100}$ held good with considerable accuracy for mixtures of benzene and ethylene dichloride, and the data for this pair of liquids are therefore given in full.

TABLE 24

$A =$ Ethylene dichloride ; $B =$ Benzene ; $c = 1·134$;
Temperature $= 49·99°$; P_A / P_B at $50° = 1·135$

Molecular percentage of A in liquid.	Vapour pressures.			Molecular percentage of A in vapour.		
	Observed.	Calculated.	△.	Observed.	Calculated.	△.
0·0	268·0	268·0	0·0	...	...	...
7·16	265·5	265·7	+0·2	...	...	...
7·07	265·8	265·8	0·0	...	...	...
15·00	263·3	263·2	−0·1	11·52	13·47	+1·95
15·00	263·8	263·2	−0·6	12·72	13·47	+0·75
29·27	258·8	258·7	−0·1	26·38	26·73	+0·35
29·27	259·3	258·7	−0·6	27·06	26·73	−0·33
29·79	259·0	258·5	−0·5	27·22	27·23	+0·01
41·56	254·7	254·8	+0·1	38·72	38·68	−0·04
41·65	255·0	254·8	−0·2	38·90	38·63	−0·27
52·15	251·3	251·4	+0·1	49·00	49·01	+0·01
52·34	252·0	251·4	−0·6	49·42	49·20	−0·22
65·66	247·3	247·1	−0·2	62·66	62·77	+0·11
65·66	247·4	247·1	−0·3	62·61	62·77	+0·16
75·42	244·1	244·0	−0·1	72·96	73·02	+0·06
75·42	243·9	244·0	+0·1	73·07	73·02	−0·05
92·06	238·7	238·7	0·0	91·00	91·09	+0·09
91·89	238·3	238·8	+0·5	90·72	90·90	+0·18
100·0	236·2	236·2	0·0	...	...	...

With the exception of the first two percentages the agreement is good, and in the case of these two it will be seen that the percentage of A in the liquid is the same, while the observed percentages of A in the vapour differ by 1·2. It seems clear that little weight can be attached to the first two observations in this series.

For this pair of liquids the constant c has almost exactly the same value, 1·134, as the ratio of the vapour pressures, 1·135.

Linebarger's Results.—It is unfortunate that the method employed by Linebarger [1]—passing a known volume of air through the mixture at constant temperature—did not give more trustworthy results, for he examined some mixtures of fairly closely related substances. It may be well to give the results obtained with one pair of liquids, chlorobenzene and benzene, for Linebarger found that there was no appreciable heat change on mixing these substances in several different proportions.

TABLE 25

A = Chlorobenzene ; B = Benzene ; $c = 7\cdot3$; $t = 34\cdot8°$;
$P_A/P_B = 7\cdot16$ at $34\cdot8°$

Molecular percentage of A in liquid.	Vapour pressures.			Molecular percentage of A in vapour.		
	Observed.	Calculated.	△.	Observed.	Calculated.	△.
15·18	126·3	126·4	+0·1	1·33	2·39	+1·06
29·08	107·9	109·0	+1·1	6·11	5·30	−0·81
65·06	63·6	64·0	+0·4	19·37	20·32	+0·95
79·21	47·0	46·3	−0·7	35·15	34·29	−0·86

Here the difference between c and P_B/P_A is probably within the limits of experimental error, but the number of observations is too small, and the errors are too large to allow of the definite statement that c is quite constant.

Lehfeldt's Results.—The experiments of Lehfeldt [2] show that the vapour pressures of mixtures of toluene and carbon tetrachloride are slightly lower than those calculated from the formula $P = \dfrac{\text{M}P_A + (100 - \text{M})P_B}{100}$, while those of Lehfeldt, of Zawidski and of Young and Fortey show that the vapour pressures of mixtures of benzene and carbon tetrachloride are somewhat higher.

For toluene and carbon tetrachloride c appears to be a constant, but with benzene and carbon tetrachloride there is a small but distinct variation.

---

[1] Linebarger, "The Vapour Tensions of Mixtures of Volatile Liquids," *Journ. Amer. Chem. Soc.*, 1895, **17**, 615.

[2] Lehfeldt, "On the Properties of Liquid Mixtures," Part II., *Phil. Mag.*, 1898, [V.], **46**, 42.

[TABLE

TABLE 26

A = Toluene ; B = Carbon tetrachloride ; $c = 2·76$; $t = 50·0°$;
P_B/P_A at 50° = 3·33

Molecular percentage of A in liquid.	Vapour pressures.			Molecular percentage of A in vapour.		
	Observed.	Calculated.	$\triangle$.	Observed.	Calculated.	$\triangle$.
0·0	310·2	310·2	0·0	...	...	...
9·01	288·8	290·6	+1·8	...	...	...
26·02	248·5	253·7	+5·2	...	...	...
29·2	...	...	...	12·8	13·0	+0·2
36·49	226·5	230·9	+4·4	...	...	...
48·3	...	...	...	25·8	25·3	-0·5
48·58	197·7	204·7	+7·0	...	...	...
60·30	174·8	179·2	+4·4	...	...	...
65·0	...	...	...	40·3	40·2	-0·1
75·99	140·8	145·1	+4·3	...	...	...
76·0	...	...	...	53·0	53·4	+0·4
83·7	...	...	...	65·6	65·1	-0·5
86·49	117·9	121·3	+3·4	...	...	...
92·7	...	...	...	81·9	82·1	+0·2
96·01	99·0	101·7	+2·7	...	...	...
100·0	93·0	93·0	0·0	...	...	...

In this case the difference between the values of c and P_B/P_A is much greater than in the previous ones, but the agreement between the observed and calculated percentages is good.

TABLE 27

A = Benzene ; B = Carbon tetrachloride ; $c = 0·984 + 0·003M$;
$t = 50·0°$; P_B/P_A at 50° = 1·145

Molecular percentage of A in liquid = M.	Vapour pressures.			Molecular percentage of A in vapour.		
	Observed.	Calculated.	$\triangle$.	Observed.	Calculated.	$\triangle$.
0·0	310·2	310·2	0·0	...	...	...
17·0	...	...	...	16·5	16·5	0·0
34·0	...	...	...	32·3	32·2	-0·1
38·69	302·3	295·0	-7·3	...	...	...
62·4	...	...	...	58·8	58·6	-0·2
69·56	290·0	282·9	-7·1	...	...	...
80·3	...	...	...	76·7	76·9	+0·2
84·22	281·0	277·1	-3·9	...	...	...
95·7	...	...	...	94·6	94·6	0·0
100·0	270·9	270·9	0·0	...	...	...

Modification of Brown's Formula.—Brown's formula is not applicable to this pair of liquids, but by taking c = const. $(c_0) + a$M instead of c = const., a good agreement between the observed and calculated values is obtained.

Mixtures of benzene with carbon tetrachloride have also been investigated by Zawidski with very similar results. He finds, however, somewhat lower vapour pressures for the pure substances, and his pressure differences are rather larger. The formula deduced from his data would be $c = 0.961 + 0.0036\text{M}$, and there is a very fair agreement between the calculated and observed percentages of A in the vapour, though not quite so good as in the table on the previous page. The first formula for c would indicate the existence of a mixture of minimum boiling point containing 6.7 molecules per cent of benzene, the second a mixture containing 10.8 molecules per cent.

Distillations under Constant Pressure and at Constant Temperature.—

Of other pairs of liquids, the vapour pressures of which do not differ very greatly from those calculated from the formula $P = \dfrac{\text{M}P_A + (100 - \text{M})P_B}{100}$, there are two, carbon disulphide with carbon tetrachloride and carbon disulphide with benzene, which have been investigated by Brown, the latter also by Carveth, but their distillations were carried out in the usual manner under constant pressure, not at constant temperature.

Brown observed, however, that the relation between m'_A/m'_B and m_A/m_B was the same for mixtures of carbon disulphide with carbon tetrachloride whether the distillation was carried out under a pressure of 432 mm., or under atmospheric pressure. If the values of c for other substances also are independent of, or vary only slightly with, the pressure, they can be calculated from the results of a distillation carried out either at constant temperature or under constant pressure.

It would not be safe to conclude, without further evidence, that the same values of c would be obtained in other cases by both methods, but we may at any rate assume that the differences would not be great enough to invalidate the general conclusions deduced from a comparison of the behaviour of different pairs of liquids.

In the cases so far considered, the composition of the vapour was determined experimentally. Such direct determinations have not yet been carried out for mixtures of chlorobenzene and bromobenzene, nor for those of any of the five pairs of closely related liquids mentioned on p. 32. Rosanoff, Bacon and Schulze,[1] however, showed how the ratio of the partial pressures p_A/p_B and, in the case of substances of normal molecular weight, the molar percentages in the vapour could be calculated with accuracy from the vapour pressures of the pure components and the total pressures, P (p. 93).

They determined the vapour pressures of mixtures of benzene and toluene with great care at 79.7°, and their results, like those of Fortey and Young, show that these pressures are represented without serious error by the formula $P = \text{M}P_A + (1 - \text{M})P_B$, the maximum difference between the calculated and observed values of P being $+1.9$ mm. [Molar percentage of benzene in the liquid $= 54.51$; P (observed) $= 537.5$ mm.]

From their results and by means of their formulae, Rosanoff, Bacon

[1] " A Method of finding the Partial from the Total Vapour Pressures of Binary Mixtures, and a Theory of Fractional Distillation," J. Amer. Chem. Soc., 1914, 36, 1993.

and Schulze calculated the molar percentages of benzene, and these, which may be accepted with great confidence, are given in the table below.

TABLE 28

A = Toluene ; B = Benzene ; c = 2·589 ; t = 79·7° ; P_B/P_A at 79·7° = 2·5951

Molecular percentage of A in liquid.	Vapour pressures.			Calculated molecular percentage of A in vapour.		
	Observed.	Calculated. $P = \dfrac{M P_A + (100 - M) P_B}{100}$	$\triangle$.	R. B. & S.	Brown c = 2·589.	$\triangle$.
100·0	748·7	748·7	0·0	...	...	...
95·65	729·0	728·7	−0·3	98·27	98·27	0·0
91·89	711·4	711·4	0·0	96·72	96·70	−0·02
82·43	668·0	667·9	−0·1	92·49	92·40	−0·09
73·27	624·9	625·7	+0·8	87·82	87·65	−0·17
63·44	579·2	580·4	+1·2	81·97	81·79	−0·18
54·51	537·5	539·4	+1·9	75·74	75·62	−0·12
43·52	487·0	488·8	+1·8	66·56	66·61	+0·05
33·83	443·1	444·2	+1·1	56·76	56·97	+0·21
22·71	392·8	393·0	+0·2	42·95	43·21	+0·26
11·61	341·5	341·9	+0·4	25·30	25·38	+0·08
0·0	288·5	288·5	0·0	...	...	...

The chemical relationship between benzene and toluene is not quite so close as that between the members of the other five pairs of liquids, and the volume and temperature changes on admixture are noticeably greater, but, even so, it is found that Brown's formula is applicable without serious error to mixtures of these two substances and that the best value of the constant c differs but slightly from the ratio P_B/P_A.

Cases to which Brown's Modified Formula is inapplicable.—When, for a given pair of substances, the deviations of the vapour pressures from those calculated from the formula $P = \dfrac{M P_A + (100 - M) P_B}{100}$ are great, the values of c are not only far from constant, but their relation to M cannot be represented by such a simple formula as $c = c_0 + a$M. This is well seen in the case of carbon disulphide and methylal, which form a well-defined mixture of maximum vapour pressure, and of chloroform and acetone, which form a mixture of minimum vapour pressure (Zawidski).

[TABLE

TABLE 29

A = Carbon disulphide ; B = methylal ; t = 35·17°

Molecular percentage of A in liquid.	Vapour pressures.			$c = \dfrac{m'_B}{m'_A} \cdot \dfrac{m_A}{m_B}.$
	Observed.	Calculated.	$\triangle$.	
0	587·7	587·7	0·0	...
10	637·3	580·4	− 56·9	0·554
20	670·0	573·1	− 96·9	0·643
30	690·9	565·7	−125·2	0·744
40	700·7	558·4	−142·3	0·875
50	701·9	551·1	−150·8	1·041
60	696·0	543·8	−152·2	1·253
70	682·0	536·5	−145·5	1·530
80	658·9	529·1	−129·8	1·926
90	612·3	521·8	− 90·5	2·531
100	514·5	514·5	0·0	...

TABLE 30

A = Chloroform ; B = Acetone ; t = 35·17°

Molecular percentage of A in liquid.	Vapour pressures.			$c = \dfrac{m'_B}{m'_A} \cdot \dfrac{m_A}{m_B}.$
	Observed.	Calculated.	$\triangle$.	
0	344·5	344·5	0·0	...
10	324·0	339·4	+ 15·4	2·080
20	304·0	334·2	+ 30·2	1·878
30	284·6	329·1	+ 44·5	1·693
40	266·9	323·9	+ 57·0	1·470
50	253·9	318·8	+ 64·9	1·257
60	247·8	313·7	+ 65·9	1·034
70	251·4	308·5	+ 57·1	0·848
80	262·1	303·4	+ 41·3	0·681
90	277·1	298·2	+ 21·1	0·580
100	293·1	293·1	0·0	...

Influence of Molecular Association.—The variability of c is even more marked when the molecules of one of the substances are associated in the liquid state. This is the case, for example, with mixtures of benzene and ethyl alcohol, which have also been investigated by Zawidski.

At 50° the vapour pressure of benzene is higher than that of alcohol, but at 80° it is the alcohol which has the higher vapour pressure.

TABLE 31

$A=$ Ethyl alcohol ; $B=$ Benzene ; $t-50°$

Molecular percentage of A in liquid.	Vapour pressures.			$c=\dfrac{m'_B}{m'_A}\cdot\dfrac{m_A}{m_B}.$
	Observed.	Calculated.	$\triangle$.	
0	270·9	270·9	**0·0**	...
8·8	350·4	266·4	− **84·0**	0·247
12·1	369·0	264·7	−**104·3**	0·290
21·5	397·0	259·8	−**137·2**	0·493
35·5	406·0	252·7	−**153·3**	0·857
44·4	404·4	248·1	−**156·3**	1·116
56·1	397·6	242·1	−**155·5**	1·556
69·7	378·4	235·1	−**143·3**	2·184
88·6	315·0	225·4	− **89·6**	3·863
100·0	219·5	219·5	**0·0**	...

In Fig. 32, c is plotted against M, the molecular percentage of A in the liquid, and it will be seen that the curvature is very marked.

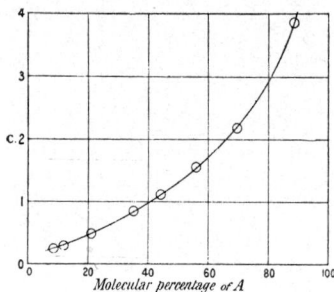

FIG. 32.—Ethyl alcohol and benzene.

Mathematical Investigations. —The whole question of the relations between the composition of liquid mixtures and (a) the partial pressures of the vapours of the components, (b) the composition of the vapour, has been discussed mathematically by Duhem,[1] Margules,[2] Lehfeldt,[3] and Zawidski[4] and others. In this connection " The Phase Rule," by Bancroft,[5] may also be consulted.

Formula of Duhem and Margules.—The formula arrived at by Duhem, and, later, by Margules, may be written :—

$$\frac{d\log p_1}{d\log \text{M}}=\frac{d\log p_2}{d\log (1-\text{M})},$$

where p_1 and p_2 are the partial pressures of the vapours of the two liquids A and B, and M and $(1-\text{M})$ their molecular fractional amounts in the liquid mixture, taking the normal molecular weights as correct.

[1] Duhem, " On the Vapours emitted by Mixtures of Volatile Substances," *Ann. de l'École Normale Sup.*, 1887, [3], **4**, 9 ; " Some Remarks on Mixtures of Volatile Substances," *ibid.*, 1889, [3], **6**, 153 ; " Solutions and Mixtures," *Trav. et Mém. de la faculté de Lille*, 1894, III. D. ; *Traité élémentaire de méchanique chimique*, 1899.

[2] Margules, " On the Composition of the Saturated Vapours of Mixtures," *Sitzungsber. der Wiener Akad.*, 1895, **104**, 1243.

[3] Lehfeldt, " On the Properties of Liquid Mixtures," 3 Parts, *Phil. Mag.*, 1895, [V.], **40**, 397 ; 1898, *loc. cit.* ; 1899, [V.], **47**, 284.

[4] *Loc. cit.*

[5] Bancroft, " The Phase Rule," 1897.

Lehfeldt's Formula.—Starting from this equation, Lehfeldt adopts the formula,

$$\log t = \log K + r \log q$$

or

$$t = Kq^r,$$

where $t =$ the ratio of the masses of the two substances in the vapour, q the ratio in the liquid and K and r are constants.

For r, Lehfeldt gives the equation :—

$$r = \frac{\log S - \log \dfrac{\pi_A}{\pi_B}}{\log \dfrac{Bq}{A}},$$

where π_A and π_B are the vapour pressures of the pure liquids at the temperature of experiment ; S is the ratio of the number of molecules of the two substances in the vapour ; A and B are the normal molecular weights of the components.

Margules has pointed out that, when $r < 1$, the equation $t = Kq^r$ leads to infinite values of $dp/d\text{M}$ when $\text{M} = 0$ or $\text{M} = 1$, which does not agree with the facts, but Lehfeldt finds that the equation holds very well for mixtures which do not contain a very small proportion of either component, provided that the molecular weights of both substances are normal in the liquid as well as in the gaseous state. For associating liquids the formula does not hold good at all.

Benzene and Carbon Tetrachloride. — As an example we may consider the case of mixtures of benzene and carbon tetrachloride, for which Lehfeldt gives the formula :—

$$\log t = 0\cdot065 + 0\cdot947 \log q.$$

[In Tables 32 and 33 the molecular composition is given instead of the composition by weight.]

TABLE 32

Liquid.	Molecular percentage of A (Benzene).		
	Vapour.		
	Observed.	Calculated.	$\triangle$.
17·0	16·5	16·5	**0·0**
34·0	32·3	32·2	**− 0·1**
62·4	58·8	59·2	**+ 0·4**
80·3	76·7	77·1	**+ 0·4**
95·7	94·6	94·4	**− 0·2**

In the above equation $\log t = \log q$ when $\log q = 1\cdot2264$, that is to say, when mass of CCl_4/mass of $C_6H_6 = 16\cdot84$. In other words, there would be a mixture of minimum boiling point, containing 5·6 per cent by weight or 10·5 molecules per cent of benzene.

Generally, if K is positive and r is less than unity, there must be a particular value of q for which $\log t = \log q$, and there must therefore be a possible mixture of constant boiling point.

Relation between Brown's and Lehfeldt's Formulae.—In the equation $t = Kq^r$, when $r = 1$, $t = Kq$, or Brown's law holds good. Lehfeldt himself gives for mixtures of toluene and carbon tetrachloride the formula

$$\log t = 0\cdot440 + 1\cdot0 \log q,$$

or $$t = 2\cdot755q,$$

which agrees almost exactly with that given on p. 84, $\dfrac{M'_B}{M'_A} = 2\cdot76\dfrac{M_B}{M_A}$.

Carbon Disulphide and Methylal.—As an example of two substances, the molecular weights of both of which are presumably normal, but which form a mixture of maximum vapour pressure considerably higher than that of either component, we may take carbon disulphide and methylal, examined by Zawidski.

In Table 33 are given the observed molecular percentages of carbon disulphide and those calculated by means of the formula $\log t = 0\cdot036 + 0\cdot619 \log q$.

TABLE 33

Molecular percentages of A (Carbon disulphide).							
Liquid.	Vapour.			Liquid.	Vapour.		
	Observed.	Calculated.	$\triangle$.		Observed.	Calculated.	$\triangle$.
4·96	8·98	12·9	+3·9	60·60	54·76	54·6	−0·2
10·44	17·39	19·6	+2·2	68·03	59·21	59·5	+0·3
16·51	24·44	25·2	+0·8	73·53	62·74	63·4	+0·7
27·19	34·39	33·3	−1·1	79·27	66·76	67·8	+1·0
34·80	39·97	38·4	−1·6	84·21	70·92	72·2	+1·3
39·04	42·63	41·1	−1·5	85·73	72·83	73·6	+0·8
45·42	46·34	45·1	−1·2	91·30	80·00	80·3	+0·3
49·42	48·52	47·6	−0·9	95·76	89·23	86·4	−2·8
53·77	50·99	50·3	−0·7				

The agreement is not very good, and there appears to be evidence of curvature.

Oxygen and N.trogen.—Baly[1] found that, for mixtures of liquid oxygen and nitrogen, Lehfeldt's formula gave very good results, but Brown's did not.

Associating Substances.—As regards mixtures of ethyl alcohol, an associating substance, with benzene and with toluene, Lehfeldt points out that the relations between $\log t$ and $\log q$ are far from linear, and he did not attempt to find an equation for the curves.

Zawidski's Formula.—Zawidski adopts the following equations to express the relations between the partial pressures, p_1 and p_2, of

[1] Baly, " On the Distillation of Liquid Air and the Composition of the Gaseous and Liquid Phases," *Phil. Mag.*, 1900, [V.], **49**, 517.

the components in the mixture, the vapour pressures, P_1 and P_2, of the pure components, and the molecular fractional amounts, M and $1 - $M, of the two substances in the mixture.

$$p_1 = P_1 \text{M} . e^{\frac{a_2}{2}(1-\text{M})^2 + \frac{a_3}{3}(1-\text{M})^3}$$

$$p_2 = P_2(1 - \text{M})e^{\frac{\beta_2}{2} . \text{M}^2 + \frac{\beta^3}{3} . \text{M}^3},$$

and
$$\beta_2 = a_2 + a_3 \; ; \; \beta_3 = -a_3,$$

where a_2 and a_3 are constants, the values of which can be ascertained from the partial pressure curves ; or, by Margules's method, from the tangents to the total pressure curve at the extreme points where M $= 0$ and M $= 1$ by means of the equations :

$$\left(\frac{a_2}{2} + \frac{a_3}{3}\right) \log e = \log \left[\left(\frac{d\pi}{d\text{M}}\right)_0 + P_2\right] - \log P_1,$$

$$\left(\frac{a_2}{2} + \frac{a_3}{6}\right) \log e = \log \left[P_1 - \left(\frac{d\pi}{d\text{M}}\right)_1\right] - \log P_2.$$

Relation between Brown's and Zawidski's Formulae.— Zawidski points out that if, in his formula, the constants a and β vanish, the equations become

$$p_1 = P_1 \text{M} \text{ and } p_2 = P_2(1 - \text{M}),$$

whence $\qquad p_1/p_2 = P_1/P_2 . \text{M}/(1 - \text{M}),$

or $\qquad p_1/p_2 = \text{const.} \times \text{M}/(1 - \text{M}) \; ;$

and since, for substances of normal molecular weight, the partial pressures are proportional to the number of molecules present, the equation, in its final form, simply expresses Brown's law, taking the relative number of molecules instead of relative masses for both liquid and vapour.

Zawidski shows that this simple relation holds for mixtures of ethylene and propylene dibromides and of benzene and ethylene dichloride, but that for the first pair of liquids, the constant, 0.758, differs somewhat widely from the ratio P_1/P_2, 0.737, though the agreement is excellent for the second pair (0.880 and 0.881)

Zawidski shows, for these two pairs of substances, not only the correctness of the relation

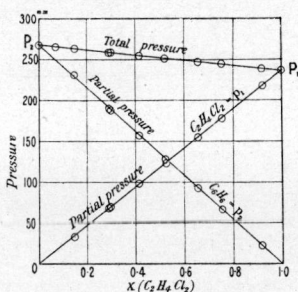

FIG. 33.
Benzene and ethylene dichloride.

$$P = p_1 + p_2 = P_1 \text{M} + P_2(1 - \text{M}),$$

which is equivalent to $P = \dfrac{\text{M}P_A + (100 - \text{M})P_B}{100}$, but also of the formulae

$$p_1 = P_1 \text{M} \text{ and } p_2 = P_2(1 - \text{M}),$$

as will be seen from the diagram (Fig. 33) which is taken from his paper.

In the case of other mixtures of liquids with normal molecular weight, the constants α and β were found to have finite values, and the simple formulae $p_1 = P_1 M$ and $p_2 = P_2(1 - M)$ were not found to be applicable ; the relations between the molecular composition and the pressures, whether total or partial, are, in fact, represented by curves. Thus, even with benzene and carbon tetrachloride the curvature, though slight, is unmistakable.

Benzene and Carbon Tetrachloride. — For this pair of liquids Zawidski gives the constants α_2 and α_3, calculated by both methods ; they are—

I. From the partial pressure curve, $\alpha_2 = 0\cdot308$, $\alpha_3 = 0\cdot00733$.

II. From the tangents $(d\pi/d\text{M})_o = 90$, $(d\pi/d\text{M})_1 = -4\cdot3$, $\alpha_2 = 0\cdot312$, $\alpha_3 = -0\cdot0168$.

Now the value of the tangents $(d\pi/d\text{M})_1$ calculated from the first constants would be $-5\cdot0$, so that in either case there is a negative value for the tangent when $\text{M} = 1$. If this is correct, there must be a mixture of maximum vapour pressure containing little benzene, and there is thus additional evidence (pp. 84 and 89) that these two liquids can form such a mixture, though the difference between the maximum pressure and the vapour pressure of carbon tetrachloride is probably too small to be determined by direct experiment.

For this and other pairs of liquids of normal molecular weight, the agreement between the observed molecular composition of the vapour and that derived from the calculated partial pressures is fairly good, though there are occasionally differences amounting to 4 or 5 per cent.

Carbon Disulphide and Methylal.—In Table 34 below are given

TABLE 34

A = Carbon disulphide ; B = Methylal
From $(d\pi/d\text{M})_o = +578$ and $(d\pi/d\text{M})_1 = -1310$; $\alpha_2 = 2\cdot9$; $\alpha_3 = -1\cdot89$

Molecular percentage of Carbon disulphide.							
Liquid.	Vapour.			Liquid.	Vapour.		
	Observed.	Calculated.	$\triangle$.		Observed.	Calculated.	$\triangle$.
4·96	8·98	8·95	−0·03	60·60	54·76	53·97	−0·79
10·44	17·39	17·09	−0·30	68·03	59·21	57·87	−1·34
16·51	24·44	24·44	0·00	73·53	62·74	61·20	−1·54
27·19	34·39	34·45	+0·06	79·27	66·76	65·36	−1·40
34·80	39·97	39·97	0·00	84·21	70·92	69·88	−1·04
39·04	42·63	42·63	0·00	85·73	72·83	71·53	−1·30
45·42	46·34	46·25	−0·09	91·30	80·00	79·05	−0·95
49·42	48·52	48·36	−0·16	95·76	89·23	88·03	−1·20
53·77	50·99	50·56	−0·43				

the results for carbon disulphide and methylal, so as to compare the calculated values with those obtained by means of Lehfeldt's formula.

The agreement up to 50 molecules per cent of carbon disulphide is excellent, but for mixtures richer in that component it is not nearly so satisfactory ; it is probable, however, that by altering the constants, better results might be obtained.

Associated Substances.—Zawidski included in his investigation some pairs of substances, of which the molecules of one component are associated in the liquid state (water), or in both the liquid and gaseous states (acetic acid), and he concludes that his formula can be employed for such mixtures, provided that in calculating the values of M and $1 - $M we take the average molecular weight of the associated liquids under the conditions of the experiment. As a rule, however, the experimental data at present available are only sufficient to afford a rough estimate of the average molecular weight of an associating substance when mixed with another liquid.

Rosanoff's Formulae.—In recent years M. A. Rosanoff and his co-workers have carried out a series of excellent experimental and theoretical researches on distillation ; and in 1914 Rosanoff, Bacon and Schulze[1] pointed out that the method of Margules, which depends on the graphic measurement of the slope of the total pressure-curve at its two ends, is liable to yield inaccurate results.

They therefore sought to formulate a general relationship, even if only empirical, between the total and partial vapour pressure curves.

They found that, in the cases examined, if a set of values of $\dfrac{dP}{d\text{M}}$ were plotted against the corresponding values of log $[p_{\text{A}}(1 - \text{M})/p_{\text{B}}\text{M}]$, the result was a straight line passing through the origin of the co-ordinates, which indicated that the simplest possible relationship exists between the two quantities. This apparently general law is expressed by the equation

$$\frac{dP}{d\text{M}} = (1/\text{K}) \log [p_{\text{A}}(1 - \text{M})/p_{\text{B}}\text{M}],$$

where

$$1/\text{K} = (P_{\text{A}} - P_{\text{B}})/(\log P_{\text{A}} - \log P_{\text{B}}).$$

Therefore

$$\frac{dP}{d\text{M}} = [(P_{\text{A}} - P_{\text{B}})/(\log P_{\text{A}} - \log P_{\text{B}})] \log [p_{\text{A}}(1 - \text{M})/p_{\text{B}}\text{M}] \quad . \quad . \quad . \quad . \quad \text{(I)}.$$

The authors state that this equation is not in conflict with the thermodynamical equation of Duhem and Margules, and they show that it faithfully reproduces the experimental results in all types of cases, even when mixtures of maximum or minimum vapour pressure are formed.

From Zawidski's measurements of the vapour pressures of mixtures of carbon tetrachloride and benzene at 49·99° they obtain the equation—

$$P = 268 \cdot 075 + 80 \cdot 853\text{M} - 43 \cdot 826\text{M}^2 + 16 \cdot 531\text{M}^3 - 13 \cdot 695\text{M}^4,$$

[1] *J. Amer. Chem. Soc.*, 1914, **36**, 1993.

according to which a mixture containing 91·65 molecules per cent of carbon tetrachloride has a vapour pressure of 308·43 mm., which is a maximum, the vapour pressure of pure carbon tetrachloride being 308·0 mm. The existence of a mixture of maximum vapour pressure is thus confirmed (pp. 84, 89, 92), and the calculated molar percentage of benzene, 8·35, agrees well enough with the values previously found.

For carbon tetrachloride (A) and benzene (B)

$$\frac{\log_{10} P_A - \log_{10} P_B}{P_A - P_B} = 0 \cdot 0015103 \text{ at } 49 \cdot 99° ;$$

therefore

$$\log_{10} [p_A(1 - \text{M})/p_B\text{M}] = 0 \cdot 122115 - 0 \cdot 132383\text{M} + 0 \cdot 074900\text{M}^2 - 0 \cdot 082734\text{M}^3.$$

The observed and calculated results are given below.

TABLE 35

Molar percentage of CCl₄ in liquid.	Vapour pressure, P.			Molar percentage of CCl₄ in vapour.		
	Observed.	Calculated.	△.	Observed.	Calculated.	△.
5·07	271·8	272·1	+0·3	6·81	6·54	−0·27
11·70	277·6	277·0	−0·6	14·59	14·51	−0·08
17·58	281·5	281·0	−0·5	21·21	21·21	0·0
25·15	285·4	285·8	+0·4	29·05	29·36	+0·31
29·47	288·3	288·4	+0·1	33·65	33·81	+0·26
39·53	294·5	293·9	−0·6	43·70	43·79	+0·09
55·87	301·0	301·1	+0·1	58·61	59·08	+0·47
67·55	305·2	304·9	−0·3	69·40	69·83	+0·43
76·52	306·8	307·0	+0·2	77·74	77·66	−0·08

The agreement is good, and it is also very satisfactory in the case of ethyl iodide and ethyl acetate and of chloroform and acetone, which form mixtures of maximum and minimum vapour pressure respectively. The agreement in the last case is specially noteworthy for there can be little doubt that acetone forms complex molecules to some extent. For chloroform and acetone the equation is—

$$\log_{10} [p_A(1 - \text{M})/p_B\text{M}] = -0 \cdot 199592 - 1 \cdot 14361\text{M} + 3 \cdot 66677\text{M}^2 - 2 \cdot 07687\text{M}^3 ;$$

and the results are given on following page.

[TABLE

TABLE 36

Molar percentage of CHCl₃ in liquid.	Vapour pressure, P.			Molar percentage of CHCl₃ in vapour.		
	Observed.	Calculated.	$\triangle$.	Observed.	Calculated.	$\triangle$.
0·0	344·5	343·7	−0·8	...	...	...
6·03	332·1	333·6	+1·5	2·8	3·4	+0·6
12·03	320·1	321·6	+1·5	6·2	6·6	+0·4
12·32	319·7	320·9	+1·2	6·4	6·8	+0·4
18·18	308·0	308·3	+0·3	10·3	10·1	−0·2
29·10	285·7	285·1	−0·6	19·4	18·0	−1·4
40·50	266·9	265·1	−1·8	31·8	30·1	−1·7
50·83	252·9	253·4	+0·5	45·6	44·7	−0·9
58·12	248·4	249·7	+1·3	56·3	56·2	−0·1
66·35	249·2	250·2	+1·0	68·3	68·8	+0·5
79·97	261·9	261·3	−0·6	85·7	85·5	−0·2
80·47	262·6	261·9	−0·7	85·4	86·0	+0·6
91·79	279·5	279·0	−0·5	95·0	95·0	0·0
100·0	293·1	293·7	+0·6	...	...	...

The case of benzene and toluene has already been considered (p. 85).

General Conclusions.—The conclusions arrived at may be stated shortly as follows :—

1. The composition of the vapour from a pair of non-miscible liquids at a given temperature may be accurately calculated from the vapour pressures and vapour densities of the components.

2. The composition of the vapour from a pair of closely related miscible liquids at a given temperature may, so far as is known, be calculated by means of Brown's formula $\dfrac{m'_A}{m'_B} = c\dfrac{m_A}{m_B}$. The value of the constant, c, certainly does not differ greatly from the ratio of the vapour pressures of the components at the temperature of experiment, but the data at present available are perhaps insufficient to warrant the statement that it is always equal to this ratio, and it appears to be necessary to determine it experimentally.

3. As regards substances which are not closely related, Brown's formula is only applicable when the vapour pressure of any mixture is given by the formula—$P = \dfrac{MP_A + (100 - M)P_B}{100}$, and this is probably never the case when there is a marked volume or temperature change on mixing the pure liquids.

4. If the vapour pressures at constant temperature of mixtures of two infinitely miscible liquids—the molecular weights of which are normal—are not given by the formula $P = \dfrac{MP_A + (100 - M)P_B}{100}$, the relation between vapour pressure and composition must be determined experimentally at the required temperature. The composition of the vapour from any mixture may then be calculated with moderate accuracy by means of the formula adopted by Zawidski (p. 90), the

values of the constants a_2 and a_3 being ascertained from the pressure-composition curve by the method of Margules; but much better results are given by the equations put forward by Rosanoff, Bacon and Schulze.

5. When the molecules of either liquid are associated, the relation between the composition of the vapour and that of the liquid cannot be ascertained, even approximately, by means of Zawidski's formula unless the average molecular weights of the associating substance under the varying conditions of the experiment are known.

In the case of chloroform and acetone, however, the formula of Rosanoff, Bacon and Schulze gave satisfactory results in spite of the fact that the molecules of acetone are associated to some extent.

6. If, for two miscible liquids, a sufficient number of determinations of the relative composition of liquid and vapour at constant temperature have been made to allow of a curve being constructed—the molecular percentages of the vapour being mapped against those of the liquid, m'_A/m'_B against m_A/m_B, the logarithms of these ratios against each other, or the partial pressures of each component separately against the molecular fractional amount of one of them—other values may be read from the curve, or the constants for an interpolation formula may be calculated.

If the vapour pressures of mixtures of the two substances differ but little from those given by the formula $P = \dfrac{M P_A + (100 - M) P_B}{100}$, a modification of Brown's formula may be used; $m'_A/m'_B = c' m_A/m_B$, where $c' = c_0 + a M$.

Better results are, however, generally given by Lehfeldt's formula, $\log t = K + r \log q.$, and this may be used even when the observed vapour pressures differ somewhat considerably from the calculated. If, however, these differences are great, the formula of Zawidski is to be preferred.

Most of the investigations of the relation between the composition of liquid and of vapour have been carried out at constant temperature, but in practice a liquid is almost always distilled under constant pressure. Brown, however, whose distillations were carried out in the usual manner, found, in the case of carbon tetrachloride and carbon disulphide, that when a mixture was boiled the composition of the vapour was independent of the pressure under which ebullition took place, and, if this were generally true, a curve constructed from results obtained at constant temperature could be used to ascertain the vapour composition in a distillation under constant pressure. It is, however, to be noticed that the ratio of the vapour pressures, even of two closely related liquids, is not the same at different temperatures, and if the relation $m'_A/m'_B = P_A m_A/P_B m_B$ is really true for such liquids, P_A/P_B would be a constant for a distillation at constant temperature, but would vary slightly if the distillation were carried out under constant pressure.

Lehfeldt found that, in the case of Brown's distillations, the logarithms of the ratios of the masses of the components in the liquid and vapour phases had a linear relation, and Baly [1] obtained a similar result with the distillation of mixtures of oxygen and nitrogen under constant pressure. Lehfeldt's formula could therefore be used for interpolation in these cases.

The question has been discussed by Rosanoff, Bacon and Schulze,[1] who determined not only the vapour pressures of mixtures of benzene and toluene at a constant temperature, but also the boiling points of mixtures of the same substances under a constant pressure (750 mm.). It had been shown by Rosanoff and Easley [2] that the composition of vapours from binary mixtures may, in general, be accurately represented by an expression of the form—

$$ln\frac{p_A(1-\text{M})}{p_B\text{M}}=ln\frac{P_A}{P_B}+a_2[(1-\text{M})-1/2]+\frac{a_3}{2}[(1-\text{M})^2-1/3]+\frac{a_4}{3}[(1-\text{M})^3-1/4],$$

and that the coefficients a_2, a_3 and a_4 are, in all cases in which the heat of dilution is moderate, practically independent of the temperature, so that changes of temperature influence the vapour composition only by affecting the value of P_A/P_B. If, therefore, from an expression found for some given temperature the logarithm of P_A/P_B corresponding to that temperature is subtracted, and to the remainder is added an expression representing the logarithm of P_A/P_B as a function of the temperature, a more general expression would be obtained, from which the vapour composition could be calculated for any temperature or temperatures within the given range.

In the case of benzene and toluene the heat of dilution is very small and the authors therefore felt justified in applying the principle just stated within the temperature range involved, and they thus arrived at a formula which enabled them to calculate the molar percentage of benzene in the vapour from mixtures boiling under a pressure of 750 mm.

[1] *Loc. cit.*
[2] *J. Amer. Chem. Soc.*, 1909, **31**, 957.

CHAPTER VII

DIRECTIONS FOR CARRYING OUT A FRACTIONAL DISTILLATION

Mixtures separable into Two Components

It has been stated (p. 62) that when a mixture of two substances is heated, the vapour is richer than the liquid[1] in the more volatile of the two components into which the mixture tends to separate, whether these components are the original substances which were mixed together or a mixture of constant boiling point and one of the original substances.

If, then, we distil the mixture in the usual manner until, say, one half of the total quantity has passed over, the distillate will be richer than the residue in the more volatile component. If we were to redistil the distillate and again collect the first half, the new distillate would be still richer in the more volatile component, and by repeating the operation several times we might eventually obtain some of the more volatile component in a pure, or nearly pure, state. The amount of distillate would, however, become smaller each time, and, if a large number of distillations were required, it would be relatively very small indeed. In order to obtain a fair quantity of both components in a sufficiently pure state, systematic fractional distillation is necessary.

Let us suppose that we have 200 grams of a mixture of equal weights of benzene and toluene. The relation between the vapour pressures and the molecular composition of mixtures of these liquids is expressed very nearly by a straight line and it is probable that Brown's law, $\dfrac{m'_B}{m'_A} = c\dfrac{m_B}{m_A}$, is very nearly true, and that c does not differ greatly from 2·47, the mean ratio of the vapour pressures at equal temperatures between 80° and 110°. At any rate the liquid tends to separate on distillation into the original components, benzene and toluene, no mixture of constant boiling point being formed.

Collection of Distillate in "Fractions."—The mixture should be first distilled and the distillate collected in a convenient number of fractions, the receivers being changed when the boiling point reaches certain definite temperatures to be arranged beforehand. In order to trace the course of the separation as clearly as possible we will,

---

[1] Unless a mixture of constant boiling point, of the same composition as that of the original mixture, is formed, in which case the composition of the vapour would be the same as that of the liquid.

in the first place, make the range of temperature nearly the same for most of the fractions. The boiling point of benzene is 80·2° and of toluene, 110·6°, the difference being 30·4° and we might take 10 small flasks to provide for 10 fractions, 8 with a range of 3° and 2 with a range of 3·2°, but it is better to take two fractions each for the first and last 3°. It is, as a rule, convenient to have the same number of fractions above and below the middle temperature between the two boiling points. Suitable temperature ranges for the twelve fractions are given in the second column of Table 37.

TABLE 37

Number of receiver.	Temperature ranges.		
	760 mm.	745 mm.	To be read on thermometer.
1	80·2— 81·2°	79·6— 80·6°	79·8— 80·8°
2	81·2— 83·2	80·6— 82·6	80·8— 82·8
3	83·2— 86·2	82·6— 85·6	82·8— 85·8
4	86·2— 89·2	85·6— 88·6	85·8— 88·8
5	89·2— 92·2	88·6— 91·6	88·8— 91·8
6	92·2— 95·4	91·6— 94·8	91·8— 95·0
7	95·4— 98·6	94·8— 97·9	95·0— 98·2
8	98·6—101·6	97·9—100·9	98·2—101·2
9	101·6—104·6	100·9—103·9	101·2—104·2
10	104·6—107·6	103·9—106·9	104·2—107·2
11	107·6—109·6	106·9—108·9	107·2—109·2
12	109·6—110·6	108·9—109·9	109·2—110·2

The first three receivers are, however, not required for the preliminary distillation.

Correction of Temperature.—Before distilling the mixture we must read the barometer, because 80·2° and 110·6° are the boiling points of benzene and toluene respectively under normal pressure, and we must find what they would be under the actual barometric pressure, and alter the temperatures accordingly. If the thermometer does not register true temperatures, the necessary corrections must be ascertained and taken into account.

Let us suppose that the height of the barometer, corrected to 0° (p. 229), is 745 mm. and that the thermometer reads 0·2° too high at 80° and 0·3° too high at 110°. Referring to p. 14 we find that the value of c $\left(=\dfrac{dt}{dp} \cdot \dfrac{1}{T} \right)$ for benzene is 0·000121 and for toluene, 0·000120. The corrections will therefore be $\Delta t = (760 - 745)$ $(273 + 80) \cdot 0·000121 = 0·6°$ for benzene and $\Delta t = (760 - 745)$ $(273 + 111) \cdot 0·000120 = 0·7°$ for toluene, and the boiling points under a pressure of 745 mm. will therefore be $80·2° - 0·6° = 79·6°$ and $110·6° - 0·7° = 109·9°$ respectively. The temperatures of the fractions under 745 mm. pressure are given in the third column and the actual readings on the thermometer in the fourth column.

When it is desired to separate the components of a mixture in the purest state attainable, the above corrections must be made with the greatest possible care, and it may not be sufficient to estimate the temperatures to 0·1° but to 0·05° or less ; it may also be necessary to read the barometer from time to time during the course of the distillation and to recalculate the corrections if there is any change in the pressure.[1]

Rate of Distillation.—The mixture must now be slowly distilled : for laboratory purposes the drops of distillate should fall at the rate of about 1 per second, but on the large scale a much greater rate would be necessary. The slower the distillation the better is the separation and, although each distillation takes longer, time will on the whole be saved, because the same result will be attained with a smaller number of distillations. Unless otherwise stated, all the laboratory distillations recorded in this book have been carried out at the rate of 1 drop of distillate per second.

Systematic Fractional Distillation.—The results of the fractionation [2] of benzene and toluene with an ordinary distillation bulb, now being described, are given in Tables 38 and 39 ; the read temperatures are not stated but only the true temperatures under normal pressure.

A flask of about 270 c.c. capacity was used for the first distillation and the bulb of the thermometer was covered with a little cotton-wool to prevent super-heating (p. 13).

First Fractional Distillation. — In the first distillation the temperature rose almost at once to 86° and the first portion of the distillate was therefore collected in the 4th receiver. On the other hand the temperature reached 110·6°, the boiling point of toluene, before the whole of the liquid had come over. The distillation was therefore stopped, the apparatus allowed to cool and the residue in the flask weighed ; this residue, amounting to 10·9 grams, consisted of pure toluene and did not require to be distilled.

---

[1] When a mixture of unknown composition is to be distilled, we cannot decide upon the temperature ranges of the fractions beforehand. In that case, the height of the barometer must be noted and the thermometer must be read when fractions of suitable bulk have been collected. The thermometer readings may be corrected afterwards.

[2] For the results obtained with an improved still-head, see Table 55 (p. 158).

[TABLE

TABLE 38

Number of fraction.	Temperature range.	Weight of fraction = $\triangle W$.			
		I.	II.	III.	IV.
1	80·2— 81·2°	...	...	12·95	31·55
2	81·2— 83·2	...	3·8	24·8	23·9
3	83·2— 86·2	...	33·85	22·75	16·2
4	86·2— 89·2	9·75	22·3	13·5	9·55
5	89·2— 92·2	51·8	19·65	11·8	8·0
6	92·2— 95·4	28·85	13·6	9·15	5·8
7	95·4— 98·6	21·2	12·95	7·3	5·35
8	98·6—101·6	12·8	9·05	6·75	4·65
9	101·6—104·6	11·45	8·9	6·3	3·85
10	104·6—107·6	14·15	10·8	7·95	5·85
11	107·6—109·6	13·45	9·6	8·95	7·4
12	109·6—110·6	24·9	30·75	33·05	30·5
Pure toluene	110·6	10·9	22·95	31·35	42·1
Total weight		199·25	198·2	196·6	194·7
Percentage weight of distillate below middle temperature, 95·4°		45·4	47·0	48·3	48·8

Second Fractionation.—The flask of 270 c.c. capacity was now replaced by a smaller one—about 80 c.c. and the first fraction, that in receiver No. 4, was redistilled. A small quantity came over below 83·2° and was collected in flask No. 2 ; this receiver was removed and No. 3 substituted as soon as the temperature reached 83·2°, and No 4 was put in the place of No. 3 when the temperature had risen to 86·2°. At 89·2° the flame was removed and, after the cotton-wool on the thermometer had become dry, the second fraction from the first distillation—that in No. 5—was added to the residue in the still. On recommencing the distillation it became clear that the temperature would not reach 86·2° for a considerable time and the first portion of the distillate was therefore collected in receiver No. 3. The process was continued as before ; flask No. 4 was substituted for No. 3 at 86·2°, No. 5 for No. 4 at 89·2° and the distillation was stopped at 92·2°. The contents of receiver No. 6 were now added to the residue in the still and the distillate was collected in Nos. 4, 5 and 6 ; the remainder of the fractionation was carried out in this manner except that, after the 11th fraction from the first distillation had been placed in the still, no distillate was collected in No. 9 because the temperature rose at once to 104·6° ; so also, after the addition of the last fraction, no distillate came over below 107·6°.

Third Fractionation.—The third fractionation was carried out in the same manner as the second, the first portion of the first distillate, and also the first portion of the distillate which came over after the contents of receiver No. 3 had been added to the residue in the still, being collected in No. 1.

Fourth Fractionation.—In the fourth fractionation, the first fraction was redistilled, the second being placed in the still when the temperature reached 81·2°. In other respects the procedure was the same as in the third fractionation.

Procedure dependent on Number of Fractions. — If the range of temperature for each fraction had been larger, say 5° instead of 3° for the majority of them, then on recommencing the distillation after addition of any fraction to the residue in the still, the temperature would have risen at once, or very rapidly, to the initial temperature of the fraction below ; thus, on adding any fraction, say No. 4, no distillate would have been collected in No. 2, and receiver No. 3 might have been left in position while fraction No. 4 was being placed in the still and the distillation was recommenced. This method is usually adopted, but by taking a large number of receivers there is a greater amount of separation of the components in a complete distillation and, on the whole, time is saved.

Loss of Material.—On the first distillation there was a loss of 0·75 gram of material (200·0 – 199·25), due entirely to evaporation. In the second complete distillation the total weight fell from 199·25 to 198·2, and in the subsequent distillations the loss was in nearly every case more than 1 gram. The greater loss was due partly to increased evaporation, and partly to the small amount of liquid left adhering each time to the funnel through which the fraction was poured into the still. The total loss after fourteen fractionations was 18·1 grams, a very appreciable amount.

Relative Rate of Separation of Components.—It will be observed that, in the first distillation, 10·9 grams of pure toluene were obtained, but that only 9·75 grams of distillate came over below 89·2°, a temperature 9° above the boiling point of benzene ; and at the end of the fourth complete distillation the weight of pure toluene recovered was 42·1 grams, while only 31·55 grams of benzene boiling within 1 degree had been obtained. It is thus evident that the separa tion of the less volatile component is much easier than that of the more volatile, and this is, indeed, always found to be the case.

It will also be noticed that the weights of the middle fractions steadily diminished, while those of the lowest and highest fractions increased ; this also invariably occurs when a mixture of two liquids is repeatedly distilled.[1]

[1] An account has been given by Kreis * of the fractional distillation of a mixture of 25 grams of benzene and 25 grams of toluene with an ordinary distillation bulb ; the data for the first distillation are such as might be expected, but the results given for the second are quite impossible, for while in the first distillation 17 c.c. of liquid are stated to have been collected above 108°, in the second complete distillation the volume of the corresponding fraction is given as only 3 c.c., and it was not until the 8th fractionation that the volume again

---

* Kreis, " Comparative Investigations on the Methods of Fractional Distillation," *Liebigs Annalen*, 1884, **224**, 259.

Weight of Distillate below Middle Temperature.—The percentage weight of distillate, coming over below the middle temperature between the boiling points of the pure components, was not far below that of the benzene in the original mixture; the percentage was actually 45·4, 47·0, 48·3 and 48·8 respectively in the four distillations, while the original mixture contained 50 per cent of benzene. Similar results are always obtained when a mixture of two liquids, which separates normally and easily into the original components, is distilled; and, as will be seen later, when an improved still-head is used, the weight of distillate below the middle temperature should, in general, be very nearly equal to that of the more volatile component in the original mixture even in the first distillation.

Alteration of Temperature Ranges.—The results of the first four complete distillations are given in Table 38, and it will be seen that if the fractionation were continued in the same manner as before, the middle fractions would soon become too small to distil, and the first fraction would never consist of pure benzene.

It is therefore preferable gradually to increase the temperature ranges of the middle fractions, and to diminish those of the fractions near the boiling points of the pure substances.[1] This was accordingly done in the subsequent fractionations, which were carried to their extreme limit. The results, together with those of the fourth fractionation, are given in Table 39.

Ratio of Weight of Fraction to Temperature Range. —So long as the fractions, or most of them, are collected between

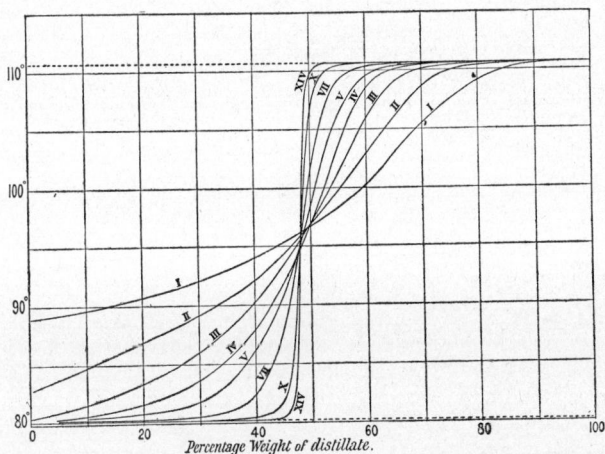

Fig. 34.—Results of fractional distillation of mixture of benzene and toluene.

reached 17 c.c. Assuming the correctness of the data for the first distillation, the last fraction in the second should have been at least 6 times as great as that stated; it is difficult to understand how the mistake can have arisen.

[1] Mendeléeff, A paper, without title, on the fractional distillation of Baku Petroleum, *Journ. Russ. Phys. Chem. Soc.*, Protok., 1883, 189.

equal intervals of temperature, the weights of distillate indicate clearly enough the progress of the separation, but when the temperature ranges are gradually altered this becomes less evident, and it is advisable either to plot the percentage (or total) weight of distillate against the temperature (Fig. 34), or to divide the weight of each fraction, Δw, by its temperature range, Δt, and to tabulate these ratios as well as the actual weights. The purer the liquid in any fraction, the higher is the ratio $\Delta w/\Delta t$, that for a pure liquid being, of course, infinitely great ; and it will be seen from Table 39 that in the later fractionations, while the weights of the middle fractions diminish, and those of the lowest and highest increase much more slowly than before, the ratios $\Delta w/\Delta t$ continue to change rapidly.

Separation of Pure Benzene.—When the liquid collected below 80·45° in the seventh fractionation was distilled at the beginning of the eighth, the temperature remained at the boiling point of benzene, 80·2°, for a short time, and a considerable amount of distillate came over below 80·25°. As the range of this fraction was reduced to 0·15°, it was concluded that the first portion of the first distillate in the ninth fractionation would consist of pure benzene ; it was therefore collected separately as there was no necessity to redistil it, and in each of the subsequent fractionations the first portion of the first distillate was taken to be pure benzene.

Elimination of Fractions.—In the tenth fractionation the weights of the 11th and 12th fractions had become very small, and their temperature ranges were reduced to 0·2° and 0·05° respectively. In the next fractionation, therefore, after No. 12 distillate had been added to the residue in the still, the 11th receiver was left in position until the temperature reached 110·6° ; the distillation was then stopped, and the residue taken as pure toluene. In this way the number of fractions was reduced to eleven by the elimination of No. 12. In the 12th and 13th fractionations the last fraction was in each case similarly eliminated, and, in the latter, the number of fractions was further reduced by the exclusion of No. 1, and in the 14th by that of No. 2.

In the 14th fractionation, after No. 6 had been added to the residue in the still, it was found that the temperature rose at once above 81°, so that nothing was collected in No. 5, and also that when the distillation had been carried as far as possible, the temperature had not risen higher than 91·4°. The distillation was therefore stopped at this point, and the residue in the still, after cooling, was placed in a separate receiver R. Fraction No. 7 was then placed in the still, and the distillate collected in R until the temperature reached 100°, and then in No. 7. The rest of the fractionation was carried on as before.

Final Fractionations.—In the remaining distillations the fractions below and above the middle point were treated separately, the residue from the last of the lower fractions and the first portion of distillate from the first of the higher ones being collected each time in R.

Four additional operations were required for the separation of the benzene and two for that of the toluene. The total weight of pure benzene recovered by pushing the series of fractionations to their extreme limit was 81·4 grams out of 100 taken ; the weight of toluene was 88·8 grams. The amount of liquid collected in R, and rejected, was 7·8 grams, and the loss by evaporation, and by transference from flasks to still, was 22·0 grams. The amount of time required for the whole operation was a little over thirty hours.

˙MIXTURES SEPARABLE INTO THREE COMPONENTS

We have seen that when a mixture tends to separate on distillation into two components, it is the less volatile component which is the easier to obtain in a pure state. When a mixture tends to separate into three components the least volatile of them can, as a rule, be most readily isolated and, of the other two, the more volatile is the easier to separate.

Methyl, Ethyl and Propyl Acetates.—An illustration of this is afforded by the fractionation of a mixture of 200 c.c. of methyl acetate (b.p. 57·1°), 250 c.c. of ethyl acetate (b.p. 77·15°) and 200 c.c. of propyl acetate ʼ(b.p. 101·55°) carried out with a plain vertical still-head one metre in length. Full details of this fractionation have been published in the *Philosophical Magazine*[1] and in Table 40 the results of the first eight fractionations and of 12th, 16th, 20th and 24th are given.[2] It will be noticed that the method adopted differed from that already described in so far that the number of fractions was small at first and was gradually increased. For convenience, the fractions are so numbered that they fall into their proper places in the later distillations. The data given are as follows :—F, the number of the fraction ; t, the final temperature, corrected and reduced to 760 mm. pressure, for each fraction, but in the first four fractionations the lowest temperature (in brackets) is that at which the distillate began to come over ; Δw, the weight of each fraction, and the ratio $\Delta w/\Delta t$.

First and Second Fractionations.—The first fractionation, I., requires no comment ; the second, II., was carried out in the following manner : —the first fraction from I. (No. 5) was distilled and the distillate was collected in receiver No. 4 until the temperature rose to 63·8°, when receiver No. 5 was substituted for it and the distillation was continued until the temperature reached 71·0°. The gas was then turned out and the second fraction from I. (No. 8) was added to the residue in the still. Heat was again applied and the distillate was collected in No. 5 until the temperature rose to 71·0°, when this receiver was replaced by No. 8, and the distillation was continued until the temperature reached 77·1°. The third fraction from I. (No. 11) was then added to the residue and the distillate was collected in No. 8 until the

---

[1] Barrell, Thomas and Young, " On the Separation of Three Liquids by Fractional Distillation," *Phil. Mag.*, 1894, [V.], **37**, 8.

[2] For the results obtained with an improved still-head, see Table 57, p. 160.

TABLE 39

No. of fraction, F	IV — Final temperature	IV — Weight of fraction	IV — Δw/Δt	V — t	V — Δw	V — Δw/Δt	VI — t	VI — Δw	VI — Δw/Δt	VII — t	VII — Δw	VII — Δw/Δt	VIII — t	VIII — Δw	VIII — Δw/Δt	IX — t	IX — Δw	IX — Δw/Δt	X — t	X — Δw	X — Δw/Δt	XI — t	XI — Δw	XI — Δw/Δt
Pure Benzene (init)	...	...	...	...	...	...	...	...	...	...	...	...	...	...	...	80.2°	10.2	∞	80.2°	23.35	∞	80.2°	44.4	∞
1	81.2°	31.55	31.55	80.9°	40.05	57.2	80.6°	42.95	107.4	80.45°	52.4	209.6	80.35°	54.45	363.0	80.3	45.0	450.0	80.25	36.25	725.0	80.25	17.1	342.0
2	83.2	23.9	11.95	82.5	23.3	14.6	81.8	23.15	19.3	81.25	17.2	21.5	80.85	15.6	31.2	80.6	17.55	58.5	80.45	12.25	61.0	80.35	10.25	102.5
3	86.2	16.2	5.4	85.1	10.9	4.2	84.0	10.2	4.6	83.0	8.0	4.5	82.0	8.6	7.5	81.4	7.25	9.0	80.95	8.75	17.5	80.6	8.0	32.0
4	89.2	9.55	3.2	88.1	7.85	2.6	87.0	6.65	2.2	86.0	5.85	2.0	84.7	5.5	2.0	83.5	3.3	2.1	82.4	4.0	2.8	81.4	3.75	4.7
5	92.4	8.0	2.7	91.3	5.75	1.8	90.5	4.75	1.4	90.0	4.3	1.1	89.0	3.55	0.8	88.0	2.75	0.7	86.4	3.2	0.8	85.0	3.65	1.0
6	95.4	5.35	1.7	95.4	5.2	1.3	95.4	3.8	0.9	95.4	3.2	0.7	95.4	3.2	0.5	95.4	2.6	0.4	95.4	2.3	0.3	95.4	2.4	0.2
7	98.6	5.35	1.7	99.5	4.15	1.0	100.3	3.65	1.0	100.8	3.6	0.6	101.8	3.0	0.5	102.8	2.85	0.6	104.4	2.65	0.6	105.8	2.25	0.2
8	101.6	4.65	1.5	102.7	3.8	1.2	103.8	4.0	1.3	104.8	3.6	0.8	106.1	3.5	0.7	107.3	3.15	1.5	108.4	3.05	2.1	109.4	2.8	0.8
9	104.6	3.85	1.3	105.7	4.5	1.5	106.8	4.85	1.2	107.8	4.9	1.2	108.8	4.4	1.3	109.4	3.45	4.3	109.85	3.4	6.8	110.2	2.8	3.5
10	107.6	5.85	1.9	108.3	5.2	2.0	109.0	8.5	7.1	109.55	8.7	2.8	109.95	5.3	3.8	110.2	7.0	23.3	110.35	5.5	27.5	110.45	3.05	12.2
11	109.6	7.4	3.7	109.9	7.25	4.5	110.2	20.4	51.0	110.35	13.05	10.9	110.45	11.45	10.6	110.5	5.75	57.5	110.55	4.15	83.0	110.6	5.25	35.0
12	110.6	30.5	30.5	110.6	28.1	40.1	110.6			110.6		52.2	110.6		76.3	110.6			110.6					
Pure Toluene	42.1			46.8			54.05			61.75			67.15			72.1			75.2			79.6		
Total weight	194.7			192.85			191.45			189.9			188.55			187.3			186.6			185.3		
Percentage weight of distillate below middle point =	48.8			48.3			48.2			48.2			48.2			48.3			48.3			48.3		

TABLE 39—*continued*

No. of fraction F	XII Final temperature	XII Weight of fraction	XII $\frac{\Delta w}{\Delta t}$	XIII t	XIII Δw	XIII $\frac{\Delta w}{\Delta t}$	XIV t	XIV Δw	XIV $\frac{\Delta w}{\Delta t}$	XV t	XV Δw	XV $\frac{\Delta w}{\Delta t}$	XVI t	XVI Δw	XVI $\frac{\Delta w}{\Delta t}$	XVII t	XVII Δw	XVII $\frac{\Delta w}{\Delta t}$	XVIII t	XVIII Δw	XVIII $\frac{\Delta w}{\Delta t}$
Pure Benzene	80·2°	52·75	∞	80·2°	62·25	∞	80·2°	73·3	∞	80·2°	75·7	∞	80·2°	77·7	∞	80·2°	79·0	∞	80·2°	81·4	∞
1	80·25	10·85	217·0	80·25	11·1	222·0	80·25	5·45	109·0	80·3	3·85	38·5	80·3	2·45	24·5	80·3	3·5	35·0			
2	80·3	7·6	152·0	80·3	5·45	109·0	80·4	3·55	23·7	80·65	2·6	7·4	80·6	3·65	12·2						
3	80·4	6·35	63·5	80·55	2·3	9·2	80·9	1·05	2·1	81·85	3·25	2·7									
4	80·8	5·15	12·9	81·4	3·4	4·0	91·4	3·85	0·4												
5	83·4	3·6	1·4	95·4	3·9	0·3															
6	95·4	2·55	0·2																		
7	107·4	2·35	0·2	109·4	2·8	0·2	109·8	2·95	0·3	{107·5 to 110·2}	1·95	0·7	{110·45 to 110·6}	1·5	10·0						
8	110·0	2·6	1·0	110·35	2·65	2·8	110·5	2·05	2·9	110·6	2·7	6·8									
9	110·4	3·15	7·9	110·6	4·4	17·6	110·6	2·1	21·0												
10	110·6	4·2	21·0																		
11																					
12																					
Pure Toluene		82·95	∞		85·0	∞		86·3	∞		87·55	∞		88·8	∞						
Total weight		184·1			183·25			(181·9)													
Percentage weight of distillate below middle point		48·3			48·3																

TABLE 40

F.	I. t.	I. $\Delta w.$	I. $\frac{\Delta w}{\Delta t}$	II. F.	II. t.	II. $\Delta w.$	II. $\frac{\Delta w}{\Delta t}$	III. F.	III. t.	III. $\Delta w.$	III. $\frac{\Delta w}{\Delta t}$	IV. F.	IV. t.	IV. $\Delta w.$	IV. $\frac{\Delta w}{\Delta t}$	V. F.	V. t.	V. $\Delta w.$	V. $\frac{\Delta w}{\Delta t}$	VI. F.	VI. t.	VI. $\Delta w.$	VI. $\frac{\Delta w}{\Delta t}$
A	(63·8°)				(60·9°)				(58·4°)			A	(57·7°)								57·3°		
1																					1	57·3°	6·3	31·5
2													2	58·6	17·2	19·1	2	58·2°	19·4	17·6	2	58·2	28·5	31·7
3									3	60·95	36·5	14·3	3	61·0	58·1	23·7	3	60·7	69·9	28·0	3	60·1	55·0	29·0
4				4	63·8	59·7	20·6	4	63·85	77·2	26·6	4	63·9	53·1	18·3	4	63·85	49·0	15·3	4	63·85	54·2	14·3
5	71·3			5	71·0	134·2	18·6	5	71·0	100·6	14·0	5	71·05	85·1	11·8	5	71·0	69·0	9·6	5	70·75	47·4	6·9
6																								
7													7								7	75·7	52·0	10·4
8	77·8	141·4	18·9	8	77·1	98·4	16·1	8	77·15	75·4	12·4	8	77·15	66·6	10·9	8	77·15	71·1	11·6	8	77·15	33·4	22·3
9																								
10																								
11	89·2	145·2	22·3	11	84·4	67·2	9·2	11	84·4	67·7	9·3	11	84·3	84·7	11·8	11	83·35	85·1	13·6	11	78·4	34·8	29·0
12																					12	82·5	47·9	11·7
13				13	91·7	51·0	7·0	13	91·65	59·6	8·2	13	91·6	40·8	5·6	13	90·75	33·7	4·6	13	90·2	28·1	3·6
14 {	above 89·2	146·9	12·8	14	98·4	86·1	12·7	14	98·4	46·7	6·9	14	98·4	33·8	5·0	14	98·45	33·5	4·4	14	98·45	27·8	3·3
15	} 143·3		..	15	101·5	63·7	20·6	15	101·45	78·3	26·1	15	100·95	44·1	17·6	15	100·95	29·4	11·8	15	100·95	25·6	10·2
16												16	101·45	38·7	77·4	16	101·45	36·7	73·4	16	101·45	30·3	60·6
17				17	(101·55)	14·3		17	101·55	19·7	197·0	17	101·55	24·2	242·0	17	101·55	31·1	311·0	17	101·55	34·0	340·0
X																								
Y																								
Z								Z	..	10·0	∞	Z	..	20·2	∞	Z	..	34·7	∞	Z	..	50·6	∞
		576·8				574·6				571·7				566·6				562·6				555·9	

TABLE 40—continued

VII.

F.	t.	Δw.	Δw/Δt
1	57.3°	13.3	**66.5**
2	58.2	42.9	**47.7**
3	60.0	47.5	**26.4**
4	63.8	43.5	**11.4**
5	70.75	45.9	**6.7**
7	75.8	42.7	**8.4**
8	77.2	45.0	**32.1**
11	78.3	37.8	**34.4**
12	82.1	41.1	**10.8**
13	90.15	23.4	**2.9**
14	98.45	22.3	**2.7**
15	100.95	17.6	**7.0**
16	101.45	24.3	**48.6**
17	101.55	23.7	**237.0**
Z	..	80.2	∞
		551.2	

VIII.

F.	t.	Δw.	Δw/Δt
A	57.1°	6.0	∞
1	57.3	11.7	**58.5**
2	58.2	49.1	**54.6**
3	60.1	45.4	**23.9**
4	63.8	36.0	**9.7**
5	70.7	37.8	**5.5**
7	75.8	41.9	**8.2**
8	77.15	43.8	**32.4**
11	78.2	51.1	**48.7**
12	81.1	34.1	**11.8**
13	90.05	22.8	**2.5**
14	98.5	17.8	**2.1**
15	101.0	16.7	**6.7**
16	101.45	19.4	**43.3**
17	101.55	12.6	**126.0**
Z	..	101.5	∞
		547.7	

XII.

F.	t.	Δw.	Δw/Δt
A	57.1°	41.1	∞
1	57.2	17.0	**170.0**
2	57.65	30.5	**67.8**
3	58.9	30.3	**24.2**
4	63.5	31.0	**6.7**
5	71.9	23.8	**2.8**
7	76.45	32.9	**7.1**
8	77.0	30.5	**55.5**
9	77.15	27.6	**184.0**
10	77.3	36.2	**241.3**
11	77.6	35.6	**85.3**
12	78.85	18.7	**14.9**
13	88.9	17.9	**1.8**
14	99.55	9.5	**0.9**
15	101.15	8.1	**5.1**
X	..	7.8	
Y	..	8.1	
Z	..	132.6	∞
		529.2	

XVI.

F.	t.	Δw.	Δw/Δt
A	57.1°	79.9	∞
1	57.2	14.6	**146.0**
2	57.6	17.3	**43.2**
3	58.6	14.6	**14.6**
4	65.35	25.0	**3.7**
5	73.9	15.6	**1.8**
7	76.8	26.6	**9.2**
8	77.1	26.7	**89.0**
9	77.15	39.0	**780.0**
10	77.2	39.9	**798.0**
11	77.35	24.4	**162.6**
12	77.9	11.6	**21.1**
U	..	6.6	
V to Z	..	172.4	
		514.2	

XX.

F.	t.	Δw.	Δw/Δt
A	57.1°	105.5	∞
1	57.15	5.9	**118.0**
2	57.55	10.7	**26.7**
3	58.55	8.9	**8.9**
4	67.25	17.6	**2.0**
5	75.95	16.6	**1.9**
7	76.95	16.0	**16.0**
8	77.15	13.4	**67.0**
9	77.15	45.0 ⎫	
10	77.2	49.4 ⎬	**1888.0**
11	77.25	13.0	**260.0**
S	..	8.6	
T to Z	..	192.3	
		502.6	

XXIV.

F.	t.	Δw.	Δw/Δt
A	..	105.5	∞
B & C	..	21.7	**0.8**
D	67.2°	10.0	**1.9**
4	75.3	6.8	**5.2**
5	76.8	15.1	
6	77.1	7.8	**40.7**
7	77.15	12.2	**292.2**
8	77.15	14.6 ⎫	very high
9	77.15	31.9 ⎬	
10		26.9 ⎭	
Q	..	13.1	
R to Z	..	228.8	
		494.4	

temperature again rose to 77·1°, when No. 11 was put in its place. At 84·4° a new receiver, No. 13, was substituted for No. 11 and the temperature was allowed to rise to 91·7°. The last fraction from I. was then added to the residue in the flask and the distillate collected in No. 13 up to 91·7°, after which fractions were collected in No. 14 from 91·7° to 98·4° and in No. 15 from 98·4° to 101·5°, when the distillation was stopped and the residue was poured into No. 17.

Separation of Propyl Acetate.—The third fractionation, III., was carried out in a similar manner, a new fraction, No. 3, being collected at the beginning. As the temperature rose to 101·55°, the boiling point of propyl acetate, before the end of the last distillation, the residue was placed in a separate flask, z, and was not redistilled. The residues from subsequent fractionations up to the 10th were collected in z, but after this, as a large amount of propyl acetate had been removed, the temperature did not reach 101·55° and the residue from the 11th fractionation (b.p. above 101·45°) was placed in a new flask, y. At the end of the 12th fractionation the temperature rose only to 101·15° and the residue was placed in a third flask, x, and subsequently the residues were placed in w, v . . . . q, as shown in Table 40.

Separation of Methyl Acetate.—It was not until the 5th fractionation that the first fraction began to boil at 57·1°, the boiling point of methyl acetate, and it was not thought advisable to separate the first portion of the first distillate until the 8th fractionation. This portion and also the corresponding ones up to the 20th fractionation were collected together in a flask, A, after which the first portions were collected in B, C, D and E.

Accumulation of Ethyl Acetate in Middle Fractions.—The presence of the middle substance, ethyl acetate, is not clearly indicated until the 4th fractionation, when the value of $\Delta w/\Delta t$ for the 6th distillate (No. 11) is somewhat higher than for those above and below it ; but the gradual accumulation of the ethyl acetate in the middle fractions in subsequent fractionations is clearly shown by the rise in the value of $\Delta w/\Delta t$ for the fractions 8 and 11 and, after the 10th fractionation, for fractions 9 and 10. The range of temperature for Nos. 9 and 10 was gradually diminished from 0·2° each in the 11th fractionation until no rise could be detected ; there was, indeed, no perceptible rise of temperature during the collection of No. 11 in the 22nd and 23rd fractionations. It was therefore certain that after the 26th fractionation the tenth fraction was free from propyl acetate, and that the remaining fractions B to E and 5 to 10 contained only methyl and ethyl acetates. Similarly, it is safe to conclude that the fractions Q to z were free from methyl acetate. The preliminary series of fractionations was therefore completed, no fraction now containing more than two components.

Graphical Representation of Results.—The progress of the

separation is well seen by mapping the temperatures against the
percentage weights of distillate collected, and the curves for the first
twelve fractionations are shown in Fig. 35 (*a* and *b*). The horizontal
lines at the extremities of the later curves represent the methyl and
propyl acetates removed in the first portion of the first distillates and
in the residues respectively. The presence of ethyl acetate is clearly
indicated in the fourth curve but not in the earlier ones.

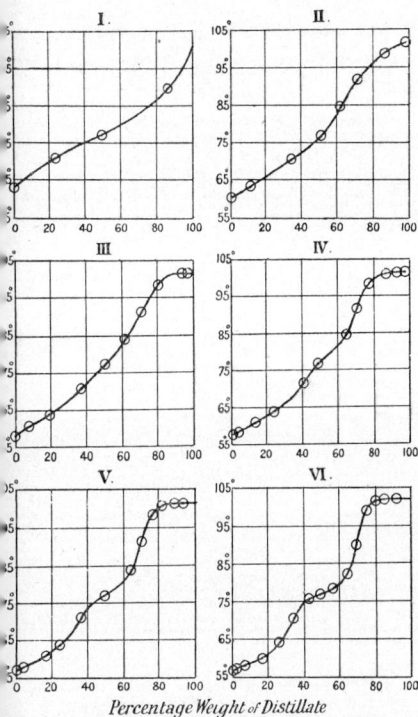

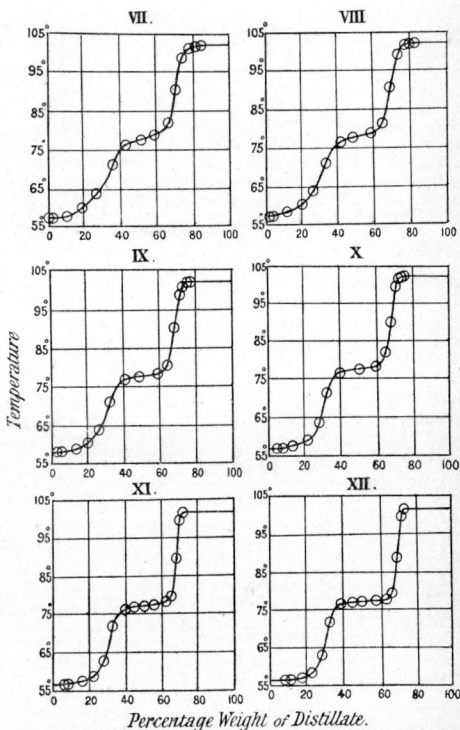

FIG. 35 (*a*).—Results of fractional distillation of mixture
of methyl, ethyl and propyl acetates.

FIG. 35 (*b*).—Results of fractional distillation of mixture
of methyl, ethyl and propyl acetates (*continued*).

Final Fractionations.—The fractions into which the esters had been
separated at the end of the 26th fractionation are shown in Table 41.
The total weight was 490·8 grams, and therefore 86 grams had been
lost by evaporation and by transference from flask to still. The final
separation of methyl and ethyl acetates and of ethyl and propyl acetates
was carried out in the manner described for the later fractionations of
mixtures of benzene and toluene, but it was necessary to treat the
methyl acetate with phosphorus pentoxide to remove moisture, and the
propyl acetate with potassium carbonate to remove free acid due to
slight hydrolysis. The loss was thus much greater than it would
otherwise have been.

[TABLE

TABLE 41

Methyl and Ethyl acetates.			Ethyl and Propyl acetates.		
Fraction.	Temperature range.	Weight.	Fraction.	Temperature range.	Weight.
A	57·1°	105·5	Q	77·15°	32·3
B	57·1 —57·15°	11·6	R	77·15— 77·2°	19·7
C	57·15—57·55	10·1	S	77·2 — 77·3	16·8
D	57·55—58·55	10·0	T	77·3 — 77·65	9·7
E	58·55—68·9	11·9	U	77·65— 79·6	10·2
5	68·9 —75·7	7·1	V	79·6 — 99·7	15·8
6	75·7 —76·9	10·4	W	99·7 —101·15	8·1
7	76·9 —77·1	15·4	X	101·15—101·45	7·8
8	77·1 —77·15	16·4	Y	101·45—101·55	8·1
9 & 10	77·15°	31·3	Z	101·55°	132·6
		229·7			261·1

The final results are given below.

TABLE 42

	Weights.			Specific gravity at 0° 4°.*	
	Taken.	Recovered.	Percentage recovered.	Before mixing.	After fractionation.
Methyl acetate .	183	88·0	48·1	0·95932	0·95937
Ethyl acetate .	222	{ 56·1 62·7 } 118·8	53·5	0·92436	{ 0·92438 0·92437
Propyl acetate .	175	126·8	72·5	0·91016	0·91008

* Specific gravity at 0° compared with that of water at 4°.

The esters were redistilled over phosphorus pentoxide before their specific gravities were determined.

COMPLEX MIXTURES

The procedure in the case of complex mixtures is similar to that adopted when there are three components, but the number of fractions must be increased and the time required is longer.

If, in a complex mixture, there are two liquids boiling at temperatures not very far apart, together with others which have boiling points relatively much lower and much higher, the early fractionations do not appear, as a rule, to indicate the presence of the two liquids, but of a single substance with a boiling point between those of the two which are actually present, and it is only after some progress has been made with the fractionations that the presence of both liquids is clearly shown.

Separation of Pentanes from Petroleum.—A striking instance of this is afforded by the separation of isopentane (b.p. 27·95°) and normal pentane (b.p. 36·3°) from the light distillate from American petroleum.[1] After treatment with a mixture of concentrated nitric and sulphuric acids to remove impurities, the " petroleum ether " consists chiefly of butanes and probably a little tetramethyl methane, all boiling below 10°, the two pentanes referred to, and hexanes and other hydrocarbons boiling higher than 60°, with very much smaller quantities of penta-methylene and trimethylethyl methane, both of which boil at about 50° ; there are no substances present except the two pentanes with boiling points between about 10° and 50°.

In order to effect the separation of the pentanes it is necessary to employ a very efficient still-head (Chapter XII.), but the method of arranging the fractions is the same as when an ordinary still-head is used.

Graphical Representation and Interpretation of Results.
—The fractionations will be referred to more fully later on (pp. 149 and 205), but it may be well here to consider the curves showing the results of the 1st, 4th, 7th, 10th and 13th fractionations (Fig. 36).

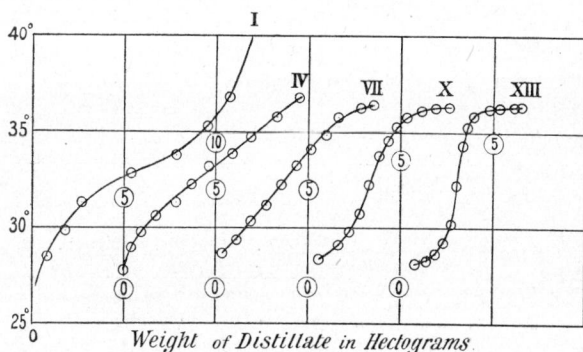

FIG. 36.—Separation of normal and isopentane from American petroleum.

The first curve, I., seems to indicate [2] that a single substance boiling at about 33° is being separated from others boiling considerably lower (butanes) and higher (hexanes), but, if that were really the case, the curves for subsequent fractionations should become more horizontal (indicating greater purity) at about that temperature, and more vertical at, say, 25° and 40°.

Instead of this, they become less horizontal at about 33° and the next curve, IV., approximates to a straight line between about 30° and 37°. This change is a sure sign that there are at least two substances present boiling at temperatures not very far apart, but little or no light is thrown on the actual boiling points.

[1] Young and Thomas, "Some Hydrocarbons from American Petroleum. I. Normal and Iso-pentane," *Trans. Chem. Soc.*, 1897, **71**, 440.
[2] Young, "Experiments on Fractional Distillation," *Journ. Soc. Chem. Industry*, 1900, **19**, 1072.

Curve VII. is fairly straight from about 28·5° to about 35·5° but then becomes much more horizontal, terminating at 36·4°. This seems to indicate that the less volatile component boils at a temperature not far from 36° (the hexanes having now been almost completely eliminated).

Curve X. is distinctly more vertical in the middle; it becomes nearly horizontal above and terminates at 36·3°, showing that the higher boiling point is a little over 36·0°. As the curve is more horizontal below than No. VII. there is evidence that the second component must boil not far from 28°.

The upper extremity of the curve XIII. is perfectly horizontal at 36·3° and the true boiling point of the less volatile component (normal pentane) is thus established; the form of the lower part of the curve indicates that the boiling point of the more volatile component (isopentane) must be very close to 28° and further fractionation showed that it is really 27·95°.

Hexamethylene in American Petroleum.—When, in a complex mixture, one component is present in relatively very small quantity, it may very easily be overlooked, and it is only by keeping a careful record of the results of the fractionations and especially of the values $\Delta w/\Delta t$, or by plotting the weights of distillate against the temperature, that the presence of such substances can be detected.

TABLE 43

V.				VI.				VII.				XII.			
$F.$	$t.$	$\triangle w.$	$\frac{\triangle w}{\triangle t}$	$F.$	$t.$	$\triangle w.$	$\frac{\triangle w}{\triangle t}$	$F.$	$t.$	$\triangle w.$	$\frac{\triangle w}{\triangle t}$	$F.$	$t.$	$\triangle w.$	$\frac{\triangle w}{\triangle t}$
	(66)				(66·0)				(66·0)						
1	67	85·9	85·9	1	67·0	74·9	74·9	1	66·9	49·8	55·3				
2	68	94·1	94·1	2	68·0	82·6	82·6	2	67·7	62·3	77·9				
								3	68·4	70·0	100·0				
4	69	81·1	81·1	4	69·0	103·0	103·0	4	69·0	68·8	114·7				
5	70	73·3	73·3	5	70·0	78·3	78·3	5	69·8	81·9	102·4				
6	72	64·6	32·3	6	71·5	52·2	34·8	6	71·0	46·9	39·1				
								7	72·5	45·4	30·3				
8	78	81·4	13·6	8	76·0	69·7	15·5	8	76·0	20·5	5·9				
												9	(74·0) 78·8	12·7	2·7
10	84	62·5	10·4	10	82·0	64·8	10·8	10	79·5	38·1	10·9	10	80·0	16·4	13·7
												11	80·5	24·3	48·6
												12	81·1	12·1	20·2
												13	83·0	11·5	6·1
14	89	66·0	13·2	14	87·0	58·9	11·8	14	85·0	72·9	13·3	14	85·0	7·1	3·5
15	91	39·9	20·0	15	90·0	35·3	11·8	15	88·0	23·4	7·8				

As an example, consider the distillate from American petroleum coming over between 66° and 91°.[1] In this case, again, no satisfactory result could be obtained without the use of a very efficient still-head. The results of the 5th, 6th, and 7th fractionations between the above-named limits of temperature and of the 12th fractionation between 74° and 85° are given in Table 43. The 5th fractionation gave high values of $\Delta w/\Delta t$ from 61 5° to 70°, with a maximum at 67° to 68°,

---

[1] Young, "Composition of American Petroleum," *Trans. Chem. Soc.*, 1898, **73**, 905.

Above 70° the ratios fell to a minimum at 78° to 84° and then rose again. Now, if there were no substance present between the hexanes and heptanes, the ratio $\Delta w/\Delta t$ would become smaller each time at the intermediate temperatures, and notably at about 80°; but it will be seen that in the next fractionation the ratio was slightly higher instead of lower. The number of fractions in the neighbourhood of 80° was therefore increased, and in the 7th fractionation a maximum value of $\Delta w/\Delta t$ was observed at about this temperature in addition to that due to normal hexane near 69°.

Further fractionation of the distillates coming over between 72° and 88° showed that there was really a substance present boiling not far from 80°, as will be seen from No. 12. It might be supposed that this substance was benzene (b.p. 80·2°); but that could not have been the case, for a 10 or 20 per cent solution of benzene in hexane boils almost constantly at nearly the same temperature as hexane itself, and, when American petroleum is distilled, almost the whole of the benzene comes over below 70°, mostly from about 63° to 68°. Moreover, in this case the aromatic hydrocarbons had been removed by treatment with nitric and sulphuric acids.

The quantity of liquid was too small to allow of a fraction of quite constant boiling being obtained, though considerable improvement was effected.

Hexamethylene in Galician Petroleum.—Later on, a large quantity of Galician petroleum was fractionated by Miss E. C. Fortey,[1] who obtained a considerable amount of liquid boiling quite constantly at 80·8°. A chemical examination of the liquid led to the conclusion that it consisted of pure hexamethylene [2][3]; but it was afterwards found [4] that it could be partially, but not completely, frozen in an ordinary freezing mixture, and it was necessary to resort to fractional crystallisation to separate the hexamethylene (b.p. 80·85°) in a pure state. It is evident that there is another hydrocarbon present in small quantity, no doubt a heptane, and that the two substances cannot be separated by fractional distillation.

Mixtures of Constant Boiling Point.—Reference has been made in Chapter IV. to the formation of mixtures of constant (minimum or maximum) boiling point. When a liquid contains two components which are capable of forming a mixture of constant boiling point, it is not possible to separate both components; all that can be done is to separate that component which is in excess from the mixture of constant boiling point, and even this is not possible when the boiling

[1] Fortey, "Hexamethylene from American and Galician Petroleum," *Trans. Chem. Soc.*, 1898, **73**, 932.

[2] Baeyer, "On the Hydro-derivatives of Benzene," *Berl. Berichte*, 1893, **26**, 229; "On the Reduction Products of Benzene," *Liebigs Annalen*, 1893, **278**, 88.

[3] Markownikoff, "On the Presence of Hexanaphthene in Caucasian Naphtha," *Berl. Berichte*, 1895, **28**, 577; "On Some New Constituents of Caucasian Naphtha," *ibid.*, 1897, **30**, 974.

[4] Young and Fortey, "The Vapour Pressures, Specific Volumes and Critical Constants of Hexamethylene," *Trans. Chem. Soc.*, 1899, **75**, 873.

points are very near together. Thus, no amount of fractionation with the most perfect apparatus would make it possible to separate either pure normal hexane or pure benzene from a mixture containing, say, 2 per cent of benzene,[1] because the boiling points of hexane and of the binary mixture differ, in all probability, by less than 0·1°. On the other hand, if the original mixture contained, say, 50 per cent of benzene, a small quantity of that component could be separated in a pure state from the mixture of constant boiling point.[2]

American petroleum contains a relatively small amount of benzene, and the whole of it comes over with the hexanes; but Russian petroleum is much richer in aromatic hydrocarbons, and, consequently, some of the benzene comes over at its true boiling point, only a portion of it distilling with the hexanes.

When a liquid contains three components which are not closely related to each other it may happen that both a ternary and a binary mixture of constant boiling point are formed on distillation. In that case it is only possible to separate one of the original components in a pure state. These points will be considered more fully in Chapters XIII. and XV.

[1] *Loc. cit.*

[2] Jackson and Young, "Specific Gravities and Boiling Points of Mixtures of Benzene and Normal Hexane," *Trans. Chem. Soc.*. 1898. **73** 923.

CHAPTER] VIII

THEORETICAL RELATIONS BETWEEN THE WEIGHT AND COMPOSITION OF DISTILLATE

Application of Brown's Formula.[1]—It has been pointed out in previous chapters—

1. That the vapour pressures of mixtures of two closely related compounds—and rarely of others—are represented with small error by the formula $P = \dfrac{\text{M}P_A + (100 - \text{M})P_B}{100}$;

2. That when this formula holds good, the composition of the vapour from any mixture is given, approximately at any rate, by Brown's formula $\dfrac{m'_B}{m'_A} = c\dfrac{m_B}{m_A}$, where the constant c does not differ greatly from the mean ratio of the vapour pressures of the pure substances at temperatures between their respective boiling points. We may conclude, then, that Brown's formula can be used without much error for mixtures of two closely related substances, and it is probable that the two formulae referred to above, when suitably modified, are applicable also to mixtures of three or more closely related substances.

Mixtures of Two Components.—Taking first the case of mixtures of two liquids, Brown's formula may be written

$$\frac{d\xi}{d\eta} = c\frac{\xi}{\eta},$$

where ξ = residue of liquid B at any instant, η = residue of liquid A at the same instant and $d\xi$ and $d\eta$ = the weights of B and A, respectively, in the vapour.

Taking L and M as the weights of B and A originally present and $L + M = 1$, we obtain by integration

$$(M^c/L)y\{c + (1 - c)y\}^{c-1} = c^c(1 - x)^{c-1}(1 - y)^c,$$

where y = quantity of the more volatile liquid B in unit weight of the distillate coming over at the instant when x is the quantity of liquid distilled. By means of this equation the changes of composition that

---

[1] Barrell, Thomas and Young, "On the Separation of Three Liquids by Fractional Distillation," *Phil. Mag.*, 1894, [V.], **37**, 8.

take place in the course of a distillation may be traced, and the variation in the composition of the distillate represented graphically.

To take a very simple case, suppose that $c=2$ and that $L=M=\frac{1}{2}$.

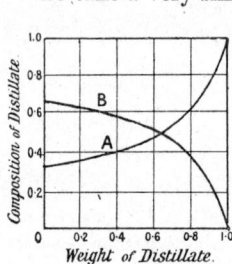

In the diagram (Fig. 37), the amounts of distillate that have been collected are represented as abscissae, and the relative quantities of the two liquids, A and B, in the distillate as ordinates.

It will be seen that the composition of the distillate alters slowly at first, then more and more rapidly, also that while the first portion of the distillate contains a considerable amount of the less volatile substance, A, the last portion is very nearly free from the lower boiling component, B. These points are fully confirmed by experiment, and an explanation is afforded of the fact that it is much easier to separate the less volatile than the other component in a pure state.

FIG. 37.—$B=0.5$, $A=0.5$, in original mixture.

By fractionating a few times in the ordinary way, collecting the distillates in six or eight fractions, we shall have a large excess of B in the first fraction and a still larger excess of A in the last.

Suppose, now, that two of these fractions, one containing B and A in the ratio of $9:1$ and the other in the ratio $1:9$, are distilled separately and completely; the results will then be represented by Figs. 38 and 39. Again, it will be seen that the first tenth of the

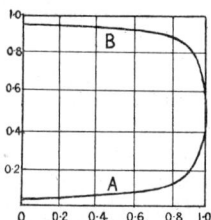

FIG. 38.—$B=0.9$, $A=0.1$, in mixture distilled.

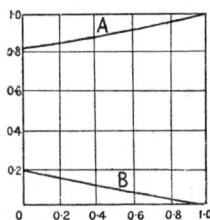

FIG. 39.—$B=0.1$, $A=0.9$, in mixture distilled.

distillate from the first of these fractions is not so rich in B as the last tenth from the second fraction is in A; the purification of A still proceeds more rapidly than that of B.

In making use of this formula it is assumed that no condensation (and therefore no fractionation) goes on in the still-head, but that the vapour reaches the condenser in the same state as when first evolved from the liquid in the still. By using an improved still-head (Chapters X. to XII.) a more rapid separation would be effected.

Mixtures of Three Components.—If we have a mixture of three closely related substances, C, B and A, it may be conjectured that the proportion of the three substances in the vapour at any instant is the same as that of the weights of the three substances in the residue in the still, each weight being multiplied by a suitable constant, which is approximately proportional to the vapour pressure of the corresponding liquid. Here again formulae may be obtained by integration, which enable us to follow the course of a distillation.

Taking the three constants as $c=4$, $b=2$, $a=1$—which are roughly proportional to the vapour pressures of methyl, ethyl and propyl acetates at the same temperature—and the original weights of the three liquids C, B and A as $L=M=N=\frac{1}{3}$, Fig. 40 a represents the first distillation.

If the distillate were collected in five equal fractions they would have the following composition—

TABLE 44

	C or L.	B or M.	A or N.
IIa	0·543	0·300	0·157
IIβ	0·47	0·33	0·20
IIγ	0·37	0·365	0·265
IIδ	0·22	0·39	0·39
IIϵ	0·047	0·265	0·687

It will be seen that while the last fraction is more than twice as rich in A as the original mixture and is nearly free from C, and the amount of C in the first fraction is 0·543, as against 0·333 in the original mixture, the fraction richest in B, the fourth, contains only 0·39 of that substance. It is evident, therefore, that the middle substance is the one that is the most difficult, and the least volatile substance the one that is easiest to separate. If these five fractions were separately distilled we should get the results indicated in the curves, Fig. 40, b, c, d, e, f.

It is to be noted that in IIa, which is richest in the most volatile component, C, the amount of C rises from 0·543 to about 0·72 in the first fifth of the distillate. In IIϵ, which is richest in

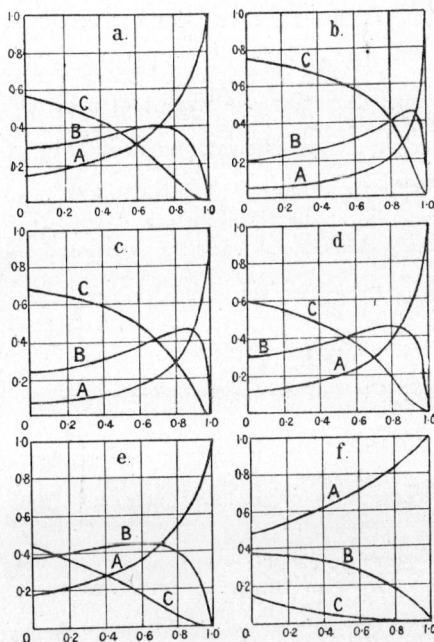

FIG. 40 (a to f).—Distillation of mixtures containing three components, A, B and C.

the least volatile liquid A, the amount of A rises from 0·687 to about 0·9 in the last fifth of the distillate. On the other hand, in IIδ, which is richest in B, the middle liquid, the improvement is merely from 0·39 to rather less than 0·45 in the third fifth of the distillate.

If the fractions richest in C, B and A respectively were redistilled we should have the following result for the best distillates :—

Improvement in C from .	.	0·7	to	0·82
,,	B ,,	.	.	0·45	,, 0·50
,,	A ,,	.	.	0·9	,, 0·98

Thus, the improvement in B continues to be very slow, and even when a fraction very rich in B is redistilled the improvement is very slight. Thus, on redistilling a mixture containing 0·02 of C, 0·96 of B and 0·02 of A, the amount of B advances only from 0·96 to 0·97 in the most favourable part of the distillate ; it is thus clear that there must be far greater difficulty in separating the middle substance than either of the others from a mixture.

It is, however, not difficult to obtain fractions rich in C but free from A, and also rich in A but free from C, and the best method of obtaining B in a pure state is to carry out a series of preliminary fractionations in the manner described for methyl, ethyl and propyl acetates (p. 105) until all the fractions coming over below the boiling point of B contain only C and B, and all those coming over above that temperature contain only B and A. We have, then, only to deal with the separation of two components from the mixtures, and these separations will already be far advanced.

In 1902 Lord Rayleigh [1] brought forward another theory of distillation. Denoting the weight of a given liquid mixture by w and the weight of the first component in it by y, $\dfrac{y}{w} = \xi =$ the fractional amount of the first component in the liquid.

Calling the total weight of vapour dw and the weight of the first component in it dy, the fractional amount of the first component in the vapour $= \dfrac{dy}{dw} = f(\xi)$.

Therefore $\qquad \dfrac{d(w\xi)}{dw} = f(\xi)$,

and if w_0 and ξ_0 denote the initial values of w and ξ

$$ln\frac{w}{w_0} = \int_{\xi_0}^{\xi} \frac{d\xi}{f(\xi) - \xi}.$$

Lord Rayleigh makes the assumption that in mixtures containing very little of the first component the fraction of that component in the vapour is proportional to its fraction in the liquid, or $\dfrac{dy}{dw} = \kappa\xi$, and therefore $\qquad \xi/\xi_0 = (w/w_0)^{\kappa-1}$.

Rosanoff, Bacon and Schulze [2] point out that Lord Rayleigh's

[1] *Phil. Mag.*, 1902, [5], **4**, 527.
[2] "A Theory of Fractional Distillation without Reflux Condensation," *J. Amer. Chem. Soc.*, 1914, **36**, 2000.

theory is thus confined to very dilute mixtures, and that the solution of the problem proposed by Barrell, Thomas and Young is based on Brown's law, which is only of limited applicability. They themselves bring forward a more general theory based on the relationship expressed by their equation,[1]

$$\frac{dP}{d\text{M}} = [(P_A - P_B)/(\log P_A - \log P_B)] \log [p_A(1-\text{M})/p_B\text{M}] \quad . \quad . \quad \text{(A)},$$

which has been found to hold good in all types of cases.

Let ξ and η denote the weights of the two components in the liquid and $d\xi$ and $d\eta$ their weights in the vapour, p_A and p_B, as before, the partial pressures of the components, and M the molar fraction of the first component in the liquid. Then p_A/p_B, the ratio of the partial pressures, equals the ratio of the molar fractions in the vapour, the molar fractions being, of course, based on the actual molecular weights of the components. It is assumed that the molecular weights have the same value in the vapour as in the liquid, otherwise the thermodynamic equation of Duhem and Margules would not hold good. It may therefore be stated that the quotient of the ratios of the molar fractions in vapour and liquid is equal to the quotient of the ratios of the weights, that is to say—

$$\frac{p_A(1-\text{M})}{p_B\text{M}} = \frac{\eta \, d\xi}{\xi \, d\eta}$$

or

$$\frac{d \ln \xi}{d \ln \eta} = \frac{p_A(1-\text{M})}{p_B\text{M}}.$$

It follows from equation A that

$$\log \frac{p_A(1-\text{M})}{p_B\text{M}} = \kappa \frac{dP}{d\text{M}}$$

or

$$\frac{p_A(1-\text{M})}{p_B\text{M}} = e^{\kappa \, dP/d\text{M}},$$

therefore

$$\frac{d \ln \xi}{d \ln \eta} = e^{\kappa \, dP/d\text{M}} \qquad . \qquad . \qquad . \qquad . \qquad \text{(B)},$$

and all conclusions regarding the course of a fractional distillation based on this formula will be as reliable as equation A itself.

Denoting by ξ_0 and M_0 the initial weight and the initial molar fraction of the first component in the liquid the authors finally deduce the equation

$$\ln \frac{\xi}{\xi_0} = \int_{\text{M}_0}^{\text{M}} \frac{e^{\kappa \, dP/d\text{M}}}{(e^{\kappa \, dP/d\text{M}} - 1)(1-\text{M})\text{M}} \cdot d\text{M} \quad . \quad . \quad . \quad \text{(C)},$$

and for the second component, similarly

$$\ln \frac{\eta}{\eta_0} = \int_{\text{M}_0}^{\text{M}} \frac{d\text{M}}{(e^{\kappa \, dP/d\text{M}} - 1)(1-\text{M})\text{M}} \quad . \quad . \quad . \quad \text{(D)}.$$

---

[1] P. 93.

But as ξ and η are connected by the equation

$$\eta = \frac{1-\text{M}}{\text{M}} \cdot \frac{M_\text{B}}{M_\text{A}} \cdot \xi,$$

where M_B and M_A are the molecular weights of B and A respectively, only one integration need be carried out.

The authors point out that equations C and D express the relation between the changes of weight and composition accompanying isothermal fractional distillation in terms of $dP/d\text{M}$, the slope of the total pressure curve.

When the total and partial pressure curves are straight lines, equation B becomes identical with Brown's formula, or

$$\frac{dln\xi}{dln\eta} = \frac{P_\text{A}}{P_\text{B}} = \text{constant},$$

and the integrals in C and D assume the values

$$\frac{\xi}{\xi_0} = \left[\frac{\text{M}(1-\text{M}_0)}{\text{M}_0(1-\text{M})} \right]^{P_\text{A}/(P_\text{A}-P_\text{B})}$$

and

$$\frac{\eta}{\eta_0} = \left[\frac{\text{M}(1-\text{M}_0)}{\text{M}_0(1-\text{M})} \right]^{P_\text{B}/(P_\text{A}-P_\text{B})}.$$

The authors have tested the theory, in the form of equations C and D, by direct experiment with satisfactory results.

See also W. K. Lewis, " Theory of Fractional Distillation," *J. Ind. and Eng. Chem.*, 1909, **1**, 522.

CHAPTER IX

RELATION BETWEEN THE BOILING POINTS OF RESIDUE
AND DISTILLATE

Data required.—If, for mixtures of any two substances, the curve representing the relations between boiling point and molecular composition has been constructed, and if the relation between the composition of the liquid mixture and of its vapour is known, the boiling point of the distillate may be read from the curve. We have seen that if the two substances are closely related we may safely assume that the boiling points of mixtures may be calculated from the vapour pressures of the components, and that Brown's formula, taking the mean ratio of the vapour pressures for the value of the constant, c, gives the relation between the composition of liquid and vapour with fair accuracy.

Benzene and Toluene.—In the case of benzene and toluene, the boiling points of mixtures have been found to agree very closely with those read from the theoretical curve, but the relation between the composition of liquid and vapour has only been indirectly determined. The mean ratio of the vapour pressures at temperatures between 80° and 110° is roughly 2·5, and for our present purpose we may assume that the relation is expressed with sufficient accuracy by the formula $m'_B/m'_A = 2 \cdot 5 m_B/m_A$.

In Table 45 are given—

1. The molecular percentages of benzene in mixtures which boil at the temperatures in the first column.

2. The corresponding molecular percentages of benzene in the distillate, calculated by means of Brown's formula.

3. The boiling points of these distillates read from the curve.

4. The differences (Δt) between the boiling points of liquid and distillate.

Isopentane and Normal Pentane.—The corresponding data for mixtures of isopentane and normal pentane are given in Table 46, but in this case neither the boiling points of mixtures nor the relation between the composition of liquid and of vapour have been directly determined. The theoretical boiling point curve differs very slightly from a straight line, the maximum deviation being 0·28° when the molecular percentage of normal pentane is 51·5 (p. 46). The ratio

123

of the vapour pressures at 30° is 1·33, and this has been taken as the constant in Brown's formula.

TABLE 45

Benzene and Toluene

Boiling point of liquid.	Molecular percentage of benzene.		Boiling point of distillate.	Δt.	Boiling point of liquid.	Molecular percentage of benzene.		Boiling point of distillate.	Δt.
	Liquid.	Distillate.				Liquid.	Distillate.		
81°	96·3	98·5	80·55°	0·45°	96°	38·3	60·8	89·35°	6·65°
82	91·5	96·4	81·0	1·0	98	32·2	54·3	91·15	6·85
84	82·4	92·1	81·95	2·05	100	26·4	47·3	93·2	6·8
86	73·8	87·6	82·85	3·15	102	21·0	39·9	95·5	6·5
88	66·0	82·9	83·9	4·1	104	15·8	31·9	98·1	5·9
90	58·4	77·8	85·05	4·95	106	10·8	23·2	101·2	4·8
92	51·3	72·5	86·3	5·7	108	6·0	13·8	104·75	3·25
94	44·6	66·8	87·8	6·2	110	1·4	3·4	109·1	0·9

TABLE 46

Isopentane and n-Pentane.

	Isopentane.					Isopentane.			
28·5°	92·5	94·25	28·38°	0·12°	33°	36·35	43·15	32·40°	0·60°
29	85·95	89·05	28·76	0·24	34	25·0	30·70	33·49	0·51
30	72·9	78·15	29·59	0·41	35	13·95	17·75	34·65	0·35
31	60·3	66·90	30·48	0·52	36	3·15	4·15	34·91	0·08
32	48·0	55·10	31·41	0·59					

Application to Distillation of Benzene and Toluene.—Let us consider first the case of benzene and toluene. In the first four fractionations, details of which are given on p. 100, the range of temperature for most of the fractions was 3°. For fraction 6, collected from 92·2° to 95·4°, the middle temperature is 93·8°, and from Table 45 we see that the distillate from a mixture boiling at 93·8° would have a boiling point about 6·2° lower, or 87·6°. That is to say, if this fraction were distilled in such a manner that no condensation could take place in the still-head, it would begin to boil at 87·6°, but, as a matter of fact, there was some condensation which would lower the temperature to some extent at first. On the other hand, in the actual fractionation, the fractions, with the exception of the first, were not distilled separately but were added to the residues left in the still. Thus the fraction that came over from 92·2° to 95·4° was added to the residue from No. 5, which was boiling at 92·2°, and was therefore richer in toluene, and the boiling point would therefore be somewhat higher than if the fraction were distilled alone. We may perhaps suppose that the two disturbing factors would about counterbalance each other, and that the mixture would actually begin to boil at about 87·5°. The temperature ranges below 92·2° to 95·4° were as follows : No. 5, 89·2° to 92·2° ; No. 4, 86·2° to 89·2°, and it is clear that a considerable amount of distillate would be collected below 89·2° in receiver No. 4.

In the sixth fractionation (p. 106) the corresponding temperature ranges were—No. 4, 84·0° to 87·0° ; No. 5, 87·0° to 90·5° ; No. 6, 90·5° to 95·4°. The middle temperature for No. 6 would be 92·95°, and the distillate from this would begin to boil at about 87·0° or 5·95° lower. In this case none of the distillate would be collected in No. 4, and receiver No. 5 might be (and actually was) left in position when No. 6 fraction was added to the residue in the still and the mixture was redistilled.

In the thirteenth fractionation the range of No. 6 had been increased to 14°, from 81·4° to 95·4°. The middle temperature would be 88·4°, and the boiling point of the distillate would be, roughly, $88·4° - 4·2° = 84·2°$, so that no distillate would be collected in No. 5, and there would be no object in continuing the fractionation in the same way as before. This was, in fact, found to be the case, the temperature rising at once above 81°. The fractions above and below 95·4° were therefore treated separately as described on p. 104.

Application to Distillation of Pentanes.—Let us now consider the behaviour of a mixture of isopentane and normal pentane (Table 46). The difference between the boiling points is 8·35° and the middle temperature is 32·6°. If the distillate were collected in the same number of fractions as in the case of benzene and toluene, the range of the middle ones would be about 0·84°, say, 0·8°. We should then have for the corresponding fractions—No. 5, 31·0° to 31·8° ; No. 6, 31·8° to 32·6°.

The middle temperature for No. 6 would be 32·2°, and the boiling point of the distillate from this would be $32·2° - 0·6 = 31·6°$. Very little would therefore be collected even in No. 5. The temperature range would have to be reduced to 0·4° [No. 5, 31·8° to 32·2° ; No. 6, 32·2° to 32·6°] in order that the distillate from No. 6 might begin to boil at the initial temperature of No. 5, and that would mean that the number of fractions would have to be doubled to begin with, and that it would not be possible to increase the temperature range of the middle fractions beyond a very small amount.

It will thus be seen that, even without considering the excessive loss by evaporation of such volatile liquids, the separation of isopentane from normal pentane with an ordinary still-head would be practically impossible. It is only by using a greatly improved still-head that such a difficult separation is rendered possible.

CHAPTER X

MODIFICATIONS OF THE STILL-HEAD

Object of Modifications.—It is evident from what has been stated in Chapter VII. that the process of fractional distillation with an ordinary still-head is frequently an exceedingly tedious one, and many attempts have been made to modify the still-head in such a manner as to bring about a more complete separation in a single operation.

The idea of improving the still-head is by no means a new one, for in a book entitled *Philosophorum, seu liber de Secretis Naturæ*, by Philip Ulstadius, 1553, there is an illustration of a five-headed still (Fig. 41). When dilute spirit was distilled through this "alembic" it is clear that the alcohol collected in the uppermost receiver must have been considerably stronger than that in the lowest.

Great progress was made in the improvement of the still-head in commerce—notably for the better separation of alcohol from weak spirit—before any advance was made in the laboratory, and, indeed, the improved still-heads first employed in the laboratory were, for the most part, merely adaptations of those already in use on the large scale.

It may, however, be convenient to consider first those forms of apparatus which are suitable for laboratory purposes.

Fig. 41.—Five-headed still described by Ulstadius, 1553.

The object of the modifications which have been made in the still-head is to bring about a more complete separation of the components of the mixture in a single distillation, a process of fractionation—more or less effective according to the form of apparatus and the rate of distillation—taking place in the still-head.[1]

---

[1] Young, "On the Relative Efficiency and Usefulness of Various Forms of Still-head for Fractional Distillation, with a Description of some New Forms possessing Special Advantages," *Trans. Chem. Soc.*, 1899, **75**, 679.

Inefficiency of Plain Vertical Still-head.—Of all possible forms, the plain vertical still-head is the least efficient. During the distillation, as the vapour rises up the cylindrical tube, the outer parts of it come in contact with the condensed liquid flowing down the sides of the tube. Since the tube is constantly losing heat by radiation and by conduction to the surrounding air, this liquid is slightly cooled and fresh condensation constantly goes on, the outermost layer of vapour probably condensing almost completely without much change of composition. The central portion of vapour rises rapidly up the tube and can only reach the liquid by diffusion or by convection currents, and much of it may pass through the still-head without reaching the liquid at all ; the condensed liquid, on the other hand, flows rapidly down the vertical walls of the tube, back to the still. In order, however, that a satisfactory separation of the components of the mixture may take place, as much of the vapour as possible should be brought into intimate contact with the condensed liquid, so that a state of equilibrium, as regards composition, may be brought about.

Brown's Formula.—Let us suppose, to facilitate the consideration of what occurs, that we are dealing with a mixture of two closely related compounds, to which Brown's formula $\dfrac{m'_B}{m'_A} = c\dfrac{m_B}{m_A}$ is applicable, and that m' and m refer to the weights of vapour and liquid respectively. Also let us suppose that, in the first place, equal weights of the two substances are taken, so that $m_A = m_B$.

Then the composition of the first small quantity of vapour formed will be given by the equation $\dfrac{m'_B}{m'_A} = c$ and the percentage of B in the vapour will be $\dfrac{c \times 100}{c+1}$.

If a distillation bulb with exceedingly short still-head be employed, this will practically be the percentage of B in the first portion of distillate.

Theoretically Perfect Still-head.—Suppose, now, that in the lower part of a long and theoretically perfect still-head condensation goes on, and that the condensed liquid remains in this part of the tube until its weight is, say, 100 times as great as that of the residual uncondensed vapour in the same part of the tube, the total weight of liquid and vapour being, say, 1/1000 of that of the mixture taken. We may then assume

1. That the composition of the liquid and vapour (taken together) in the still-head will be practically the same as that of the vapour first formed, and

2. That the composition of the condensed liquid will not differ sensibly from this, or $\dfrac{m_{1B}}{m_{1A}} = c$.

If the residual vapour is brought into thorough contact with the condensed liquid, its composition will practically be given by the equation $\dfrac{m'_{1B}}{m'_{1A}} = c \times c = c^2$, and the percentage of B in it will be $\dfrac{c^2 \times 100}{c^2+1}$.

If we suppose, as a theoretical case, that the still-head is divided into n sections, and that the distillation goes on in such a manner that perfect equilibrium between vapour and condensed liquid is established in each section, and, lastly, that the condensed liquid in any section has the same composition as the vapour in the section below, then we should have, for the composition of the vapour in the nth section, when the distillate begins to come over, $\dfrac{m'_{nB}}{m'_{nA}} = c^{n+1}$.

If we take $c = 2\cdot5$, as in the case of benzene and toluene, and start with equal weights of the components, the composition of the vapour in the still and in the first nine sections would be as follows :—

TABLE 47

No. of Section.	Percentage of B.	No. of Section.	Percentage of B.
Still	71·43	5	99·59
1	86·21	6	99·84
2	94·55	7	99·93
3	97·51	8	99·97
4	98·99	9	99·99

Such an arrangement is not realisable in practice, but the attempt should be made to approach as closely to it as posssible.

Characteristics of a good Still-head.—In choosing or constructing a still-head, the first point to be considered is its efficiency in separating the components of a mixture. It frequently happens, however, that the quantity of liquid available is small and, in any case, when a series of fractional distillations has to be carried out, some of the fractions eventually become very small : in such a case, of two still-heads of equal efficiency, that one is the more useful which allows of the distillation of the smaller quantity of liquid. Now there is necessarily, at any moment, a certain amount of condensed liquid in every efficient still-head and, obviously, the smallest quantity of a substance that can be usefully distilled must be considerably greater than that of the liquid and vapour in the still-head. It is therefore of great importance to construct the still-head in such a manner that —consistently with efficiency—the quantity of condensed liquid in it at any moment shall be as small as possible. It is also important that, after the removal of the source of heat, this liquid shall return as completely as possible to the still.

Among other points to be considered are (a) ease of construction, (b) freedom from liability to fracture, (c) convenience in handling.

Comparison of Still-heads. Mixture distilled.—In comparing the efficiency and usefulness of different still-heads it is necessary always to distil a mixture of the same composition. Equal weights of pure benzene and toluene may conveniently be employed.

Rate of Distillation.—It is necessary also that the rate of distillation should always be the same, and it is best to collect the distillate at the rate of one drop per second. A good plan is to have a seconds pendulum—a weight attached to a string 39·1 inches or, say, 1 metre long, serves the purpose very well—swinging behind the receiver, so that the drops may be easily timed.

That the rate of distillation does greatly influence the separation was pointed out by Brown,[1] and is well shown by Table 48, in which

TABLE 48

Temperature range.	Number of drops of distillate per minute.		
	30	60	120
	Percentage weight of distillate.		
80·2— 83·2° . . .	0·8	0·6	0·6
83·2— 86·2	29·6	21·8	10·8
86·2— 89·2	9·8	14·3	20·2
89·2— 92·3	5·6	7·8	12·6
92·3— 95·4	3·8	5·0	7·0
95·4— 98·5	3·0	4·7	5·4
98·5—101·6	2·8	4·0	5·0
101·6—104·6	3·0	3·6	5·0
104·6—107·6	4·3	5·2	5·6
107·6—110·0	6·2	7·3	9·2
110·0—110·6	11·6	9·8	11·1
Pure toluene by difference .	19·5	15·9	7·6*
	100·0	100·0	100·0

* Temperature barely reached 110·6°; residue not quite pure toluene.

are recorded the results of three distillations, each of 50 grams of the benzene-toluene mixture, at the rate of 30, 60 and 120 drops per minute. An improved still-head was used, and the distillate was collected in eleven fractions, the twelfth consisting of the residue (calculated by difference) after the temperature had reached the boiling point of pure toluene. For convenience, the results are stated as percentages.

It will be seen that the separation is greatly improved by diminishing the rate of distillation.

Explanation of Tables.—In Tables 49 to 53, which show the relative efficiency and usefulness of different still-heads, the following data are given.

1. The vertical height from the bottom of the still-head to the side delivery tube.

2. The final temperature for each fraction.

3. The percentage weight of distillate.

4. The weight of vapour and liquid in the still-head during the

[1] Brown, "The Comparative Value of Different Methods of Fractional Distillation," *Trans. Chem. Soc.*, 1880, **37**, 49.

K

distillation when pure or nearly pure toluene was coming over. This was arrived at by continuing the distillation as nearly as possible to the last drop and weighing the liquid, when cold, in the flask (still) and—if necessary and possible—in the still-head. From this amount was subtracted the constant weight of liquid and vapour in the flask itself, which was estimated to be 0·85 gram.

For the very long still-heads, which could not conveniently be weighed, the estimated loss by evaporation +0·85 gram was subtracted from the difference between the original weight of the mixture and the sum of the weights of the fractions. The actual loss by evaporation, when measured, varied from 0·00 to 0·65 gram, and was greater for the large than the small still-heads.

The Plain Vertical Still-head.—Although the plain vertical still-head is less efficient than any other, yet a certain amount of fractionation does take place in it, and it may be well to consider the effect of altering (a) its length, (b) its internal diameter. The influence of such alterations is shown in Table 49.

TABLE 49

PLAIN VERTICAL STILL-HEADS

Vertical height in cm.	12	62	120	From 62 to 70 cm.				
Internal diameter in mm.	From 14 to 15·5 mm.			5·1	8·0	14·0	19·4	25·7
Final temperature.	Percentage weight of distillate.							
83·2°	*							
86·2	..	0·1	1·0	0·5	0·4	0·1	1·3	0·2
89·2	2·1	12·8	19·4	22·2	16·2	12·8	15·2	20·0
92·3	28·4	21·5	16·6	16·5	18·3	21·5	19·4	18·8
95·4	18·0	11·0	10·6	9·2	11·7	11·0	11·2	9·2
98·5	10·9	8·9	7·2	7·2	8·7	8·9	8·6	7·2
101·6	8·6	7·4	6·5	5·8	7·5	7·4	6·1	4·7
104·6	7·2	6·2	6·0	5·5	5·8	6·2	5·6	5·9
107·6	7·2	7·3	6·1	5·9	6·9	7·3	6·6	6·4
110·0	6·8	8·4	7·8	7·1	8·2	8·4	9·5	10·0
110·6	5·9	9·0	8·1	8·9	9·5	9·0	8·0	7·2
Pure toluene by difference	4·9	7·4	10·7	11·2	6·8	7·4	8·5	10·4
	100·0	100·0	100·0	100·0	100·0	100·0	100·0	100·0
Weight of liquid and vapour in still-head	0·3	1·55	3·55	1·05	1·15	1·55	2·35	3·15

Influence of Length.—In the first three distillations the diameter of the still-head was nearly the same, but the third tube was ten times as long as the first. As might be expected, the efficiency is improved by this alteration, but the weight of liquid and vapour also increases and is, roughly, proportional to the length.

Influence of Width.—For tubes of approximately equal length the efficiency is smallest when the diameter is rather less than 14 mm., and rises when it is either increased or diminished, as will be seen from the results of the last five distillations. On the other hand, the weight of liquid and vapour in the still-head increases with the diameter of the tube, and it is therefore clearly more advantageous, in making a plain still-head, to use very narrow rather than very wide tubing. It will be noticed also that the narrowest tube gives a better result than the tube of medium diameter nearly twice as long, though the weight of liquid and vapour is only 1·05 as compared with 3·55 grams.

The diameter of the tubing cannot, however, be diminished beyond a certain amount, depending on the nature of the liquid distilled, especially on its boiling point, for if the tube is too narrow, the condensed liquid will unite into columns which will be driven bodily upwards. This blocking is much less liable to occur if a short piece of wider tube is sealed to the bottom of the narrow one. There must also be a wider piece at the top to admit the thermometer.

Condensation in the still-head may be diminished by covering it with cotton wool or any other non-conducting material, but while the amount of liquid in the tube is thus diminished, so also is the efficiency. The device is, however, occasionally useful, for if the liquid in the tube is just on the point of joining up into columns, the blocking may be prevented by slightly diminishing the amount of condensation.

Modifications of the Still-head.—It has been pointed out that the inefficiency of the plain vertical still-head is due to the want of thorough contact between the vapour and the condensed liquid, owing, firstly, to the central portion of the vapour passing rapidly up the tube and possibly never meeting with the liquid at all, and, secondly, to the condensed liquid flowing very rapidly back to the still. Consequently, any modification that brings about better admixture of the ascending vapour or that retards the down-flow of condensed liquid should increase its efficiency.

Sloping Still-head.—The simplest and most obvious alteration that can be made is to change the slope of the tube so as to retard the down-flow of liquid. This may be done by bending the tube near the top and bottom so that its ends remain vertical, while the middle part slopes very gently upwards.

It will be seen from Table 50 that, by altering a tube in the manner described, a notable improvement is effected, especially as regards the separation of the benzene, and it is remarkable that, although the down-flow of liquid is retarded, the quantity of liquid in the still-head is slightly diminished.

Spiral Still-head.—The efficiency is further improved by bending the sloping portion of the tube into the form of a spiral, probably because a better admixture of the vapour is thus produced ; by this device the amount of liquid in the still-head is still further reduced.[1]

"Rod and Disc" Still-head.—In the still-head shown in Fig. 42a the down-flow of part of the condensed liquid is greatly retarded by

---

[1] G. Berlomont, "New Tube for Fractional Distillations, modified by Lebel," *Bull. Soc. Chim.*, 1895, [III], **13**, 674 ; *J. Soc. Chem. Ind.*, 1895, **14**, 821.

TABLE 50

Nature of still-head.	Vertical.	Sloping.	Spiral.	Plain vertical.	With rod and 20 discs.	Same as last with constrictions.
Vertical height in cm.	70	28·5	32	62 to 63		
Internal diameter.	8 mm.			14 mm.		
Final temperature.	Percentage weight of distillate.					
83·2°	...	...	...	...	1·4	2·0
86·2	0·4	3·5	8·0	0·1	18·1	20·5
89·2	16·2	24·8	22·6	12·8	15·8	14·4
92·3	18·3	14·0	11·6	21·5	9·7	8·6
95·4	11·7	9·3	8·8	11·0	6·1	5·8
98·5	8·7	6·7	5·5	8·9	4·2	3·6
101·6	7·5	5·6	5·0	7·4	3·8	3·4
104·6	5·8	5·4	5·5	6·2	4·2	4·2
107·6	6·9	6·3	6·0	7·3	4·3	4·2
110·0	8·2	8·2	7·4	8·4	8·0	7·4
110·6	9·5	9·0	10·7	9·0	6·6	7·5
Pure toluene by difference .	6·8	7·2	8·9	7·4	17·8	18·4
	100·0	100·0	100·0	100·0	100·0	100·0
Weight of liquid and vapour in still-head	1·15	1·10	0·95	1·55	2·35	2·2

the discs on the central glass rod, and this liquid is protected from the cooling action of the air ; at the same time eddies and cross currents are produced in the ascending vapour. The increase in efficiency with this apparatus is very marked, as will be seen from the second part of Table 50 ; it is easily constructed and is very convenient to handle. The quantity of liquid and vapour in the still-head is the same as that in the plain tube of 19·4 mm. diameter, but the efficiency is very much greater. On removing the source of heat, the liquid returns almost completely to the still, but when it is of special importance to avoid all loss, it is advisable with this and many other forms of still-head to disconnect the condenser and to tilt the still and still-head, while hot, from side to side so as to facilitate the back-flow of liquid.

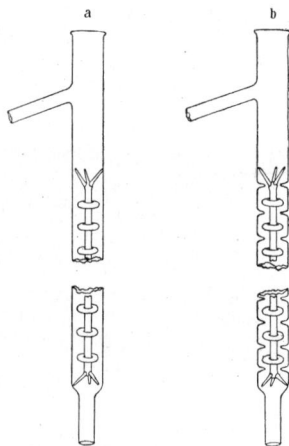

FIG. 42.—The "rod and disc" still-heads ; (a) without, (b) with constrictions in outer tube.

If that is done, the whole of the liquid usually returns to the still.

A slight further improvement, both in efficiency and as regards the quantity of liquid in the still-head, is effected by constricting the outer tube between the discs (Fig. 35*b*). Better contact between the vapour and the condensed liquid on the outer tube is thus ensured.

Bulb Still-heads.—A vertical tube with a series of bulbs blown on it was recommended by Wurtz.[1] For a given diameter of tube and bulbs, the greater the number of bulbs the higher is the efficiency, but the greater also is the quantity of liquid and vapour in the still-head (Table 51).[2]

The tube with thirteen bulbs was somewhat more efficient than the slightly shorter tube with rod and discs, but the quantity of liquid and vapour in the still-head was half as large again.

TABLE 51

Nature of still-head.	3 bulbs	7 bulbs	13 bulbs	" Pear " still-head, 13 bulbs
Vertical height in cm.	· 26	42	66	62
Final temperature.		Percentage weight of distillate.		
83·2°	...	0·2	· 1·4	3·0
86·2	2·2	14·2	24·4	26·2
89·2	18·8	18·0	10·2	11·0
92·3	18·4	12·0	7·9	5·8
95·4	9·4	5·1	6·0	5·2
98·5	7·8	6·4	3·7	2·4
101·6	6·7	4·6	3·6	2·4
104·6	5·8	4·8	3·4	2·8
107·6	6·4	5·6	3·4	3·6
110·0	8·0	6·2	6·2	5·0
110·6	10·0	10·4	8·0	11·0
Pure toluene by difference	6·5	12·5	21·8	21·6
	100·0	100·0	100·0	100·0
Weight of liquid and vapour in still-head	0·95	1·75	3·55	2·6

The " Pear " Still-head.—The " Wurtz " still-head may be improved, to some extent in efficiency, but chiefly as regards the quantity of condensed liquid, by blowing pear-shaped instead of spherical bulbs on the tube (Fig. 43).

As a result of this alteration, the condensed liquid in any bulb, after flowing past the constriction, instead of spreading itself over the inner surface of the bulb below, mixing with the liquid condensed

---

[1] Wurtz, " Memoir on Butyl Alcohol," *Ann. Chim. Phys.*, 1854, [III], **42**, 129.

[2] Neither the experimental results obtained by Kreis * nor the conclusions he deduces from them can be accepted, for he obtained a very much worse separation of benzene with four bulbs than with two, though the separation of toluene was better. It was, however, according to his figures, not so good as with an ordinary distillation bulb.

---

* Kreis, " Comparative Investigations on the Methods of Fractional Distillation," *Liebigs Annalen*, 1884, **224**, 259.

in that bulb and flowing down the sides with increasing velocity, collects on the depression in the bulb below and falls in drops near the middle of the bulb. The liquid on the inner surface of each bulb is merely the small amount condensed in that bulb, and its velocity of down-flow is no greater in the bottom bulb than in the top one. The liquid, on the other hand, that collects on a depression (that is to say, the total quantity condensed in the part of the still-head above it) is brought well in contact with the ascending vapour in a part of the bulb that is less exposed to the cooling action of the air than any other. It is probable also that the eddies in the vapour are greater than in the ordinary bulb tube.

FIG. 43.—
The "pear"
still-head.

The "pear" still-head is more efficient than the "rod and disc" tube of the same length and possesses the same advantages except that it is somewhat more difficult to construct. It may be especially recommended for liquids of high boiling point.

The "Evaporator" Still-heads.—Greater efficiency, for a given vertical height, and less condensation, for a given efficiency, is attained by the "evaporator" still-heads.

Original Form of "Evaporator" Still-head.—The general form of the apparatus, as originally designed, is shown in Fig. 44. Each section consists of three separate parts :—

1. The outer tube, A, of 22 to 24 mm. internal diameter, connected above and below with other sections, the length of each being about 10 or 10·5 cm. ;

2. An inner thin-walled tube, B, of 7·5 to 8 mm.[1] internal diameter and 60 mm. long, open at each end and widened below into the form of a funnel which rests on the constricted part of A and may be prevented from fitting it too accurately by fusing three or four minute beads of glass to the rim of the funnel. Two large holes (B') are blown on the sides of B near the top ;

3. An intermediate tube, C, of about 14 mm.[1] internal diameter and 50 mm. long, like a small inverted test tube. Above the tube C and attached to it by three glass legs, shown in Fig. 45, is a small funnel, C_1, which must be a little wider than the depression in the tube A just above it. The tubes B and C are centred and kept in position by the little glass projections shown in the same figure.

When the vapour first reaches a section from below, a large amount of condensation takes place, and the narrow passage, D, where the inner tube rests on the constriction in the outer one, becomes at once blocked by the condensed liquid. The vapour therefore rises up the inner tube, then passes down between the inner and middle tubes and finally up again between the middle and outer tubes and so into the section above.

---

[1] It has been found advisable to use slightly wider tubing for the inner and intermediate tubes; the internal diameter of the former should be 8 to 9 mm. and of the latter 15 to 16 mm.

The condensed liquid in any section collects together and falls in drops from the depression in the section below into the funnel ; from this it falls on to the top of the middle tube and spreads itself over its surface, falling again in drops from the bottom of this tube and finally flowing through the passage D.

Owing to gradual removal of the less volatile component, the condensing point of a mixed vapour becomes lower and lower as the vapour rises through the still-head ; thus, in any section, the vapour that rises through the inner tube will be hotter than that which reaches the section above. The condensing point of the vapour in the inner tube must, indeed, be higher than the boiling point of the liquid that falls from the little funnel, and when the two components differ considerably in volatility and neither of them is in great excess, evaporation of the liquid on

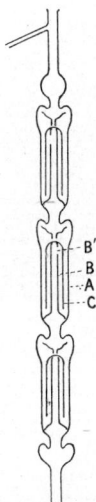

FIG. 44. — The "evaporator" still-head ; original form.

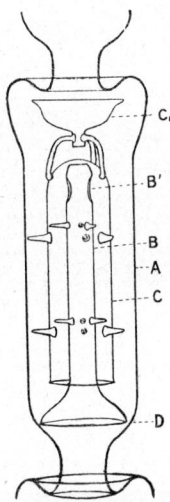

FIG. 45.—One section of the original "evaporator" still-head.

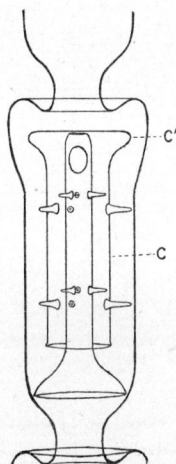

the middle tube may be easily observed. Under such conditions there is a tendency for the drops of liquid from the funnel to assume the spheroidal state, and the progress of a distillation is, indeed, rendered evident by the appearance or disappearance of such drops.

With a pure liquid, the spheroidal drops are never seen unless the quantity of liquid in the still is very small, when, owing to superheating of the vapour, they may be formed in the lowest section.

Modified " Evaporator " Still-head.—A modification of the " evaporator " still-head is shown in Fig. 46. In this, the rather fragile funnel on the three legs is done away with, and the top of the middle tube C is blown into a flattened bulb C', on which the drops of liquid from the depression above fall and collect into a shallow pool which soon overflows, and the liquid then spreads itself as before over the surface of the

FIG. 46.—One section of the modified " evaporator " still-head.

tube. Spheroidal drops are not nearly so readily formed in this apparatus, but when rapid separation is taking place, the liquid flowing down the sides of the inverted tube breaks up into separate streams.

It will be seen from Table 52 that very good results are obtained with the evaporator still-heads of either form. The sections in the modified apparatus are shorter than in the original one, and for a given height of still-head the efficiency is greater, while, for a given efficiency, the weight of liquid and vapour in the still-head is usually slightly less.

TABLE 52

Nature of still-head.	"Evaporator" still-heads.						"Hempel," 200 large beads.
	Original.		Modified.				
Number of sections. .	3	5	3	5	8	13	
Vertical height in cm.	57	77	46	62	78	131	58
Final temperature.	Percentage weight of distillate.						
81·2° . . .	} 15·4	6·5	} 12·8	3·5	12·0	42·5[1]	} 20·6
83·2 . . .		24·1		22·5	23·85	2·6	
86·2 . . .	19·6	10·0	21·15	12·35	6·5	1·7	15·4
89·2 . . .	6·2	3·8	7·5	5·6	2·9	1·1	7·2
92·3 . . .	5·1	2·3	4·8	3·25	2·15	0·65	3·4
95·4 . . .	3·5	1·6	3·35	1·95	1·4	0·5	3·2
98·5 . . .	2·9	1·7	2·7	1·3	1·15	0·55	2·2
101·6 . . .	2·6	1·5	2·3	1·8	1·05	0·5	1·8
104·6 . . .	2·4	1·8	2·5	1·5	1·15	0·45	2·9
107·6 . . .	3·8	2·9	3·5	2·45	1·6	0·9	3·2
110·0 . . .	5·8	4·8	6·5	4·2	3·95	1·95	6·8
110·6 . . .	10·8	8·2	10·5	11·4	9·6	2·85	8·4
Pure toluene by } difference .	21·9	30·8	22·4	27·9	32·7	43·75	24·9
	100·0	100·0	100·00	100·00	100·00	100·00	100·0
Weight of liquid } and vapour in } still-head .	3·45	5·0	2·7	4·55	6·25	16·25	7·85

When there are many sections, the inverted tube should be made somewhat shorter in the lower sections, so as to increase the vertical distance from the bottom of that tube to the bottom of the inner tube ; in the modified still-head the vertical distance from the depression to the flattened top of the inverted tube should be somewhat increased in the lower sections. It is important that the holes near the top of the inner tubes should be as large as possible.

[1] The specified dimensions were unfortunately not adhered to in constructing the still-head of thirteen sections, with the result that there was a tendency for two or three of the sections to become blocked with condensed liquid. The efficiency of the apparatus was thus increased, but its usefulness was greatly diminished and the quantity of liquid in the still-head was much greater than it should have been.

The Kubierschky Column.—The Kubierschky column [1] belongs to this type of still-head, but has only been used for large-scale work. See p. 307 and Fig. 115, p. 308.

The " Hempel " Still-head.—The great advantages of the " Hempel " apparatus [2] are simplicity and efficiency ; on the other hand the amount of liquid in the still-head is excessive, and it is therefore unsuitable for the distillation of small quantities of liquid.

The still-head consists simply of a wide vertical tube, filled with glass beads of special construction, and constricted below to prevent the beads from falling out. A short, narrower, vertical tube with side delivery tube is fitted by means of an ordinary cork into the wide tube.

From the following table it will be seen that in efficiency the Hempel tube, which contained 200 of the large beads now used, came about midway between the original " evaporator " of 3 sections, the length of which was nearly the same, and the modified " evaporator " of 5 sections, which was a little longer. But the weight of liquid in the 3-section " evaporator " still-head was considerably less than half, and in the 5-section " evaporator " not much more than half as great as in the Hempel tube.

Length for length, the modified " evaporator " still-head would be more efficient, and would contain only about half the amount of condensed liquid.

Many other fillings for the still-head to replace Hempel's heavy glass beads have been suggested. Partington and Parker [3] employ four pointed stars. Raschig [4] recommends thin sheet-iron rings one inch long and one inch in diameter for large still-heads, smaller rings being employed for laboratory purposes. Goodwin [5] uses rings each of which consists of two hollow truncated cones united at their narrower ends. Lessing [6] has modified the Raschig rings in such a manner that they are easier to manufacture and give a larger surface of contact between liquid and vapour. Fig. 47 shows the form of the Lessing ring, which may be made of any suitable metal or of earthenware. For laboratory purposes Lessing recommends rings of $\frac{1}{4}$-inch diameter, 1000 of which would occupy 400 c.c. The larger rings have been found very suitable for the distillation of

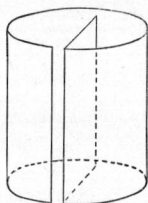

FIG. 47.

large quantities of liquid ; for example, a ring still of 18 inches diameter was found to give a better separation than a column still of the Coffey type of the same height but 2 feet in diameter.

I am indebted to Dr. Lessing for 1500 of the $\frac{1}{4}$-inch rings, and have carried out distillations of 200 grams of the mixture of equal weights

[1] E. Graefe, " Use of the Kubierschky Column in the Distillation of Mineral Oils," Petroleum, 1913, **9**, 303 ; *J. Soc. Chem. Ind.*, 1914, **33**, 1146 ; Borrmann, *Zeitschr. ang. Chem.*, 1915, **28**, 377, 381 ; *J. Soc. Chem. Ind.*, 1915, **34**, 1232.

[2] Hempel, " Apparatus for Fractional Distillation," *Fersenius' Zeitschr. für Anal. Chem.*, 1882, **20**, 502.

[3] *J. Soc. Chem. Ind.*, 1919, **38**, 75 T.   [4] Eng. Patent, 1914, No. 6288.
[5] *Ibid.*, 1917, No. 110,260.   [6] *Ibid.*, 1920, No. 139,880.

of benzene and toluene at the usual rate of 60 drops of distillate per minute. The tube was 3·7 cm. in diameter and the height of the column of rings 60 cm.

The separation of benzene was distinctly better than that obtained with an evaporator still-head of 8 sections but not nearly so good as with an evaporator of 13 sections. It would be about equal to that given by a Young and Thomas still-head of 10 sections. The separation of toluene was about equal to that with the evaporator of 8 sections, but the amount of toluene actually recovered was far less owing to the large quantity left adhering to the rings, in spite of the fact that as soon as the distillation was over the still-head was disconnected from the condenser and was tilted from side to side and was then left standing for an hour to allow the liquid to flow back to the flask as completely as possible. The average weight of toluene lost in this way was 12·5 grams as against 0·2 gram in the case of the evaporator still-head. The amount of liquid in the still-head during distillation was more than 28 grams as great as with the evaporator still-head.

On account of its ease of construction and convenience the Lessing ring still-head may be strongly recommended for the distillation of large quantities of liquid, especially if the residual liquid is not of value ; but for the distillation of small quantities and for the recovery of the least volatile component it is not to be compared with the evaporator still-head.

Lessing states that the principal point is the adjustment of the amount of reflux obtainable over the whole tube. For a still-head of 40 cm. length he recommends a reflux condenser, but for the longer still-heads he states that the lower part of the column—up to about 10 cm. from the top—should be lagged with asbestos. The reflux condenser would, of course, increase the amount of liquid in the still-head and would make it still less suitable for the distillation of small quantities of liquid. The lagging would have the reverse effect, but the amount of liquid in the still-head and the loss of material would remain excessive.

In the experiments referred to above the still-head was not lagged, nor was a reflux condenser used.

A simple and efficient still-head has been described by Foucar.[1] It consists essentially of a helical septum which traverses the annular space between two concentric cylinders. The vapour from the still passes up the long spiral passage before reaching the condenser. If desired, the inner cylinder can be used as a thermostat, the column then acting as a regulated temperature still-head (Chapter XII.). The Foucar apparatus may be described as a greatly improved spiral still-head.

An easily constructed spiral still-head, suitable for laboratory work, has been described by Dutton.[2] A spiral of copper wire is wound round an inner closed tube which fits quite loosely into an outer tube, thus leaving a spiral passage up which the vapour ascends. The amount of condensed liquid in the still head increases from top to bottom

[1] Eng. Patent, 1908, No. 19,999.   [2] J. Soc. Chem. Ind., 1919, 38, 45 T.

and, to allow for this, tubes of increasing size may be used—either the diameter of the outer tube increasing downwards or that of the inner tube upwards, the thickness of the wire increasing downwards in either case. It was found necessary to protect the still-head from too rapid cooling by packing it in a thickness of about two inches of cotton-wool wadding covered with asbestos yarn. Very good results are claimed with a still-head 120 cm. in height, but as the distillation was carried out with extreme slowness—10 drops of distillate a minute—comparison with other still-heads is hardly possible with the data at present available.

M. Robert[1] has obtained very good results with a vacuum jacketed column, with a reflux condenser above it to act as a dephlegmator.

[1] M. Robert, "Colonne à distiller pour Laboratoire," *Ann. de Chim. analytique*, 1919, Series II. **1**, 372.

CHAPTER XI

MODIFICATIONS OF THE STILL-HEAD (*continued*)

BUBBLING STILL-HEADS [1]

IN many of the still-heads employed on the large scale, for example the Coffey still (Fig. 68, p. 167) the condensed liquid is made, by means of suitable obstructions, to collect into shallow pools, and the ascending vapour has to force its way through these pools ; very good contact is thus brought about at definite intervals between vapour and liquid. The excess of liquid is carried back from pool to pool and finally to the still by suitable reflux tubes.

The "Linnemann" Still-head.—The first still-head constructed on this principle for use in the laboratory was devised by Linnemann.[2] A number of cups of platinum gauze, A, were placed at different heights in the vertical tube (Fig. 48) and the liquid collected in these, but, as no reflux tubes were provided, the liquid gradually accumulated in the still-head until the quantity became unmanageable, when the distillation had to be discontinued until the liquid flowed back to the still. There was thus much waste of time, and increased loss of material by evaporation, and it was impossible to make an accurate record of the temperature.

In the more recent forms of bubbling still-head, reflux tubes are provided, and it is on the size and arrangement of these tubes that the efficiency and usefulness of the still-heads chiefly depends.

The "Glinsky" Still-head.—The Glinsky still-head [3] has only one reflux tube, which carries the excess of liquid from the large bulb to the tube below the lowest obstruction, practically back to the still.

---

[1] When a description of the still-head devised by G. L. Thomas and the author was first published in the *Chemical News* (1895, **71**, 177) the term " dephlegmator " was adopted to distinguish this form of still-head from others such as the " Wurtz " and " Hempel," in which no bubbling through pools of condensed liquid takes place. The term was retained in the paper on still-heads published in 1899 in the *Trans. Chem. Soc.* (**75**, 679) and in *Fractional Distillation* (Macmillan, 1903). The word is, however, very loosely used in the literature, and Mr. Kewley has kindly looked up a considerable number of references, from which it would appear that it is most correctly employed to denote the small condenser placed above a fractionating column to return a sufficient amount of condensed liquid to the uppermost section of the column. It is therefore thought best to adopt the term " bubbling column," used by Mariller in *La Distillation Fractionnée*, in place of " dephlegmator."

[2] Linnemann, " On a Substantial Improvement in the Methods of Fractional Distillation," *Liebigs Annalen,* 1871, **160**, 195.

[3] Glinsky, " An Improved Apparatus for Fractional Distillation," *Liebigs Annalen,* 1875, **175**, 381.

In its original form (Fig. 49), the dephlegmator otherwise resembled the Linnemann still-head very closely, but, as now constructed, there

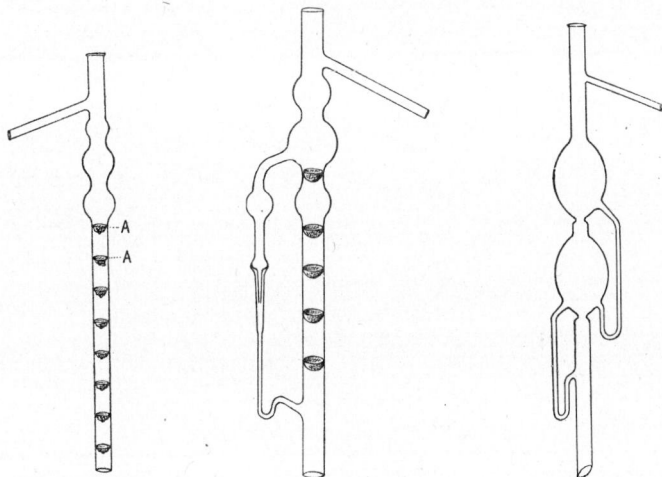

FIG. 48.—The " Linnemann " still-head.

FIG. 49.—The " Glinsky " still-head.

FIG. 50.—The " Le Bel-Henninger " still-head.

are bulbs on the vertical tube, and spherical glass beads, instead of platinum cups, rest on the constrictions between the bulbs.

The " Le Bel-Henninger " Still-head.—In the Le Bel-Henninger apparatus [1] the obstruction is usually caused by placing platinum cones on the constrictions between bulbs blown on the vertical tube ; each bulb is connected by a reflux tube with the one below it (Fig. 50) so that the liquid is carried back from bulb to bulb and not straight to the still.

In these still-heads, unlike that of Coffey, the reflux tubes are external, and the returning liquid is thus exposed to the cooling action of the air. The still-heads of Brown,[2] and of Young and Thomas,[3] [4] follow the principle of the Coffey still more closely, the reflux tubes being much shorter and being heated by the ascending vapour.

The " Young and Thomas " Still-head.—The Young and Thomas still-head [5] is shown in Fig. 51 *a*, *b*, and *c*. It consists of a long, glass tube of about 17 mm. internal

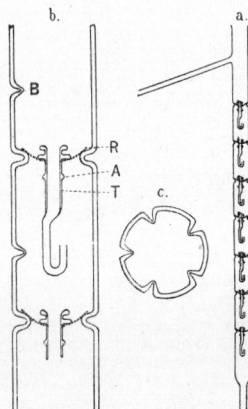

FIG. 51.—The " Young and Thomas " dephlegmator.

[1] Le Bel and Henninger, " On Improved Apparatus for Fractional Distillation," *Berl. Berichte*, 1874, **7**, 1084.

[2] Brown, " The Comparative Value of Different Methods of Fractional Distillation," *Trans. Chem. Soc.*, 1880, **37**, 49.

[3] Young and Thomas, " A Dephlegmator for Fractional Distillation in the Laboratory," *Chem. News*, 1895, **71**, 117.

[4] Young, " The Relative Efficiency and Usefulness of Various Forms of Still-head for Fractional Distillation," *Trans. Chem. Soc.*, 1899, **75**, 679.   [5] *Loc. cit.*

diameter, with the usual narrow side delivery tube.   In the wide tube are sharp constrictions, on which rest concave rings of platinum gauze, R, previously softened by being heated to redness, and these support small, glass reflux tubes, T, of the form shown in Fig. 51 *b*.   The upper and wider part of the reflux tube has an internal diameter of 4·5 mm., the narrow U-shaped part, which serves as a trap, an internal diameter of 3 mm. ; if, however, the number of constrictions exceeds 10 or 12, the traps for the lower reflux tubes should be slightly wider, say, 3·5 mm. The length of the reflux tubes should be about 45 mm., and the distance between two constrictions about 60 to 65 mm.   The enlargement, A, on the reflux tube prevents it from slipping through the ring if the tube is inverted, and the reflux tube and ring together are prevented from falling out of position by the five internal projections (made by heating the glass with a fine blow-pipe flame and pressing it inwards with a carbon pencil), one of which, B, is shown in Fig. 51 *b*.   A horizontal section through the tube at B is shown in Fig. 51 *c*.

Comparison of Bubbling Still-heads.[1] — On comparing the efficiency of the three still-heads—each of three sections—it was found that when a large quantity (400 grams) of the mixture of benzene and toluene was distilled, the Le Bel-Henninger still-head gave slightly better results than the Young and Thomas, and both of these distinctly better results than the Glinsky.   On distilling 50 grams of the mixture, the Young and Thomas still-head was found to be the best ; with the Le Bel-Henninger tube the residual toluene was not quite pure, and with the Glinsky it contained a quite appreciable amount of benzene, although the temperature reached 110·6° in all three cases.

With 25 grams of the mixture the differences in efficiency were much accentuated, and it was only with the Young and Thomas still-head that the temperature reached 110·6°, though even in this case the toluene was not quite pure.   The highest temperature reached with the Glinsky still-head was 107·6°, and with the Le Bel-Henninger, 107·35°, but the residual toluene was far less pure in the former case, for, on distillation from a small bulb, the Glinsky residue came over between 102·2° and 110·4°, and the Le Bel-Henninger from 105·7° to 110·6°, and the Young and Thomas from 110·4° to 110·6°.

With the " rod and disc," the " pear " and the " evaporator " (3 and 5 sections) still-heads, nearly as good results were obtained with 25 as with 50 or more grams of the mixture, and in all these cases the residual toluene was quite pure.

It will thus be seen that, for small quantities of liquid, the Young and Thomas still-head gives better results than the Glinsky or Le Bel-Henninger, but that none of them are so satisfactory as the other forms of still-head.

The relative efficiency of the three bubbling still-heads when 400 and 25 grams, respectively, of the mixture were distilled, and the effect of increasing the number of sections in the case of the Young and Thomas still-head, are shown in Table 53.   The results in the last

---

[1] Young, *loc. cit.*

TABLE 53

Nature of still-head.	Glinsky	Le Bel-Henninger	Young and Thomas	Glinsky	Le Bel-Henninger	Young and Thomas	Young and Thomas	Young and Thomas	Young and Thomas	Young and Thomas	Young and Thomas
Number of sections.				3	3	3	3	6	12	18	18
Vertical height in cm.	30	43	51[1]	30	43	51[1]	51[1]	78[1]	122[1]	130	130
Weight of mixture distilled.		400		25	25				100		
Final temperature.											
Percentage weight of distillate.											
80·7°	..	..	..	..	..	..	..	..	} 21·1	} 32·4	33·7
81·2	..	..	..	..	..	..	..	0·6			7·2
83·2	1·2	0·6	0·3	0·6	0·4	0·6	0·4	22·8	18·6	9·6	3·4
86·2	19·1	25·8	20·3	13·2	13·4	18·6	20·2	14·5	4·3	3·0	2·0
89·2	14·2	12·1	13·9	15·0	15·2	15·2	14·5	6·3	1·9	1·5	1·2
92·3	8·7	9·2	10·4	8·8	10·0	7·6	8·6	3·1	1·8	1·1	0·9
95·4	6·6	4·3	4·5	6·8	7·4	6·0	5·2	2·5	1·3	0·9	0·7
98·5	5·5	3·1	4·2	6·0	5·0	4·6	4·5	2·1	0·8	0·8	0·3
101·6	6·1	4·3	3·3	6·4	5·2	3·8	3·7	1·7	0·8	1·0	0·4
104·6	3·7	2·9	3·4	5·2	5·2	4·6	3·9	2·0	1·0	1·1	0·6
107·6	6·2	4·8	3·8	20·8	9·6	5·6	4·9	3·9	1·8	1·4	1·2
110·0	7·1	6·0	8·3			9·8	8·5	4·6	4·1	2·8	2·8
110·6	10·0	10·4	10·5			7·6	9·3	8·0	5·5	6·0	3·0
Pure toluene by difference.	11·6	16·5	17·1	(17·2)	(9·87)	16·0	16·3	27·9	37·0	38·3	42·6
	100·0	100·0	100·0	100·0	100·0	100·0	100·0	100·0	100·0	100·0	100·0
Weight of liquid and vapour in still-head	2·8	5·85	2·8	2·8	5·85	2·8	2·8	5·3	10·6	38·3	12·1

[1] These were unnecessarily long.

column were obtained by taking the distillation at only half the usual rate. In this case, 42·6 out of the 50 grams of toluene were recovered in a pure state in the single distillation, and 33·7 grams of benzene were obtained with a temperature range of only 0·5°.

General Remarks on the Construction of Bubbling Still-heads.—The results of the experiments which have been made serve to indicate the requirements which should be fulfilled in order that a still-head of this kind may give the best possible results.

1. **Number of Sections.**—It should be possible to greatly increase the number of sections without seriously adding to the difficulty of construction or to the fragility of the apparatus. This requirement is best fulfilled by the " Brown " and the " Young and Thomas " still-heads.

2. **Size of Constrictions, etc.**—As the amount of condensed liquid flowing back at any level is greatest at the bottom of a still-head and least at the top, it follows that, in order to retard the flow sufficiently for a pool to be formed, more complete obstruction is necessary at the top of the tube than at the bottom. In the Le Bel-Henninger still-head, however, the constrictions are frequently made widest, and the platinum cones largest, at the top of the tube. There is thus a tendency, on the one hand, for the liquid to flow past the upper cones without forming a pool, and, on the other hand, for the quantity of liquid in the pools in the lower bulbs to be unnecessarily large. In the Young and Thomas apparatus it is advisable to make the upper constrictions somewhat deeper than the lower ones.

3. **Width of Reflux Tubes and Depth of Traps.**—It has been pointed out that, consistently with efficiency, the amount of condensed liquid in the still-head during distillation should be as small as possible. The reflux tubes should, therefore, not be made wider—in that part which is filled with liquid during the distillation—than is necessary freely to carry back the condensed liquid ; also the U-shaped parts, acting as traps, should be no deeper than is required to prevent the ascending vapour from forcing its way through them. Additional width or depth simply means waste liquid in the still-head, but if the number of sections is very large, the lower traps should be made rather wider than the upper ones. On the other hand, the upper part of the reflux tubes may be advantageously made fairly wide so as to facilitate the entrance of the condensed liquid, and also to prevent bubbles or columns of vapour from being caught and carried down with the liquid through the traps. Such columns of vapour, when formed, are liable to drive out the liquid, and the ascending vapour may then pass more easily through the traps than through the pools formed by the obstructions.

The poor results obtained with the Le Bel-Henninger still-head, when only 25 grams of the mixture was distilled, were probably largely due to the excessive width and depth of the reflux tubes. The weight of liquid and vapour in the still-head was more than twice as great as in those of Glinsky or Young and Thomas.

4. **Flow of Liquid through the Reflux Tubes.**—It is of the utmost importance that there should be a rapid flow of condensed liquid through the reflux tubes, especially if they are outside the main tube and are not heated by the ascending vapour. To take an extreme case, suppose that there were no back flow at all through the reflux tubes and that the traps simply became filled with the first portions of condensed liquid. This most volatile liquid would thus remain lodged in the traps until the end of the distillation, and would then form part of the residue. In the distillation, for instance, of a mixture of benzene and toluene, the last fraction might consist of pure toluene, while the residue at the end of the fractionation would be very rich in benzene.

Of the three bubbling still-heads compared, the Young and Thomas is the best in this respect ; in the Glinsky apparatus, on the other hand, the flow of liquid was exceedingly slow, and it is for this reason that, when only 25 grams of the mixture was distilled, the residue was so much richer in benzene than the last portions of distillate. The Glinsky apparatus is, indeed, quite unsuited for the distillation of very small quantities of liquid. The same fault is to be observed with the Le Bel-Henninger still-head, but it is not nearly so marked.

In fractionating tubes of the Glinsky or Le Bel-Henninger type, the upper end of the reflux tube should be wide and the junction with the bulb should be low down, in order that the quantity of liquid in the pool may not become unnecessarily large. This is a point that is frequently overlooked.

5. **Arrangement of the Reflux Tubes.**—To get the best results, there should be a reflux tube connecting each section with the one below it, so that the change in composition may be regular from bottom to top of the still-head. That is the case with the still-heads of Le Bel-Henninger, Brown, and Young and Thomas, but in the Glinsky apparatus there is only one reflux tube connecting the top bulb practically with the still. Thus the condensed liquid, which is returned to the still through the reflux tube, is richer in the more volatile component of the mixture than the liquid in the lower pools.

6. **Return of Liquid from Still-head to Still after the Distillation is completed.**—When the residual liquid is valuable, it is of importance that it should return as completely as possible from the still-head to the still. The weights of liquid actually left in the still-heads after cooling were as follows :—Glinsky, 0·2 gram ; Le Bel-Henninger, 1·4 grams ; Young and Thomas, 0·55 gram. In this respect, and this only, the Glinsky still-head gave the best results, but the amounts of liquid left in the "Rod and Disc," the "Pear," and the "Evaporator" tubes were far smaller.

Of the various still-heads that have been described, it may be concluded that, when only moderate efficiency is required, the "Rod and Disc " or " Pear " is to be most strongly recommended, but for great efficiency the " evaporator " still-heads give the best results.

Comparison of Improved with Plain Vertical Still-head.
—The relative efficiency of the different still-heads is well shown by the following comparison of the number of fractional distillations with a plain vertical tube of 30 cm. height, which give the same result as a single distillation with the improved apparatus.

Description of still-head.	No. of fractionations.
1. " Rod and Disc " (20 discs)	More than 2.
2. " Pear " (13 bulbs)	Nearly 3.
3. " Hempel " (200 large beads)	Nearly 4.
4a. " Evaporator," original form (3 sections) . .	More than 3.
4b. ,, ,. ,, (5 ,,) . .	About 5.
5a. ,, modified ,, (3 ,,) . .	More than 3.
5b. ,, ,, ,, (5 ,,) . .	Nearly 5.
5c. ,, ,, ,, (8 ,,) . .	About 6.
6a. Young and Thomas Still-head (3 sections) . .	Between 2 and 3.
6b. ,, ,, ,, ,, (6 ,,) . .	About 4.
6c. ,, ,, ,, ,, (12 ,,) . .	About 7.
6d. ,, ,, ,, ,, (18 ,,) . .	Nearly 8.
6d. ,, ,, ,, ,, (18 ,,) ⎱ (half rate) ⎰ . .	Nearly 9.

In general, the improvement in the separation of toluene is better than in that of benzene.

Comparisons of the relative efficiency of various still-heads have been made by Rittman and Dean, *J. Ind. Eng. Chem.*, 1915, **7**, 754 ; *Bureau of Mines*, Washington, 1916, *Bull.* 125, " Petroleum Technology," 34 ; Friedrichs, *Zeit. f. angew. Chem.*, 1919, **32**, 341, and others.

Fractionating columns are in many respects analogous to the scrubbing towers employed for the absorption of soluble gases. The theory of scrubbing towers has been considered mathematically by Donnan and Masson,[1] and the conclusions arrived at as regards the conditions required for high absorption efficiency are applicable with respect to efficiency of separation in distillation.

The efficiency in each case depends on—

1. High interfacial area between gas (vapour) and liquid.
2. High relative motion of gas (vapour) and liquid within limits.
3. High degree of turbulent motion in one or both phases.

[1] Donnan and Masson, " Theory of Gas Scrubbing Towers with Internal Packing," *J. Soc. Chem. Ind.*, 1920, **39**, 237 T.

Other laboratory still-heads of various types have been described by—

R. Rempel, *Chem. Zeit.*, 1886, **10**, 371.
P. Monnet, *Monit. Scient.*, 1887, (IV.), **1**, 335.
G. E. Claudon and E. C. Morin, *Bull. Soc. Chim.*, 1888, **48**, 804.
M. Ekenberg, *Chem. Zeit.*, 1892, **16**, 958.
E. Varenne, *Bull. Soc. Chim.*, 1894, **11**, 289.
M. Otto, *ibid.*, 1894, **11**, 197.
C. W. Volney, *J. Amer. Chem. Soc.*, 1894, **16**, 160.
A. Tixier, *Bull. Soc. Chim.*, 1897, **17**, 392.
H. Vigreux, *ibid.*, 1904, **31**, 1116.
A. Golodetz, *Chem. Ind.*, 1912, **35**, 102, and 141.

4. Sufficient rate of flooding to secure the maximum drip effect. [As regards distillation this requirement is only applicable in certain cases.]

To put the matter shortly, what is required for both absorption and fractionation is the most perfect possible contact between liquid and gas (vapour).

CHAPTER XII

MODIFICATIONS OF THE STILL-HEAD (*continued*)

"REGULATED" OR "CONSTANT TEMPERATURE" STILL-HEADS

BY surrounding the still-head with water or any other liquid, the temperature of which is kept as little above the boiling point of the more volatile component as will allow of vapour passing through, a considerable improvement in the separation is effected. The temperature of the bath, however, requires very careful regulation if the boiling points of the components are near together, or if one component is present in large excess, for, in either case, a fall of a fraction of a degree would cause complete condensation of the vapour, while a rise of temperature to a similar extent would prevent any condensation from taking place and there would be no fractionation at all.

When a regulated temperature still-head is employed, it is better, for two reasons, to bend the tube into the form of a spiral; in the first place, the effective length of the still-head may be thereby greatly increased without unduly adding to the height of the bath, and in the second place, as has been already pointed out, the spiral form is more efficient than the vertical. Rosanoff and his co-workers have found, however, that a double-walled metallic cylinder (Fig. 31, p. 75), open at both ends and immersed in liquid kept at a constant temperature and vigorously stirred, gives better results than the ordinary spiral form of still-head.

Warren's Still-head.—The employment of an elongated spiral still-head, kept at a constant or slowly rising temperature, was first recommended by Warren.[1] The spiral tube was heated in a bath of water or oil; its length varied from $1\frac{1}{2}$ to 10 feet and its internal diameter from $\frac{1}{4}$ to $\frac{1}{2}$ inch.

Warren carried out fractional distillations of petroleum and other complex mixtures, and observed that, as the fractions became purer, the temperature of the bath had to be brought nearer to the boiling point of the liquid in the still and required more careful regulation.

Brown's Still-head.—A modification of this apparatus, devised with a view to the better control of the temperature of the still-head,

---

[1] Warren, "On the Employment of Fractional Condensation," *Liebigs Annalen*, 1865, Suppl., **4,** 51.

is described by Brown.[1] The liquid to be distilled is boiled in the
vessel A (Fig. 52) ; the vapour rises through the coil C, the temperature
of which is that of a liquid boiling in the vessel E. The vapour from
the liquid in E passes through the tube D to the worm condenser F,
and the condensed liquid returns by the tube K to the bottom of the
vessel E. The pressure under which the jacketing liquid boils is
regulated and measured by a pump and gauge connected with the
tube M. The liquid in E is heated by the ring burner B.

The vapour of the liquid which is being distilled passes through
the side delivery tube G and is condensed and collected in the usual
manner. The temperature of the
vapour as it leaves the still-head
is registered by a thermometer at
a ; that of the jacketing liquid is
regulated by the pressure and may
be read from the vapour pressure
curve, or a second thermometer
may be placed in the central tube.

Brown obtained very good
results with his apparatus, and
arrived at the important conclusion
(p. 73) that " in distillations with
a still-head maintained at a constant temperature, the composition
of the distillate is constant, and
is identical with that of the vapour
evolved by a mixture whose boiling point equals the temperature
of the still-head." As already
stated (p. 75) this conclusion has
been confirmed by Rosanoff and
Bacon.

In Brown's apparatus the temperature of the still-head can be
kept constant for any length of
time or it can be altered rapidly

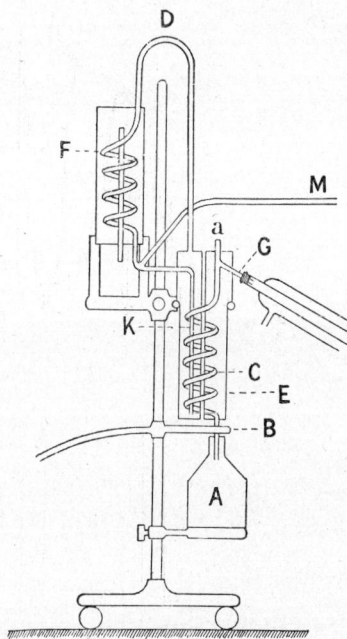

FIG. 52.—Brown's "regulated temperature"
still-head.

and easily by altering the pressure under which the jacketing liquid is
boiling, but, owing perhaps to the fact that the number of observations
and the amount of attention required during a distillation are greater
than when the still-head is merely exposed to the cooling action of the
air, it has not come into general use.

Separation of Pentanes from Petroleum.—A regulated tem-
perature still-head, combined with a bubbling still-head (Fig. 53), has
been found very useful for the separation of the lower paraffins from

[1] Brown, "The Comparative Value of Different Methods of Fractional Distillation,"
Trans. Chem. Soc., 1880, **37**, 49 ; "Fractional Distillation with a Still-head of Uniform
Temperature," *ibid.*, 1881, **39**, 517. A still-head similar in principle to Brown's has been
described by Hahn, *Ber.*, 1910, **43**, 419 ; *J. Soc. Chem. Ind.*, 1910, **29**, 300.

American petroleum.[1] The vapour from the boiling " petroleum
ether " passed first through a six column Young and Thomas still-
head, on leaving which its temperature was read on a thermometer, A.
It then passed upwards through a spiral tube in a large bath, the water
in which was either cooled by adding ice, or warmed by a ring burner
below, and was kept constantly stirred by an arrangement similar to
that employed by Oswald, in which a propeller, with four blades of
thin sheet copper, was kept rotating by a windmill actuated by a ring
of gas jets, the efficiency of the windmill being greatly increased by

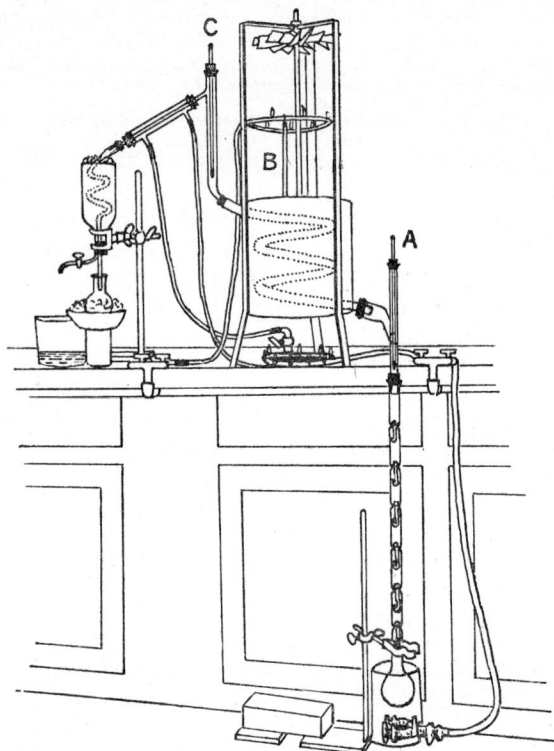

Fig. 53.—Combined " regulated temperature " still-head and " Young and Thomas "
still-head.

the use of a chimney. The temperature of the bath was registered by
the thermometer B.

After leaving this part of the apparatus the vapour passed through
a vertical tube, where its temperature was read on the thermometer C,
then into a spiral condenser cooled by ice ; the condensed liquid was
collected in ice-cooled flasks.

The general course of this separation has already been referred to
(p. 113), and is indicated by the curves in Fig. 36. Details of a single

[1] Young and Thomas, " Some Hydrocarbons from American Petroleum. 1. Normal and
Isopentane," *Trans. Chem. Soc.*, 1897, **71**, 440. Cf. F. M. Washburn, " Constant Temperature Still-head for Light Oil Fractionation," *J. Ind. and Eng. Chem.*, 1920, **12**, 73.

fractionation, the eleventh, are given in Table 54 ; the temperatures are the final ones for the fractions and they are all corrected to 760 mm., and for the thermometric errors. Since the temperature of the vapour is affected by changes of the barometric pressure, that of the water bath has been altered in each case to the same extent. The lowest fractions consist chiefly of isopentane ; the middle ones are mixtures of iso- and normal pentane ; the highest consist of nearly pure normal pentane.

TABLE 54

ELEVENTH FRACTIONATION

Fraction.	D.	B.	T.	$D-T$.	$\triangle T$.	$\triangle W$.	$\dfrac{\triangle W}{\triangle T}$.
1	28·42°	28·25°	28·30°	0·12°	(1·15°)	78	(70)
2	29·15	28·85	28·90	0·25	0·60	101	168
3	30·17	29·65	29·65	0·52	0·75	58	77
4	31·75	30·55	30·55	1·20	0·90	44	49
5	33·10	32·15	32·20	0·90	1·65	43	26
6	34·60	33·90	33·85	0·75	1·65	48	29
7	35·35	34·65	34·75	0·60	0·90	37	41
8	35·65	35·20	35·40	0·25	0·65	40	61
9	35·90	35·80	35·85	0·05	0·45	43	95
10	36·11	36·00	36·10	0·01	0·25	80	320
11	36·25	36·30	36·23	0·02	0·13	81	623
12	36·32	36·30	36·31	0·01	0·08	71	890

D = temperature of vapour on leaving the bubbling still-head.
B = temperature of bath.
T = temperature of vapour before entering the condenser.

The bath was kept at such a temperature, and the source of heat was so regulated, that the drops of distillate fell as nearly as possible at the rate of 60 per minute, and it will be noticed that the temperature of the vapour before condensation was in all cases nearly the same as that of the bath, generally very slightly higher. It will also be seen that the fall in temperature of the vapour, during its passage through the regulated temperature still-head ($D-T$), is greatest for the middle and least pure fractions and smallest for the highest and purest ; indeed, for the last fraction the three temperatures are almost identical, while for the fourth there is a difference of 1·2° between D and T.

Very careful regulation of the temperature of the bath was necessary when the liquid distilled was nearly pure.

Suitability for very Volatile Liquids. — For very volatile liquids there can be no question that the regulated temperature still-head is the most suitable and the principle has been applied by Ramsay, Dewar and others to the purification of gases. Thus, if a mixture of helium with any other gases is passed at the ordinary pressure through a spiral tube cooled by liquid hydrogen boiling under reduced pressure, all other known gases would be condensed in the tube and the helium alone would pass through.

Separation of Oxygen and Nitrogen from the Air.—Large quantities of pure oxygen and also of nitrogen are now prepared by the fractional distillation of liquid air or by the combined fractional condensation of air and distillation of the resulting liquids.

Claude's apparatus may be regarded as a combination of a constant temperature differential liquefier with a rectifying column ; by means of it not only pure oxygen but also pure or nearly pure nitrogen are obtained.   The air, under a pressure of five atmospheres, enters the chamber A (Fig. 54) through the tube B.   From A it passes upwards through the vertical tubes, immersed in liquid oxygen, into the small

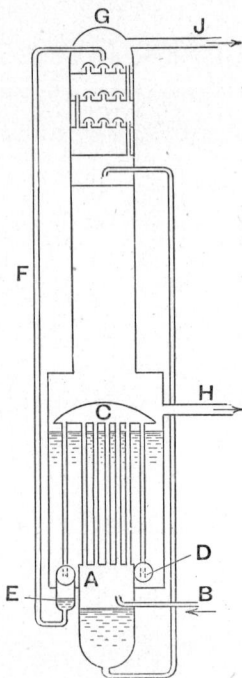

FIG 54.

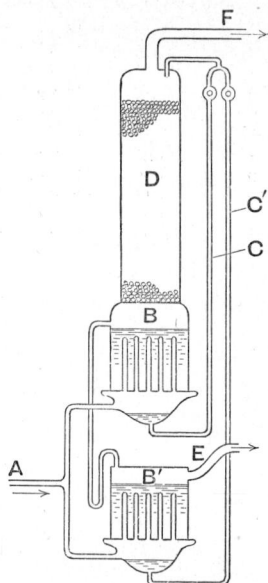

FIG. 55.

chamber C.   Partial liquefaction and fractionation takes place in these vertical tubes, the condensed liquid which collects in A containing about 47 per cent of oxygen.   The remaining air, rich in nitrogen, which escaped condensation in the upward passage through the vertical tubes, passes downwards through the outer vertical tubes and is there finally condensed.   The crude liquid nitrogen collects in a sort of hydraulic main D, flows into the vessel E, and is forced up through the tube F into the top of the rectifying column G.   The liquefied air, rich in oxygen, which collects in A is forced up through the other vertical tube and enters the column at a lower level.   In its passage downwards the liquid from E and A undergoes fractionation, so that pure liquid oxygen leaves the bottom of the column and fills the vessel in which the vertical tubes are placed. Pure oxygen gas, formed by evaporation of the liquid, passes out by the tube H, and pure nitrogen leaves the top of the rectifying column through the tube J.

In Linde's apparatus (Fig. 55) there is nothing in the nature of a differential condenser, the two baths of liquid oxygen causing the complete condensation of the compressed air. The air entering the apparatus through the tube A is liquefied in B and B', and the liquid air is forced up the tubes C and C' and is delivered at the top of the rectifying column D. In its descent through the column fractionation takes place and the liquid which reaches the bottom and flows into B is pure oxygen : the excess, together with any oxygen gas, passes into B', and the pure oxygen gas formed by evaporation of the liquid is carried off through the tube E. The gas which passes away from the top of the column through the tube F contains about 7 per cent of oxygen.

Helium.—It has been found that helium is present in nearly all natural gases, those of the Western States of America containing from 1 to 2 per cent of the element, whilst the Bow Island gas in Alberta, supplied to Calgary, contains about 0·35 per cent. On account of its lightness and non-inflammability helium would be an ideal gas for aeronautical purposes if it could be obtained in sufficient quantity at not too great a cost. From preliminary calculations as to cost, etc., carried out by Sir Richard Threlfall, he was led to believe that helium might be successfully obtained from natural gas. The necessary investigations were entrusted to and were successfully carried out by J. C. M'Lellan, who has recently given a full description[1] of the plant actually used at Calgary for the production of helium of at least 99 per cent purity, and of a commercial plant designed to deal with the whole of the natural gas supplied from Bow Island. This plant should yield more than 10,000,000 cubic feet of 97 per cent helium annually, assuming that the supply of natural gas showed no diminution.

In order to isolate the helium it is necessary to condense all the less volatile gases, chiefly methane and nitrogen. The principle of the method is similar to that on which the production of pure oxygen and nitrogen from the air is based, but the difficulties are much greater inasmuch as it is the most volatile component which has to be separated and as this component is present in relatively very small quantity.

[1] J. C. M'Lellan, "Helium: its Production and Uses," *Trans. Chem. Soc.*, 1920, **117**, 923.

CHAPTER XIII

CONTINUOUS distillation of wort has long been carried out on the large scale by means of the Coffey still (p. 167) and others of the same type, but it is only comparatively recently that any attempt has been made to devise a process of the kind suitable for laboratory purposes.

Carveth's Apparatus.—Carveth [1] suggests that by maintaining two parts of a system at different temperatures, corresponding to the boiling points of the two components of a mixture, it should be possible to effect continuous separation, especially if use were made

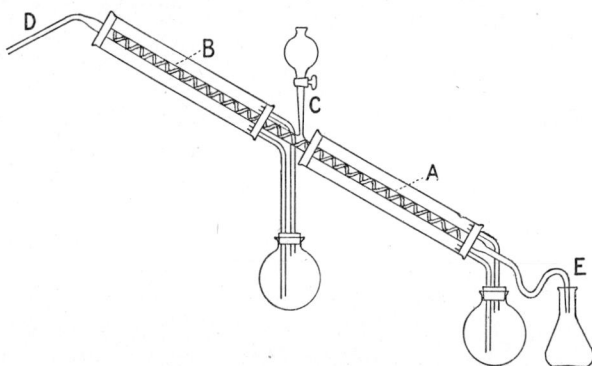

FIG. 56.—Carveth's still for continuous distillation.

of dephlegmating intercepts. He describes an apparatus used by Derby, in 1900, for mixtures of alcohol and water. It consists (Fig. 56) of a long block-tin tube surrounded at its lower end, A, by the vapour of boiling water, at its upper end, B, by vapour from boiling alcohol, and filled with intercepts. The mixture of alcohol and water to be distilled was slowly dropped in at C, the vapour passing to the condenser at D, and the residue through the trap at E. Carveth, however, gives no details regarding the length or diameter of the tubes or the nature of the intercepts.

So far as the lower half of the still-head is concerned, there is no

---

[1] Carveth, " Studies in Vapour Composition," Part II. *Journ. Phys. Chem.*, 1902, **6**, 253.

objection to the use of steam as a jacket, the object aimed at being
to keep the temperature up to the boiling point of water in order that
the alcohol may be vaporised as completely as possible ; the great
amount of heat evolved by the condensation of steam makes that
substance a very efficient heating agent. But the use of the *vapour*
of alcohol for the upper part of the still-head is wrong in principle,
and it is a mistake to suppose that this part would thus be maintained
" at a temperature corresponding to the boiling point of alcohol."
What is here required is to keep the temperature as far as possible
from rising above the boiling point of alcohol, in order that the vapour
of water may be condensed as completely as possible, while that of
the alcohol passes on to the condenser. This can be done by surrounding
the still-head with a *liquid* which is kept at the required temperature
by suitable means, but if a vapour, such as that of the more volatile
component, is used as a jacket, it easily becomes superheated and the
still-head is not prevented from rising in temperature at all ; indeed
the effect is very similar to that produced by covering the tube with
cotton-wool—condensation is somewhat diminished and so also is the
efficiency.

Carveth makes the general statement that the percentage of alcohol
by weight in the residue was found to average about 0·5 when working
carefully, but in five cases for which he gives details the percentage
varied from 1·5 to 2·0. These five distillations were carried out with
extreme slowness, the average rate being only 7·5 grams of distillate
per hour.

Again, the percentage of alcohol in the distillate is stated to have
varied from 90·6 to 93·9 in four cases, but in the five very slow distillations referred to above the percentages varied from 76·7 to 89·9. It
does not appear possible to form any very definite idea of the efficiency
of the apparatus from the data given.

Lord Rayleigh's Apparatus.—Lord Rayleigh[1] describes an
apparatus similar in principle to Carveth's except that the temperatures of both parts of the still-head are regulated by liquids and not
by vapours.

The apparatus consists of a long length (12 metres) of copper tubing,
15 mm. in diameter, arranged in two spirals which are mounted in
separate pails. For the distillation of mixtures of alcohol and water
the lower and longer spiral was heated by boiling water, the upper
one by water maintained at a suitable temperature, usually 77°. The
spirals were connected by a straight glass or brass tube of somewhat
greater bore, provided with a lateral junction through which the mixture could be introduced. With the exception of the two extremities,
the whole length of tubing sloped gently and uniformly upwards from
near the bottom of the lower pail to the top of the upper pail, where
it turned downwards and was connected with an ordinary Liebig's
condenser. The lower end of the tubing was, if necessary, connected
with an air-tight receiver heated to 100°.

---

[1] Rayleigh, " On the Distillation of Binary Mixtures," *Phil. Mag.*, 1902, [VI.], **4**, 521.

The mixture was introduced at such a rate that it fell in a rapid but visible succession of drops and, when rich in alcohol, it was previously heated. Mixtures containing 20, 40, 60 and 75 per cent of alcohol were distilled, and in all cases the water, collected in the lower receiver, was nearly pure, never containing more than 0·5 per cent of alcohol. The distillate varied but little in strength and contained from 89 to 90·3 per cent of alcohol.

It seems probable that such a continuous process may prove very useful for mixtures which separate into only two components and when the quantity to be distilled is large.

Continuous Separation of Three Components.—It would, moreover, be possible to devise an arrangement by which three components

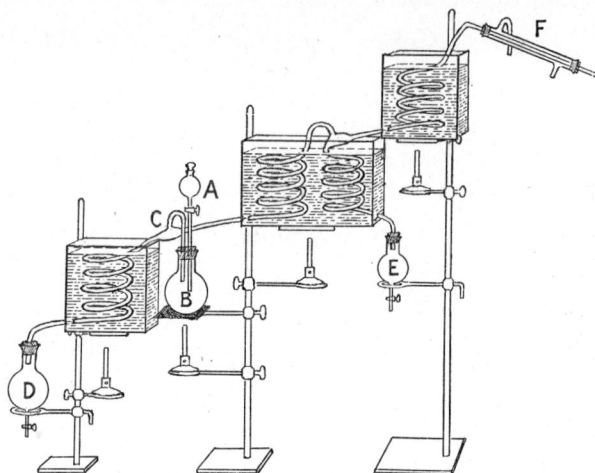

FIG. 57.—Still for continuous distillation of three components.
(Adapted from Lord Rayleigh's still.)

could be separated in a similar manner, but it would be necessary to have two coils for the middle substance ; the form of apparatus required is shown diagrammatically in Fig. 57. Suppose, for example, that the liquids to be separated were methyl, ethyl and propyl acetates (b.p. 57·1°, 77·15° and 101·55° respectively). The highest bath would be kept at 57·1°, the middle one at 77·15° and the lowest at 101·55°. The mixture would be introduced slowly through the funnel at A, into the still B, where it would be boiled, and the mixed vapour would enter the still-head at C. The propyl acetate would be collected in the receiver D, which might be heated by a ring burner. The vapour in passing upwards through the first coil in the middle bath would be freed, more or less completely, from propyl acetate and the mixture of ethyl and methyl acetates would enter the top of the second coil, chiefly in the form of vapour. The condensed liquid which reached the bottom of the coil, and was collected in E, would be nearly pure ethyl acetate ; while the vapour that reached the top of the coil in the highest bath, and was condensed in F, would be nearly pure methyl acetate.

It should thus be theoretically possible to separate all three liquids in a pure state, and the components of a still more complex mixture should be separable, if there were as many baths as components kept at the boiling points of those components, and two coils in each bath except the lowest and highest.

Messrs. E. Barbet et Fils et Cie, of Paris, supply an efficient laboratory apparatus which can be used for either continuous or discontinuous distillation.

CHAPTER XIV

FRACTIONAL DISTILLATION WITH AN IMPROVED STILL-HEAD

Benzene and Toluene.—That much time is saved by the use of an improved still-head is seen by comparing the results of a fractional distillation of 200 grams of a mixture of equal weights of benzene and toluene with a modified " evaporator " still-head of 5 sections with those obtained with a plain vertical tube 30 cm. in height.

TABLE 55

	I.			II.		
No. of fraction.	Temperature range.	$\triangle w$.	$\dfrac{\triangle w}{\triangle t}$.	Temperature range.	$\triangle w$.	$\dfrac{\triangle w}{\triangle t}$.
1	...	...	...	80·2— 80·4°	39·25	**196 2**
2	81·5— 83·2°	51·6	**30·4**	80·4— 81·4	43·55	**43 5**
3	83·2— 95·4	47·1	**3·9**	81·4— 95·4	16·5	**1 2**
4	95·4—110·0	30·1	**2·1**	95·4—110·0	8·9	**0 6**
5	110·0—110·6	19·8	**33·0**	110·0—110·6	7·9	**13 2**
Toluene .	110·6°	51·0	∞	110·6°	82·65	∞
		199·6			198·75	

	III.		IV.		V.	
No. of fraction.	Temperature range.	$\triangle w$.	Temperature range.	$\triangle w$.	Temperature range.	$\triangle w$.
Benzene .	80·2°	21·9	80·2°	56·3	80·2°	81·8
1	80·2—80·3°	44·35	80·2—80·25°	30·7		
2	80·3—81·7	27·0				15·0
Residue .	...	4·95		10·3		
		98·2		97·3		96·8
...	Rejected below 110·6°	} 7·25				
Toluene .	110·6°	91·95				
		197·4				

When the " evaporator " still-head was used it was not necessary to divide the distillate into more than five fractions. In the first distillation (Table 55), more than half the total quantity of toluene was recovered in a pure state, and it will be noticed that the values of $\Delta w/\Delta t$ for the middle fractions are relatively very low.

In the second fractionation, a little liquid came over at the boiling point of benzene, but it was not considered advisable to collect this separately. The value of $\Delta w/\Delta t$ for the first fraction was more than 300 times as great as for the fourth. More than 82 per cent of the toluene was now obtained in a pure state.

In the third fractionation 21·9 grams of pure benzene was obtained, and when fraction No. 3, collected from 81·4° to 95·4°, was added to the residue in the still, and the distillation was continued, the temperature did not rise above 81·7°; the whole of the distillate was therefore collected in receiver No. 2, and the small residue (4·95 grams) was rejected. The apparatus was then dried, and the higher fractions were distilled, when only 7·25 grams came over below 110·6°, the residue (9·3 grams) consisting of pure toluene. The distillate was too small to allow of further fractionation.

Two more distillations of the lower fractions sufficed to complete the recovery of the benzene.

Comparative Results with Plain and Improved Still-heads.—The great improvement effected by employing an efficient apparatus is clearly shown in Table 56.

TABLE 56

	Evaporator still-head of 5 sections.	Plain still-head 30 cm. in height.
Weight of pure benzene recovered	81·8 grams	81·4 grams.
„ „ toluene „	91·95 „	88·8 „
Time required for actual distillation	6½ hours	About 30 hours.
Loss by evaporation, &c.	4·0 grams	22·0 grams.
Mixture left undistilled	22·25 „	7·8 „

The chief gain was in the time occupied, which was reduced nearly to one-fifth, but the recovery also was somewhat better, and in many cases would be much better; there was also much less actual loss of material, though the amount left undistilled was greater.

Methyl, Ethyl and Propyl Acetate.—The separation of a mixture of 100 grams of methyl acetate, 120 grams of ethyl acetate and 100 grams of propyl acetate was carried out by distillation with a modified evaporator still-head of 8 sections. Details of the fractionations are given in Table 57 (p. 160).

[TABLE

TABLE 57

No. of fraction	I. Temperature range	Δw	Δw/Δt	II. Temperature range	Δw	Δw/Δt	III. Temperature range	Δw	Δw/Δt
1	..	..	..	57·1 — 57·4°	26·25	**87·5**	Methyl acetate 57·1 —57·25°	9·45	cc
2	Below 60·85°	49·65	?	57·4 — 59·4	49·15	**24·6**	57·25—58·05	42·2	**281·3**
3	60·85— 67·1	48·3	**7·7**	59·4— 67·1	24·1	**3·3**	58·05—67·1	30·5	**38·1**
4	..	..	..	67·1 — 76·15	15·75	**1·7**	67·1 —76·65	15·45	**1·7**
5	67·1 — 77·15	55·85	**5·6**	76·15— 77·15	43·4	**43·4**	76·65—77·15	11·3	**1·2**
6	..	..	..	77·15— 77·65	45·65	**91·3**	77·15—77·3	46·9	**82·0**
7	77·15— 89·35	64·85	**5·3**	77·65— 89·35	13·45	**1·1**	77·3 —77·7	12·65	**312·7**
8	89·35—101·35	34·45	**2·9**	89·35—101·55	14·2	**1·2**	Residue . .	6·3	**31·6**
9	101·35—101·55	11·45	**57·2**						
								215·75	
							Rejected below 101·55°	5·15	
	Propyl acetate	54·65		Propyl acetate	85·35	cc	Propyl acetate	94·2	cc
		319·2			317·3			315·1	

No. of fraction	IV. Temperature range	Δw	V. Temperature range	Δw	VI. Temperature range	Δw
1	Methyl acetate . 57·1—57·2° . .	34·45 32·0	Methyl acetate . . 57·1—57·15° . .	58·9 24·0	Methyl acetate . .	76·8
2	57·2—58·0 . .	23·1	Total residue . .	12·55	Total residue . .	18·15
	Residue . .	6·55				
		96·1		95·45		94·95
	Rejected below 74·0°	3·35				
4	74·0—77·1° . .	12·7				
5	77·1—77·15 . .	36·05	Rejected below 76·5°	5·5	Rejected below 77·15°.	8·2
6	Ethyl acetate .	52·55	76·5—77·15° . .	10·6	Ethyl acetate . .	95·8
		200·75	Ethyl acetate . .	88·1		198·95
	Total residue .	12·6		199·65		
		213·35				

In the second fractionation, when fraction No. 7 from I. had been distilled over as completely as possible, the temperature had risen only to 81·5°. Fraction 8 from I. was, however, added to the residue in the still as usual and, on distillation, a considerable amount was collected below 89·35°, but the quantity that came over between this temperature and 100° was very small, and it was clearly not worth while to attempt to recover any more ethyl acetate in the next fractionation from the eighth fraction of II. This last fraction was therefore merely redistilled to recover as much propyl acetate as possible, and all that came over below 101·55 (5·15 grams) was rejected.

Similarly, in the third fractionation, when fraction 3 from II. had been distilled as completely as possible, the temperature had only risen to 65°, but fraction 4 from II. was added, and the distillation continued as usual. In fractionation IV., after fraction 3 from III. had been added to the residue and the distillation continued, the quantity of liquid left in the still was extremely small when the temperature reached 58° ; the distillation was therefore stopped, and the residue was rejected. The fractionations numbered III. to VI. were therefore not continuous, but in each case consisted of two parts. Thus in III. the distillation was stopped when the temperature had reached 77·7° ; the still and still-head were dried, and the last fraction from II. was separately distilled, when the recovery of propyl acetate was completed.

In IV. the first part of the fractionation ended when the temperature reached 58°, and was recommenced by the distillation of fraction 4 from III. In this case the temperature did not rise above 77·15°, even when the highest fraction from III. was distilled ; fraction 6 was therefore taken to be pure ethyl acetate.

Comparative Results with Plain and Improved Still-head.—The great superiority of the " evaporator " still-head is clearly shown by the results given in Table 58.

TABLE 58

	Long plain still-head.	" Evaporator " still-head of 8 sections.
Percentage weight of methyl acetate recovered	48·1	76·8
„ „ ethyl acetate „	53·5	79·8
„ „ propyl acetate „	72·5	94·2
Number of fractionations required	32	6
Time required in hours	About 70	17
Percentage weight of material—		
(1) Lost by evaporation and transference	21·0	2·8
(2) Lost by treatment with reagents	5·6	0·0
(3) Left undistilled	17·1	13·8

In the long-continued fractionation with the plain still-head, a considerable amount of moisture was absorbed by the esters and some hydrolysis took place. Before the fractionations were completed it was necessary to treat the propyl acetate with potassium carbonate and the methyl acetate with phosphorus pentoxide, and the loss was thus increased.

With the " evaporator " apparatus, on the other hand, the propyl acetate showed no acid reaction whatever, and the methyl acetate was only slightly moist.

The specific gravities were again determined and were found to agree well with those given on p. 112.

Fractional Distillation under Reduced Pressure.—The improved still-heads may be employed for distillation under reduced pressure. A 12-column Young and Thomas still-head, for example, was used by Francis[1] for the separation of isoheptyl and normal heptyl bromides from the products formed by the action of bromine on the distillate from American petroleum coming over between 93·5° and 102°. The

---

[1] Francis and Young, " Separation of Normal Isoheptane from American Petroleum," *Trans. Chem. Soc.*, 1898, **73**, 920.

M

pressure in this case was 70 mm. Wade and Merriman [1] employed an "evaporator" still-head for the distillation of mixtures of ethyl alcohol and water under pressures both lower and higher than that of the atmosphere (pp. 17, 59).

[1] Wade and Merriman, "Apparatus for the Maintenance of Constant Pressures above and below the Atmospheric Pressure. Application to Fractional Distillation," *Trans. Chem. Soc.*, 1911, **99**, 984.

CHAPTER XV

Plant required.—The plant required for the distillation of alcohol, acetone, coal-tar and its products, petroleum and its products, glycerine and essential oils on the large scale is fully described in the sections dealing with those substances, but a brief sketch of the chief forms of plant may be useful.

When only a rough separation of the constituents of a complex mixture is required the vapour passes directly from the still to the condenser, but for the better separation of the components modified still-heads are employed.

Mansfield's Still. — A simple contrivance, which was adopted by Mansfield, is to cool the still-head with water, the temperature of which is allowed to rise to a suitable extent. The head A (Fig. 58) is surrounded by water which becomes heated to its boiling point. Liquids which boil at temperatures higher than 100° are for the most part condensed in the still-head and return to the still; afterwards the stop-cock B is opened and the vapour

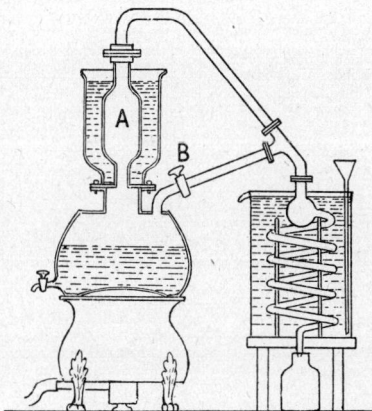

FIG. 58.—Mansfield's still.

then passes directly to the condenser. With this still, Mansfield was able to separate benzene in a fairly pure state from coal-tar.

Dephlegmators.—A constant or regulated temperature still-head, such as that employed in Mansfield's still, is now usually termed a dephlegmator; it now forms a more or less important part of many plants. It has two functions, (a) fractionation, (b) the providing a sufficient amount of condensed liquid for the satisfactory working of the bubbling or other still-head. In Mansfield's still it performs the first function only, but in many modern plants the second function is by far the more important.

Coupier's Still.—In Coupier's still (Fig. 59), a more elaborate cooling arrangement is combined with a bubbling still-head or column.

163

The vapour from the still A passes first through the rectifier or column B, and then by the pipe C into a series of bulbs placed in a cistern D containing brine, which may be warmed by steam from the pipe E. The liquid condensed in the bulbs returns by the pipes F, F, to the

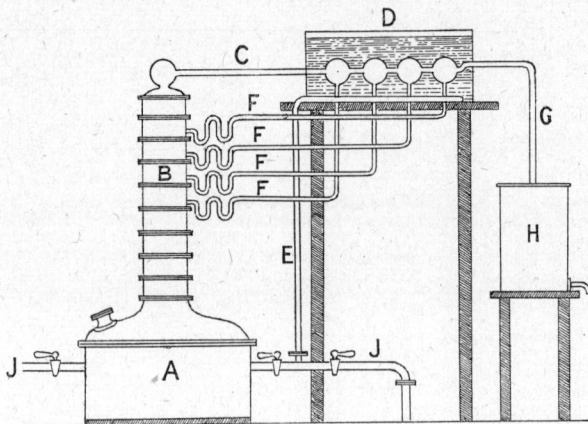

FIG. 59.—Coupier's still.

column, the less volatile vapours being for the most part condensed in the first bulb and the liquid returning to a low part of the column, while that from the other bulbs reaches it successively at higher levels. The vapour, freed from substances boiling at higher temperatures than that of the water in the tank, then passes by the pipe G to the condenser H. The contents of the still are heated by the steam-pipe J.

French Column Apparatus. — A very similar arrangement is seen in the French column apparatus (Fig. 60). The liquid to be distilled is heated by a steam-pipe in the still A; the vapour rises through the rectifier B, passes through the series of pipes in the tank C, and then enters the condenser D. Cold water from the cistern E enters the condenser at the bottom, and in its passage upwards is warmed by the condensation of the vapour; the warm water passes through the pipe F into the tank, which is divided into sections by the vertical partitions shown in the figure. Here the

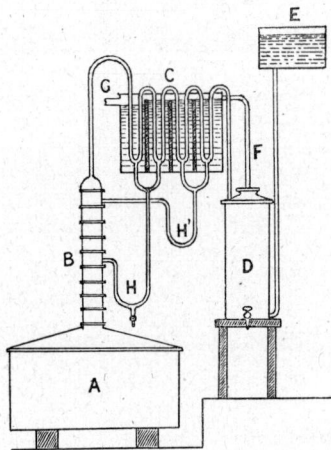

FIG. 60.—French column apparatus.

water receives more heat owing to condensation of vapour in the pipes, so that it is hottest where it leaves the tank by the pipe G. Thus the vapour, in passing from the rectifier to the condenser through the dephlegmator, is cooled by successive stages, and the liquid which returns to the lower part of the rectifier through the pipe H is much richer in the less volatile components than that which reaches the top by the pipe H'.

In the case of very complex mixtures, from which it is only necessary to separate fractions of different volatility, the successive condensates from such a dephlegmator may be collected separately instead of being returned to the rectifier. Fractionation is then the sole function of the dephlegmator, as in Mansfield's still.

Bubbling Still-heads.—In many still-heads part of the vapour is condensed and forms pools of liquid through which the rising vapour

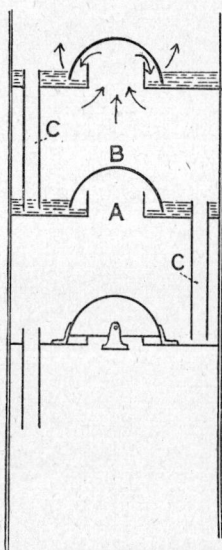

a

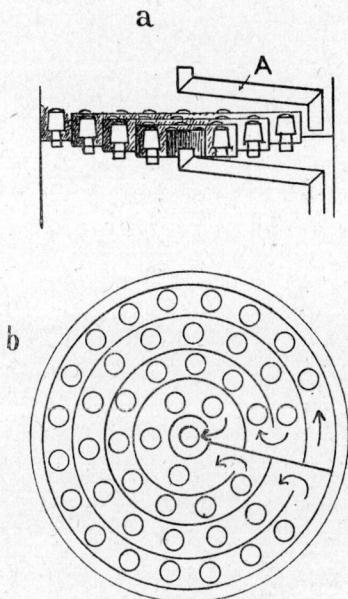

FIG. 61.—Dubrunfaut's still-head.   FIG. 62.—Egrot's still-head.

has to force its way, so that very complete contact between vapour and liquid is ensured. A dephlegmator is very frequently placed at the top of the still-head in order to provide sufficient liquid for the pools. The details of the various bubbling still-heads differ considerably ; a few of them are shown in Figs. 61 to 65 ; others are described in the special sections (pp. 302, 303, 331, 402, 403).

The condensed liquid collects on trays which in some cases are perforated ; in others they are plain and the vapour then rises through a central pipe A (Fig. 61) in the tray, but its ascent is barred by a dome B, and it bubbles through the condensed liquid in the tray. The dome may be serrated round its lower edge (Fig. 134, p. 331) or may have vertical slits (Fig. 109, p. 303) so as to break up the bubbles more completely. In some cases, as in the Egrot still-head (Fig. 62), there are numerous domes or similar contrivances on a tray, and the condensed liquid is made to follow a zigzag course from the circumference to the centre of each tray. The liquid reaches the tray by pipe A (Fig. 62a), and follows the course shown by the arrows in Fig. 62b.

In all cases the excess of liquid flows back through a reflux pipe, the lower end of which is trapped by liquid in various ways, as shown in Figs. 63 to 65, C, Fig. 61, and A, Fig. 62a.

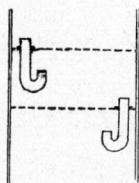

FIG. 63.
Savalle's still-head.

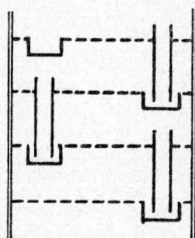

FIG. 64.
Savalle's still-head.

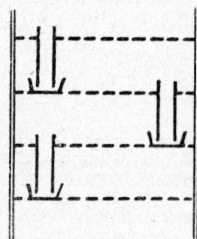

FIG. 65.
Coffey's still-head.

Other Still-heads.—Various forms of still-head have been devised in which good contact between vapour and liquid is brought about without the actual passage of the vapour through the liquid in the form of bubbles. These still-heads have the advantage that there is very little, if any, rise in pressure from the highest to the lowest section of the column. An early form of still-head of this kind is that of Pistorius (Fig. 66). Here the vapour entering a section of the still-head is deflected from the centre to the circumference by the flat dome, A ; it then passes back to the centre above the dome and is partially condensed by the water in B, above it. It is probable that a considerable improvement in the efficiency of this apparatus would be effected by the simple modification shown in Fig. 67. The condensed liquid,

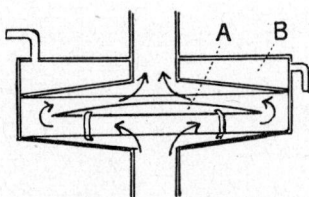

FIG. 66.—The " Pistorius " still.

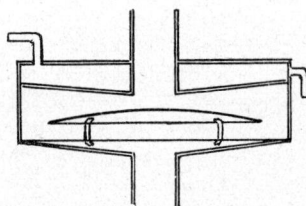

FIG. 67.—Modified " Pistorius " still.

instead of flowing down the outer walls of the section, would drop from the central tube on to the flat dome as in the " evaporator " still-head (p. 134), and the ascending vapour would come into better contact with this hot liquid and would cause some of it to evaporate again.

Among the still-heads of this class may be mentioned the various " Ring " columns, such as Raschig's and Lessing's (Figs. 164, 47, pp. 400, 137), Foucar's spiral still-head, and the columns of Kubierschky, Perrier, Guillaume, Ilges, and others (Figs. 115, 111, pp. 308, 304), in which the condensed liquid is caused to fall in drops or spray.

Foucar's apparatus may be regarded as a greatly improved form of spiral still-head; it is simple, compact, and efficient, and is well suited for vacuum distillation.

Kubierschky's still-head contains perforated plates like some of the bubbling columns, but it is the liquid and not the vapour that passes through the perforations; the vapour rises without obstruction alternately through central pipes and circumferential rings nearly to the top of the section in each case, leaving the section near the bottom. This column is very efficient, but, according to Mariller,[1] the other "spray" still-heads do not give very satisfactory results.

CONTINUOUS DISTILLATION

When large quantities of liquid are to be dealt with there are obvious advantages in making the process of distillation continuous; the

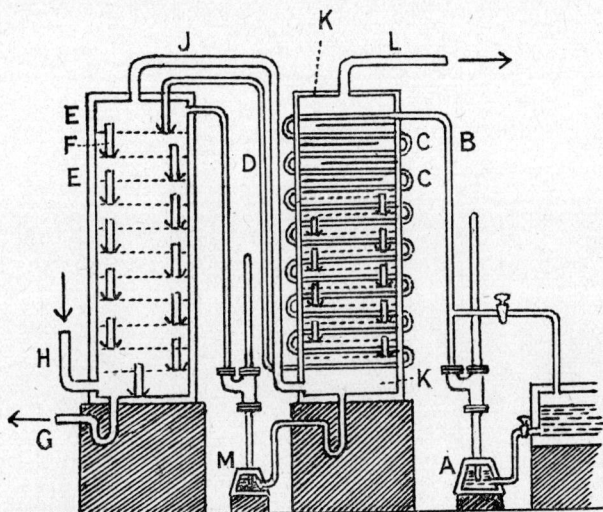

FIG. 68.—The "Coffey" still.

advantages, as well as certain disadvantages, are discussed in the sections on Petroleum and Coal-tar (pp. 319, 359).

The continuous process was first applied to the production of ethyl alcohol from fermented liquors. In the Coffey plant, which may be regarded as typical of those in general use for the continuous production of alcohol, there is no still, but the fermented liquor is heated by live steam which enters the bottom of the still-head. The plant, in its simplest form, is shown in Fig. 68. The wort is pumped from a reservoir A up the pipe B, and passes down the zigzag pipe C C, where it is heated by the ascending vapour in the rectifier; then up the pipe D into the highest section of the analyser E. It then descends

[1] Mariller, *La Distillation fractionnée*, Paris, 1917.

from section to section through the tubes F, and is finally allowed to escape through the trapped pipe G.

Steam is passed into the analyser by the pipe H, and causes the wort to boil, so that by the time it has reached the bottom it is completely deprived of alcohol. The ascending vapours pass through the perforations in the plates and bubble through the liquid on them, a portion of the aqueous vapour being thus condensed, and the descending wort heated, by each washing.

On reaching the top of the column, the concentrated alcoholic vapour passes through the pipe J into the bottom of the rectifier K, and then ascends through perforated plates similar to those in the analyser; the ascending vapour is, however, not washed by wort in the rectifier, but by the liquid formed by partial condensation of the vapour. In the upper part of the rectifier there are usually only shelves which compel the vapour to take the same zigzag course as the pipes which convey the wort downwards. The purified vapour then passes through the pipe L to the condenser.

The mixture of weak spirit and fusel oil, condensed in the rectifier, flows into a reservoir M, from which (usually after separation of the fusel oil) it is pumped into the top of the analyser, where it mixes with the descending wort.

When such a still is working regularly the composition of liquid and vapour in each section of the rectifier should remain practically constant. The fusel oil tends, however, to accumulate in the lower part of the rectifier, and in modern plants liquid is run off continuously at the level of greatest oil concentration at such a rate that this concentration remains constant.

The vapour also contains impurities more volatile than alcohol, and these would, of course, be contained in the spirit produced with the simple plant described. The alcohol at some little distance from the top of the rectifier is much freer from these impurities than that at the very top, and, at the present time, the alcohol is run off at a suitable rate from the level of greatest purity.

Part of the vapour which leaves the top of the rectifier is condensed in a dephlegmator, the liquid being returned to the rectifier, the rest of it, richest in the volatile impurities, passes on to a condenser; or the whole of the vapour may be condensed and a part of the liquid returned to the rectifier.

Other still-heads of this class are described in the sections on alcohol, acetone, and coal-tar products.

Such complex mixtures as coal-tar or petroleum could not be distilled in this manner, but various methods of continuous distillation are employed.

The petroleum or coal-tar, usually previously heated in a pre-heater, may, for example, pass through a series of stills at successively slightly lower levels, the temperature rising as the liquid passes from still to still (p. 377). The vapour evolved from each still—and in some cases from the pre-heater—is separately condensed, and it is clear that the vapour from the first still (or the pre-heater if hot enough) will contain the most volatile components and that from the last still the least volatile. The residue from the last still (pitch in the case of coal-tar) is run off continuously and is usually passed through pre-heaters, so that its heat may be utilised (pp. 336, 378). In some cases the number of fractions is increased by the fractional condensation in dephlegmators of the vapour from each of the stills (p. 338).

Occasionally the crude liquid is strongly and rapidly heated so as to expel all volatile matter at once. Partial separation is then effected by fractional condensation in a series of dephlegmators (pp. 343, 373).

Full details of these and other methods, also methods of vacuum and steam distillation, are given in the different sections dealing with large scale processes.

CHAPTER XVI

Determination of Composition of Mixture.—The composition of a mixture of liquids which are not difficult to separate may, as a rule, be ascertained with a fair degree of accuracy from the results of a single distillation with an efficient still-head, or, if the components are more difficult to separate, from the results of two or three distillations.

It will be well to consider first the case of mixtures which tend, on distillation, to separate normally into the original components.

Taking first the simplest case, that of a mixture of two liquids, it is found that the weight of distillate that comes over below the " middle point " is, as a rule, almost exactly equal to that of the more volatile component, even when the separation is very far from complete.

By " middle point " is to be understood in all cases the temperature midway between the boiling points of the two components, whether single substances or mixtures of constant boiling point, into which the original mixture tends to separate ; or, in the case of more complex mixtures, the temperature midway between the boiling points of any two consecutive fractions of constant boiling point.

If the original mixture tends to separate on distillation into more than two, say n, components, the weights of these components will be very nearly equal respectively to (No. 1) the weight of distillate below the first middle point, (No. 2 to $n-1$) the weights of distillate between the successive middle points, (No. n) the weight above the last middle point.

Loss by Evaporation.—It is obvious that there must be some loss by evaporation, which always makes the weight of distillate somewhat too low. This loss will be greater as the initial boiling point of the liquid is lower and as the temperature of the room is higher. It is not, however, proportional to the amount of liquid distilled, for a great part of it is caused by the saturation of the air in the flask and still-head with vapour when the liquid is first heated ; since this vapour is mixed with much air its partial pressure is low, and a large proportion of it escapes condensation when cooled in the condenser. Under otherwise similar conditions the loss is, therefore, roughly proportional to the volume of

---

[1] Young, " Experiments on Fractional Distillation," *Journ. Soc. Chem. Industry,* 1900, **19,** 1072. Young and Fortey, " Fractional Distillation as a Method of Quantitative Analysis," *Trans. Chem. Soc.,* 1902, **81,** 752.

air in the still and still-head, and it is advantageous to use as small a flask as possible and to employ a still-head of as small a capacity as is consistent with efficiency.

Choice of Still-head.—A plain wide tube or one with spherical bulbs is the least satisfactory, but the " pear " still-head, owing to the diminished capacity of the bulbs and the increased efficiency, gives much better results.   Of all forms, the " evaporator " is the best, because the capacity is very small relatively to the efficiency, and the amount of condensed liquid in it is smaller than in any other equally efficient still-head ; moreover, almost the whole of the condensed liquid returns to the still at the end of the distillation.   With a liquid of low viscosity, like one of the lower paraffins, the quantity left in the still-head is almost inappreciable, and in other cases it may be reduced to a very small amount by disconnecting the apparatus, while hot, from the condenser, tilting the tube from side to side to facilitate the flow of liquid back to the still, and, if the original form of " evaporator " still-head is used, shaking out any liquid remaining in the funnels.

Estimation of Loss by Evaporation.—The following may be taken as an example of the estimation of loss by evaporation.   Mixtures of benzene and methyl alcohol, one with benzene, the other with methyl alcohol in excess (these liquids form a mixture of minimum boiling point), were distilled through an " evaporator " still-head of five sections of the original form, the distillation being stopped in each case when the middle point was reached.   The following results were obtained :—

TABLE 59

	Component in excess.	
	Benzene.	Methyl alcohol.
Weight of distillate	128·7	132·0
Weight of liquid in still	24·9	27·2
Total	153·6	159·2
Weight of mixture taken	154·2	160·1
Loss by evaporation and left in still-head .	0·6	0·9

When the benzene was in excess, it is certain that the amount of it left in the still-head did not exceed 0·1 gram, and the loss by evaporation was therefore estimated as 0·5 gram ; in calculating the composition this amount was added to the observed weight of distillate.   With methyl alcohol in excess the total loss was greater, but this more viscous liquid does not flow back so completely to the still, and the loss by evaporation was taken to be the same, 0·5 gram.   With an " evaporator " apparatus of five sections the loss by evaporation is usually from 0·3 to 0·5 gram.

Mixtures containing Two Components.—The following are examples of the distillation of mixtures of two liquids which separate normally into the original components.

TABLE 60

METHYL ALCOHOL AND WATER

Boiling points—Methyl alcohol, 64·7° ; water, 100° ; middle point, 82·35°.

I.—*Methyl alcohol in large excess*

Mixture taken.	Weight of distillate below middle point.	Percentage composition of mixture.		
		Found.		Taken.
		Uncorrected.	Corrected.	
Alcohol　90·9 Water　　24·4	Observed　90·5 Corrected　90·8	Alcohol　78·5 Water　　21·5	78·7 21·3	78·8 21·2
115·3		100·0	100·0	100·0

II.—*Water in large excess*

Mixture taken.	Weight of distillate below middle point.	Found Uncorrected.	Corrected.	Taken.
Alcohol　39·7 Water　161·5	Observed　33·9 Corrected　34·2	Alcohol　16·9 Water　　83·1	17·0 83·0	19·7 80·3
201·2		100·0	100·0	100·0

The first result is quite satisfactory, while the second is not, but it must be remembered that it is always difficult to separate the more volatile component of a mixture when present in relatively small quantity, and, in such a case, a second distillation is usually necessary. The first distillation was therefore continued until the temperature reached 100°, and the whole of the distillate, weighing 66·8 grams, was then redistilled and the double correction for loss by evaporation was applied.

The weight below the middle point was now 38·9, corrected 39·5, giving the percentage composition.

	Uncorrected.	Corrected.	Taken.
Alcohol	19·3	19·6	19·7
Water	80·7	80·4	80·3
	100·0	100·0	100·0

It will thus be seen that, by repeating the distillation, the result was as satisfactory as that given by a single distillation when the alcohol was in excess. Even without correcting for loss by evaporation the agreement is fairly good, but it is much improved by introducing the correction.

TABLE 61

ISOAMYL ALCOHOL AND BENZENE

Boiling points—Benzene, 80·2° ; isoamyl alcohol, 132·05° ; middle point, 106·1°.

Mixture taken.	Weight below middle point.	Percentage composition of mixture.		
		Found.		Taken.
		Uncorrected.	Corrected.	
Alcohol 26·6	Observed 85·55	Alcohol 23·8	**23·6**	**23·7**
Benzene 85·7	Corrected 85·85	Benzene 76·2	**76·4**	**76·3**
112·3		100·0	**100·0**	**100·0**

Here the agreement is very satisfactory.

One Component in Large Excess.—That a single distillation may be sufficient when the more volatile of two components is present in large excess, while two or more distillations are necessary when it is present in relatively small amount, is further shown by the following results.

A mixture containing 90 grams of benzene and 10 grams of toluene was distilled through an evaporator still-head of three sections.

Weight below middle point : observed,   89·6 ; corrected, 89·9.
Percentage of benzene in mixture : taken **90 0** ; found,   **89·9.**

When 100 grams of a mixture containing only 10 per cent. of benzene was distilled through the same still-head, very little came over below the middle point, 95·4°, and the quantity was too small to admit of fractional distillation. By twice redistilling all that came over below 110·6°, however, the weight below the middle point rose to 9·0 grams. Allowing a loss of 0·3 gram for each distillation the corrected weight would be 9·9 instead of 10 grams.

With a larger quantity, 250 grams, a fairly satisfactory result was obtained even with a " pear " still-head. In this case the distillate was divided into three fractions, and the following results were obtained :—

TABLE 62

	I.	II.	III.	IV.	V.
1. Below 95·4°	0	16·2	21·1	22·7	23·2
2. 95·4—104·7°	43·7	21·0	10·9	6·4	3·5
3. 104·7—110·5°	76·0	39·4	19·4	9·6	6·3
	119·7	76·6	51·4	38·7	33·0

The weight below the middle point is clearly approaching a limit, the increase being smaller each time ; allowing 0·3 gram for loss by evaporation in each distillation the last weight would be 23·2 + 1·5 = 24·7, and the percentage 9·9 instead of 10.

Advantages of Efficient Still-head.—A great saving of time is, however, effected and a more certain result is obtained by the use of a very efficient still-head. Thus, on distilling 300 grams of the above mixture through an 18-column Young and Thomas still-head, the first distillation gave 21·4 grams below the middle point, and a total of 76·1 below the boiling point of toluene. On redistillation of the 76·1 grams the weight below the middle point was 29·2 grams, so that with no correction for loss by evaporation the calculated percentage of benzene would be 9·7, and with the larger still-head it would be fair to allow 0·4 gram for each distillation, which would bring up the weight to 30·0 grams, and the percentage of benzene to 10·0.

Mixtures of Three Components.

—The fractionation of a mixture of methyl, ethyl, and propyl acetates with a plain vertical still-head one metre in length has already been referred to (p. 105), and some details have been given. With so large numbers of fractions and of fractional distillations, it would not be possible to distribute the loss by evaporation, and by transference from receiver to still, between the different fractions, but the total loss was ascertained in each fractionation, and we may calculate the percentages on the total quantity of material left at the end of each operation instead of on the original quantity taken.

The boiling points of the three esters are 57·1°, 77·15°, and 101·55° respectively, and the two middle points are therefore 67·1° and 89·35°. The percentage weights of distillate below 67·1°, from 67·1° to 89·35°, and above 89·35° were read from the curves (Fig. 35, p. 111) and are given below.

TABLE 63

Number of fractionation.	Percentage weight of distillate.		
	Below 67·1°.	From 67·1° to 89·35°.	Above 89·35°.
1	11·5	74·5	14·0
2	22·5	45·5	32·0
3	28·5	41·5	30·0
4	32·5	36·5	31·0
5	31·0	38·5	30·5
6	30·0	39·5	30·5
7	31·5	38·5	30·0
8	30·0	39·0	31·0
9	29·0	40·0	31·0
10	31·0	39·0	30·0
11	31·0	37·7	31·3
12	30·5	38·3	31·2
Mean of last 9 percentages . .	**30·7**	**38·6**	**30·7**
Percentage taken . .	**31·7**	**38·2**	**30·1**

It will be seen that the numbers remain nearly constant after the first three fractionations, and that the mean percentages calculated from the last nine distillations agree fairly well with those in the original mixture. The actual loss by evaporation must have been greatest for methyl acetate and least for propyl acetate, and the calculated percentages are too low for the first and too high for the second of these esters.

Advantages of Efficient Still-head.—With an evaporator still-head of eight sections the following results were obtained (p. 159).

Weight of methyl acetate taken   100
„   ethyl   „   „.   .   .   .   120
„   propyl   „   „   .   .   .   100
                                                     ———
                                                     320
                                                     ═══

Weight of distillate below first middle point, 97·95 ; corrected, 98·45.
Weight of distillate between first and second
  middle points   .   .   .   .   .   120·7 ;   „   120·8

Percentage composition of mixture.	Found.	Taken.
Methyl acetate . . .	**30·77**	**31·25**
Ethyl „ . . .	**37·75**	**37·50**
Propyl „ . . .	**31·48**	**31·25**
	100·00	100·00

The distillation was continued until the temperature reached the boiling point of propyl acetate. The residue in the still was weighed when cold, and the loss, due partly to evaporation and partly to the minute amount of liquid left in the still-head, was found to be 0·8 gram. It was assumed that of this loss 0·7 gram was due to evaporation, and that 0·5 gram was lost below the first middle point, 0·1 between the two middle points, and 0·1 above the second middle point.

The fractionation of the mixture was continued, and the results obtained in the second complete distillation are given below.

Weight below first middle point   .   .   .   .   .   99·5
Weight between first and second middle points .   . 118·25
Weight above second middle point   .   .   .   .   99·55
                                                            ———
                                                            317·3
        Total loss   .   .   .   .   2·7
                                                            ———
                                                            320·0
                                                            ═══

Here, again, it is impossible to distribute the loss correctly between the different fractions, and it is best to calculate the percentages on the total amount of material left at the end of the distillation.

Percentage composition of mixture.	Found.	Taken.
Methyl acetate . . .	**31·36**	**31·25**
Ethyl „ . . .	**37·27**	**37·50**
Propyl „ . . .	**31·37**	**31·25**
	100·00	100·00

The agreement is better than after a single distillation, and is very satisfactory.

Complex Mixtures.—The separation of isopentane and normal pentane (b.p. 27·95° and 36·3° ; middle point 32·15°) from a mixture containing also butanes, hexanes, and a very little pentamethylene has been already referred to (p. 113). Taking the weight of distillate between 27·95° and 36·3° as 100 in each case, the percentage coming over between 27·95° and 32·15° became roughly constant after the first distillation. The variation in this case was greater than with the esters (40—46, mean 42 per cent in 12 fractionations), but the difference between the boiling points of the components is only 8·35° against 20·15° and 22·4°.

It would appear, then, that in American petroleum, of the two paraffins, about 42 per cent consists of isopentane and 58 per cent of normal pentane. At the end of the fractionations 101 grams of pure isopentane and 175 grams of pure normal pentane were obtained, or 36·6 per cent of the isoparaffin ; but the loss by evaporation must have been greater, and there is never so good a recovery of the more volatile component.

Mixtures of Constant Boiling Point.—For the sake of brevity, an azeotropic mixture containing two components will be referred to in this chapter simply as a " binary " mixture, and an azeotropic mixture containing three components as a " ternary " mixture.

The quantity of a mixture of constant boiling point may be estimated by the distillation method in exactly the same way as that of a single substance. The methods of experiment and of calculation are similar in all respects.

Two examples may be given to prove this point ; binary mixtures of isopropyl alcohol and tertiary butyl alcohol respectively with water were mixed with excess of water and distilled with the following results :—

TABLE 64

ISOPROPYL ALCOHOL AND WATER

Boiling points : Binary mixture, 80·37° ; water, 100·0° ; middle point, 90·2°.,

Mixture taken.		Weight below middle point.	Percentage composition of mixture.	
			Found.	Taken.
Binary mixture .	57·7	Observed 57·3	Binary mixture . 74·05	74·15
Water . . .	20·1	Corrected 57·6	Water . . . 25·95	25·85
	77·8		100·00	100·00

[TABLE

TABLE 65

TERTIARY BUTYL ALCOHOL AND WATER

Boiling points : Binary mixture, 79·9° ; water, 100·0° ; middle point, 89·95°.

Mixture taken.	Weight below middle point.	Percentage composition of mixture.	
		Found.	Taken.
Binary mixture . 58·8	Observed 58·2	Binary mixture . 66·25	66·6
Water . . . 29·5	Corrected 58·5	Water . . . 33·75	33·4
88·3		100·00	100·0

The found percentages have been calculated in both cases from the corrected weights of distillate. The agreement is very good in the first case and satisfactory in the second.

DETERMINATION OF THE COMPOSITION OF AZEOTROPIC MIXTURES BY DISTILLATION

Binary Mixtures.—Since a mixture of constant boiling point behaves like a single substance on distillation, it is possible, if we know the composition of the mixture distilled, to calculate that of the binary mixture.

For a mixture of minimum boiling point, the ratio of the weight of the component not in excess in the original mixture to the corrected weight of distillate below the middle point is equal to the proportion of that component in the binary mixture.

In the case of a mixture of maximum boiling point, the ratio of the weight of the component not in excess to that of the residue after the middle point has been reached is equal to the proportion of that component in the binary mixture.

The following examples may be taken :—

TABLE 66

I. *Normal propyl alcohol and water, with the latter in excess.*

Boiling points : Binary mixture, 87·72° ; water, 100·0° ; middle point, 93·85°.

Mixture taken.	Weight below middle point.	Percentage composition of binary mixture.	
		Distillation method.	From specific gravity.
Alcohol . . 76·6	Observed 106·4	Alcohol . 71·8	71·69
Water . . 50·0	Corrected 106·7	Water . 28·2	28·31
126·6		100·0	100·00

The calculation is carried out as follows :—

Weight of propyl alcohol, 76·6 grams.

Weight of binary mixture = corrected weight of distillate = 106·7 grams.

Percentage of alcohol in binary mixture $= \dfrac{76·6 \times 100}{106·7} = 71·8$.

N

In the calculation of the composition from the specific gravity of the redistilled binary mixture, the necessary correction has been introduced for the contraction that occurs on mixing the components. In many cases the boiling point of the binary mixture is too near that of one of the components to allow of a determination of composition being made with that component in excess, but if the boiling point is greatly depressed one may frequently determine the composition even when the more volatile of the two original components is in excess.

Thus with methyl alcohol and benzene two separate determinations were made with the following results :—

TABLE 67

I. *Benzene in excess.*

Boiling points : Binary mixture, 58·34° ; benzene, 80·2° ; middle point, 69·25°.

II. *Methyl alcohol in excess.*

Boiling points : Binary mixture, 58·34° ; methyl alcohol, 64·7° ;
middle point, 61·5°.

Mixture taken.			Weight below middle point.			Percentage composition of binary mixture.		
	I.	II.		I.	II.		I.	II.
Alcohol	51·2	79·9	Observed	128·7	132·0	Alcohol	**39·6**	**39·5**
Benzene	103·0	80·2	Corrected	129·2	132·5	Benzene	**60·4**	**60·5**
	154·2	160·1					**100·0**	**100·0**

Non-miscible and Partially Miscible Liquids.—The method is applicable to liquids which are non-miscible or miscible within limits.

Thus, with isoamyl alcohol and water, which are partially miscible, the following results were obtained :—

TABLE 68

I. *Water in excess.*

Boiling points : Binary mixture, 95·15° ; water, 100·0° ; middle point, 97·6°.

II. *Isoamyl alcohol in excess.*

Boiling points : Binary mixture, 95·15° ; alcohol, 132·05° ; middle
point, 113·6°

Mixture taken.			Weight below middle point.			Percentage composition of binary mixture.		
	I.	II.		I.	II.		I.	II.
Alcohol	38·8	68·3	Observed	76·4	85·65	Alcohol	**50·5**	**50·3**
Water	69·5	42·7	Corrected	76·9	85·95	Water	**49·5**	**49·7**
	108·3	111·0					**100·0**	**100·0**

Ternary Mixtures.—When a mixture of three liquids gives rise, on distillation, to the formation of a ternary mixture of minimum boiling point, the separation may, theoretically, take place in twelve different ways, and, in addition to these, if the original mixture has the same composition as the ternary mixture, its behaviour on distillation would be precisely that of a pure liquid.

Ethyl Alcohol—Benzene—Water.—As an example we may consider mixtures of ethyl alcohol, benzene, and water. For convenience, the components are represented by the initial letters A, B, and W. The possible cases are as follows :—

	First fraction.	Second fraction.	Residue.		First fraction.	Second fraction.	Residue.
1	$A.B.W.$	$A.W.$	$W.$	7	$A.B.W$	...	$A.$
2	,,	$B.W.$	$W.$	8	,,	...	$B.$
3	,,	$A.W.$	$A.$	9	,,	...	$W.$
4	,,	$A.B.$	$A.$	10	,,	...	$A.B.$
5	,,	$B.W.$	$B.$	11	,,	...	$A.W.$
6	,,	$A.B.$	$B.$	12	,,	...	$B.W.$

13   Distillate $= A.B.W.$

The first six cases, and on redistillation of the first fraction the thirteenth, would be those commonly met with. Of the first six, the third is unrealisable in practice, owing to the very small difference between the boiling points of the second fraction ($A.W.$) and the residue (ethyl alcohol).

Mixtures, however, tending to separate in the other five ways specified were distilled in order to determine the composition of the ternary mixture.

For the calculation it is necessary to know not only the composition of the original mixture, but also that of the binary mixture forming the second fraction.

Data required.—The boiling points of all possible components and the percentage composition of the three binary mixtures are given in Table 69.

TABLE 69

	Boiling points.	Percentage composition.		
		$A.$	$B.$	$W.$
$W.$	$100\cdot0°$	...	...	100
$B.$	$80\cdot2$	...	100	...
$A.$	$78\cdot3$	100	...	...
$A.W.$	$78\cdot15$	$95\cdot57$	...	$4\cdot43$
$B.W.$	$69\cdot25$	...	$91\cdot17$	$8\cdot83$
$A.B.$	$68\cdot24$	$32\cdot36$	$67\cdot64$	...
$A.B.W.$	$64\cdot86$	...	...	...

The middle points are therefore as follows :—

TABLE 70

		Middle points.	
Fractions.		First.	Second.
I. *A.B.W.* ; *A.W.* ; *W.*.		71·55°	89·1°
II. *A.B.W.* ; *B.W.* ; *W.*.		67·05	84·6
IV. *A.B.W.* ; *A.B.* ; *A.*.		66·55	73·3
V. *A.B.W.* ; *B.W.* ; *B.*.		67·05	74·7
VI. *A.B.W.* ; *A.B.* ; *B.*.		66·55	74·2

Experimental Results.—In Table 71 are given—(*a*) the actual weight of the components in the mixtures distilled, (*b*) the weights of

TABLE 71

(*a*) *Mixtures taken.*

	I.	II.	IV.	V.	VI.
Alcohol	66·0	18·4	75·0	18·5	35·0
Benzene	74·2	120·0	108·0	160·1	148·3
Water	50·5	52·1	7·5	12·1	7·6
	190·7	190·5	190·5	190·7	190·9

(*b*) *Weights below and between the middle points.*

	I.	II.	IV.	V.	VI.
First.—Observed . .	99·5	94·9	100·6	97·1	111·6
Corrected . .	99·9	95·3	101·0	97·5	112·0
Second.—Observed . .	51·7	54·0	47·5	52·5	42·6
Corrected . .	51·8	54·1	47·6	52·6	42·7

(*c*) *Percentage composition of ternary mixture.*

	I.	II.	IV.	V.	VI.	Mean.
Alcohol . .	16·5	19·3	17·5	19·0	18·9	**18·2**
Benzene . .	74·3	74·2	75·1	73·4	74·3	**74·3**
Water . .	9·2	6·5	7·4	7·6	6·8	**7·5**
	100·0	100·0	100·0	100·0	100·0	**100·0**

distillate below and between the middle points, (*c*) the calculated percentages of the components in the ternary mixture.

Method of Calculation.—In calculating the composition of the ternary mixture it is assumed, as before, that the corrected weights of the two distillates are equal to those of the ternary and binary mixtures respectively which would be obtained if the separation were perfect. That being so, in case I., the weight of benzene in the ternary mixture is simply that in the original mixture ; the weight of alcohol is that taken, less the amount in the binary mixture, which can be calculated ; the weight of water is given by difference.

The composition of the ternary mixture was also directly determined, and it will be seen that the agreement with the mean value obtained by the distillation method is very satisfactory.

	Direct determination.	Distillation method.
Alcohol	18·5	18·2
Benzene	74·1	74·3
Water	7·4	7·5
	100·0	100·0

On the other hand, some of the individual values, notably those of alcohol and water in I., differ somewhat widely from the mean. The explanation of the rather large errors in this distillation is given below.

Cases to which the Distillation Method is not applicable.— When small quantities of alcohol are successively added to water, the boiling point is rapidly lowered ; the middle temperature between the boiling points of the pure components is, in fact, reached when the mixture contains 6·5 molecules per cent of ethyl alcohol. On the other hand, water must be added to alcohol until the mixture contains 25 molecules per cent before the boiling point rises 0·1° above that of pure alcohol, and with equal molecular proportions the rise of temperature is only 1·5°.[1]

Very similar results are obtained with normal hexane and benzene ; a mixture containing 16 molecules per cent of benzene boils only 0·1° higher than normal hexane, and the mixture which has the boiling point 74·6°, midway between those of the pure components, contains 79 molecules per cent of benzene.[2]

In both cases mixtures of minimum boiling point, very rich in the more volatile component, are formed so that the separation would be that of the mixture of constant boiling point from that component which is in excess.

Form of Boiling Point Composition Curve.—It is, however, found practically to be impossible in either case to separate the mixture of minimum boiling point even from the less volatile component, although the difference between their boiling points is considerable. In both cases the boiling point composition curve is very flat where the more volatile component is in large excess, and it is in such cases—when the curve is very flat at either one end or the other—that one at least of the components is exceedingly difficult to separate, and that the distillation method cannot be relied on for the determination of composition.

Ethyl Alcohol and Water.—Thus on distilling ethyl-alcohol-water

[1] Noyes and Warfel, " The Boiling Point Curves of Mixtures of Ethyl Alcohol and Water," *Journ. Amer. Chem. Soc.*, 1901, **23**, 463.

[2] Jackson and Young, " Specific Gravities and Boiling Points of Mixtures of Benzene and Normal Hexane," *Trans. Chem. Soc.*, 1898, **73**, 923.

mixtures, containing from 15 to 25 per cent by weight of water, through an 18-column dephlegmator and calculating the percentage of water in the mixture of constant boiling point in the usual way from the weight of distillate below the middle point, values from 7·6 to 8·0 instead of 4·43 per cent were obtained. Referring back to the calculation of the composition of the ternary ethyl-alcohol-benzene-water mixture from the first distillation (p. 180), if we take 7·8 as the percentage of water in the binary *A.W.* mixture, the calculated composition of the ternary mixture becomes :

Alcohol	18·2
Benzene	74·3
Water	7·5
	100·0

which agrees very well indeed with that observed.

General Conclusions.—In the great majority of cases the distillation method may be safely employed for the determination of the composition of a mixture which separates normally into its components, provided that a very efficient still-head be employed and that the distillation be carried out slowly. But it must be borne in mind that from a mixture of two liquids it is almost always more difficult to separate the more volatile component than the other, and therefore, if the original mixture contain a relatively very small amount of that component, a second distillation may be necessary, and a large quantity of the original mixture will be required in order to give a sufficient amount of distillate for a second operation. As regards the separation of three or more components from a mixture, it must be remembered that, as a general rule, the least volatile component is the easiest, while the intermediate components are the most difficult, to separate.

If a binary mixture of constant boiling point is formed, the composition of the original mixture may be determined if that of the mixture of constant boiling point is known ; or if the composition of the original mixture is known, that of the mixture of constant boiling point may be determined. The method may even be applied to the determination of the composition of a ternary mixture of constant boiling point.

It appears to be only when the separation of the components (either simple substances or mixtures of constant boiling point) by distillation is exceedingly difficult that the method is inapplicable.

The " middle point " method was carefully investigated by M[lle] J. Reudler[1] with good results ; it has been used for the determination of the composition of binary and ternary mixtures of constant boiling point by Wade,[2] Atkins and Wallace,[3] Hill,[4] and Merriman[5] ; it has frequently been employed by Atkins, and Lecat[6] made use of it in the

---

[1] " Eenige Opmerkingen over Sydney Young's Distillatieregel," *Versl. Amst.*, 1903–1904, **12**, 968 [Eng. trans. *Proc. Amst.*, **6**, 807].
[2] *Trans. Chem. Soc.*, 1905, **87**, 1656.   [3] *Ibid.*, 1912, **101**, 1179 and 1958.
[4] *Ibid.*, 1912, **101**, 2467.   [5] *Ibid.*, 1913, **103**, 1790.
[6] " La Tension de vapeur des mélanges de liquides, l'azéotropisme," Brussels, 1918.

case of more than 350 out of the 1100 azeotropic mixtures examined by him. Lecat states on p. 56 that it is only when the separation of the fractions by distillation is very difficult that the method becomes inapplicable.

The method has also been found useful for the analysis of commercial products such as crude benzene, toluene, etc.

In the case of very complex mixtures, such as petrol, no attempt is made to estimate the relative amounts of individual hydrocarbons present, but for commercial purposes 100 c.c. of the " spirit " is slowly distilled under specified conditions usually without an efficient still-head. The initial and final temperatures are noted and the weights or volumes of distillate coming over between these limits, the range of temperature for each fraction being usually 10°. Some chemists, however, consider that an improved still-head should be used, and Washburn recommends a more complete separation of light oils by distillation through a combined Hempel and constant temperature still-head.[1]

[1] Rittman and Dean, "The Analytical Distillation of Petroleum," *U.S. Bureau of Mines*, Washington, 1916, *Bull*, 125, *Petroleum Technology*, 34; Lomax, "Testing and Standardisation of Motor Fuel," *J. Inst. Pet. Tech.*, 1917-18, p. 6; Anflogoff, "Distillation Test of Petrol," *J. Soc. Chem. Ind.*, 1918, **37**, 21 T; Phillips, "Some Laboratory Tests on Mineral Oils," *J. Inst. Pet. Tech.*, 1919; *J. Soc. Chem. Ind.*, 1919, **38**, 393 R; Dean, "Motor Gasoline Testing," *U.S. Bureau of Mines Technical Paper*, 214, *Petroleum Technology*, 52; Luynas-Bordas Apparatus as used in France, *J. Soc. Chem. Ind.*, 1920, **39**, 220 A.; F. M. Washburn, "Constant Temperature Still-head for Light Oil Fractionation," *J. Ind. and Eng. Chem.*, 1920, **12**, 73.

CHAPTER XVII

METHODS BY WHICH THE COMPOSITION OF MIXTURES OF CONSTANT BOILING POINT MAY BE DETERMINED

1. Distillation Method.—In the last chapter it has been shown how the composition of an azeotropic mixture may be ascertained from the weight of distillate that comes over below the middle point, when a mixture of known composition is distilled. This method is generally but not universally applicable.

There are several other methods by which the composition of a mixture of constant boiling point may be determined.

2. By Separation of Pure Mixture.—The most accurate method —applicable, however, only to those mixtures for which the first method can be employed—is to separate the mixture of constant boiling point in a pure state by fractional distillation and to determine its composition either (*a*) by chemical analysis, (*b*) by the removal of one component, (*c*) from its specific gravity, (*d*) from its refractive power, (*e*) from its rotatory power or from some other physical property.

(*a*) If one of the substances is an acid, or a base, the ordinary methods of volumetric analysis may be conveniently employed, or if one component contains a halogen, sulphur, etc., the amount of that element may be determined; but this method is not, as a rule, to be recommended.

(*b*) When one component is easily soluble in water, and the other insoluble, or nearly so, for example alcohol and benzene, a fairly accurate result may be obtained by shaking the mixture with water in a separating funnel and washing the insoluble component once or twice with water. The volume of this component at a known temperature may then be ascertained or its weight determined, but there is inevitably some loss by evaporation and by adhesion to the sides of the separating funnel and to the solid dehydrating agent, if this is added. As a rule, also, a little of the component that is insoluble in water remains dissolved by the aqueous solution of the other constituent, and to obtain an accurate result it would be necessary to distil this solution, and to treat the first small portion of the distillate with more water in order to separate the remainder of the insoluble component.

This method was employed for the direct determination of the composition of the ternary mixture of ethyl alcohol, benzene, and water

(p. 181), the benzene being determined in the manner described above and the alcohol from the specific gravity of the aqueous solution.[1] It was frequently employed by Lecat.

(c and d) As the specific gravities and refractive powers of mixtures are not usually strictly additive properties, it is almost always necessary to determine the values for a prepared mixture of about the same composition as that which boils at a constant temperature or, better, to determine the values for a series of mixtures in order to find what correction must be applied. Such series of determinations of specific gravity have been made by different observers in the case of mixtures of the lower alcohols with water[2] and by Brown[3] for some other pairs of liquids. The specific refractive powers have been determined for several series of mixtures by Lehfeldt,[4] by Zawidski,[5] and others.

(e) The composition of several binary mixtures was determined by Lecat from their rotatory power.

3. Method of Successive Approximations. — Mixtures of different composition may be distilled, and, by successive approximations, that mixture may finally be made up which distils (a) at a constant temperature, or (b) without change of specific gravity.

(a) This method was employed by Roscoe and Dittmar[6] in the case of mixtures of strong acids with water, and by Ryland[7] to ascertain the approximate composition of the large number of mixtures of constant boiling point examined by him.

(b) If the boiling point of the mixture differs only slightly from that of either of the pure components, observations of the temperature would be useless, but we may find what mixture gives a distillate of the same specific gravity (or refractive power) as itself, or, better, collecting the distillate each time in three or four fractions, we may proceed until the first and last fraction have the same specific gravity. The last method has been employed in the case of ethyl alcohol and water,[8] and the following results were obtained with the two last mixtures :—

TABLE 72

I. Weight of fraction.	Sp. gr. at 0°/4°.	II. Weight of fraction.	Sp. gr. at 0°/4°.
23·6	0·81936	21·2	0·81946
73·4	...	55·0	...
27·6	0·81927	26·0	0·81953
		15·0	...
		26·1	0·81954

[1] Young and Fortey, "The Properties of Mixtures of the Lower Alcohols with Benzene and with Benzene and Water," *Trans. Chem. Soc.*, 1902, **81**, 739.
[2] Young and Fortey, "The Properties of Mixtures of the Lower Alcohols with Water," *ibid.*, 1902, **81**, 717.
[3] F. D. Brown, "Theory of Fractional Distillation," *ibid.*, 1879, **35**, 547; "On the Distillation of Mixtures of Carbon Disulphide and Carbon Tetrachloride," *ibid.*, 1881, **39**, 304.
[4] Lehfeldt, "On the Properties of Liquid Mixtures," Part II, *Phil. Mag.*, 1898 [V] **46**, 42.
[5] Zawidski, "On the Vapour Pressures of Binary Mixtures of Liquids," *Zeitschr. physik. Chem.*, 1900, **35**, 134.
[6] Roscoe and Dittmar, *Quart. Journ. Chem. Soc.*, 1860, **12**, 128 ; Roscoe, *ibid.*, 1861, **13**, 146 ; 1862, **15**, 270 ; *Proc. Roy. Soc.*, 1862, **11**, 493.
[7] Ryland, "Liquid Mixtures of Constant Boiling Point," *Amer. Chem. Journ.*, 1899, **22**, 384.   [8] Young and Fortey, *loc. cit.*

In the first case, the last fraction has a lower specific gravity than the first, showing that alcohol was in excess ; in the second case it is the first fraction which has the lower specific gravity and therefore there was excess of water in the still. It is clear that the specific gravity of the mixture that distils without change of composition must

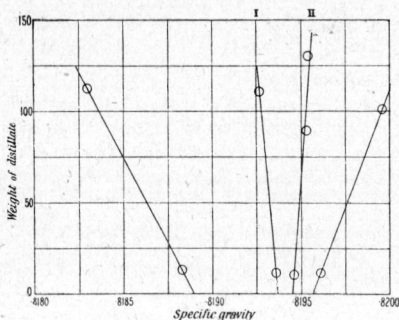

FIG. 69.—Ethyl alcohol an water.

be between those (0·81936 and 0·81946) of the first fractions in these distillations.

Mapping the specific gravities as abscissae against the weights of distillate as ordinates in each case, it is found that the lines slope almost equally, the first to the left and the second to the right (I. and II., Fig. 69), and it may therefore be assumed that the required specific gravity is 0·81941, the mean of the other two. If the two lines are produced, they intersect each other at a point between 0·81941 and 0·81942.

According to Mendeléeff's tables, the percentage of alcohol in a mixture which has the specific gravity 0·81941 at 0°/4° is 95·57.

Wade and Merriman [1] adopted the same method, the only modification being the substitution of algebraical for graphical interpolation. From their distillations under normal atmospheric pressure they found that the mixture of minimum boiling point contained 95·59 per cent of alcohol.

In calculating the composition from Mendeléeff's data they used a differential method of interpolation which is probably more accurate than the graphical method employed by Fortey and Young. They find that the specific gravity 0·81941 corresponds to 95·62 per cent of alcohol. In any case the agreement is very satisfactory. The method of successive approximatiors was employed by Lecat in the majority of cases, the results, however, being usually verified by other methods.

4. Graphically from Vapour Pressures or Boiling Points.—

If the vapour pressures at constant temperature, or the boiling points under constant pressure, of a series of mixtures of known composition have been determined, these values may be mapped against the percentages of one of the components, and the percentage corresponding to the maximum or minimum pressure or temperature can then be read off. The pressure- (molecular) composition curve for carbon disulphide and methylal [2] is shown in Fig. 70, but it will be seen that while the maximum pressure can be read with considerable accuracy the corre-

---

[1] " Influence of Water on the Boiling Point of Ethyl Alcohol at Pressures above and below the Atmospheric Pressure," *Trans. Chem, Soc.*, 1911, **99**, 997.

[2] The molecular weights of these two substances are equal and the molecular percentages are therefore equal to the percentages by weight.

sponding percentage of carbon disulphide can only be roughly estimated.
The same objection applies to the boiling-point composition curve.

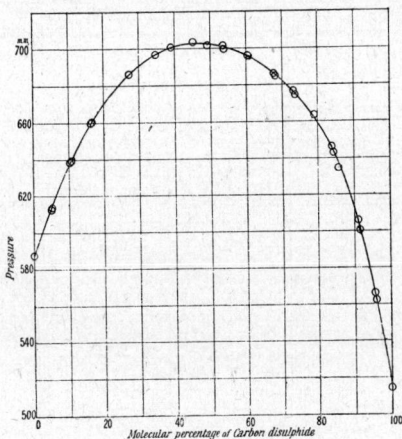

FIG. 70.—Carbon disulphide and methylal.

5. Graphically from Composition of Liquid and Vapour.—
If the relative composition of vapour and liquid has been determined
for a series of mixtures, the composition of the mixture of constant
boiling point may be ascertained in various ways.

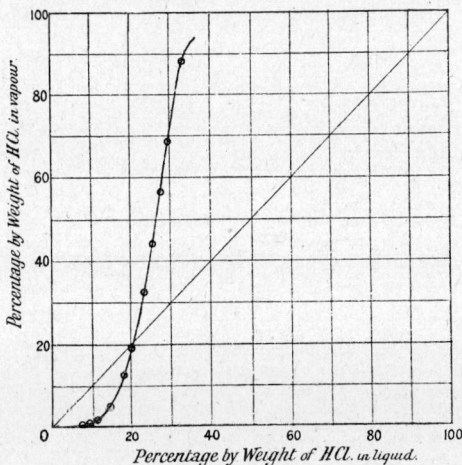

FIG. 71.—Hydrogen chloride and water.

(a) The percentages by weight m or the molecular percentages M
of one component in the liquid may be plotted against the percentages
m' or M' of the same component in the vapour.[1]

(b) The ratios R of the weights, or of the number of grammolecules,
of the two components A and B in the liquid may be mapped against
the corresponding ratios R' in the vapour.

[1] Rayleigh, "On the Distillation of Binary Mixtures," *Phil. Mag.*, 1902 [VI], **4**, 521.

(c) The logarithms of these ratios may be plotted in the same way.[1] Whichever method is adopted, the composition will be given by that point on the curve at which the ordinate and abscissa have the same value or, in other words, by the point of intersection of the curve with the straight line corresponding to equal values of m and m', M and M', R and R', or log R and log R'.

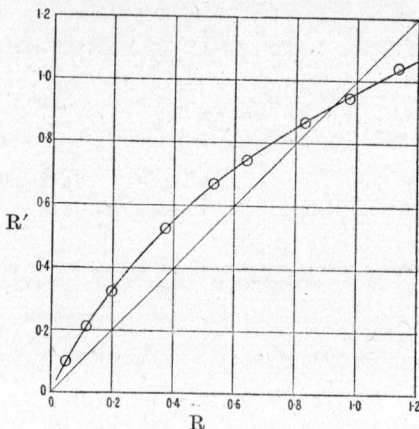

FIG. 72.—Carbon disulphide and methylal.

As an example of the first method, Lord Rayleigh's determinations of the composition of liquid and vapour for mixtures of hydrogen chloride and water (Fig. 71) may be mentioned ;[2] Vrevskij [3] employed this method in the case of aqueous solutions of some of the alcohols. For the second and third methods we may take the

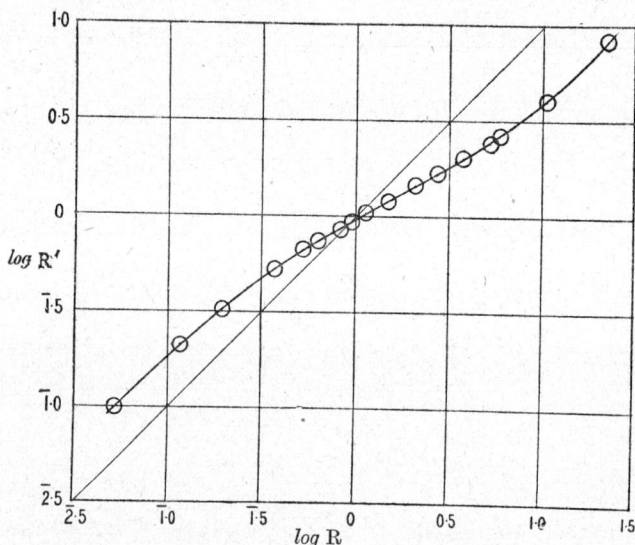

FIG. 73.—Carbon disulphide and methylal.

results obtained by Zawidski for mixtures of carbon disulphide and methylal (Figs. 72 and 73).

6. Graphically by means of Brown's Formula.—The relative number of molecules (or the relative weights) of the components in the

[1] Lehfeldt, loc. cit.   [2] Rayleigh, loc. cit.
[3] Vrevskij, Zeitschr. physik. Chem., 1913, **83**, 551.

liquid M_A and M_B and in the vapour M'_A and M'_B may be calculated from the experimental observations and the values of $\dfrac{M'_B}{M'_A} \cdot \dfrac{M_A}{M_B}$ plotted against the percentage number of molecules (or percentages by weight) of one component.

The percentage corresponding to $\dfrac{M'_B}{M'_A} \cdot \dfrac{M_A}{M_B} = 1$ is that required.

Here, again, we may make use of Zawidski's data for carbon disulphide and methylal (Fig. 74).

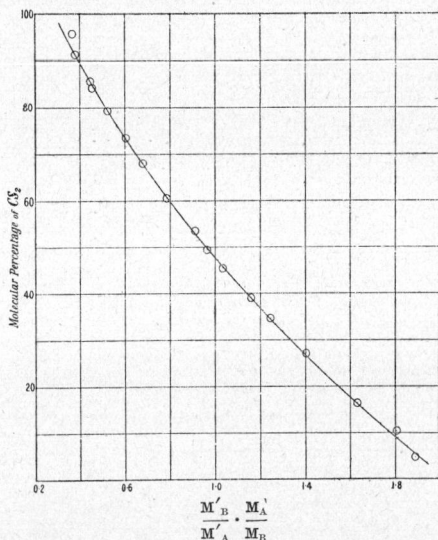

FIG. 74.—Carbon disulphide and methylal.

Results obtained.—For benzene and ethyl alcohol the results given in Table 73 have been obtained.

TABLE 73

Method.	Observers.	Percentage of benzene by weight.	Temperature.
1.	Young and Fortey	67·55	Boiling point under normal pressure.
2c.	,, ,,	67·64	
2b.	Ryland	68·1	
2b.	,,	72·1	50–51°
5c.	Lehfeldt	71·3	50°
6.	,,	71·3	50°

When the mixture that distils without change of composition boils at almost exactly the same temperature as one of the components, it is probable that methods 5, 6, and 3b are the only ones that can be relied upon, and it is on the first and second of these methods that the conclusion is based (p. 92) that benzene forms a mixture of constant boiling point with carbon tetrachloride, while the exact composition of the ethyl-alcohol-water mixture has been determined by method 3b.

CHAPTER XVIII

INDIRECT METHOD OF SEPARATING THE COMPONENTS OF A MIXTURE OF
CONSTANT BOILING POINT

Distillation after the Addition of a Third Substance.—It has
been pointed out that there are many cases in which the two components
of a mixture cannot be separated by fractional distillation, owing to the
formation of a mixture of constant boiling point. It is, however, sometimes possible to eliminate one of them by adding a third substance and
then distilling the mixture.[1]

Formation of Binary Mixture of Minimum Boiling Point.
—Suppose, for example, that we have a mixture of isobutyl alcohol and
benzene containing 10 per cent by weight of the alcohol. The components cannot be separated by fractional distillation, because a mixture
of minimum boiling point (79·93°), containing 9·3 per cent of isobutyl
alcohol, is formed. Neither can they be satisfactorily separated by
treatment with water, owing to the fact that the solubility of isobutyl
alcohol in benzene is greater than in water.

If, however, a little water be added and the mixture be distilled,
the first fraction will consist of the binary mixture of benzene and water,
boiling at 69·25° and containing 91·2 per cent of benzene.

In order to remove the benzene, all that is necessary is to add water
to the mixture in the ratio of 8·7 to 100 parts by weight and to distil
with an efficient still-head. The mixture will then tend to separate
into (a) the binary benzene-water mixture (b.p. 69·25°), and (b) pure
isobutyl alcohol (b.p. 108·05°). Here the difference between the boiling
points is considerable, and the separation is an easy one.

If too much water were added we should have, as an intermediate
fraction, the binary alcohol-water mixture which boils at 89·8°, and
contains 66·8 per cent of the alcohol.

If, on the other hand, too little water were added, only a part of the
benzene would be carried over with it and the remainder would form the
binary benzene-alcohol mixture ; the residue in either case would consist of pure isobutyl alcohol.

The method has been employed in a number of cases by Golodetz.[2]

---

[1] Young, " The Preparation of Absolute Alcohol from Strong Spirit," *Trans. Chem. Soc.*,
1902, **81**, 707 ; Young and Fortey, " The Properties of Mixtures of the Lower Alcohols with
Benzene and with Benzene and Water," *ibid.*, 1902, **81**, 717.

[2] Golodetz, *J. Russ. Phys. Chem. Ges.*, 1911, **43′**, 1041 ; *Chem. Zentr.*, 1912, **1**, 69.

Formation of Ternary Mixture of Minimum Boiling Point.

—The substance added sometimes forms a ternary mixture of constant boiling point with the two components of the original mixture, but the relative weights of these components in the ternary mixture differ from those in the binary mixture of constant boiling point which they themselves form.

Tertiary Butyl Alcohol and Water with Benzene.—Take, for example, the case of tertiary butyl alcohol and water. This alcohol is a crystalline solid, melting at 25·53° and boiling at 82·55° ; when liquefied, it is miscible with water in all proportions and forms with it a mixture of constant boiling point (79·91°) containing 88·24 per cent of the alcohol.

The difference between the boiling point of the binary mixture and of the alcohol is only 2·64°, and the separation of the last traces of water, or rather of the binary mixture, from the alcohol by distillation is difficult. It was not, in fact, found possible to obtain the alcohol quite pure by this method, the highest melting point of the distilled alcohol observed being 25·25°, and the boiling point 82·45°.

Fractional crystallisation gave a better result, the melting point of the recrystallised alcohol being 25·43°.

Eventually, however, it was found that the last traces of water could best be removed by distillation with benzene, both the melting and boiling points (given above) of the residual alcohol being higher than those of the product purified by recrystallisation. Table 74 gives the boiling points and the composition of the binary and ternary mixtures and the boiling points of the single components.

TABLE 74

	Boiling point.	Percentage composition.		
		Alcohol.	Benzene.	Water.
Water	100·0°	...	...	100
Tertiary butyl alcohol . .	82·55	100	...	...
Benzene	80·2	...	100	...
Alcohol-water	79·9	88·24	...	11·76
Alcohol-benzene . . .	73·95	36·6	63·4	...
Benzene-water	69·25	...	91·17	8·83
Alcohol-benzene-water . .	67·30	21·4	70·5	8·1

It will be seen that the ratio of water to alcohol in the binary mixture $=\dfrac{11·76}{88·24}=0·133$ and in the ternary mixture $=\dfrac{8·1}{21·4}=0·379$, so that the latter contains nearly three times as much water, relatively to the alcohol, as the binary mixture does.

Not only can we remove the last traces of water from the nearly pure alcohol by means of benzene ; but it is also possible to obtain the pure alcohol from the binary alcohol-water mixture.

Suppose that we start with 100 grams of this mixture ; the water in it, 11·76 grams, would require $\dfrac{11\cdot76 \times 70\cdot5}{8\cdot1} = 102$ grams of benzene for conversion into the ternary mixture, and if a perfect separation could be effected by a single distillation of the mixture, 145 grams of the ternary mixture and 57 grams of alcohol would be obtained. In practice, however, we should get a rather smaller amount of the ternary mixture, a little binary alcohol-benzene mixture, and a residue of alcohol, the second and third fractions still containing a little water.

It is better to add a larger amount of benzene, say 125 grams, in the first place, when the quantity of the second fraction (alcohol-benzene) will be increased and the smaller amount of residual alcohol will be obtained free from water in a single operation.

Results of Distillation.—In an actual experiment, 117·5 grams of the alcohol-water mixture (containing 103·7 grams of tertiary butyl alcohol and 13·8 grams of water) with 145 grams of benzene were distilled through a 5-section " evaporator " still-head of the original form, and the following fractions were collected :—

TABLE 75

	Temperature range.	Weight.	Theoretical composition.		
			Alcohol.	Benzene.	Water.
1.	67·3—70·6° . . .	169·3	36·2	119·4	13·7
2.	70·6—78·2 . . .	39·9	14·6	25·3	
3.	78·2—82·55 . . .	18·9 ⎫	52·7		
	Residue, alcohol . .	33·8 ⎭			
	Loss	0·6			
		262·5			

The residue solidified on cooling.

If the separation had been complete, the fractions would have had the composition stated in Table 75, but the first fraction must really have contained rather more alcohol and less benzene and water, and the second rather more benzene and a very little water. The third fraction must have contained some benzene with the dry alcohol.

Treatment of " Fractions."—The first fraction separated into two layers, the lower one consisting chiefly of alcohol and water with a little benzene, the upper one of benzene containing some alcohol and a little water. By adding more water in a separating funnel, running off the aqueous alcohol and washing the benzene repeatedly with small quantities of water to extract the remaining alcohol, a dilute aqueous solution of the alcohol could be obtained almost free from benzene, and by fractional distillation almost the whole of the alcohol could be recovered in the form of the binary alcohol-water mixture of constant boiling point.

Theoretically, the weight of this mixture would be 41·0 grams, and to remove the water 41·8 grams of benzene would be required. But the second fraction contains about 25 grams of benzene, and it would be necessary, in practice, to add about 35 grams more, making altogether about 60 grams.

On mixing together the recovered alcohol-water mixture, the second fraction, and the additional benzene we should have a liquid of approximately the following composition :—

Alcohol	50·8
Benzene	60·3
Water	4·8
	115·9

The mixture should now be distilled, the first two fractions collected as before, and the distillation stopped when the second middle point is reached ; fraction 3 from the first distillation should then be added to the residue in the still and the distillation continued. The results would, theoretically, be those given in Table 76.

TABLE 76

Temperature range.	Weight.	Theoretical composition.		
		Alcohol.	Benzene.	Water.
1. 67·3—70·6°	59·3	12·7	41·8	4·8
2. 70·6—78·2	29·2	10·7	18·5	
3. 78·2—82·55	18·9 ⎫	46·3		
Residue of alcohol	27·4 ⎭			

Amount of Alcohol recovered.—No doubt the quantity of alcohol actually recovered would be somewhat less than this, but it should be at least 20 grams, giving a total of, say, 54 grams out of 103·7. Moreover, the remainder of the alcohol, except the small amount actually lost by evaporation, could be recovered in the form of the alcohol-water mixture of constant boiling point.

Advantage of Larger Quantities.—Working with larger quantities the result would be much better, for fraction 3 remains nearly the same whatever the quantity distilled, and fraction 2 need not be made much larger.

Thus, on distilling 300 grams of the alcohol-water mixture, containing 264·7 grams of alcohol, with 340 grams of benzene, the weights of the fractions would be 435·6, 52·1 and 21 respectively ; leaving 131·3 grams of pure alcohol, or nearly half the total quantity. Moreover, the second distillation would give a much more satisfactory result. Recovering the alcohol as the binary mixture with water, adding fraction 2 and 95 grams of benzene, and distilling in the same manner as with the smaller quantity, the weights of the fractions would be 153·1, 31·7, and 21·0 respectively, with a residue of 68 grams of alcohol. The first fraction would also be large enough for a third operation which would yield, say, an additional 20 grams of pure alcohol. It would, in fact, be possible to recover about 215 out of the 264·7 grams of alcohol in the pure state, with very little actual loss.

Alcohols—Water—Benzene.—It has been pointed out that the monhydric aliphatic alcohols may be regarded, on the one hand, as alkyl

O

derivatives of water and, on the other, as hydroxyl derivatives of the paraffins ; and that, as the complexity of the alkyl group increases, the properties of the alcohols recede from those of water and approach those of the paraffins or of benzene.

Methyl Alcohol.—Methyl alcohol and water may be separated from each other without difficulty by distillation with an efficient still-head, because their properties are so similar, and their boiling points so far apart that a mixture of minimum boiling point is not formed, and, indeed, the boiling point-molecular composition curve is nowhere nearly horizontal.

Methyl alcohol and benzene cannot be separated from each other by distillation, because their properties are so dissimilar that a mixture of minimum boiling point is formed, and, as its boiling point is much lower than that of the binary benzene-water mixture, no separation can be effected by adding water and distilling. A ternary mixture. does not come over, but the first fraction still consists of the benzene-alcohol mixture of constant boiling point.

On the other hand the methyl alcohol may readily be extracted from its mixture with benzene by shaking with water, because, although the alcohol is miscible in all proportions both with benzene and with water, it resembles water much more closely, and its solubility in that substance may perhaps be said to be greater than in benzene.

The greater the molecular weight of an alcohol, or, in the case of isomers, the higher the boiling point, the more difficult is the extraction of the alcohol from a solution in benzene by means of water ; with isobutyl alcohol the process is very slow indeed.

Ethyl, Isopropyl, Normal Propyl, and Tertiary Butyl Alcohol.—Ethyl, isopropyl, normal propyl, and tertiary butyl alcohol all form binary mixtures of minimum boiling point both with water and with benzene. Pure ethyl alcohol cannot be obtained, even from its very strong aqueous solution, by distillation because its boiling point is so very little higher than that of the alcohol-water mixture ; and, on account of the similarity of the properties of the two substances, dehydrating agents act in a very similar manner on them, and it is only under very special conditions and, apparently, only with one dehydrating agent— freshly ignited lime—that the last traces of water can be removed.

Each of the four alcohols, however, forms a ternary mixture of minimum boiling point with benzene and water, and the latter substance may be removed from the strong alcohols by distillation with benzene.

The method has been employed since 1908 by Kahlbaum of Berlin for the production of absolute ethyl alcohol from strong spirit.

Normal hexane gives very good results, but is less easily obtainable than benzene. Lecat has tried cyclohexane as a substitute, and Chavanne [1] has investigated the behaviour of a number of substances when distilled with ethyl alcohol and water and has observed the formation of several ternary mixtures of minimum boiling point.

---

[1] Chavanne, " Sur une application de la méthode de Young pour la préparation de l'alcool absolu," *Bull. Soc. Chim. de Belg.*, 1913, **27**, 205. See also *Compt. Rend.*, 1914, **158**, 1698.

Isobutyl Alcohol.—Isobutyl alcohol does not form a ternary mixture of minimum boiling point with benzene and water, and when a mixture of these three liquids is distilled, the first portion of the distillate consists, as has been stated, of the benzene-water mixture which boils at 69·25°.

Higher Alcohols.—Benzene may, in fact, be removed from isobutyl alcohol or any other of higher boiling point by adding sufficient water to form the binary benzene-water mixture and distilling with an efficient still-head ; or, conversely, water may be removed by adding the requisite amount of benzene and distilling the mixture.

General Statement.—Thus, it is possible to extract water from the alcohols—except methyl alcohol—by adding benzene and distilling, and to extract benzene from the higher alcohols by adding water and distilling.

Atkins [1] has used this method successfully for the dehydration of certain solid substances such as laevulose. Absolute alcohol is first added to the moist laevulose in a flask and then benzene, preferably in excess. The flask, provided with an efficient still-head, is then heated over a water-bath and the distillate is collected. Almost the whole of the water comes over in the first turbid fraction (ternary alcohol-water-benzene mixture, b.p. 64·85°) ; the next fraction consists of the alcohol-benzene binary mixture boiling at 68·25°, and the rest of the water is carried over with this ; the residual liquid is benzene free from water and alcohol. The greater part of this is distilled off, and the small amount remaining in the flask is removed by means of a current of dry air. The method is especially useful when the solid is liable to soften, melt, or decompose at a temperature sufficiently high for the water to be expelled by direct heating.

[1] Atkins, " Preparation of Anhydrous Solids," *Trans. Chem. Soc.*, 1915, **107**, 916.

CHAPTER XIX

GENERAL REMARKS

Purposes for which Fractional Distillation is required. Interpretation of Experimental Results. Choice of Still-head,
Number of Fractions, etc.

Purposes for which Fractional Distillation is required.
—Fractional distillation may be employed for various purposes, of which
the following are among the most important :—

1. The isolation of a single substance from a mixture with the
smallest possible loss.

2. The separation of the components from a mixture of known
qualitative composition.

3. The determination of the quantitative composition of a mixture
of which the qualitative composition is already known, or the determination of the quantitative composition of a mixture of constant
boiling point.

4. A general study of the qualitative and quantitative composition
of a complex mixture when only the general nature of the chief components is known, or when it is definitely known that certain substances
are present but it is not known what others there may be.

1. The Separation of a Single Substance from a Mixture with the Smallest Possible Loss

When a pure substance is to be isolated from a mixture, the procedure will be somewhat simplified if the true boiling point of the chief
component is already known.

In any case a careful record should be made of the temperature
range and weight of each fraction, and we may then either calculate the
values of $\Delta w/\Delta t$ (p. 103) or plot the total weights of distillate against
the final temperatures of the fractions.

Much time will be saved if an improved still-head is employed ; for
liquids of high boiling point, the " pear," and for volatile liquids the
" evaporator " apparatus may be especially recommended.

Interpretation of Results. First Case.—If the results obtained by the distillation of the liquid under examination give a curve

similar in form to Fig. 75, we may conclude that there is present a relatively small quantity of some impurity of much higher boiling point, but that impurities more volatile than the chief component are absent.

As the temperature shows no appreciable rise until towards the end of the distillation, the first fraction will have a perfectly constant boiling point when redistilled, and we may conclude that it will most probably consist of the pure substance required.   In the second and subsequent fractionations the greater part of the distillate from the first fraction may, in each case, be taken as most probably pure and will not require to be redistilled.   The separation of the liquid of constant boiling point will in such a case as this be an easy one.

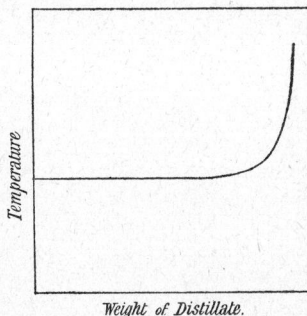

Weight of Distillate.

FIG. 75.

Proofs of Purity.—Even, however, if the true boiling point of the pure substance is known, and if the observed boiling point agrees with it, we cannot be absolutely sure of the purity of the liquid unless we know—from the method of preparation—that there is no possibility of the formation of a mixture which distils without change of composition at practically the same temperature as the pure liquid.   If we can be sure that the original mixture contains only substances which are very closely related to each other, such as members of a homologous series, the problem will be much simplified, for it may then be concluded with certainty that no mixture of constant boiling point can be formed.

If any doubt exists and if the specific gravity of the pure liquid is accurately known, it is best to determine that of the distillate : if both the boiling point and the specific gravity of the distillate agree with those of the pure substance, we may conclude with confidence that the distillate is really pure.   Instead of the specific gravity, the refractive index, the vapour density, the melting point (if the liquid can be easily solidified), or some other physical constant may be determined for comparison, or a chemical analysis of the liquid may be made.

Hexane and Benzene.—Suppose, for example, that the original liquid consisted chiefly of normal hexane, and that we were unacquainted with the previous history of the specimen ; there would be the possibility of the presence of some benzene.   A mixture of normal hexane and benzene containing as much as 10 per cent of the aromatic hydrocarbon boils at almost exactly the same temperature as pure hexane, and therefore the presence of the benzene could not be detected, and the impurity could not be removed, by fractional distillation.   The high specific gravity of the distillate would, however, show that some other substance besides normal hexane was present.

If, on the other hand, it were known that the hexane had been prepared synthetically from pure propyl iodide by the action of sodium, there would be no possibility of the presence of benzene.

Ethyl Alcohol and Water.—Ethyl alcohol containing, say, 15 or 20 per cent water would behave in the manner indicated by the same curve (Fig. 75), but, in this case, even with the most efficient still-head, we should not have pure alcohol in the first part of the distillate, nor even the pure mixture of constant boiling point, but a mixture containing at least 5 and probably as much as 7 or 8 per cent of water, for the mixture of constant boiling point which contains 95·6 per cent of alcohol is extremely difficult to separate from water, although there is a wide difference between the two boiling points.

In this case, again, the boiling point of the binary alcohol-water mixture is so slightly lower than that of pure alcohol that the reading of the thermometer—unless very accurate—would hardly be sufficient to distinguish with certainty between the two.

Isopropyl Alcohol and Water.—With isopropyl alcohol and water, similar results would be obtained except that the difference between the boiling points of the pure alcohol and of the binary mixture is sufficiently great for the two to be distinguished without difficulty by the observed temperature, and also that the binary mixture can be separated in a pure state from water.

Second Case.—Figure 76 represents the separation of a liquid of constant boiling point from a mixture which contains only much more

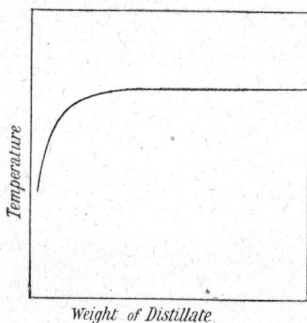

Weight of Distillate

Fig. 76.

volatile impurities. The separation is an easier one than the last since it is the least volatile component that is to be isolated. If the temperature remains constant for a considerable time and shows no rise whatever at the end of the distillation, the last portion need not be redistilled, and in subsequent fractionations the distillation may be stopped as soon as the maximum temperature has been reached, and the residue may be taken as pure.

Here, again, it is possible that we may be dealing with a mixture of maximum boiling point as, for example, in the separation of the binary mixture of chloroform and methyl acetate from a slight excess of the ester, but it is less probable, for mixtures of maximum boiling point are not met with so frequently as those of minimum boiling point. It is also possible, though less likely, that we might have a mixture of minimum boiling point containing neither substance in excess but contaminated with a more volatile impurity.

Third Case.—A curve like that in Fig. 77, would result from the distillation of a liquid containing an impurity, the boiling point of which was not much higher than that of the chief component.

The separation in such a case would be much more difficult, and several fractionations would be required before the boiling point of the first fraction became quite constant. When the boiling points of the components of a mixture are very near together and the chemical relationship is not very close, mixtures of constant boiling point are not unlikely to be formed ; that is the case, for example, with carbon tetrachloride and benzene, and it would be practically impossible to separate either pure tetrachloride or the mixture of constant boiling point from any mixture of the two substances, although benzene, if present in large excess in the original mixture, could be separated in a pure state by repeated fractionation.

Fourth Case.—A curve such as that in Fig. 78, would represent the behaviour on distillation of any mixture—such as that last-named with benzene in large excess—in which the chief component was the less volatile, but in which the difference between the boiling points of the

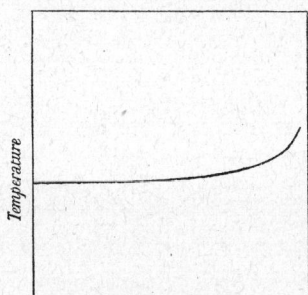

Weight of Distillate.

FIG. 77.

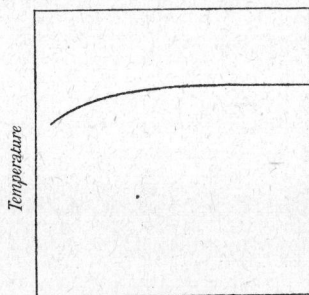

Weight of Distillate

FIG. 78.

components was small. The separation of the chief component would almost invariably be easier than in cases represented by Fig. 77. Indeed, Fig. 78 bears the same relationship to Fig. 77 as Fig. 76 to Fig. 75.

Here, again, the highest fraction might consist of a mixture of maximum boiling point such as chloroform-methyl acetate ; (b.p. 64·5°) with chloroform (b.p. 60·5°) in excess ; or, less probably, of a mixture of minimum boiling point containing a more volatile impurity, as, for example, the isopropyl alcohol-water mixture (b.p. 80·35°) with a little ethyl alcohol (b.p. 77·3°).

Fifth and Sixth Cases. — The curves in Figs. 79 and 80 represent the distillation of liquids which contain impurities of both greater and less volatility. In the case of Fig. 79, the boiling points of these impurities are far removed from that of the substance to be separated ; in the case of Fig. 80 they are near it. Such separations as these have very frequently to be carried out.

When, for example, benzene is nitrated by treatment with nitric and sulphuric acids, some of the benzene is usually unacted on, while a certain amount of dinitrobenzene is formed. Here the boiling point of the chief component, nitrobenzene, is very much higher than that of benzene and far below that of dinitrobenzene, and the separation is therefore an easy one.   If pure benzene is used for the preparation, and the mixture is distilled through a " pear " still-head with, say, 12 bulbs, the collection of pure nitrobenzene may be commenced after the second or third fractionation.

When, however, the boiling points are not so far apart, as, for example, in the separation of ethyl acetate from methyl and propyl acetate, the process is tedious if an ordinary apparatus is used (p. 105), and even with a 5-column " evaporator " still-head several fractionations are required (p. 159), for the middle substance is always more difficult to separate than the others.

If the boiling points of the components are very near together (Fig. 80) the separation of the middle substance is extremely trouble-

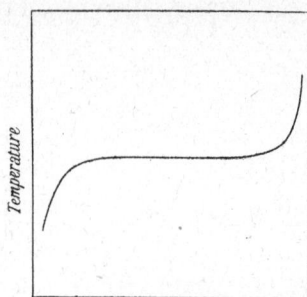

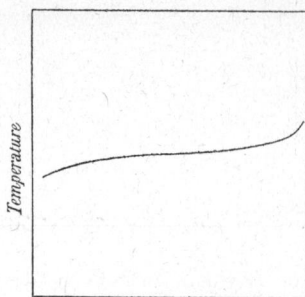

Weight of Distillate.	Weight of Distillate
FIG. 79.	FIG. 80.

some, and there is greater probability that mixtures of constant boiling point may be formed than in the cases previously considered.

Other Cases.—Other cases than those referred to may be met with, for example a liquid may contain only impurities of greater volatility, but the boiling points of some of these may be near, those of others far below that of the chief component.   The most volatile impurities in such a case will be easy to remove, while the less volatile will only be eliminated with difficulty and the distillation will correspond to those represented by the curve in Fig. 78 rather than in Fig. 76.

2. THE SEPARATION OF THE COMPONENTS FROM A MIXTURE OF KNOWN QUALITATIVE COMPOSITION

Two Components. — The chief points to be considered are (a) the boiling points of the pure components, and the difference between them ; (b) the chemical relationship of the components ; (c) the form of the boiling point-molecular composition curve, if it can be ascertained.

Closely related Substances.—By far the simplest cases are those in which the substances present are chemically closely related to each other, for the form of the boiling-point molecular composition curve must then be normal or nearly so (Chap. IV.) ; there is no possibility of the formation of mixtures of constant boiling point, and the only points to be considered are the actual boiling points of the components and the difference between them.

Boiling Points of Components.—On the actual boiling points will depend the nature of the still-head that can be used ; the greater the difference between them, the more easily can the separation be effected.

Substances not Closely Related ; Mixtures of Constant Boiling Point. —If the substances are not closely related to each other, mixtures of constant boiling point may be formed and the list of such mixtures which are at present known (p. 49) may be consulted. It will be sufficient to remark here that water forms mixtures of minimum boiling point with the majority of organic compounds ; that, generally, such mixtures are most frequently met with when the molecular weight of one of the components is abnormal in the liquid state, and that this is usually the case with compounds which contain a hydroxyl group.

It should also be remembered that when the boiling points of any two substances are near together, a small deviation from the normal boiling point-molecular composition curve is sufficient to give rise to the formation of a mixture of constant boiling point, or, at any rate, to make the curve nearly horizontal over a large part of its course starting from one extremity, usually that corresponding to the lowest temperature. It has been pointed out (p. 181) that when the curve is of this form it is very difficult, and may be impossible, to separate that component which is in excess where the curve is nearly horizontal.

When a mixture which distils without change of composition is formed, and its boiling point is far below that of the more volatile component, as in the case of normal propyl alcohol and water, or of methyl alcohol and benzene, or far above that of the less volatile component, as with nitric acid and water, it is usually possible to separate in a pure state both the mixture of constant boiling point and that component which is in excess. But, if the boiling point of the mixture is very near that of one of the two components, as in the case of ethyl alcohol and water (Fig. 81), or of normal hexane and benzene, it is practically impossible to separate that component in a pure state, and it may be impossible to separate the mixture of constant boiling point even when the component which boils at a widely different temperature is in excess. Thus, in the cases referred to, whatever the composition of the original mixture, it is impossible to obtain normal hexane, ethyl alcohol, the binary hexane-benzene mixture or the alcohol-water mixture in a pure state by fractional distillation, and it is only the less volatile component, benzene or water, which can be so obtained.

The composition of a mixture of constant boiling point, however, depends on the pressure, and it has been shown by Wade and Merriman [1]

---

[1] Wade and Merriman, ' Influence of Water on the Boiling Point of Ethyl Alcohol at Pressures above and below the Atmospheric Pressure," *Trans. Chem. Soc.*, 1911, **99**, 997 ; also Merriman, " The Vapour Pressures of the Lower Alcohols and their Azeotropic Mixtures with Water," *ibid.*, 1913, **103**, 628.

in the case of ethyl alcohol and water that as the pressure falls the percentage of water in the azeotropic mixture diminishes, and that at pressures lower than about 75 mm. no such mixture is formed. It would, therefore, theoretically be possible to separate both ethyl alcohol and water from a mixture of these substances by distillation under sufficiently reduced pressure. In practice, however, the separation of pure alcohol under low pressures would be as difficult as that of the azeotropic mixture under atmospheric pressure.

Separation of Components.—The manner in which the composition of the distillate is related to the total amount collected has been discussed in Chap. VIII., and full details of the separation of benzene and

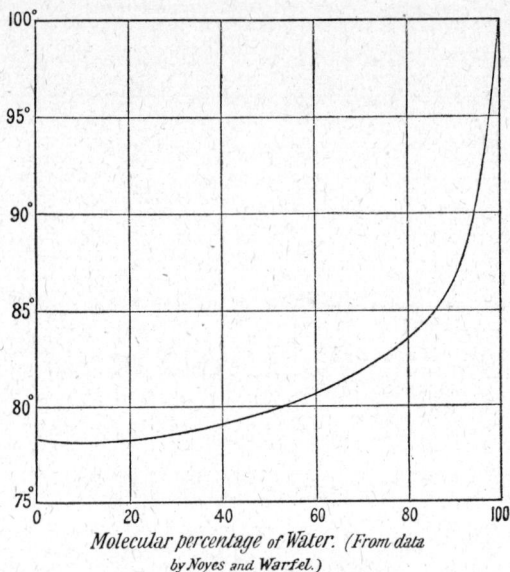

Molecular percentage of Water. (From data by Noyes and Warfel.)

FIG. 81.—Boiling points of mixtures of ethyl alcohol and water.

toluene by fractional distillation both with an ordinary and an improved still-head have been given (pp. 100 and 158).

Number of Fractions required and Choice of Still-head.—It is difficult to formulate any definite rules regarding the number of fractions in which the distillate should be collected, so many points have to be taken into consideration. The most important of these are the following :—

(*a*) The efficiency of the still-head employed.
(*b*) The approximate quantity of each component.
(*c*) The difference between the boiling points of the components.
(*d*) The form of the boiling point-molecular composition curve.

Efficiency of Still-head.—It may be stated quite generally that, for a given mixture, the more efficient the still-head the smaller is the number of fractions that will be required.

Amount of each Component present.—The approximate quantity of each component must be taken into consideration, not only in deciding upon the number of fractions but also in choosing the still-head to be used.

Suppose, for example, that it was desired to separate the benzene as completely as possible from a mixture consisting of 30 grams of that hydrocarbon with 270 grams of toluene.

The best plan would be to employ for the first distillation the most efficient still-head available, and if the recovery of the toluene need not be regarded as of special importance, the complete return of the liquid from the still-head to the still at the end of the distillation would not be essential, and a " bubbling " still-head of many sections might be used. A very slow distillation would be advantageous. It would probably be best to collect all the distillate that came over below 110·0° in one fraction and that from 110·0° to 110·6° in a second. The large residue would consist of pure toluene. The same still-head might be used for the second distillation and the distillate might be collected in three or four fractions, the temperature ranges of which would depend on the efficiency of the still-head. With an 18 section Young and Thomas still-head it is probable that nearly 20 grams of distillate would come over below 81·2°. The fractions would now be small, and it would be necessary to employ a still-head of fewer sections for the remaining distillations. It would be important that the amount of condensed liquid in the still-head should be as small as possible, and an evaporator still-head of 5 sections would probably be the most convenient.

If the original mixture contained 270 grams of benzene and 30 grams of toluene, and it was desired to recover the toluene, it would again be best to employ a very efficient still-head, and as, in this case, a most essential point would be that the liquid should return very completely from the still-head to the still at the end of the distillation, an evaporator still-head of many sections would be the best for the purpose.

A mixture of this composition, distilled through a very efficient still-head, would give a considerable amount of nearly pure benzene ; so far as the recovery of toluene is concerned, the first 100 grams, or even more, would not be worth redistilling. After rejecting this, fractions might be collected below 80·5°, from 80·5° to 95·6°, and above 95·6°. If the temperature reached 110·6° before the end of the distillation the residue would consist of pure toluene ; if not, it would require to be redistilled. In the second fractionation, the distillate from the first fraction of I. would consist of nearly pure benzene and might be rejected. The remaining fractions would now be small, and an evaporator still-head of not more than 5 sections would have to be used for the remaining distillations.

Boiling Points of Components.—For mixtures of closely related substances, or others which behave normally on distillation, the greater the difference between the boiling points the more readily can the components be separated by distillation. Thus, it has been shown (p. 125) that a mixture of normal and isopentane (b.p. 36·3° and 27·95° respectively) would require a much larger number of fractions than one of benzene and toluene if a still-head of the same efficiency were employed for both.

Boiling Point-Composition Curve.—When the form of the boiling point-molecular composition curve is normal it is steeper in the region of high temperature than of low, and in general the temperature ranges above the middle point may be somewhat greater than below ; in other words, the number of fractions may be somewhat smaller. When the actual boiling points of mixtures of the two substances are lower than those given by the normal curve, this difference above and below the middle point becomes accentuated ; and when the curve is very flat near its lower extremity, the number of fractions in the low temperature region must be considerably increased while those above the middle point may be diminished in number. In most cases, however, data for the construction of the boiling point-molecular composition curve are wanting, and, unless we can judge from the nature of the substances in the mixture whether the deviation of the actual from the normal curve is likely to be large or not, we cannot decide on the number of fractions that may ultimately be required until after the fractionation has made some progress.

Three Components.—If there are three substances present and the mixture behaves normally on distillation, the least volatile component will be the easiest to separate and the substance of intermediate boiling point the most difficult.

If, however, the components are not closely related to each other, one or more mixtures of constant boiling point may be formed and the problem becomes more complicated. Thus, when a mixture of ethyl alcohol, benzene, and water is distilled it tends to separate into (*a*) the ternary mixture of constant boiling point, (*b*) one of the three possible binary azeotropic mixtures, (*c*) that pure component which is in excess. It may, however, happen that the quantities in the original mixture are such that we have only the two fractions *a* and *b*, or *a* and *c*, or the fraction *a* alone. There are, in fact, 12 different ways in which separation may take place, or the mixture may distil unchanged (p. 179).

Again, the substances present may be capable of forming one or two binary mixtures but no ternary mixture of constant boiling point. That would be the case, for example, with isoamyl alcohol, benzene, and water, for the only azeotropic mixtures that can be formed are those of benzene and water (b.p. 69·25°) or of water and amyl alcohol (b.p. 95·15°). There are, therefore, five different ways in which separation may occur, but under no conditions can the mixture distil without change of composition. Employing the initial letters *A*, *B*, and *W* for the components—alcohol, benzene, and water—we have the following possible separations :—

First fraction.	Second fraction.	Third fraction.
$B.W.$	$A.W.$	$A.$
$B.W.$	$A.W.$	$W.$
$B.W.$	$B.$	$A.$
$B.W.$	$A.W.$	...
$B.W.$	$A.$	...

If only a single binary mixture of constant boiling point can be formed, as would be the case, for example, with a mixture of ethyl alcohol (A_1), benzene, and isoamyl alcohol (A_2), the following three separations would be possible :—

First fraction.	Second fraction.	Third fraction.
A_1B	A_1	A_2
A_1B	B	A_2
A_1B	A_2	...

A detailed description of the separation of the three closely related liquids, methyl acetate, ethyl acetate, and propyl acetate (both with an ordinary and an improved still-head) is given in Chapters VII and XIV.

3. The use of FRACTIONAL DISTILLATION as a METHOD OF QUANTITATIVE ANALYSIS has been fully discussed in Chapter XVI.

4. A GENERAL STUDY OF THE QUALITATIVE AND QUANTITATIVE COMPOSITION OF A COMPLEX MIXTURE WHEN ONLY THE GENERAL NATURE OF THE CHIEF COMPONENTS IS KNOWN, OR WHEN IT IS DEFINITELY KNOWN THAT CERTAIN SUBSTANCES ARE PRESENT, BUT IT IS NOT KNOWN WHAT OTHERS THERE MAY BE.

Rough Estimate of Composition.—When a complex mixture, the qualitative composition of which is only partly known, is distilled through an efficient still-head and the distillate is collected in fractions of either equal temperature range or of approximately equal weight, a rough estimate of the boiling points and of the amounts of the components may frequently be obtained from the quantities of distillate in the first case, or from the temperature ranges in the second, or, generally, from the values of $\Delta w/\Delta t$.

Causes of Confusion.—But confusion is apt to arise when (a) there are two substances present with boiling points very near together, (b) one of the components is present in relatively very small amount, or (c) two or more of the substances form mixtures of constant boiling point.

a. Two Components boiling at nearly the same Temperature.—If there are isomeric members of a homologous series present in the mixture, it is very likely that their boiling points may in some cases be very near together.

Pentanes in Petroleum.—An instance of this has already been referred to [1] in the case of the light distillate from American petroleum consist-

[1] Young and Thomas, " Some Hydrocarbons from American Petroleum, I, Normal and Isopentane," Trans. Chem. Soc., 1897, 71, 440.

ing chiefly of butanes, pentanes, and hexanes. Even with the very efficient still-head shown in Fig. 53 (p. 150), the results of the first distillation seem to indicate the presence of only a single substance between the butanes and hexanes, boiling at about 33° ; and it is only after repeated fractionation that the presence of both normal and isopentane (boiling points 36·3° and 27·95°) is clearly shown. The fact, however, that for the middle fractions the values of $\Delta w/\Delta t$ diminish, while for those below and above them they increase, is a clear indication that we are not dealing with a single substance.

This may be seen from Table 77, in which the results of the first three fractionations are given ; it will be seen that, in the first fractionation, the highest values of $\Delta w/\Delta t$ are those for fractions 5 and 6 ; in the second, the fractions that have the highest values are Nos. 5 and 7, and in the third they are Nos. 4 and 7.

TABLE 77

	I.		II.		III.	
Number of fraction.	Final temperature $=t$.	$\dfrac{\Delta w}{\Delta t}$.	t.	$\dfrac{\Delta w}{\Delta t}$.	t.	$\dfrac{\Delta w}{\Delta t}$.
1	...	...	28·05°	?	27·95°	?
2	28·5°	?	29·15	68	29·15	35
3	29·9	72	30·55	71	30·5	74
4	31·3	58	31·7	123	31·5	172
5	32·85	179	32·45	204	32·3	135
6	33·85	242	33·5	140	33·3	110
7	...	...	34·45	168	34·05	193
8	35·25	136	35·4	142	34·9	142
9	36·8	74	36·8	81	36·1	117
10	41·1	17	37·8	54	37·1	88

Considering 5 and 6 as one fraction the values of $\Delta w/\Delta t$ for the first three fractionations would be 204, 167, 121, respectively ; the values for No. 4 in the three fractionations are 58, 123, and 172, and for No. 7 they are 168 and 193 in the second and third fractionations.

The whole course of the separation is, however, better seen in the diagram (Fig. 36, p. 113). Both the pentanes may be obtained in a pure state by distillation only, or, at any rate, after removal of small quantities of impurity by means of a mixture of nitric and sulphuric acids. It does not appear, however, than any hydrocarbon that boils at a higher temperature than normal pentane can be so separated.

Hexanes in Petroleum.—Let us consider, for example, the hexanes.[1] A preliminary distillation through an ordinary or fairly efficient still-head appears to indicate the presence of a single substance boiling at about 66°, and after a few fractionations with a very efficient still-head it is seen that a further separation, apparently into two components, as in the case of the pentanes, is taking place, but after long continued

[1] Young, " Composition of American Petroleum," *Trans. Chem. Soc.*, 1898, **73**, 905.

fractionation it is found that the separation does not end there. Two series of fractionations were carried out, one of American, the other of Galician petroleum.

Material employed.—The American petroleum at first came over for the most part between 28° and 95°, and was richest in hexanes ; the weight of distillate between 56° and 74° was about 800 grams. Aromatic hydrocarbons were not removed before the fractionation by treatment with nitric and sulphuric acids.

From the Galician petroleum, the hydrocarbons boiling below 40° and above 72° had previously been for the most part separated by distillation, and benzene was removed before the fractionation was commenced. The weight of the Galician petroleum was about 5 times as great as that of the American.

Still-heads used.—The fractionation of the American petroleum was carried out with a 12-section Young and Thomas bubbling still-head, that of the Galician with the combined bubbling and regulated temperature still-heads which were employed for the separation of the pentanes (Fig. 53, p. 150).

The Galician petroleum was thus freer from pentanes and heptanes to begin with than the American, and the still-head employed was more efficient.

Description of Results.—In Table 78 are given the temperature ranges and the values of $\Delta w/\Delta t$ for the fourth fractionation of the American petroleum, and it will be seen that at this early stage the fraction from 65° to 66° has the highest value, but in the subsequent fractionations the ratio for the corresponding fraction (above 65°) steadily fell, and after the tenth fractionation this fraction had the lowest value.

TABLE 78

IV.		IV.	
Temperature range.	$\dfrac{\Delta w}{\Delta t}$.	Temperature range.	$\dfrac{\Delta w}{\Delta t}$.
57 —60°	9·1	66 —67°	80·5
60 —61·5	31·8	67 —68	87·1
61·5—63	54·1	68 —69	90·2
63 —64	56·1	69 —70·5	60·7
64 —65	94·7	70·5—73	24·3
65 —66	102·0		

Graphical Representation.—So far as the middle fraction is concerned, the gradual change in the value of $\Delta w/\Delta t$ closely resembles that observed in the case of the two pentanes, but in other respects the fractionations show marked differences, which become apparent when the curves (Fig. 82) representing the separation of the hexanes are compared with those for the pentanes (Fig. 36, p. 113).

As the quantity of Galician petroleum was much greater than that of American, the weight of distillate coming over between 60° and 70° has been taken as 100 for both of them in each fractionation, and as the tenth and sixteenth fractionations of American petroleum correspond very closely with the fourth and seventh respectively of the Galician, only one curve has been drawn for each of these two pairs, the experimental results for American petroleum being indicated in these and other cases by crosses, and for Galician by circles. The fractionations of American petroleum are referred to as IV_A, X_A . . . and of Galician as IV_G, VII_G . . .

After the fifth fractionation of the pentanes, the values of $\Delta w/\Delta t$ for the lowest and highest fractions show a steady rise, until, in the thirteenth fractionation, the last portion of liquid boiled quite constantly at 36·3°, and, in the eighteenth, the first portion of distillate came over

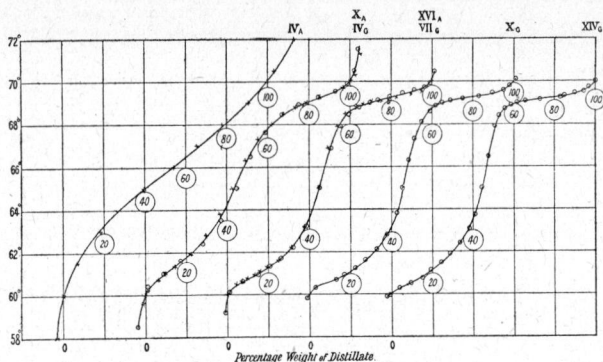

FIG. 82.—Results obtained by fractional distillation of American and of Galician petroleum.   (Fractions from 58° to 72°.)

quite constantly at 27·95°; in other words, the curves became quite horizontal at these two temperatures.

Highest Fractions.—With the hexanes, on the other hand, no fraction was obtained with perfectly constant boiling point even though, in the case of American petroleum, the liquid which came over above 66° was fractionated thirty-one times. Indeed, after the sixteenth fractionation, the maximum value of $\Delta w/\Delta t$ (at about 69·1°) showed very little further rise, and after the twenty-first fractionation it diminished slightly.

With the Galician petroleum, no further improvement in the value of $\Delta w/\Delta t$ for the fraction a little above 69° was noticeable after the eleventh fractionation.

Lowest Fractions.—Again, at the lower temperatures, in the case of American petroleum, the highest individual value of $\Delta w/\Delta t$ was reached in the twelfth fractionation (189 for the fraction coming over between 60·85° and 61·0°). In the sixteenth fractionation, the highest value had fallen to 141 for the fraction from 60·75 to 60·9, but, on the other hand, the values for the fractions above and below these temperatures showed a tendency to rise.

With Galician petroleum the highest individual value of $\Delta w/\Delta t$ at the lower temperatures was reached in the eleventh fractionation (790 for the fraction from 60·6° to 60·85°). In the last fractionation, the seventeenth, the maximum was only 626 for the fraction coming over between 60·55° and 60·75°, but in this case also the values for the higher and lower fractions appeared to be increasing slightly.

Interpretation of Results.—These facts may be explained by assuming the presence of at least four substances in both American and Galician petroleum boiling at temperatures between, say, 59° and 72°, two of them with boiling points not far from 61° and two not far from 69·5°. We should then have at first the apparent separation of a single substance boiling at about 66°, just as with the pentanes we have in the first place apparently the separation of a single component boiling at about 33°. Later on there would appear to be a further separation into two components with the approximate boiling points 61° and 69·5° corresponding to the two pentanes which boil at 27·95° and 36·3° respectively.

But while the two pentanes can be separated in a pure state, the two components from the hexane fractions cannot ; and, in the later fractionations, we have the beginning of a further subdivision. But each of the final separations must necessarily be far more difficult than the earlier ones, for the boiling points of the components of each pair are much closer together than the apparent boiling points of the pairs themselves, and although a partial separation may be effected, it is extremely doubtful whether, with the most efficient still-head, any one of the components could be separated in a pure state by fractional distillation alone.

Isohexane and Diethyl-methyl Methane.—As regards the components boiling not far from 61°, there is the possibility of the presence of the two isomers with the formulae $(CH_3)_2CH$—CH_2—CH_2—CH_3 and $(C_2H_5)_2CH$—CH_3, which would certainly resemble each other very closely indeed in their physical properties. It is doubtful, however, whether either of these paraffins has yet been prepared in a perfectly pure state, and sufficient reliance cannot be placed on the accuracy of the determinations of their boiling points and specific gravities which have been made up to the present time to warrant the formation of any estimate as to the relative quantities of the two hydrocarbons from the temperature ranges or specific gravities of the fractions.

Di-isopropyl.—It is noticeable, however, that the observed specific gravities (minimum 0·6728 at 0°/4° for the fraction from 60·55° to 60·75°) are slightly higher than the most probable value for isohexane (about 0·6721 at 0°/4°), and this may be due to the presence of a little di-isopropyl, $(CH_3)_2CH$—$CH(CH_3)_2$, which boils at 58·1° and has a somewhat higher specific gravity (0·67948 at 0°/4°), or it is just possible that the pentamethylene (b.p. about 50·5°, sp. gr. 0·7506 at 20·5°/4°), which is certainly present in petroleum, may not have been completely removed.

P

Normal Hexane and another Component.—With regard to the fractions boiling near to 69·5°, we have the following facts to guide us. The boiling point of pure normal hexane is 68·95° and its specific gravity at 0°/4° is 0·67697 ; but in the last fractionation of the Galician petroleum the fraction from 69·12° to 69·20° had the highest value of $\Delta w/\Delta t$, and its specific gravity at 0°/4° was about 0·685.

The less volatile portion of the American petroleum was fractionated thirty-one times ; the fractions were then separately treated several times with mixed nitric and sulphuric acids, and were subsequently fractionated many times with the object of separating pure normal hexane. The best specimen that could be obtained boiled at 69·05° and had the specific gravity 0·67813 at 0°/4°.

In both cases the specific gravities of the fractions above 66° were higher than those of pure normal hexane. For the fractions from 66° to 69° the change was slight, but above 69° the specific gravity rose rapidly. The following determinations were made in the course of the fractionations.

TABLE 79

Temperature range.	Sp. gr. at 0°/4°.	Temperature range.	Sp. gr. at 0°/4°.
66·4 —67·85° . . .	0·6793	69·4 —69·55° . . .	0·6898
68·6 —68·85 . . .	0·6802	69·5 —69·7 . . .	0·6962
68·95—69·03 . . .	0·6803	69·7 —69·95 . . .	0·7095
69·0 —69·1 . . .	0·6815	69·95—70·15 . . .	0·7157
69·25—69·4 . . .	0·6856	70·2 —74·0 . . .	0·7306

The fact that the fractions which showed the highest values of $\Delta w/\Delta t$ came over at temperatures so little higher than the boiling point of pure normal hexane (American petroleum about 69·1° ; Galician, about 69·2°) shows that there may be either a considerable quantity of some other substance present with a boiling point only a fraction of a degree higher than that of normal hexane, or there may be a smaller quantity of a substance which boils not more than 2° or 3° higher. That the separation is such a difficult one shows that the difference of boiling point can hardly be greater than three degrees.

Evidence afforded by Specific Gravities.—Again if the specific gravities of the fractions given in Table 79 are plotted against the mean temperatures it will be seen that there is a very sudden rise of specific gravity above 69°, and this seems also to show that the boiling point of the second substance cannot be far from that of normal hexane. If, for example, the substance boiled at a temperature as high as 80° the rise would certainly be more gradual at first. Moreover, it has been shown (p. 114) that hexamethylene, which boils at 80·85°, can be separated from the hexanes by fractional distillation.

Methyl Pentamethylene.—Further light has been thrown on the problem by the researches of Markownikoff,[1] of Zelinsky,[2] and of

---

[1] Markownikoff, "On Methyl Cyclo-Pentane from different Sources and some of its Derivatives," *Berl. Berichte,* 1897, **30,** 1222.

[2] Zelinsky, "Researches in the Hexamethylene Group," *ibid.,* 1897, **30,** 387.

Aschan [1] on Russian petroleum, which has been found to contain large quantities of hexamethylene (b.p. 80·85° ; sp. gr. 0·7968 at 0°/4°) [2] and methyl pentamethylene (b.p. about 72° ; sp. gr. about 0·766 at 0°/4°), both of which substances have also been prepared synthetically. The two hydrocarbons differ widely in their behaviour with fuming nitric acid, for hexamethylene is only attacked slowly even when heated, a large amount of adipic acid being formed, whereas methyl pentamethylene is acted on rapidly at the ordinary temperature, with evolution of much heat, and acetic acid is the chief product of oxidation.

Now the fractions for the first three or four degrees above 69° are attacked in the cold by fuming nitric acid, much heat being evolved and a large amount of acetic acid formed, and it may therefore be concluded that the substance present in American and Galician petroleum with a boiling point not far above that of normal hexane is methyl pentamethylene.

Preparation of Pure Normal Hexane.—Not only methyl pentamethylene, but also the isohexanes and other hydrocarbons which contain a $>CH$—group may be removed by heating with fuming nitric acid, and it was found that when the fractions from Galician petroleum which came over between 66° and 69·2° were subjected to prolonged heating with the fuming acid, and were afterwards distilled two or three times, almost pure normal hexane was obtained ; indeed from the fractions between 66° and 68·95° the normal paraffin appeared to be perfectly pure.

Summary of Results.—It has thus been shown that in the distillation of this portion of American or Galician petroleum, the liquid which at first seems to be a single substance boiling at about 66° proves to be a mixture of four substances, two isomeric hexanes boiling at nearly the same temperature, 61°, and normal hexane and methyl pentamethylene with the boiling points 68·95° and (about) 72° respectively. The first pair of substances are very closely related to each other, the second pair are not.

Amyl Alcohols in Fusel Oil.—As another example of a pair of very closely related liquids which are frequently met with, the isomeric amyl alcohols which are present in fusel oil may be mentioned. Distilled through an ordinary still-head, long continued fractionation would be necessary even to indicate the presence of two isomeric amyl alcohols, and the boiling points are so close together that even with an exceedingly efficient still-head the separation is very difficult.

Hexamethylene and a Volatile Heptane.—Another example of the presence, in a complex mixture, of two substances which boil at nearly the same temperature is afforded by the distillation of the portion of

[1] Aschan, " On the Presence of Methyl Pentamethylene in Caucasian Petroleum Ether," *Berl. Berichte*, 1898, **31**, 1803.

[2] Prepared synthetically by hydrogenation of benzene by Sabatier and Mailhe who give the b.p. 81° under 755 mm. pressure and sp. gr. 0·7843 at 13·5° (=0·7970 at 0°), *Compt. rend.*, 1903, **137**, 240.

American or Galician petroleum that comes over between 75° and 80°. It was found possible to separate a liquid of quite constant boiling point and, as derivatives of hexamethylene could be prepared from it,[1] it was concluded that that substance had been separated in a pure state. Later on, however, it was found that the liquid could be partially but not completely frozen in an ordinary freezing mixture, and eventually, by fractional crystallisation, nearly pure hexamethylene was obtained with practically the same boiling point as before but with a definite melting point, and of notably higher specific gravity.[2] It was evident that the substance separated by fractional distillation only was a mixture of two hydrocarbons, hexamethylene and a heptane, of which the former was present in much the larger quantity, and that either the boiling points are almost identical, or else the two substances form a mixture of constant boiling point almost identical with that of hexamethylene.

Pentamethylene and Trimethyl-ethyl-methane.—Such difficulties as have been described are of common occurrence in the distillation of petroleum. For example, American, Galician, and Russian petroleum all contain a certain amount of pentamethylene which boils at about 50° ; but there is also present a hexane, trimethyl-ethyl-methane, boiling at nearly the same temperature, and it appears to be impossible to separate these hydrocarbons by fractional distillation.

Lecat finds that *n*-heptane and methyl-hexamethylene form an azeotropic mixture very rich in heptane, boiling at a temperature slightly lower than the boiling point of the pure heptane. It is therefore quite probable that azeotropic mixtures may be formed in the two cases just described, and also with *n*-hexane and methyl pentamethylene.

b. One or more Components Present in Small Quantity.—

If one or more components of a complex mixture are present in relatively very small amount, they are apt to be altogether overlooked, and it is only by keeping a careful record of the weights and temperature ranges of the fractions, and by calculating the values of $\Delta w/\Delta t$ or mapping the total weights of distillate against the temperatures, that the existence of these components can be detected.

The manner in which the presence of a relatively very small quantity of hexamethylene was recognised in American petroleum has been described in Chap. VII. (p. 114).

Pentamethylene in Petroleum.—In a similar manner, it was found that there is a small quantity of pentamethylene (and also trimethyl-ethyl-methane) in American petroleum.[3]

The first fractionation of all the available distillates from petroleum ether obtained with the combined bubbling and regulated temperature still-heads, and collected between about 37° and 60°, gave no indication

[1] Fortey, " Hexamethylene from American and Galician Petroleum," *Trans. Chem. Soc.*, 1898, **73**, 103.
[2] Young and Fortey, " The Vapour Pressures, Specific Volumes, and Critical Constants of Hexamethylene," *ibid.*, 1899, **75**, 873.
[3] Young, *loc. cit.*

of the presence of any substance boiling in the neighbourhood of 50°, as will be seen by Table 80.

TABLE 80

I.		II.		IV.		VII.	
Temperature range.	$\frac{\triangle w}{\triangle t}$.	Temperature range.	$\frac{\triangle w}{\triangle t}$.	Temperature range.	$\frac{\triangle w}{\triangle t}$.	Temperature range.	$\frac{\triangle w}{\triangle t}$.
36 —37°	47·4	36 —37°	47·8	36·8—42·0°	5·0	45·0 —49·35°	3·8
37 —40	22·2	37 —39	11·8	42·0—47·7	3·1	49·35—50·1	34·9
40 —45	3·1	39 —42	5·9	47·7—50·7	12·9	50·1 —51·3	16·0
45 —50	6·2	42 —47	4·0	50·7—53·4	12·6	51·3 —53·7	6·7
50 —54	12·2	47 —51	6·0	53·4—56·2	7·8		
54 —57·5	22·6	51 —54	16·5	56·2—58·6	10·3		
57·5—59·5	24·7	54 —57	14·4	58·6—59·6	15·7		
59·5—60·4	39·0	57 —59·2	17·8				
		59·2—60·1	38·9				

FINAL FRACTIONATION

Temperature range.	$\triangle w$.	$\frac{\triangle w}{\triangle t}$.	Specific gravity at 0°/4°.
47·3 —49·45°	11·5	5·3	0·7029
49·45—49·55	16·3	163·0	0·7035
49·55—50·35	16·0	20·0	0·6975
50·35—56·4	15·0	2·5	

In the second distillation, however, the value of $\triangle w/\triangle t$ for the fraction 51—54° was slightly higher than for the one above and much higher than for that below it.

In the fourth fractionation the maximum was quite clearly defined, and the value of $\triangle w/\triangle t$ near 50° steadily increased in the subsequent operations, rising from 12·9 in the fourth to 34·9 in the seventh and 48·0 in the ninth.

The quantity of material was too small to allow of the separation being continued with the combined bubbling and regulated temperature still-heads, but six additional fractionations were carried out with a 12-column bubbling still-head, when the value of $\triangle w/\triangle t$ rose to 163·0 for the fraction from 49·45° to 49·55°.

The distillate, amounting to 16·3 grams, was found to yield glutaric acid on oxidation with nitric acid, and therefore contained penta-methylene. The specific gravity, 0·7035 at 0°/4° was, however, too low and the vapour density, 39·2, too high for pure pentamethylene (sp. gr. 0·751 at 15°/15° ; vap. den. 35). As the calculated vapour density of hexane is 43, it would appear that the distillate consisted of a mixture of pentamethylene and trimethyl-ethyl-methane in approximately equal molecular proportions. The results are quite in accordance with those obtained by Markownikoff,[1] who showed that both these hydrocarbons

[1] Markownikoff, " On Cyclic Compounds," *Liebig's Annalen*, 1898, **301**, 154.

are present in Russian petroleum, the quantity of pentamethylene and other naphthenes as compared with that of the paraffins being, however, much larger in Russian than in American petroleum.

Chloroform.—An interesting case is that of chloroform. It had been found that chloroform prepared from acetone was distinctly inferior as an anaesthetic to that made from alcohol, and the most probable explanation appeared to be the presence of minute quantities of irritating products of decomposition, such as chlorine, carbonyl chloride or hydrochloric acid in the chloroform derived from acetone. The subject was investigated by Wade and Finnemore,[1] who found that neither ordinary distillation nor treatment with sulphuric acid, washing, drying, and subsequent rectification through a pear still-head disclosed any difference between samples of chloroform prepared respectively from acetone and alcohol.

The problem was eventually solved by them with the aid of an evaporator still-head of five sections, which was employed in all the subsequent distillations. That chloroform and alcohol form a mixture of minimum boiling point was shown by Thayer,[2] and Wade and Finnemore isolated and examined not only this binary azeotropic mixture but also a ternary chloroform-alcohol-water mixture, boiling constantly at 55·5° and containing 4·0 per cent of alcohol and 3·5 per cent of water. Now anaesthetic chloroform always contains a little alcohol and a trace of water, and it was found that with anaesthetic chloroform prepared from acetone distillation commenced at 55·5°, and there was also distinct evidence of the formation of the binary chloroform-alcohol mixture, which boils at 59·4°. With chloroform made from alcohol, on the other hand, distillation began at 54°.

It therefore appeared that the " acetone " chloroform contained no foreign substance beyond the alcohol and water which are always present, whereas the " alcohol " chloroform contained a minute quantity of a volatile impurity. It was suspected that this might be ethyl chloride, and it was found that on distilling chloroform containing 0·5 per cent of alcohol, a trace of water, and 0·2 per cent of ethyl chloride, the liquid began to boil at 51° and the indications of the binary chloroform-alcohol mixture were obliterated. With only 0·1 per cent of ethyl chloride distillation commenced at 52·65°, and there was clear indication of the formation of the binary chloroform-alcohol mixture.

Finally, direct chemical proof of the presence of ethyl chloride was obtained in the first fractions, not only from the mixtures to which minute quantities of that substance had been added, but also from the chloroform prepared from alcohol. Not the slightest indication of its presence, on the other hand, could be detected in the first fraction from " acetone " chloroform.

It was thus proved that the superior anaesthetic properties of " alcohol " chloroform are due to the presence of a trace of ethyl chloride—probably about 0·05 per cent or 2 c.c. per Winchester quart—

[1] Wade and Finnemore, "Influence of Moist Alcohol and Ethyl Chloride on the Boiling Point of Chloroform," *Trans. Chem. Soc.*, 1904, **85**, 938.
[2] Thayer, "Boiling Point Curves," *J. Amer. Chem. Soc.*, 1899, **3** 36.

and it was found by Mr. Rowell at Guy's Hospital that "acetone" chloroform to which a small quantity of ethyl chloride has been added is therapeutically identical with chloroform made from alcohol. Wade and Finnemore suggest that the addition of a further quantity of ethyl chloride would be even more beneficial in decreasing the time of induction and generally facilitating anaesthesia. Loss by evaporation is trifling even when the chloroform contains as much as 1 per cent of ethyl chloride.

c. **Mixtures of Constant Boiling Point.** — *When two or more of the components form mixtures of constant boiling point the difficulty in interpreting the experimental results is greatly increased.*

Benzene in American Petroleum.—For example, if we subject a quantity of American petroleum to distillation, collecting fractions from 30—60°, 60—90°, and 90—120°, and then treat each fraction with a mixture of strong nitric and sulphuric acids, we shall find that a considerable amount of dinitrobenzene will be obtained from the fraction from 60—90° and of dinitrotoluene from that between 90° and 120°. The obvious conclusion from these results is that benzene and toluene are present in American petroleum, and this was, in fact, stated long ago to be the case by Schorlemmer.

If, however, instead of collecting the distillate in three large fractions, we were to separate it into nine with a range of 10 degrees each, it would then be found, after repeated distillation, that on treating each fraction with the mixed acids, dinitrobenzene would be obtained almost exclusively from the distillate that came over between 60° and 70° and dinitrotoluene from that between 90° and 100°, although the boiling points of benzene and toluene are 80·2° and 110·6° respectively.

Possible Explanations of Results.—In order to explain these facts two assumptions may be made ; either (*a*) it is not benzene and toluene that are present in American petroleum, but some other more volatile compounds from which dinitrobenzene and dinitrotoluene respectively are formed by the action of nitric and sulphuric acids ; or (*b*) benzene and toluene, when mixed with the paraffins present in petroleum come over chiefly at temperatures 14° or 15° below their true boiling points.

The first assumption, so far as the conversion of a hydrocarbon other than benzene into dinitrobenzene is concerned, was actually made by one observer, but it is practically certain that the second is the correct one, for it has been found, as already stated, that when a mixture of normal hexane (b.p. 68·95°) with, say, 10 per cent of benzene, is distilled, the benzene cannot be separated but comes over with the hexane at 68·95°. Lastly, the fact that when American petroleum is distilled, the greater part of the benzene comes over below 69° (mostly at 65—66°) is explained by the presence of isomeric hexanes which boil at lower temperatures than normal hexane and also carry down the benzene with them.

Ethyl Alcohol—Benzene—Water.—The behaviour of mixtures of ethyl alcohol, benzene, and water has already been referred to (p. 179) ;

but it may be pointed out that, when a mixture of equal weights of benzene and of 95 per cent alcohol is distilled through a very efficient still-head, the substance of highest boiling point, water, comes over in the first of the three fractions into which the distillate tends to separate, the remainder of the benzene in the second fraction, while the third fraction or residue consists of the most volatile of the original components, alcohol.

Aliphatic Acids and Water.—Again, when a mixture of formic, acetic and butyric acids with water is distilled, it tends to separate into three or more of the following components.

		Boiling point.
1.	Butyric acid-water (mixture of minimum boiling point)	99·4°
2.	Water	100·0
3.	Formic acid	100·7
4.	Formic acid-water (mixture of maximum boiling point)	107·3
5.	Acetic acid	118·5
6.	Butyric acid	163·5

With a large amount of water, the whole of the butyric acid would come over in the lowest fraction, and if the amount of acetic acid was large, the last fraction would consist of that acid ; but although acetic acid does not form a mixture of minimum boiling point with water, yet it is very difficult to separate the acid from its dilute aqueous solution by distillation, and if the amount of the acid was relatively small it might all be carried over with the water at temperatures below 107°, and in that case the highest fraction would consist of the formic acid-water mixture of maximum boiling point. The acids would then come over in the reversed order of their boiling points. Hecht[1] found that on distilling a mixture of acetic, butyric, and oenanthylic acids with much water, the whole of the oenanthylic acid came over in the first portion of the distillate, the middle portion contained chiefly butyric acid and the last portion contained acetic acid nearly free from the other two. Hecht points out that acetic acid is miscible with water in all proportions with considerable heat evolution ; butyric acid is also miscible with water in all proportions but very little heat change is observable, and oenanthylic acid is nearly insoluble in water.

The determination of the qualitative and quantitative composition of mixtures of fatty acids is a difficult problem. The methods suggested for the estimation of the acids are based either (1) on the relative solubilities of the acids or of their salts, or (2) on the relative rates of distillation of the acids or of their esters. The methods depending on the rates of distillation of the acids volatile with steam have been examined and are discussed by Reilly,[2] and he has devised a process by which a single acid in dilute aqueous solution may be identified or the percentage amounts of two or more acids in dilute solution may be determined.

---

[1] Hecht, " On Isoheptoic Acid from β-Hexyl Iodide," *Liebig's Annalen*, 1881, **209**, 321.
[2] Reilly, " The Determination of the Volatile Fatty Acids by an Improved Distillation Method," *Sci. Proc. Roy. Dubl. Soc.*, 1919, **15** (N.S.), 513.

BUTTER FAT

Butter fat is a complex mixture of the glycerides of a considerable number of fatty acids. The composition is, no doubt, somewhat variable, but the table below, due to Brown, will at any rate give an idea of the nature and the relative quantity of these glycerides.

TABLE 81

Glycerides of		Per cent.
Dihydroxy stearic acid	$C_{18}H_{34}(OH)_2O_2$	1·04
Oleic acid	$C_{18}H_{34}O_2$	33·95
Stearic acid	$C_{18}H_{36}O_2$	1·91
Palmitic acid	$C_{16}H_{32}O_2$	40·51
Myristic acid	$C_{14}H_{28}O_2$	10·44
Lauric acid	$C_{12}H_{24}O_2$	2·73
Capric acid	$C_{10}H_{20}O_2$	0·34
Caprylic acid	$C_8H_{16}O_2$	0·53
Caproic acid	$C_6H_{12}O_2$	2·32
Butyric acid	$C_4H_8O_2$	6·32

In the analysis of butter fat, when both the soluble (Reichert-Wollny-Meissl method) and insoluble (Polenské method) volatile fatty acids are to be estimated, 5 grams of the fat, 20 grams of glycerol and 2 c.c. of 50 per cent aqueous (not alcoholic) caustic soda are heated in a 300 c.c. flask until saponification is complete. The soap is dissolved in 90 c.c. of hot water ; 50 c.c. of normal sulphuric acid and about 0·1 gram of finely powdered pumice are added, and the mixture is distilled at such a rate that 110 c.c. come over in about 20 minutes. The receiver (flask) is then replaced by a 25 c.c. cylinder and the distillation is stopped. The well cooled distillate in the flask, after standing, is filtered and 100 c.c. of the filtrate is used for the estimation of the soluble volatile acids. The insoluble volatile acids are carefully collected from the flask, cylinder, and condenser, dissolved in alcohol and separately estimated.

Of the fatty acids present in butter, butyric acid is the only one which is miscible with water in all proportions. It forms with water an azeotropic mixture which, according to Lecat, boils at 99·4° under normal pressure and contains 18·4 per cent of the acid.

It may therefore be fairly assumed that the whole of the butyric acid from the butter fat is contained in the 110 c.c. of distillate.

As regards the higher acids, as the molecular weight increases the solubility in water diminishes, palmitic, stearic, and oleic acids, at any rate, being practically insoluble. All these higher acids form, of course, azeotropic " mixtures "—termed by Lecat " heterogeneous mixtures "—with water, but the vapour pressures of palmitic, stearic, and oleic and dihydroxy stearic acids are so low that the quantities that distil over with steam must be exceedingly small.

V. H. Kirkham,[1] Government Analyst, East Africa Protectorate, whose laboratory at Nairobi is 5,500 ft. above sea level, having invari-

[1] V. H. Kirkham, " The Effect of Pressure upon the Polenské and Reichert-Meissl Values," *The Analyst*, 1920, **45**, 293.

ably obtained very low values by the Polenské method in the analysis of butter, was led to investigate the influence of pressure on the Reichert-Meissl and Polenské values, and found that as the pressure was increased the Reichert-Meissl value rose slowly while the Polenské value rose very rapidly.

Kirkham made determinations of both the soluble and insoluble volatile fatty acids at pressures from 100 to 1000 mm. and found that the Polenské values were satisfactorily reproduced by the formula

$V = \dfrac{v(P - \kappa)}{p - \kappa}$ where V and v are the Polenské values corresponding

respectively to the pressures P and p, and the constant $\kappa = 45$.

The Reichert-Meissl values agreed moderately well with the formula

$V - \kappa' = \dfrac{(v - \kappa') \log P}{\log p}$.

The observed and calculated values are given below.

TABLE 82

Pressure in mm.	Reichert-Meissl values.		Polenské values.	
	Observed.	Calculated.	Observed.	Calculated.
100	22·34	22·58	0·19	0·19
180	24·43	24·19	0·48	0·48
250	25·57	25·10	0·75	0·73
380	26·93	26·23	1·14	1·19
450	27·13	26·69	1·61	1·44
627	27·60	27·60	2·06	2·07
760	27·99	28·12	2·68	2·55
900	28·17	28·60	..	..
1000	28·05	28·87	3·40	3·40

In other words the Polenské values are not far from proportional to the total pressures, increasing somewhat more rapidly, whilst the Reichert-Meissl values are very roughly proportional to the logarithms of the pressures, increasing somewhat more slowly.

The vapour pressures of some of the fatty acids have been determined at a series of temperatures, and although these temperatures are generally higher than 100° they afford the means of calculating the percentages of acid in the distillates when these acids are distilled with water under different total pressures, the assumptions being made that the vapour densities of these higher acids are normal and that the acids are insoluble in water.

[TABLE

TABLE 83

Myristic acid, D=114; Water, D=9.

T.°C.	Vapour pressures. Acid.	Water.	$\dfrac{P_A D_A}{P_W D_W}$.	Percentage of acid in distillate.	Total Pressure.	$\dfrac{\text{Percentage of acid}}{\text{Total pressure}} \times 10^6$.
190·8	10·0	9609	0·01318	1·301	9619	135
207·6	20·0	13655	0·01855	1·821	13675	133
217·4	30·0	16550	0·02296	2·245	16580	135
223·5	40·0	18573	0·02728	2·656	18613	143
			Lauric acid, D=100; Water, D=9.			
164·5	8·0	5210	0·01706	1·64	5218	314
174·1	12·0	6576	0·02028	1·99	6588	302
185·9	20·0	8625	0·02576	2·51	8645	291
195·5	30·0	10633	0·03135	3·04	10663	285
203·0	40·0	12437	0·03573	3·45	12477	277
			Oenanthylic acid, D=65; Water, D=9.			
144·0	50	3040	0·1188	10·62	3090	3436
160·0	100	4652	0·1552	13·44	4752	2828
199·6	400	11582	0·2494	19·96	11982	1666
221·0	760	17722	0·3097	23·65	18482	1280

It will be seen that with myristic acid the calculated percentage of acid in the distillate is just about proportional to the total pressure. In the case of lauric acid, although the percentage of acid increases with the total pressure, the increase is relatively less rapid. With oenanthylic acid there is still an increase in the percentage of acid in the distillate, but the ratio of the percentage of acid to the total pressure shows a marked fall as the pressure rises.

Although butyric acid and the acids of still lower molecular weight are miscible with water in all proportions, one may calculate the values of $\dfrac{P_A D_A}{P_W D_W}$ to ascertain how they would behave on distillation with water if they were insoluble in it and if their molecular weights were normal. The calculation indicates that if these conditions were fulfilled the percentage weight of butyric acid in the distillate would increase with rise of total pressure far more slowly than that of oenanthylic acid, while the percentage of acetic acid would actually diminish.

One may therefore take it as a rule, with probably no exceptions, that when the insoluble aliphatic acids are distilled with water under a series of different pressures, the higher the molecular weight of the acid the more rapid is the relative increase in the percentage of acid in the distillate, with rise of total pressure. The results obtained by Kirkham are in conformity with this rule, but, as he points out, the problem is greatly complicated by the fact that the mixture distilled contains, in addition to water, a large amount of glycerol, in which the higher acids are nearly insoluble, as well as sodium sulphate and some free sulphuric acid. There are also relatively very large quantities of palmitic and oleic acids with which the lower acids are miscible in all proportions and which are themselves nearly non-volatile at the temperature of distillation. On the other hand the quantities of caprylic and capric acids are relatively very small indeed.

The presence of a large amount of non-volatile acids would certainly diminish the volatility of small quantities of the lower homologues, and Mr. Kirkham informs me that by adding margarine to butter the Polenské value and even the Reichert-Meissl value is actually lowered.

Alcohols and Water.—All the alcohols—with the exception of methyl —form mixtures of minimum boiling point with water, but as the boiling points of the binary mixtures, up to amyl alcohol at any rate, follow the same order as those of the alcohols themselves, such reversals as are observed with the acids do not occur unless the amount of water, relatively to that of the soluble alcohols, is so large that the boiling point of the mixture of soluble alcohol and water is not far below 100° (v. p. 289).

By the fermentation of sugar there are formed, in addition to ethyl alcohol, much smaller quantities of isopropyl, n-propyl, isobutyl, n-butyl, and inactive and active isoamyl alcohols.

The boiling points of these alcohols (with the exception of active isoamyl alcohol, the boiling point of which differs but slightly from that of the inactive compound) and of their azeotropic mixtures with water are given below.

TABLE 84

	Boiling point (760 mm.).	
	Alcohol.	Azeotropic mixture.
Ethyl alcohol	78·30	78·15
Isopropyl alcohol	82·44	80·37
n-Propyl alcohol	97·20	87·72
Isobutyl alcohol	108·06	89·82
n-Butyl alcohol	117·5 [1]	92·25 [2]
Isoamyl alcohol	132·05	95·15

When the product of fermentation or " mash " is distilled through a very efficient still-head, such as the Coffey still, then, leaving out of account volatile impurities such as aldehyde which are eliminated as a separate fraction, there are collected (a) the ethyl alcohol-water azeotropic mixture containing a slight excess of water, (b) fusel oil containing the higher alcohols and water with a certain amount of ethyl alcohol.

The butyl and amyl alcohols are only partially miscible with water and the amyl alcohols are present in largest quantity. On cooling and standing two layers are formed, the upper layer consisting of the alcohols with a little water, the lower containing most of the water with a certain amount of the alcohols, chiefly ethyl and isopropyl—in solution.

When the upper layer is redistilled the water soon passes over as an azeotropic mixture (or mixtures) and then the alcohols themselves can

[1] Reilly, *Sci. Proc. Roy. Dubl. Soc.*, 1919, **15** (N.S.), 597.
[2] Lecat, *loc. cit.*

be collected. The separation of the dry alcohols should be easier than that of their azeotropic mixtures with water as the differences in boiling point are so much greater, and it is only the two isoamyl alcohols which cannot be separated from each other without great difficulty. Golodetz,[1] however, states that with an efficient still-head he has obtained a better separation of substances non-miscible with water by steam distillation than without steam.

[1] Golodetz, " Fractional Distillation with Steam," Zeitschr. physik. Chem., 1912, **78**, 641 ; J. Soc. Chem. Ind., 1912, **31**, 215.

CHAPTER XX

Sublimation and Distillation.—The processes of sublimation and distillation are very closely allied ; in practice, indeed, no hard and fast line is drawn between them.

There are three ways in which the vaporisation of a stable substance may take place :—

(1) The substance is liquid under ordinary conditions ; when heated it boils at a definite temperature depending on the pressure.

(2) The substance is solid ; when heated it melts at a definite temperature ; when further heated it boils like an ordinary liquid.

(3) The substance is solid ; when heated it does not melt, but volatilises at a definite temperature depending on the pressure.

So, also, condensation of vapour may take place in three ways :—

(1) There may be liquefaction only.

(2) Liquefaction may first take place, and then solidification.

(3) There may be direct passage from the gaseous to the solid state.

The term " distillation " is applied to vaporisation and subsequent condensation of the first kind. It should also be applied to vaporisation and condensation of the second kind since it is really liquid which is converted into vapour and which is first formed by condensation.

The term " sublimation " ought, perhaps, only to be employed when the changes of state are of the third kind, but in that case it would rarely be used. Under certain conditions, however, a substance, when heated, may first melt and then boil, but, on cooling, it may pass directly from the gaseous to the solid state. The process is then usually termed sublimation.

Triple Point.—The majority of substances may, in fact, undergo either distillation or sublimation according to the pressure under which the vaporisation and condensation take place. This may be best understood by help of the diagram, Fig. 83, in which temperatures are measured as ordinates and pressures as abscissae.

The curve AC represents the boiling points of a substance, the curve AB its volatilising (subliming) points, and the curve AD its melting points under different pressures. The three curves intersect each other at the point A, called by James Thomson the " triple point." The temperature at A, measured vertically from the horizontal axis OY, gives the melting point of the substance under the pressure (triple point pressure) represented by the horizontal distance of A from the vertical axis OX ; it is practically the same as the melting point under atmospheric pressure—generally a small fraction of a degree lower, but in the case of ice 0·0075° higher. The point A also represents both the volatilising point of the solid and the boiling point of the liquid under the triple point pressure.

If a solid substance is heated under a pressure greater than the triple point pressure, its temperature rises until the melting point is reached ; then, after fusion, the temperature again rises until the liquid reaches its boiling point. The melting point and boiling point under any given pressure are indicated by the intersection of a vertical line representing that pressure with the curves AD and AC respectively.

If, however, the pressure is lower than the triple point pressure, the solid substance cannot be melted, but, when heated, it volatilises without previous fusion at a temperature given by the intersection of the vertical line of constant pressure with the curve AB. [Although a solid cannot be melted when the pressure is lower than the triple point pressure, a liquid, under certain conditions, may be cooled below its freezing point without undergoing solidification ; it may, indeed, be made to boil at a temperature lower than the freezing point. This is indicated by the dotted line AE which is simply a continuation of the curve AC.]

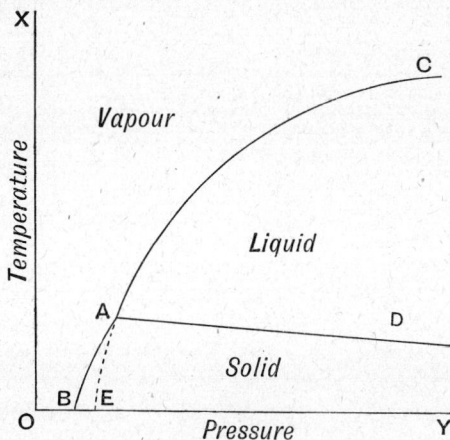

FIG. 83.

The triple point pressure differs very greatly for different substances : in the vast majority of cases it is far lower than the atmospheric pressure, and therefore, under ordinary conditions, true sublimation rarely takes place.

The following table gives the melting points, the boiling points under normal pressure, and the triple point pressures of a few substances. Complete data have not been obtained for many others.

[TABLE

TABLE 85

Substance.	Melting point.	Boiling point (760 mm.).	Triple point pressure.
Water	0°	100°	4·6 mm.
Acetic acid . . .	16·4	118·5	9·45 ,,
Benzene . . .	5·6	80·2	35·9 ,,
Bromine . . .	−7·1	58·7	44·5 ,,
Nitrogen . . .	−214·0	−193·1	60 ,,
Iodine	114·3	184·3	91 ,,
Camphor . . .	180	204	380 ,,
Mercuric chloride .	288	303	554 ,,
Carbon hexachlorine .	188 (about)	188 (about)	1 atm.
Arsenic trioxide . .	200 (about)	?	a little over 1 atm.
Aluminium chloride .	180–185	?	,, ,, ,,
Arsenic	>500	?	>1 atm.
Carbon dioxide . .	−57 (about)	(−78·2)	about 5·3 atm.

The triple point pressure of carbon dioxide is over five atmospheres ; hence when the liquid is allowed to escape from a cylinder in which it has been stored under pressure, that portion of it that escapes vaporisation solidifies at once and cannot be melted again in an open vessel, but gradually volatilises at a temperature of about −78°.

Arsenic and arsenious oxide volatilise without melting when heated under atmospheric pressure, and carbon hexachloride volatilises if the pressure is slightly lowered but melts and boils if the pressure is slightly raised. Water cannot exist as a stable liquid under a pressure lower than 4·6 mm. Below this pressure ice cannot be melted but volatilises at a temperature lower than 0° when heat is supplied.

Volatilising Point.—The curve BA represents not only the volatilising points of a solid under different pressures but also the vapour pressures of the solid at different temperatures,[1] just as the curve AC represents both the boiling points and the vapour pressures of the substance in the liquid state ; the volatilising point of a solid may therefore be defined in precisely the same way as the boiling point of a liquid.

The volatilising point of a solid is the highest temperature attainable by the solid under a given pressure *of its own vapour* when vaporising with a perfectly free surface and when heat reaches the surface from without.

Dalton's Law of Partial Pressures.—Dalton's law of partial pressures applies to the volatilising point of a solid just as it does to the boiling point of a liquid (p. 25) ; it is the partial pressure of the vapour, not necessarily the total pressure, on which the temperature depends.

Iodine.—If a quantity of iodine is heated in a porcelain basin and a cold glass funnel or plate is held over it, the iodine will melt and boil in the basin, but crystals of solid iodine will be deposited on the funnel

---

[1] Young, *Nature*, 1881, **24**, 239 ; Pettersson, *ibid.*, p. 167 ; Ramsay and Young, *Trans. Roy. Soc.*, 1884, **175**, 37 and 461 ; *Phil. Mag.*, 1887 [V], **23**, 61 ; Fischer, *Wied. Ann.*, 1886, **28**, 400 ; Ferche, *ibid.*, 1891, **44**, 265.

or plate. The heavy iodine vapour collects in the basin, so that its partial pressure in the atmosphere surrounding the substance soon rises above the triple point pressure, 91 mm., when fusion takes place ; but before reaching the cold glass surface the hot vapour becomes so diluted with air that the partial pressure falls below the triple point pressure, so that when condensation takes place the solid state is at once assumed. In the purification of crude iodine in France by " sublimation," the iodine is heated in earthenware retorts AA (Fig. 84), placed in a sand-bath B, and completely immersed in sand, so that condensation in the upper part of the retorts and necks may be prevented. The short necks of the retorts are connected with large earthenware receivers CC, so that the iodine vapour, on entering the receivers, is rapidly diluted with air. The partial pressure falls below the triple point pressure and direct passage from the gaseous to the solid state takes place. The narrow

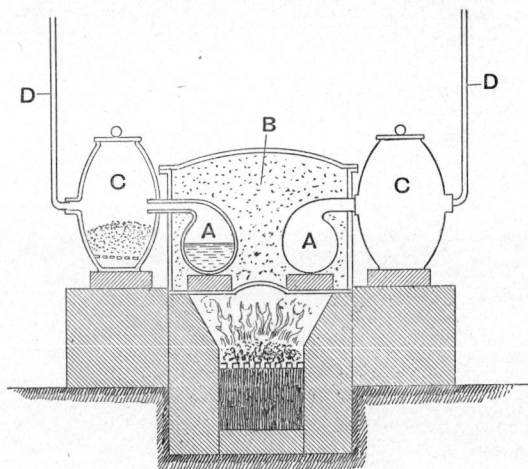

FIG. 84.

vertical tubes DD are for the escape of air or of steam if the iodine is moist.

The sublimation of the iodine obtained from " caliche " or crude sodium nitrate in Peru and Chili is carried out in a cast-iron retort to which is attached a series of wide cylindrical earthenware receivers or udells connected together so that the condenser as a whole resembles a wide drain pipe. The principle is the same in both cases.

Sulphur.—The purification of sulphur by " sublimation," with formation of flowers of sulphur, is carried out in a somewhat similar manner, but the triple point pressure of sulphur is far lower than that of iodine. When sulphur is boiled in an ordinary glass retort the forma-tion of small quantities of flowers of sulphur may always be observed in the upper part of the retort before the air has been expelled, but when, after expulsion of the air, the vapour condenses in the neck of the retort, liquid sulphur is formed.

Q

On the large scale crude sulphur is melted in the pot A (Fig. 85), whence it flows, at a rate regulated by the valve B, through the pipe C into the retort D. It is there boiled and the vapour passes through the wide pipe EE into the large brick chamber F. The sulphur vapour diffuses into the air so that its partial pressure rapidly falls and, so long as the temperature of the chamber is well below the melting point of sulphur, condensation with formation of flowers of sulphur takes place. If the temperature of the chamber is allowed to rise above about 112° liquid sulphur is condensed and this is allowed to flow, as required, through a hole in the plate G and is collected in the pot H.   The

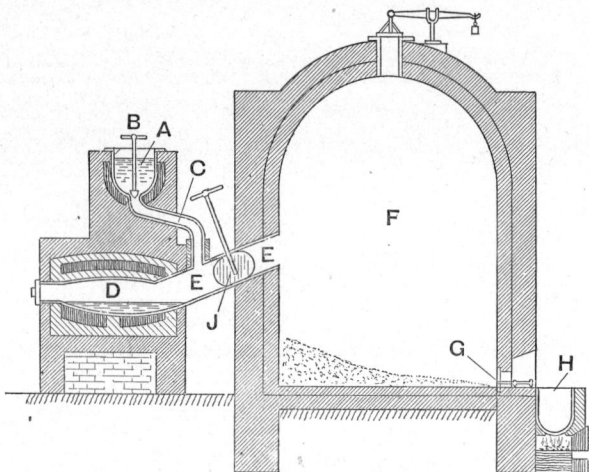

FIG. 85.

chamber can be shut off from the retort by means of the valve J, and a workman is then able to enter the chamber through a side door.

Anthracene.—In the purification of crude anthracene by passing superheated steam over the melted substance and condensing the vapour in a chamber by jets of cold water, the steam acts as an indifferent gas, and the partial pressure of the anthracene vapour is probably below the triple point pressure when condensation takes place.

Arsenious Oxide.—The triple point pressures of arsenic and of arsenious oxide are higher than the atmospheric pressure, and when these substances are heated and their vapours are condensed true sublimation takes place. The method adopted in Silesia in the case of arsenious oxide is illustrated by Fig. 86. The crude arsenious oxide, obtained by roasting arsenical pyrites or other arsenical ores, is placed in the iron pots AA, which are heated each by its own furnace BB. Above the pots are iron cylinders CC covered with caps DD. The arsenious oxide which escapes condensation in the cylinders passes through the pipes EE into the chamber or chambers F. The sublimed oxide is obtained either as a hard vitreous cake or as a loose friable mass according to the temperature at which the operation is conducted.

In Wales the vitreous " white arsenic " or arsenic glass is obtained by causing the sublimation to take place under slightly increased pressure. The crude oxide is heated in a cast-iron pan over which is placed a bell-shaped cover with a hole at the top. After the introduction of the crude oxide, the hole is closed by a plug so that the pressure rises to some extent when the pan is heated. The white arsenic condenses on the cover as a transparent glass.

In the preparation of arsenious oxide by roasting arsenical ores, or when the sublimation takes place in a furnace through which the hot

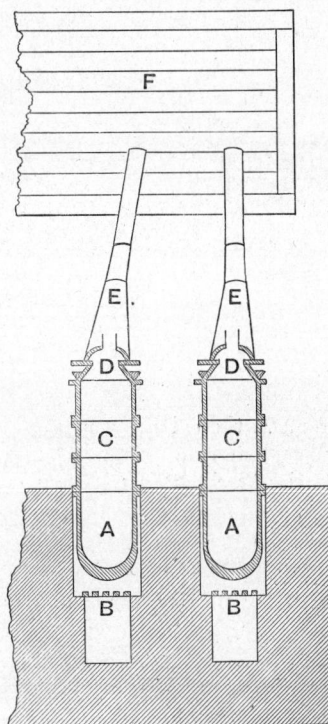

FIG. 86.

gases pass as well as the vapour of the oxide, the light sublimate is very liable to be carried away, and it is usual to conduct the gases through very long flues or through a series of condensing chambers. Electrical methods of causing the deposition of fine particles are now, however, employed for many purposes.

Ammonium Chloride.—When ammonium chloride is heated it does not melt but undergoes almost complete dissociation into ammonia and hydrogen chloride, these substances recombining when the temperature falls. The purification of ammonium chloride by sublimation differs, therefore, from that of arsenious oxide in so far that there is dissociation and recombination in place of vaporisation and condensation, but the methods employed in practice are very similar. The crude ammonium chloride is heated in large cast-iron pots A (Fig. 87), and the sublimate forms on flat or concave iron covers B, which rest on the pots.

Smaller earthenware or glass vessels are also used for the sublimation of ammonium chloride, and the crude chloride is sometimes mixed with

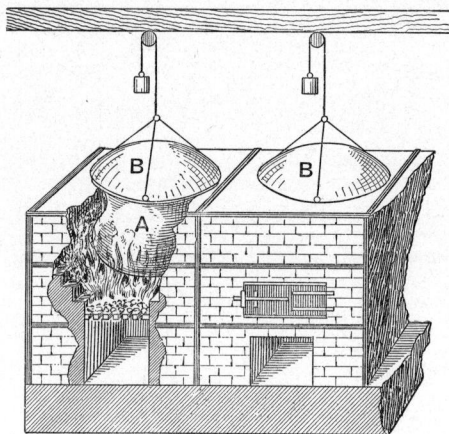

FIG. 87.

animal charcoal to remove the colouring matters derived from tar, or with calcium superphosphate or ammonium phosphate to prevent the sublimation of any ferric chloride that may be present.

It is obvious that in the process of sublimation nothing in the nature of an improved still-head can be used, and if the vapour is conveyed through tubes or pipes, they must be very wide to avoid blocking.

For the sublimation of naphthalene see p. 389.

APPENDIX

Correction of Height of Barometer for Temperature.—The atmospheric pressure is usually expressed in millimetres of mercury read at, or corrected to, $0°$. In practice the height, H, of the barometer is read at the temperature of the room, t, and the value at $0°$ calculated by means of one of the two following equations :—

1. $H_0 = H - 0 \cdot 000172 \, Ht$ if the scale is etched on the glass ; or
2. $H_0 = H - 0 \cdot 000161 \, Ht$ if a brass scale is used.

In Table 86 the values of $0 \cdot 000172 \, Ht$ are given for each degree from $10°$ to $30°$ and for pressures at intervals of 10 mm. from 720 to 780 mm. ; and in Table 87 the values of $0 \cdot 000161 \, Ht$ for the same temperatures and pressures. If the readings are smaller than 720 mm., it is convenient to plot the corrections against the pressures from 0 to, say, 800 mm. for alternate degrees from $10°$ to $30°$ on curve paper. The correction for any pressure up to 800 mm., at any temperature between $10°$ and $30°$, may then be easily ascertained from the diagram.

TABLE 86

GLASS SCALE.—VALUES OF $0 \cdot 000172 \, Ht$.

t.	Height of column of mercury.						
	720.	730.	740.	750.	760.	770.	780.
10	1·25	1·25	1·25	1·3	1·3	1·35	1·35
11	1·35	1·4	1·4	1·4	1·45	1·45	1·45
12	1·5	1·5	1·55	1·55	1·55	1·6	1·6
13	1·6	1·65	1·65	1·7	1·7	1·7	1·75
14	1·75	1·75	1·8	1·8	1·85	1·85	1·9
15	1·85	1·9	1·9	1·95	1·95	2·0	2·0
16	2·0	2·0	2·05	2·05	2·1	2·1	2·15
17	2·1	2·15	2·15	2·2	2·2	2·25	2·3
18	2·25	2·25	2·3	2·3	2·35	2·4	2·4
19	2·35	2·4	2·4	2·45	2·5	2·5	2·55
20	2·5	2·5	2·55	2·55	2·6	2·65	2·7
21	2·6	2·65	2·65	2·7	2·75	2·75	2·8
22	2·7	2·75	2·8	2·85	2·9	2·9	2·95
23	2·85	2·9	2·95	2·95	3·0	3·05	3·1
24	2·95	3·0	3·05	3·1	3·15	3·2	3·2
25	3·1	3·15	3·2	3·2	3·25	3·3	3·35
26	3·2	3·25	3·3	3·35	3·4	3·45	3·5
27	3·35	3·4	3·45	3·5	3·55	3·6	3·65
28	3·45	3·5	3·55	3·6	3·65	3·7	3·75
29	3·6	3·65	3·7	3·75	3·8	3·85	3·9
30	3·7	3·75	3·8	3·85	3·9	3·95	4·0

TABLE 87

BRASS SCALE.—VALUES OF 0·000161 Ht.

$t.$	Height of column of mercury.						
	720.	730.	740.	750.	760.	770.	780.
10	1·15	1·2	1·2	1·2	1·2	1·25	1·25
11	1·3	1·3	1·3	1·35	1·35	1·35	1·4
12	1·4	1·4	1·45	1·45	1·45	1·5	1·5
13	1·5	1·55	1·55	1·55	1·6	1·6	1·65
14	1·6	1·65	1·65	1·7	1·7	1·75	1·75
15	1·75	1·75	1·8	1·8	1·85	1·85	1·9
16	1·85	1·9	1·9	1·95	1·95	2·0	2·0
17	1·95	2·0	2·05	2·05	2·1	2·1	2·15
18	2·1	2·1	2·15	2·15	2·2	2·25	2·25
19	2·2	2·25	2·25	2·3	2·3	2·35	2·4
20	2·3	2·35	2·4	2·4	2·45	2·5	2·5
21	2·45	2·45	2·5	2·55	2·55	2·6	2·65
22	2·55	2·6	2·6	2·65	2·7	2·75	2·75
23	2·65	2·7	2·75	2·75	2·8	2·85	2·9
24	2·8	2·8	2·85	2·9	2·95	3·0	3·0
25	2·9	2·95	3·0	3·0	3·05	3·1	3·15
26	3·0	3·05	3·1	3·15	3·2	3·2	3·25
27	3·15	3·15	3·2	3·25	3·3	3·35	3·4
28	3·25	3·25	3·35	3·4	3·45	3·45	3·5
29	3·35	3·4	3·45	3·5	3·55	3·6	3·65
30	3·5	3·5	3·55	3·6	3·65	3·7	3·75

DISTILLATION OF ACETONE AND n-BUTYL ALCOHOL ON THE MANUFACTURING SCALE

JOSEPH REILLY, M.A., D.Sc., F.R.C.Sc.I., F.I.C.

CHEMIST-IN-CHARGE AT THE ROYAL NAVAL CORDITE FACTORY, HOLTON HEATH, DORSET

AND

THE HON. F. R. HENLEY, M.A., F.I.C.

CHAPTER XXI

ACETONE

ACETONE is produced on the manufacturing scale by several completely different processes. The chief of these are :—

(1) Fermentation processes.

(2) The decomposition of acetate of lime, derived from various sources.

(3) Conversion of acetic acid to acetone in presence of a catalyst.

(4) From crude wood spirit. But hitherto it has not been found possible to produce pure acetone from this source by distillation.

As the acetone in each case is mixed with considerable quantities of water, some general information on the separation of mixtures of acetone and water by distillation will first be given. The removal of the various volatile impurities will be considered separately in the more detailed description of the distillation of crude acetone from different sources, as the impurities to be removed differ in each case.

1. Separation of Acetone from Water when Mixtures of these Substances are distilled.—The vapour always contains a higher percentage of acetone than the liquid distilled. There is no general rule giving the composition of vapour and liquid in every case. But Table 88 gives the results of experiments made on a large number of mixtures.

Table 89, which is taken from E. Hausbrand, Rektifizier- und Destillier-Apparate, 1916, p. 164, shows the minimum heat abstraction required in the condenser of a column to produce a kilo of acetone from various mixtures of acetone and water. As in the case of ethyl alcohol and water a minimum is reached, in this case when the mixture contains 50 to 55 per cent of acetone.

Table 90 shows the composition of vapour and liquid on each plate of a rectifying column. A comparison of the corresponding tables for the distillation of ethyl alcohol and water shows that in the case of acetone and water, fewer plates are required in the column and that the necessary heat abstraction in the condenser is smaller.

[TABLE

233

TABLE 88

Acetone per cent by weight in liquid and vapour, H. Bergström. Boiling points [ ] H. R. Carveth, ( ) J. H. Pettit.[1]

Boiling point of liquid. °C.	Acetone in liquid. Weight per cent.	Acetone in vapour. Weight per cent.	Boiling point of liquid. °C.	Acetone in liquid. Weight per cent.	Acetone in vapour. Weight per cent.
(56·9)	100	100	[73·3]	17·5	86
	99	99·5		15	84·2
[57·2]	95	97·5	[77·16]	12·5	81·93
	90	96·3	(81·1)	11	79·95
[58·3]	89	96·18		10	78·4
[58·9]	85·5	95·76	[80·7]	9·5	77·6
	80	95·3		8	74·4
[60·7]	76·5	95·02		7	71·2
[60·45]	72	94·72		6	67·4
	70	94·6		5	62·6
(61·6)	65·5	94·33	[88·7]	4·5	59·3
(62·4)	60·5	94·03		3	47·0
[62]	55	93·6	[94·63]	2	36·0
	50	93·3		1	20·4
(64·4)	45·5	93·03		0·5	11·0
(65·3)	40·5	92·55		0·4	8·93
[65·9]	36	91·94		0·3	6·85
(67)	33·5	91·56		0·2	4·78
(70·7)	24·5	89·59	99·8	0·1	2·7
	20	87·6			

TABLE 89

Acetone and Water

Minimum amount of heat in great calories to be abstracted in the condenser to produce 1 kilo of acetone 99·75 per cent by weight from mixtures of water and acetone containing 95·5 to 1 per cent by weight of acetone.

Acetone content. Weight per cent.		Calories.	Acetone content. Weight per cent.		Calories.
Liquid.	Vapour.		Liquid.	Vapour.	
95·5	97·75	...	35	91·8	55·9
95	97·5	132·9	30	91·0	59·5
90	96·3	87·7	25	89·0	66·8
85	95·7	70·2	20	87·6	81·7
80	95·3	60·5	15	84·2	108·0
75	94·9	55·4	10	78·4	154·0
70	94·6	52·3	9	76·8	168·9
65	94·3	50·3	8	74·4	192·4
60	94	49·8	7	71·2	225·1
55	93·6	49·6	6	67·4	269·3
50	93·3	49·6	5	62·6	332·6
45	93·0	49·7	2	36·0	994·3
40	92·5	51·7	1	20·0	2188·8

[1] H. Bergström, Stockholm *Aftryk ur Bitrang til Jem-Kontoret's Annaler*, 1912; H. R. Carveth, *J. Phys. Chem.*, 1899, **3**, 193; J. H. Pettit, *J. Phys. Chem.*, 1899, **3**, 349.

Table 90

Acetone and Water

Rectifying Column

Acetone content, weight per cent of liquid and vapour on each plate of a column to produce 10 kilo. acetone 99·75 per cent by weight. The amount of heat abstracted in the condensers being 1500, 3000, 5000 great calories.

Number of plate from the top.	Acetone in liquid. Weight per cent.	Acetone in vapour. Weight per cent.	Acetone in liquid. Weight per cent.	Acetone in vapour. Weight per cent.	Acetone in liquid. Weight per cent.	Acetone in vapour. Weight per cent.
	1500 calories.		3000 calories.		5000 calories.	
	99·5	99·75	99·5	99·75	99·5	99·75
1	99	99·5	99	99·5	99	99·5
2	98·75	99·4	98·6	99·3	98·5	99·25
3	98·6	99·3	97·9	98·98	96·6	98·3
4	98·27	99·11	96·9	98·48	94·5	97·39
5	97·9	98·98	95·75	97·87	86·5	95·83
6	97·0	98·5	93·1	97·06	26	90·03
7	96·1	98·13	83	95·53	4·66	60·46
8	95·6	97·78	24·6	89·63	3·7	53·07
9	95·2	97·6	18·3	86·56	3·6	52·5
10	94·2	97·3	6·1	67·81		
11	93	97	5·5	65·13		
12	92·1	96·89				
13	88·25	96·09				
14	69·1	94·56				
15	22	88·48				
16	11	80·1				
17	10·75	79·5				
18	10·4	78·95				

CHAPTER XXII

PRODUCTION OF ACETONE AND *n*-BUTYL ALCOHOL BY THE FERMENTATION PROCESS

THE production of *n*-butyl alcohol and other alcohols from carbohydrate material, directly by fermentation, has been known for a long time. The presence of iodoform-producing substances among the products of fermentation has been recorded occasionally. The first observation of acetone as the result of fermentation of carbohydrates was made by Schardinger,[1] who obtained this substance together with acetic and formic acids. The isolation of an organism yielding acetone and *n*-butyl alcohol from amylaceous material was due to Fernbach and Strange.[2] Since this initial discovery many other processes have been described which produce acetone and *n*-butyl or ethyl alcohol by the breakdown of carbohydrates under bacterial action.

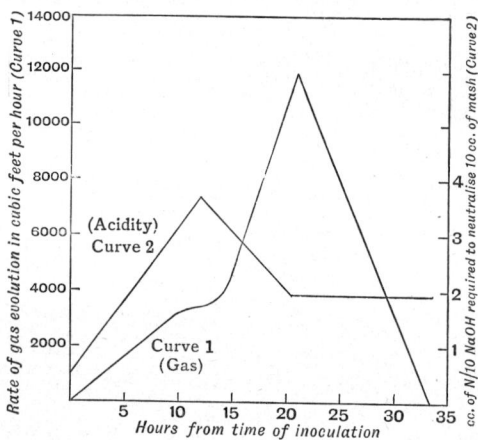

FIG. 88.

Table 91 gives the yields of products obtained in the ordinary " acetone *n*-butyl alcohol " fermentation process.

In a normal fermentation the acidity of the mash increases from a very small initial value until a maximum is reached in from 13 to 17 hours after inoculation. (The length of time taken to reach this point is influenced by the percentage of inoculant used, temperature of the mash, etc.) When the maximum acidity is reached in 6·5 per cent maize mash, 3·5 to 4·5 c.c of $N/10$ alkali are required to neutralise 10 c.c. of mash. After this point is reached a very marked acceleration in the rate of production of acetone, *n*-butyl alcohol, carbon dioxide and hydrogen takes place. The acidity gradually falls to a constant value, 1·5 to 2·5 c.c. of $N/10$ alkali being required to neutralise 10 c.c. of mash (see Fig. 88, Curve 2).

[1] Schardinger, *Centr. Bakt.*, Part II., 1907, **18**, 748.
[2] Fernbach and Strange, Eng. Patent, 1912, 21073.

The rate of gas evolution is shown in Curve 1, Fig. 88, and the composition of the gas in Table 92.[1]

TABLE 91

1000 lb. maize containing 650 lb. starch.   Volume of mash (6·5 per cent maize) = 1540 gallons.

		lb.	Carbon content.
650 lb. starch yield		70 acetone	43·5
		163 n-butyl alcohol	105·7
	3410 cu. ft. of CO_2 evolved — at 27° and 760 mm. $\Big\}$ = 390 CO_2 evolved		106·3
1 vol. of mash * at 38° dissolves 0·3 vol. CO_2. The latter being estimated at N.T.P.	74 cu. ft. CO_2 in sol. at 0° and 760 mm. $\Big\}$ = 9·2 CO_2 in solution		2·5
	2090 cu. ft. H_2 evolved at 27° and 760 mm. $\Big\}$ = 11 H_2		—
	Residual acidity † = 12 "acids" containing		5·7
		655·2	263·7

Theoretically 650 lb. of starch is equivalent to 722 lb. of hexose or 288·8 lb. of carbon.

* 1 vol. of distilled water at 38° dissolves 0·55 vols. CO_2 at N.T.P. where pressure of $CO_2 = 760$ mm.—At 38° tension of water vapour = 49·3 mm.—Actual pressure due to gas = 710·7 mm.—as gas consists of 40 per cent H_2 and 60 per cent CO_2 by vol., the partial pressure of the $CO_2 = 426·4$ mm.—At this pressure the solubility of the CO_2 is 0·31 vols. (at 0 and 760 mm. pressure) in 1 vol. of liquid. The fermentation liquors contain substances which influence the solubility and retention of carbon dioxide. Estimations of CO_2 in liquors from a series of fermentations gave results varying between 0·15 and 0·52 vols. of CO_2 dissolved in 1 vol. of mash liquor. A slightly lower value (0·3) than that of the solubility of CO_2 in water at 38° has been used.

† The residual acidity is taken as acetic acid 56·5 per cent by weight, and butyric acid 43·5 per cent by weight. The non-volatile portion is recorded as butyric acid for the purpose of calculation.

Calculated as Percentage of Starch Fermented

100 g. starch gives 111·1 g. hexose and contains 44·4 g. carbon.

		g.	Carbon.
100 g. starch gives		10·77 acetone	6·68 g.
,,	,,	25·07 n-butyl alcohol	16·21
,,	,,	61·41 carbon dioxide	16·75
,,	,,	1·60 hydrogen	—
,,	,,	1·80 residual acidity	0·85
		100·65	40·49

Rate of Gas Evolution, and Rise and Fall of Acidity throughout the Fermentation.

The curves in Fig. 88 show the changes in

(1) the rate of gas evolution, and

(2) the acidity during a fermentation. The figures are the average of 12 fermentations of 6·5 per cent maize mash (40,000 gallons) with 4·7 per cent of inoculant.

The acidity is measured in c.c. of $N/10$ alkali required to neutralise 10 c c. of mash after boiling the solution to remove carbon dioxide. Phenolphthalein is used as indicator.

[1] Reilly, Hickinbottom, Henley and Thaysen, *Biochem. Journ.*, 1920, vol. xiv. No. 2, p. 229.

It will be noted that the rate of gas evolution rises steadily with the increase of acidity for some time ; it then becomes constant (in some cases it even slackens somewhat). As the acidity falls, the rate of gas evolution rises quickly to a maximum, and then falls rapidly until the end of the fermentation. Readings of the gas evolution were taken every hour. The acidity was estimated every three hours. The acidity at the end of the fermentation is generally higher than that of the mash when inoculated.

Composition of Gas.—Estimations have been made of the composition of the gas evolved during fermentation. Table 92 shows the results of one of these experiments.

The fermenting vessel was about two-thirds full of mash, the space above the latter being occupied by air before inoculation.

TABLE 92

Time, 28. vi. '16.	Gas evolved, cubic feet per hour.	Composition of Gas evolved.	
		CO_2.	H_2.
		per cent.	per cent.
4 p.m.	—	—	—
7 ,,	253	23·0	77·0
8 ,,	834	31·9	68·1
9 ,,	822	42·2	57·8
10 ,,	660	47·4	52·6
11 ,,	760	51·6	48·4
29. vi. '16. 9.30 a.m.	1186	62·0	38.0

The percentage of carbon dioxide did not vary from 9·30 A.M. 29. vi. '16 to the end of the fermentation, which lasted 36 hours. Total gas evolved, 42,694 cubic feet. The high percentage of hydrogen in the gas evolved during the early hours is partly due to the greater solubility of the carbon dioxide.

The production of acetone and n-butyl alcohol at various stages in the fermentation is shown in Table 93.

TABLE 93

Time.	Acidity.	Ratio, acetic to butyric acid.	Amount of acetone and n-butyl alcohol mixture in 1 litre of mash.
9.0 p.m.	1·0	1 : 0·5	None
11.0 ,,	1·5	1 : 0·62	None
12.15 a.m.	2·0	1 : 0·9	Trace
5.30 ,,	3·7	1 : 1·25	1 c.c.
3.15 p.m.	2·0	1 : 0·28	4 c.c.
4.30 ,,	1·6	1 : 0·25	5 c.c.

Acidity is expressed in c.c. of $N/10$ alkali required to neutralise 10 c.c. of mash.

Wort, similar to that used in alcoholic fermentation for the production of ethyl alcohol, can also be used. The wash at the end of the fermentation of a starchy mash usually contains about 0·6 per cent of acetone, and 1·4 per cent of *n*-butyl alcohol, and a much smaller proportion of ethyl alcohol. When wort is used the concentration may be much higher than this.

The Distillation Process

The distillation of the fermented wash is carried out either in a Coffey still slightly modified for the purpose, or in one of Messrs. Blair, Campbell & M'Lean's acetone stills. Both these stills will produce almost pure acetone in one operation. But when the acetone is required for the manufacture of cordite, redistillation is necessary in order to remove carbon dioxide and traces of substances which affect the stability of cordite. This final distillation is carried out in an intermittent still with a very efficient fractionating column.

Preliminary Distillation in Coffey's Still.—See Fig. 121, p. 312.

The fermented wash is supplied to the still in the usual manner, but its temperature is from 35° to 40° C. As it passes down the rectifying column in the wash pipe its temperature must be slightly under 56° C. at the level of the spirit plate, where acetone is drawn off from the still. The difference between 35° C. and 56° C. is so small that it is difficult to keep the temperature low enough at the spirit plate. To overcome this difficulty either the wash pipe may be brought into the rectifying column lower down, and its upper coils replaced by an independent cold water coil, or the temperature of the wash may be reduced by mixing it with cold water.

After leaving the rectifying column, the wash enters the top of the analyser down which it flows, leaving the base of the column as spent wash, free from acetone and *n*-butyl alcohol.

The condensate leaving the base of the rectifier generally consists of a dilute solution of *n*-butyl alcohol in water. This is pumped back to the top of the analyser.

Provision is made for running off the *n*-butyl alcohol from one of the lower plates of the rectifying column. The method of drawing off the liquid from the plate is shown in Fig. 89. When the still is working smoothly, an azeotropic mixture of *n*-butyl alcohol and water can be run off steadily. It is then cooled in a condenser, and passes to a separator, in which two layers are formed ; the upper layer consists of 85 per cent *n*-butyl alcohol, and 15 per cent water ; the lower, of 12 per cent *n*-butyl alcohol and 88 per cent water. The lower layer is pumped back into the top of the analyser. When very concentrated wash is distilled the mixture of *n*-butyl alcohol and water is run off from the base of the rectifying column. In either case the *n*-butyl alcohol obtained contains some ethyl alcohol and acetone. It is most important

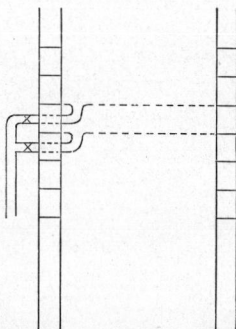

FIG. 89.

to keep up a steady flow of n-butyl alcohol from the rectifying column, so as to prevent its gradually passing up the column and contaminating the acetone distillate. Acetone is drawn off from the spirit plate. The gases dissolved in the wash leave the still by a pipe passing into the air from the top of the rectifier, carrying with them a very small amount of acetone, which can be removed by passing through a scrubber, down which water trickles. The acetone from the spirit plate flows into a cooler, where its temperature is reduced to 15° C., the specific gravity being about 0·800. Having only been in contact with the gases at a temperature of about 56° C., the amount of carbon dioxide dissolved in the acetone is much smaller than is the case when the acetone is cooled in presence of excess of the gases. This point is of some importance when the acetone is required for the manufacture of cordite.

Acetone of over 99 per cent purity can be produced with Coffey's still, and the n-butyl alcohol which is simultaneously run off contains less than 1 per cent of acetone.

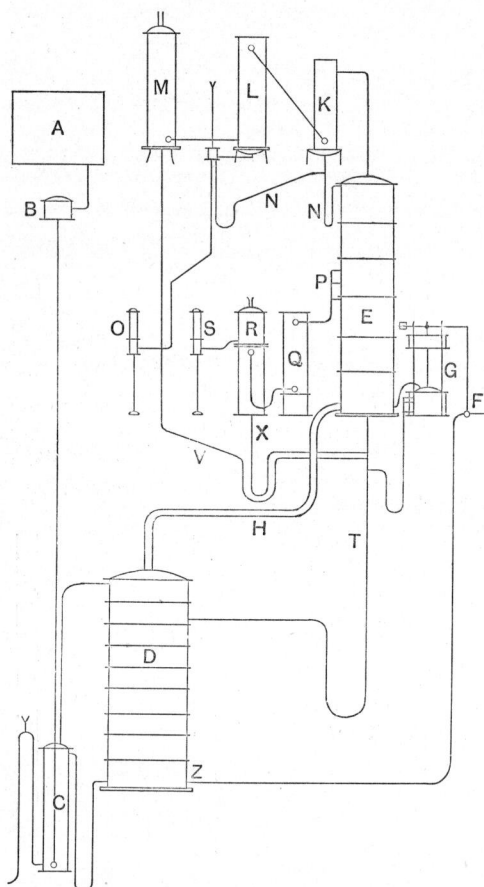

FIG. 90.

Messrs. Blair, Campbell & M'Lean's Continuous Acetone Still, Fig. 90.—The fermented wash is distilled in a boiling column, having plates of the bell type, described under alcohol distillation (p. 303). The wash flows from the reservoir A through the regulator B, and preheater C, in which it is heated by a heat exchange with the spent wash. It then enters the wash column D just above the topmost plate, and flows away as spent wash, practically free from acetone and n-butyl alcohol, by a trapped pipe at the base of the column.

Steam is introduced below the lowest plate at Z, the supply being automatically controlled by the regulator G. The rising vapour bubbles through a depth of about 1 inch of liquid on each plate. The vapour leaving the top of the boiling column by pipe H enters the base of the rectifying column, which is of the same type as the boiling column, and is also provided with plates of the bell type.

The vapour leaving the top of the rectifying column enters a water-cooled tubular dephlegmator K, which provides the necessary condensate for the rectifying column. The vapour passing uncondensed through this dephlegmator enters at the top of a water-cooled condenser L, and is there completely condensed. Part of the condensate so produced flows to the test glass O as finished acetone, and part is returned by pipe N to the top of the rectifying column with the liquid condensed in the dephlegmator K. The gas originally dissolved in the wash is led from the base of the condenser L to a gas scrubber M. This consists of a cylindrical tower filled with pieces of coke down which water is made to flow. By this means the gases are freed from acetone and the dilute aqueous solution of acetone so produced flows back through a trapped pipe V to the boiling column.

Removal of *n*-Butyl Alcohol from the Rectifying Column.—The *n*-butyl alcohol (azeotropic) mixture with water is run off from the rectifying column by one of the tapped pipes P. The *n*-butyl alcohol and water are first cooled by passing through a coil Q, immersed in water, and then flow into a separator R. The lower layer from the separator, consisting of a saturated solution of *n*-butyl alcohol in water and traces of acetone, is returned to the boiling column by pipe X. The upper layer from the separator, consisting of a saturated solution of water in *n*-butyl alcohol, is run off at the observation glass. The still runs very evenly for about 48 hours, but after that time the temperature of the vapour entering the dephlegmator tends to rise above 56° C., and it is difficult to keep it down to this normal level. The reason for this appears to be that ethyl alcohol gradually accumulates in the rectifying column above the level at which *n*-butyl alcohol is drawn off. To overcome this difficulty the supply of wash to the still must be replaced by water. The distillation is then continued till the temperature at the base of the rectifying column rises to 100° C. By this means the rectifying column is cleared of the accumulation of ethyl alcohol. The distillate collected during this process, and containing some ethyl alcohol, is treated with the rest of the crude acetone in the final distillation. A liquid rich in ethyl alcohol might also be run off from the rectifying column at a suitable level, but this has not yet been attempted.

Control of the Still.—The flow of wash is controlled by the regulator described under alcohol distillation, page 310, Fig. 117. The supply of steam to the boiling column is automatically regulated by the apparatus described under alcohol distillation, p. 310, Fig. 118 ; and the flow of acetone and *n*-butyl alcohol from the still is regulated by test glasses similar to that described on p. 311. The temperature at the base of the rectifying column and of the vapour entering the dephlegmator must be kept very steady. If acetone is not run off fast enough from the still, the temperature at the base of the rectifying column will tend to fall, and the separation of the two layers of *n*-butyl alcohol-water mixture in the separator will no longer take place properly, owing to the presence of too much acetone.   If acetone is run off too rapidly, the temperature of the vapour entering the dephlegmator will very soon rise above 56° C.   It is important to keep the pressure in the rectifying column very steady.   For this purpose pressure gauges are fixed at the base and near the top of the rectifying column.

Acetone of over 99 per cent purity can be obtained with this still, but the presence of traces of impurities make it unfit for immediate use in the manufacture of cordite.   To remove these impurities the acetone is redistilled, after the addition of small quantities of caustic soda, in a discontinuous still.

In practice the acetone was generally run off from the still at a lower strength, as this somewhat facilitates the efficient removal of *n*-butyl alcohol from the column.   But this procedure is not necessary.

Final Purification of Acetone produced by the Fermentation Process.—Two types of still used for this purpose are shown in Fig. 91 [1] and Fig. 92.   The still (Fig. 91 or 92) is charged with about 4800 lb. of crude acetone from the primary still.   The charge is slowly heated by live steam introduced into the liquid through the perforations of a ring placed in the base of the still.   By this means most of the gas dissolved in the crude acetone is driven off before any condensate is formed in the condenser.   As soon as the evolution of gas ceases, steam is shut off, and the necessary quantity of caustic soda solution, containing about 8 lb. of sodium hydroxide, is added, and the heating continued.   The crude acetone produced from Coffey's still contains so little carbon dioxide in solution that the preliminary heating is unnecessary, and the caustic soda can be added as soon as the still is charged.

The distillate is collected in three fractions :—

1. The first runnings, consisting of almost pure acetone, amount to about 125 lb., or 2·6 per cent by weight of the charge ; it contains, however, too much carbon dioxide to pass the specification and has to be submitted to a second alkaline distillation.

2. The second fraction consists of good acetone which needs no further treatment.   This fraction amounts to about 74 per cent by weight of the charge.

3. The third fraction with specific gravity over 0·802 amounts to about 23·3 per cent by weight of the charge.   The collection of this fraction ceases when the specific gravity of the distillate rises to 1·0.   This fraction is redistilled after the addition of caustic soda and permanganate.

Redistillation of Third or " High Gravity " Fraction.—The redistillation of the third fraction referred to above is carried out in the same type of still as that already described.

Three fractions are collected :—

1. Acetone specific gravity <0·802, forming nearly 13 per cent by weight of the charge.

[1] A still similar to that shown on p. 299 for the rectification of ethyl alcohol might be used.

2. " Crude Ketone," forming about 70 per cent by weight of the charge.   This fraction is redistilled.

3. *n*-Butyl alcohol and water, forming nearly 3·0 per cent by weight of the charge.   The remainder consists of water.

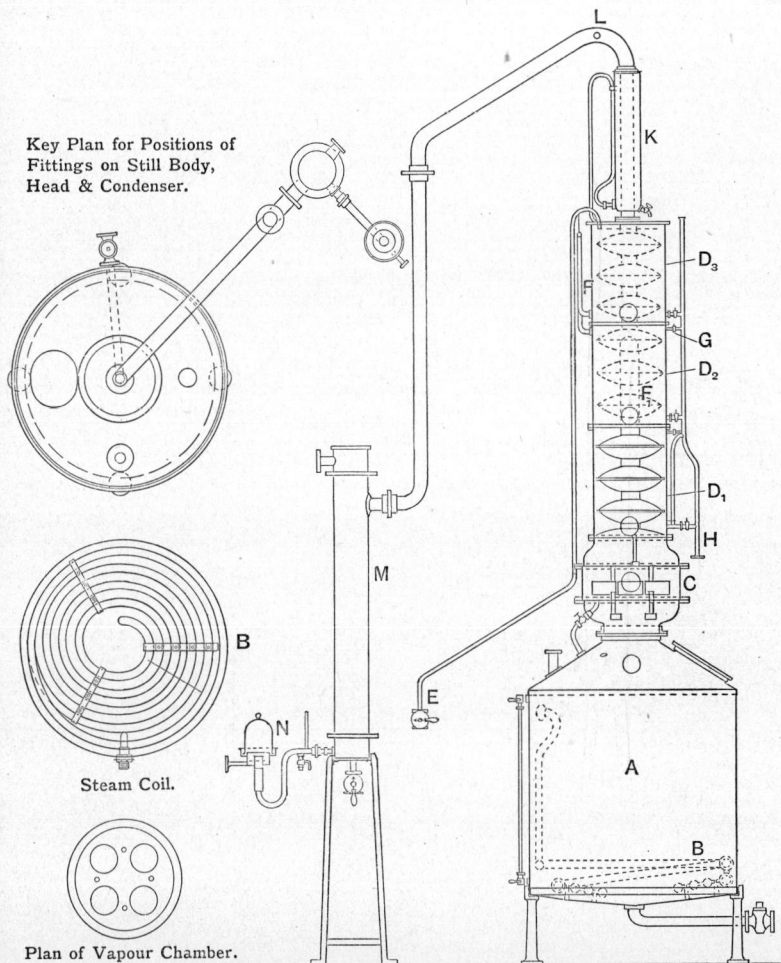

Key Plan for Positions of
Fittings on Still Body,
Head & Condenser.

Steam Coil.

Plan of Vapour Chamber.

Fig. 91.—Acetone Rectifying Still made by Messrs. J. Dore & Co., London.

A, Still; B, Steam coils; C, Vapour chamber (not water jacketed); D₁, D₂, D₃, Water-jacketed section of the column; E, Cold water supply to jackets D₁, D₂, D₃; F, Syphon from D₃ to D₂; F₁, Syphon from D₂ to D₁; G, Vent pipes from water jackets; H, Outlet pipe for water in jackets; K, Supplementary water jacket; L, Vapour pipe; M, Condenser; N, Still watcher.

Redistillation of the Second or " Crude Ketone " Fraction.

The redistillation of this fraction yields three fractions :—

1. Acetone, 3·5 per cent by weight of the charge.

2. " Finished Ketone," 87·0 per cent of the charge.

3. *n*-Butyl alcohol and water, 9·5 per cent of the charge.

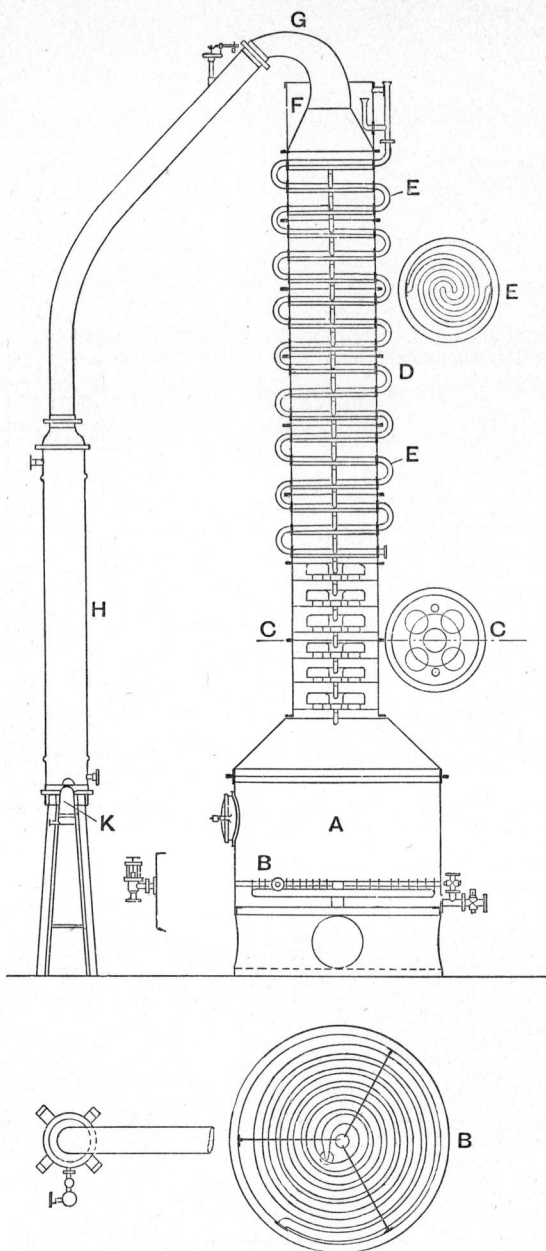

FIG. 92.—Acetone Rectifying Still, made by Messrs. J. Miller & Co., Glasgow.

A, Still; B, Steam coils; C, Lower section of still-head, not water cooled; D, Section of column provided with cooling coils E above each plate; F, Water bath; G, Vapour pipe; H, Condenser; K, Still watcher.

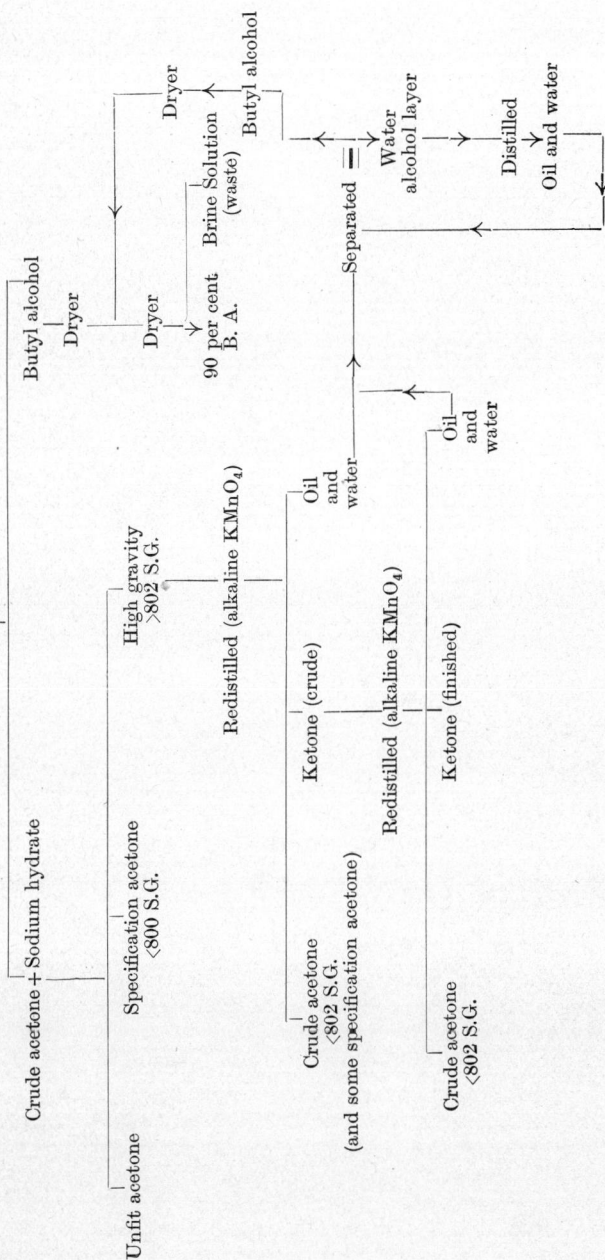

TABLE 94

FERMENTED WASH

Primary Distillation

Crude acetone + Sodium hydrate

Unfit acetone

Specification acetone
<800 S.G.

High gravity
>802 S.G.

Redistilled (alkaline KMnO$_4$)

Oil and water

Ketone (crude)

Crude acetone <802 S.G.
(and some specification acetone)

Redistilled (alkaline KMnO$_4$)

Oil and water

Ketone (finished)

Crude acetone <802 S.G.

Butyl alcohol

Dryer

Dryer

90 per cent B. A.

Brine Solution (waste)

Dryer

Dryer

Butyl alcohol

Butyl alcohol

Separated

Water alcohol layer

Distilled

Oil and water

Fraction (2) is not further treated, but appears to consist largely of ethyl alcohol, *n*-butyl alcohol and water.   A sample of this fractionated in the laboratory gave the following fractions :—

B.-pt. 65–70° C.	3·3 per cent
75–80	34·8 ,, ,,
80–85	1·6 ,, ,,
85–92	4·7 ,, ,,
n-Butyl alcohol and water . .	52·6 ,, ,,

Table 94 gives a diagrammatic representation of these distillations.

The organism *Bacillus acetoethylicum* is capable of converting carbohydrates into a mixture of ethyl alcohol, acetone, and formic acid.   The yield of acetone is 8–9 per cent and of alcohol 14–20 per cent by weight of the carbohydrate fermented.[1]   The optimum temperature for growth and for the fermentation is 43° C.   The process takes place in presence of calcium carbonate.   The process has been investigated from the commercial aspect in the plant of the Commercial Solvents Corporation at Terre Haute, Indiana.[2]

Acetone has also been prepared on an industrial scale from seaweed and kelp.[3]

---

[1] Northrop, Ashe, and Senior, *J. Biol. Chem.*, 1919, **39**, 1.
[2] Cf. Arzberger, Peterson, and Fred, *J. Biol. Chem.*, **44**, 465.
[3] *J. Ind. and Eng. Chem.*, 1918, **10**, 858, and *J. Soc. Chem. Ind.*, 1916, **35**, 565.

CHAPTER XXIII

PRODUCTION OF ACETONE BY MEANS OTHER THAN FERMENTATION

Manufacture of Acetone from Acetate of Lime, etc.—Acetone is produced commercially by the dry distillation of various acetates, calcium acetate (commercial grey acetate of lime) or barium acetate being generally used. The single acetate is usually employed, although mixtures of two acetates have been recommended (F.P. 439732/1911).

Grey calcium acetate usually contains about 80 per cent calcium acetate, the remaining 20 per cent consisting of water and various impurities, including small quantities of calcium formate and propionate as well as salts of other organic acids.

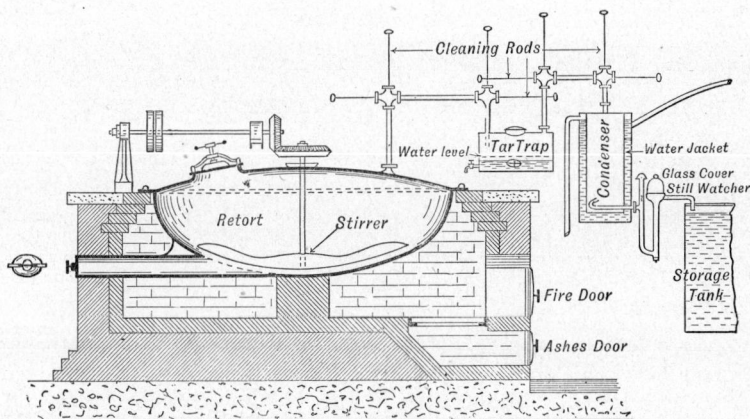

FIG. 93.

Calcium acetate when heated at a temperature of about 380° C. decomposes, giving acetone and a residue of calcium carbonate, but at the same time the accompanying calcium salts present as impurities also react and acetaldehyde and various higher ketones are formed together with the condensation products. In addition other impurities (such as dumasin) and tar-like bodies are formed.

The distillation of calcium acetate is carried out generally in a shallow circular retort taking a charge of from 300 lb. to 700 lb. Fig. 93 gives an outline drawing (kindly supplied by Mr. W. T. Thomson, F.I.C.) of the usual type of retort employed.

The heating of the retort is carried out by direct fire. Stirring gear is provided and the outlet tubes are fitted with rods passing through stuffing joints and having an iron disc fitted on the end so that the pipes may be cleaned of dust, tar, etc., at intervals during the distillation. After the retort has been charged the stirring gear is started, and heating is begun moderately at first, as acetone and water are given off at a moderate temperature. The temperature is gradually increased up to 380° C., when the decomposition of the acetate proceeds briskly. Steam is blown through at the end to sweep through the last portion of the distillate.

The vapour from the retort passes first through a tank in which the greater part of the dust and tar are retained. Acetone and water vapour mixed with volatile impurities pass on, and are condensed in a worm and collected in a galvanised tank. Towards the end of the distillation the contents of the dust and tar tank are heated by a steam coil to drive out any acetone which may have condensed with the tar.

The distillate is diluted with water till the specific gravity reaches 0·960. The liquid is then well stirred and allowed to stand for 12 hours. The tar, higher ketones, etc., which rise to the top are run off by means of a tap to an adjoining tank. To the crude liquor containing acetone is then added 3·5 lb. of sodium hydrate per 1000 gallons of liquid. The liquid is then well stirred and transferred to a rectifying still. The addition of the sodium hydrate solution prevents volatile acids from passing over into the distillate.

Two fractions are collected, the subsequent treatment of which is indicated in Table 95. The object of adding sulphuric acid is to prevent amines from passing over into the distillate. The same type of still is employed as in the final distillation of acetone produced by fermentation.

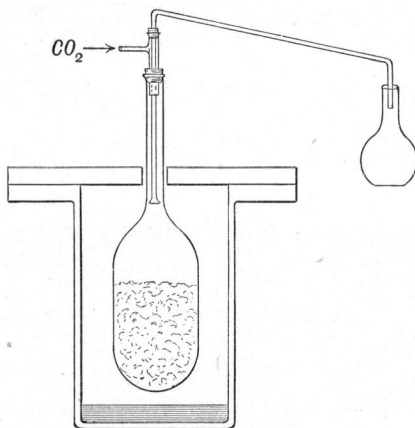

$CO_2 \rightarrow$

FIG. 94.

The presence of alkali (free lime) in the calcium acetate is considered to be a disadvantage if a high yield is wanted. Becker [1] introduces a stream of carbon dioxide into the heated vessel containing the calcium acetate to prevent the formation of free lime. For the preparation of ketones by the decomposition by heat of certain calcium salts in a current of carbon dioxide, there is an earlier reference to that of Becker. In 1891, S. Young [2] prepared dibenzyl ketone in good yield by heating dry calcium phenylacetate in a current of carbon dioxide. A short description of the apparatus (Fig. 94) and method as employed on the laboratory scale is given. The calcium salt dried at 130° was placed in a wide

[1] Zts. Ang. Chem., 1907, 20, 206.   [2] Trans. Chem. Soc., 1891, 59, 621.

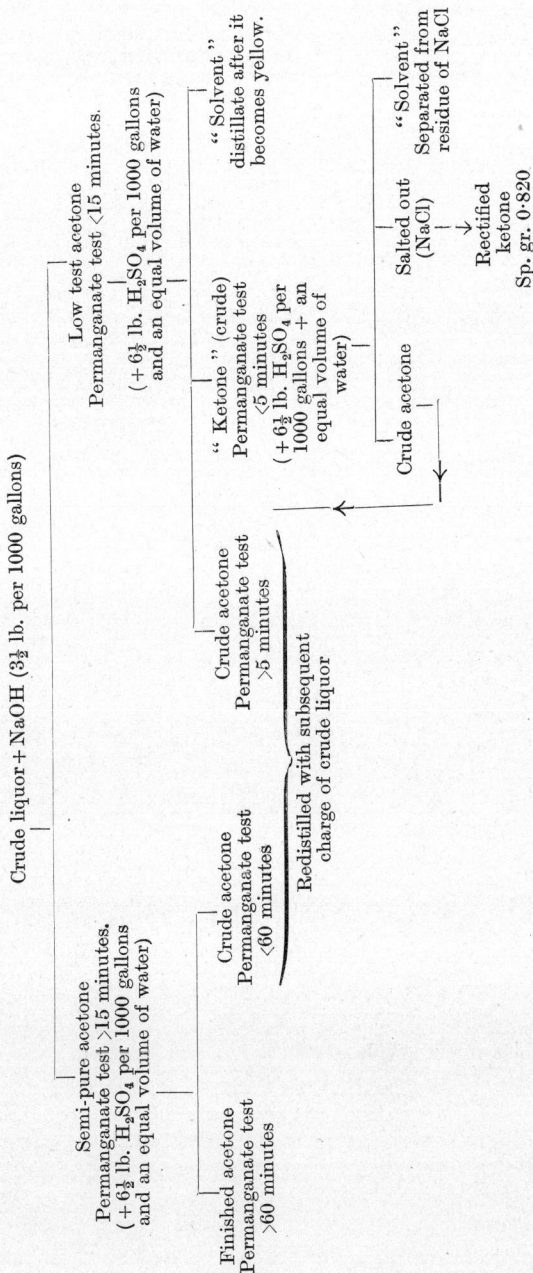

TABLE 95

Crude liquor + NaOH (3½ lb. per 1000 gallons)

Finished acetone
Permanganate test
>60 minutes

Crude acetone
Permanganate test
<60 minutes

Semi-pure acetone
Permanganate test >15 minutes.
(+ 6½ lb. H₂SO₄ per 1000 gallons
and an equal volume of water)

Crude acetone
Permanganate test
>5 minutes

Redistilled with subsequent
charge of crude liquor

"Ketone" (crude)
Permanganate test
<5 minutes
(+ 6½ lb. H₂SO₄ per
1000 gallons + an
equal volume of
water)

Low test acetone
Permanganate test <15 minutes.
(+ 6½ lb. H₂SO₄ per 1000 gallons
and an equal volume of water)

"Solvent"
distillate after it
becomes yellow.

Crude acetone

Salted out
(NaCl)

"Solvent"
Separated from
residue of NaCl

Rectified
ketone
Sp. gr. 0·820

cylindrical glass bulb sealed to a narrower tube. The bulb was heated by the vapour of sulphur boiling in a flanged iron vessel provided with a heavy iron cover. The narrow tube already referred to was provided with an india-rubber stopper, through which a T-tube passed. A piece of barometer tubing enlarged conically at its lower end passed through the T-tube and the narrow tube. It was attached to the T-tube by a cork. The barometer tubing was bent as shown in Fig. 94, after the T-tube and the cork had been placed in position. It is possible that a bath containing a mixture of equal molecular proportions of potassium and sodium nitrates might with advantage be substituted for the sulphur bath. The limit of temperature for the distillation should be 370 to 446°.

The overheating due to direct contact of the acetate with the hot

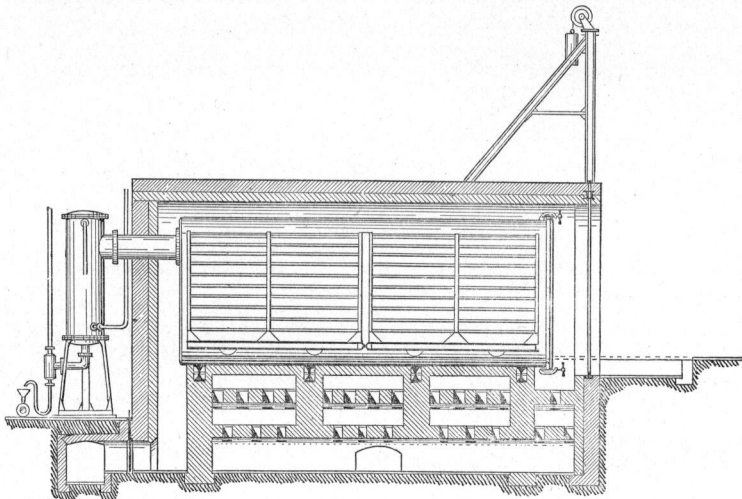

FIG. 95.

walls of the retort has been largely overcome by the employment of a retort made by F. H. Meyer of Hanover. Fig. 95 shows a retort of this type. Trays of acetate are placed upon a perforated truck which is wheeled bodily into the retort. By this method a spent charge is readily withdrawn and a fresh charge inserted with little loss of heat, and the avoidance of the dusty operation of withdrawing spent lime.

A method somewhat related to the above consists in passing a continuous current of pyroligneous acid over a heated acetate capable of forming acetone.[1]

Conversion of Acetic Acid to Acetone in Presence of a Catalyst.— When the vapours of acetic acid are passed into air-tight vessels containing some porous material saturated with lime or baryta, a good conversion to acetone is brought about.[2]

The methods developed at the Shawinigan Falls, Canada, for the

[1] *J. Soc. Chem. Ind.,* 1906, **25**, 634; 1907, **26**, 1002; 1908, **27**, 277.
[2] *J. Soc. Chem. Ind.,* 1899, **18**, 128, 828. Bauschlicker, D.R.P. 81914.

preparation of acetone from synthetic acetic acid have recently been described by Matheson.[1] In the experimental installation three tubes (four inch) 6 feet in length were employed. They were heated electrically by means of a resistance winding and the acid was

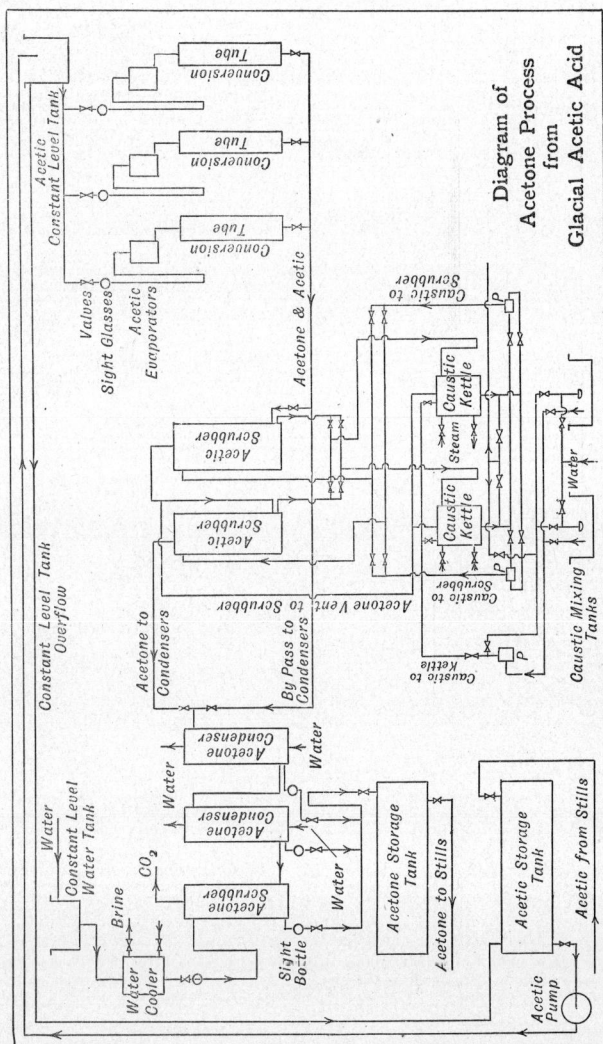

Fig. 96.

vaporised before passing into the tube. The catalyst consisted of hydrated lime mixed with a small amount of magnesia deposited on pumice stone. The yield obtained was about 95 per cent of the theoretical. When the process was developed to large scale experiment, initial difficulties were encountered, due to the difficulty of heat

[1] *Canadian Chem. J.*, Aug. 1919.

transference. Fig. 96 is a diagram of the acetone process as worked at the Shawinigan Falls. The conversion vessels consist of steel tubes, 13 feet in length, 12 inches in diameter with centrally heated core. They were filled with cast-iron balls on which the catalyst was placed by immersing the balls in a paste of the catalyst and drying them in the tube by the passage of a current of air. Seventy-two of these tubes were installed and this installation was sufficient for the production of ten tons of acetone per day. Fig. 97 shows the details of one of the conversion tubes.

The mixed vapours of acetone, water, and unconverted acetic acid were combined from 24 of these tubes to the one main leading to alkali scrubbers maintained at 98° C. by the heat from the gases ; the acetone and water vapours which passed the scrubbers were condensed to give a 20 per cent aqueous solution of acetone. This mixture was rectified in an ordinary continuous acetone still. The optimum temperature for conversion was 485° C. Each tube was supplied with three thermo-couples having leads to a central switch-board and the temperature was maintained constant by installing the resistance windings in parallel so that any of the parallel windings could be cut out as the temperature fluctuated.

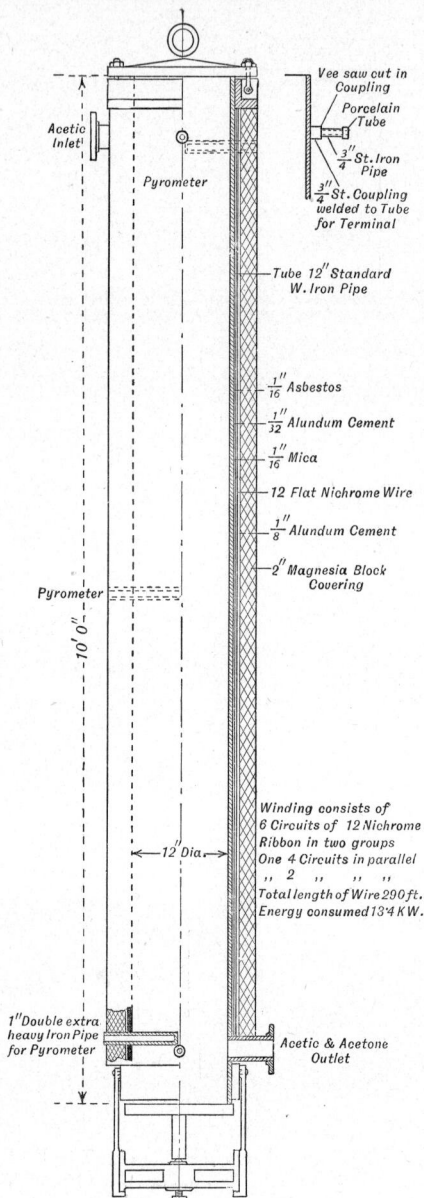

Details of 12″ Conversion Tube

FIG. 97.

Iron had a detrimental action on the conversion, but this difficulty was eliminated by coating the walls of the tube with the catalyst itself. Copper is the best metal to employ for the tubes, but its cost is very high. After every fourteen days it is necessary to renew the catalyst by recoating the balls.

The average efficiency of the process is about 85 per cent, and the acetone produced passes the strictest commercial specifications.

Acetone from Wood Spirit.—The process of removing acetone from methyl alcohol on the manufacturing scale is described by Mariller, *La Distillation fractionnée*, 1917, p. 402.

The purification of the acetone so obtained has, however, not until now been accomplished on the large scale. Attempts to do so were made in Canada by Messrs. Barbet and Fils and Co. of Paris, but they say that they failed to obtain a finished product containing more than 92 per cent of acetone. Certain impurities, particularly methyl alcohol, remained in the acetone and could neither be removed by rectification nor by chemical treatment.

Methyl alcohol and acetone give an azeotropic binary mixture boiling at 55·95° C. and containing 86·5 per cent of acetone in the distilled liquid. While the presence of azeotropic mixtures of alcohol, methyl alcohol, and such a substance as methyl acetate, has not been definitely proved, yet in practice such a ternary mixture boils at nearly constant temperature. In the distillation of wood spirit there is the possibility of the formation of various azeotropic binary and probably of ternary mixtures, and for this reason it is practically impossible to obtain a pure acetone by distillation and fractionation alone.

Acetone and carbon disulphide give an azeotropic mixture boiling at 39·25° and containing 34 per cent by weight of acetone in the mixture. On this fact has been based a process for the purification of acetone by Duclaux and Lavzenberg.[1] To the acetone is added 1·7 times its volume of carbon disulphide. The distillate between 38-40° is extracted with water and the aqueous solution distilled. The pure acetone is collected at 56·1° to 56·3°.

When methyl alcohol is present, as in the mixture from wood spirit, the process must be modified since methyl alcohol and carbon disulphide give an azeotropic mixture boiling at 37·5°. The azeotropic mixtures are left over potassium carbonate for some time before extraction with water, and copper sulphate is added to the aqueous solution extract before its fractionation. In this way it is claimed that most of the methyl alcohol is removed.

Acetone from another Source.—Another method which has been suggested for the preparation of acetone depends upon the action of a reducing agent or amine upon a monobasic acid or ester.[2]

---

[1] *Bull. Soc. Chem.*, 1920 (IV.), 779-782. Cf. also *J. Chem. Soc.* (A), 1920, **118**, 1, 818.

[2] P. de la Fresnaye and E. Cadorat de la Gabiniere, F.P. 451374/1912, abst. *J. Soc. Chem. Ind.*, 1913, **32**, 625.

CHAPTER XXIV

In processes in which acetone is used as a solvent it is often possible to recover the greater portion of the acetone from the air of the drying stoves. For its recovery from cordite drying stoves and from various other sources an economic process has been worked out by Robertson and Rintoul.[1] In the drying of the cordite, air at a temperature of 40° C. passes over the cordite. This moist air, containing a small proportion of acetone vapour, is drawn from the stoves through galvanised iron pipes to the towers, which contain a 30 per cent solution of sodium bisulphite. This solution readily absorbs the acetone, forming a compound $C_3H_6O, NaHSO_3$. A blower or exhaust-fan leads the acetone-air mixture through the towers in series or parallel as required, the air current meeting the falling stream of bisulphite liquor. The order in which the acetone-air mixture enters the various towers and the method in which the partly saturated bisulphite liquor is circulated in the towers is so adjusted as to give the best absorption in the most economic manner. Any sodium sulphate formed by oxidisation crystallises out when the bisulphite liquors are allowed to stand. The bisulphite solution when saturated with acetone is transferred to primary stills where it is gently boiled. The absorbed acetone is then liberated, without appreciable decomposition of the sodium bisulphite solution, which may therefore be used again. The aqueous acetone solution which distils off contains small quantities of sulphur dioxide. Caustic soda is therefore added to fix the sulphur dioxide before the final distillation and rectification of the acetone. These latter stages may be carried out in the ordinary acetone stills as already described.

The acetone vapours may also be recovered by absorption in water.

Various other liquids have been suggested for the absorption of vapours such as these of acetone or "alcohol-ether." These liquids include cresol,[2] glycerol,[3] and oil emulsions.[4] Various alcohols, including butyl alcohol,[5] have been suggested for the absorption of ethyl alcohol and ether, but not expressly for acetone.

A special type of tower in which the acetone is most conveniently absorbed from the acetone-air mixture on a large scale has also been

[1] E.P. 25993, 25994/1901.
[2] J, Bregeat, E.P. 128640, 131938/1917. See also E.P. 127309/1917.
[3] E.P. 9941, 23888/1907.   [4] E.P. 7098/1915,   [5] D.R.P. 207554/1909.

devised by Robertson and Rintoul (Fig. 98). This tower is square in section and for bisulphite solution is lined with lead. Inside the tower are erected frames on which are wound parallel strands on which the liquid is carried, these strands or threads being preferably arranged with interlacing fibres so as to present a large surface for contact with the vapours which are passed longitudinally through the tower. The tower is provided with a lid b through which passes a pipe c for drawing off the gases. The tower is closed at the bottom with the exception of an inlet pipe e for the gases and another pipe f for running off the absorbent liquid. The tower is divided by partitions into several chambers or cells, a branch d^1, d^2, d^3, d^4 from the main pipe supplying the absorbent liquid being led along the top of each cell. Immediately under each supply pipe is a dish or trough constructed of an inner portion h and an outer one i, so arranged one within the other that the strands or threads which pass between them pass through the absorbent liquid which has trickled over from the inner trough h, the edges of which are suitably lipped.

The strands are preferably arranged as follows. A framework adapted to fit into each division or cell of the tower is built up of crossbars k, k, passing between two end plates, the topmost bars being adapted to support the troughs h, i, while the strands are threaded between the other bars. The contents of each cell of the tower, con-

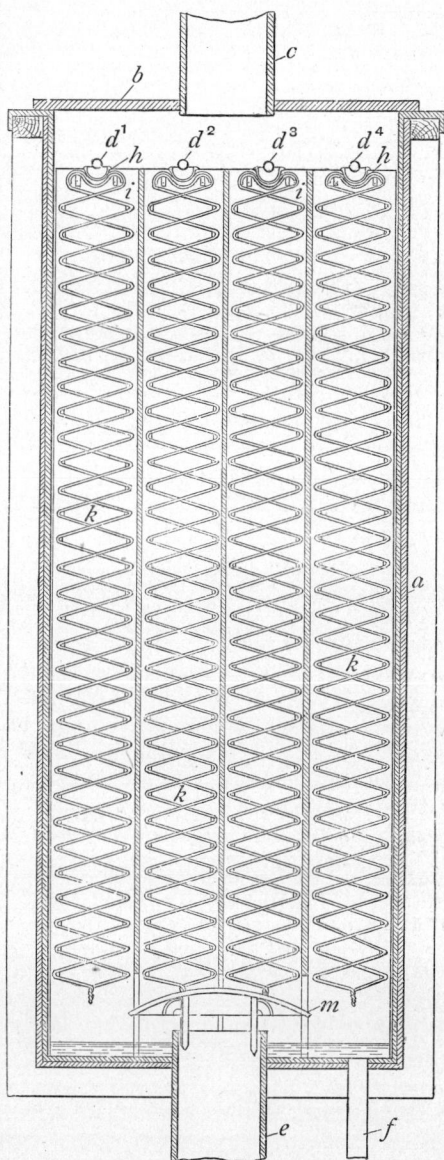

FIG. 98.

structed according to this arrangement, can thus conveniently be lifted out as a whole when necessary for overhauling or examination. The partitions separating the frames may be dispensed with, the fibres in adjacent frames being brought almost into contact with each other.

The strands after passing over the side of the lower trough i are carried downwards at an angle of about 75° to the side of the tower parallel to each other and at such a distance apart that the small fibres in adjacent strands somewhat interlace ; they now pass over one of the bars k, and are then carried downward at an angle to their former course until they reach the opposite side of the tower, where they pass round another bar k, and thus zigzag from side to side to the bottom of the tower, each layer making an angle of about 30° to the preceding one. As the parallel strands extend from side to side of the tower or section, a series of screens is thus formed through which the gases to be treated are made to pass.

A suitable baffle m is provided for the more uniform distribution of the gases through the tower, and to prevent the liquid from getting into the gas inlet pipe e. The method of using the tower is to allow an absorbent solution to flow or syphon from a reservoir near the top of the tower down the strands, and this solution, being distributed in its course over a multitude of interlacing fibres, presents a very large surface of active liquid to the ascending gas driven or aspirated through the apparatus, without causing any appreciable heading back of the gas current.

As the vapour pressure of acetone is small at low temperatures, it should also be possible to condense it by passing the vapours through vessels cooled by liquid ammonia or liquid air. In the case of acetone-air mixtures from cordite stoves this method would not be convenient owing to the possible presence of small quantities of "nitro-glycerine" in the mixed vapours unless the traces of nitro-glycerine were first decomposed. In the closed system of Nikolsky (Eng. Pat. 3661/1906) the vapours circulate through a calcium chloride refrigerating system (cooled by liquid carbon dioxide or ammonia).

Acetone may also be removed from air by drawing the mixture over dry cellulose esters which retain the acetone by absorption (Wohl, Eng. Pat. 23995/1911).

CHAPTER XXV

Acetone

Properties. — Acetone is one of the most important organic solvents employed in the chemical industry. It is a solvent (alone or in mixtures) for various explosives (*e.g.* nitrocellulose), celluloid, acetyl-cellulose, acetylene, as well as for tannin fats, and resins. Acetone is also used in the synthesis of many organic compounds (*e.g.* chloroform, sulphonal, etc.). Acetone-bisulphite is used in certain photographic developers. When reduced, acetone is converted into a mixture of isopropyl alcohol and pinacone. Dehydrating agents convert it into mesityl oxide and phorone.

Acetone is a mobile colourless hygroscopic liquid, miscible in all proportions with water, ethyl alcohol, ethyl and amyl acetates, ether and chloroform. It can be separated from aqueous solution by calcium chloride. Acetone gives a crystalline compound with sodium iodide ($NaI, 3C_3H_6O$). By means of this compound acetone may be readily purified (Shipsey and Werner).[1]

The density of pure acetone as recorded by recent workers is :—

> 0·79123 20°/4° Reilly and Ralph.[2]
> 0·79082 20°/4° Price.[3]
> 0·7912 20°/4° Bramley.[4]
>
> Its coefficient of expansion is 0·0016.

Boiling point . . 56·2° to 56·3°/760°—Approximate change of b.p. for 1 mm. change of pressure 0·03° C.

Melting point . . − 94·9°
Critical temperature 232·6° [5]
Critical pressure . 52·2 atm.[5]

The vapour pressure of acetone at various temperatures is given in Table 96.

The viscosity of acetone at 50° (n) = ·00245, c.g.s.
The refractive index at 15° μ_D = 1·360.

Palmer[6] gives the following values for the Refractive Index of Acetone for the C and F hydrogen lines :—

Hydrogen C . 1·35633   Hydrogen F . 1·36296

---

[1] *Trans. Chem. Soc.*, 1913, **103**, 1255. [2] *Sci. Proc. R. Dublin Soc.*, 1919, **15**, 598.
[3] *Trans. Chem. Soc.*, 1919, **115**, 1125. [4] *Trans. Chem. Soc.*, 1916, **109**, 455.
[5] Lecat, *L'Azéotropisme*. [6] Palmer, *Analyst*, 1920, **45**, 302.

S

TABLE 96

Temperature.	Vapour pressure of acetone in mm. of mercury.				
	Regnault.[1]	Sameshima.[2]	Taylor.[3]	Price.[4]	Paranjpé.[5]
°C.					
50	620·9	612·5	607·0	620·9	...
45	...	510·5	505·0	510·8	...
40	420·2	421·5	416·0	425·3	...
35	...	346·4	343·0	348·1	...
30	281·0	282·7	281·0	284·6	...
25	...	229·2	229·0	232·0	...
20	179·6	184·8	182·5	186·3	...
15	...	...	...	...	151·8
10	...	...	...	...	117·4
5	...	...	...	...	90·36
0	...	...	...	...	69·51
− 5	...	...	...	...	53·27
− 10	...	...	...	...	40·12
− 15	...	...	...	...	30·02
− 20	...	...	...	...	22·05
− 25	...	...	...	...	16·26
− 30	...	...	...	...	11·70
− 35	...	...	...	...	8·32
− 40	...	...	...	...	5·81
− 45	...	...	...	...	3·92
− 50	...	...	...	...	2·54

[1] Landolt and Börnstein's Tables. (An interpolation formula for vapour pressures of acetone calculated from Regnault's results is given by Ramsay and Young (*Physico-Chemical Tables*, John Castell Evans. Vol. i. p. 512).
[2] *J. Amer. Chem. Soc.*, 1918, **40**, 1482.   [3] *J. Physical Chem.*, 1900, **4**, 436.
[4] *Trans. Chem. Soc.*, 1919, **115**, 1125.   [5] *J. Indian Inst. Sci.*, 1919, **2**, Part V. 55.

Table 97 shows the propagation of flame limits obtained by White and Price [1] for acetone-air mixtures using various iron and glass tubes at 20° ± 2° C.

TABLE 97

Material and diameter of tube.	Percentage of solvent in limit mixture and direction of propagation.					
	Upwards.		Horizontal.		Downwards.	
	Upper limit.	Lower limit.	Upper limit.	Lower limit.	Upper limit.	Lower limit.
Iron 5 cm.	...	3·80	...	3·90	...	4·00
Iron 15 cm.	12·40	2·88	12·40	2·89	10·90	3·11
Glass 5 cm.	12·20	2·89	9·15	3·04	8·35	3·15

Tests.—After the separation of the acetone and *n*-butyl alcohol at the factory the purity of these compounds is tested as follows :—

The analysis of acetone includes the determination of specific gravity which in acetone for use in cordite manufacture should not exceed 0·800 at 15°/15°.

[1] *Trans. Chem. Soc.*, 1917, **115**, 1491.

It should be free from colour, and miscible with water in all proportions, showing no turbidity.

On evaporation there should be no more than a slight trace of residue.

The absence of certain compounds is shown by the *permanganate test*. In this test 1 c.c. of a 0·1 per cent solution of potassium permanganate is added to 100 c.c. of acetone. The solution is kept at 15° C., and is shielded from direct light. The characteristic pink colour produced by the permanganate should not disappear in less than 30 minutes. With the acetone produced as described above it is quite usual for this pink colour to remain for several hours.

Alkalinity is tested for by titrating a diluted solution (50 per cent of water) with N/20 sulphuric acid, paranitrophenol solution being used as indicator.

Acidity due to the presence of slight traces of carbon dioxide and volatile acids (*e.g.* acetic acid) should not exceed 0·002 per cent calculated as carbon dioxide. The acetone must be free from fixed acids. Total acidity is estimated as follows. A mixture of equal parts of acetone and neutral water is titrated with N/20 potassium hydroxide solution, phenolphthalein being used as indicator. Fixed acids are tested for in the following manner. 50 c.c. acetone are mixed with 50 c.c. neutral water to which a few drops of phenolphthalein solution have been added, and which has been made just pink by the addition of a drop of N/20 potassium hydroxide solution. The acetone is evaporated off on the water bath. The pink colour should return or remain, showing the absence of fixed acids.

Only traces of aldehydic and other substances that will reduce ammoniacal silver solutions are permitted. These impurities are tested for with a solution made by dissolving 3 grams of silver nitrate, 3 grams of caustic soda, and 20 grams of aqueous ammonia (sp. gr. 0·9) in water and making up to 100 c.c. 2 c.c. of this solution are added to 10 c.c. of acetone diluted with an equal quantity of distilled water and the mixture allowed to stand in the dark for fifteen minutes. The liquid is then decanted from the precipitated silver, and tested to ascertain whether excess of silver is still present. If there is excess of silver present then it is assumed that the amount of " aldehyde " in the acetone is less than one part in a thousand.

n-Butyl Alcohol

n-Butyl alcohol is a comparatively new industrial substance, and its commercial applications have not yet been developed. Its most important use will probably be as a solvent.[1] Even when it was only known as a laboratory product its use as a solvent for cellulose esters was suggested by R. Schüpphaus (U.S. Pat. 410204/1889. Cf. also Mosenthal, *J.S.C.I.*, 1904). Derivatives of *n*-butyl alcohol such as *n*-butyl acetate should have wide industrial application.

Another use of *n*-butyl alcohol will be in its application in synthetic organic chemistry. From *n*-butyl alcohol such compounds as *n*-amyl alcohol, *n*-valeric acid, and methyl ethyl ketone can be prepared in

[1] Cf. U.S. Pat. 1321611 and 1341745.

good yield. This latter substance has been prepared industrially from n-butyl alcohol.[1] The n-butyl anilines, n-butyl toluidines, and aminobutyl benzenes have also been investigated.[2]

The closing of many alcohol distilleries in Russia and the United States since the passing of prohibition will reduce the available supply of fusel oil, with the result that an increased demand for some alternative substance such as n-butyl alcohol will probably arise.

The n-butyl alcohol run off from the separators in both the continuous stills described consists of a saturated solution of water in n-butyl alcohol, with traces of ethyl alcohol. A considerable amount of water can be removed from this mixture by salting out with sodium chloride. The n-butyl alcohol supplied for industrial purposes has generally been prepared in this manner and contains not more than 10 per cent of water. The n-butyl alcohol is usually sold as 90 per cent butyl alcohol by weight. Salt remaining from the dehydration process is generally present as an impurity. In addition a small quantity of acetone—0·5 to 1 per cent—and some lower alcohols are often present in the commercial n-butyl alcohol.

To obtain pure n-butyl alcohol the crude liquid from the salting plant must be redistilled.

A process for the removal of the last traces of water by distilling in presence of benzene has been referred to in an earlier section of this book.

n-Butyl alcohol has a density of 0·80974 at 20°/4° and boils at 117·6/763 mm.

Refractive index	. .	$\mu_a = 1·39909$
Mol. refraction .	. .	$= 35·45$ [3]
Melting point .	. .	$-79·7°$
Critical temperature .	.	$287·1°$
Critical pressure	. .	$48·27$ atm.(?) [4]

The boiling point of the azeotropic mixture of n-butyl alcohol and water is 92·25°—the mixture being heterogeneous.

The percentage by weight of water in the azeotropic mixture is approximately 37 per cent.

n-Butyl alcohol is a colourless liquid completely soluble in hydrochloric acid (20 per cent). The purity of the alcohol can be determined by means of the temperature of critical solution of a ternary mixture of the alcohol with hydrochloric acid ; the upper critical solution temperature is 43·55° and the lower 9·6°. The c.s.t. is very sensitive to the presence of impurities.[5]

Mixture of Acetone and n-Butyl Alcohol.—The amount of water in acetone or in n-butyl alcohol may be obtained from density

[1] Weizmann and Legg, Can. Pat., 202135/1920.

[2] Reilly and Hickinbottom, *Trans. Chem. Soc.*, 1918, **113**, 99 ; 1920, **117**, 103.

[3] Brühl, *Annalen*, 1880, **203**, 16. In the literature there appears to be some confusion between n and iso butyl alcohol, and it is probable that many of the recorded values in the literature for n-butyl alcohol were not obtained with the pure substance.

[4] Ref. from Lecat. As a similar value is given for iso-butyl alcohol the value given is probably too high.

[5] Orton and Jones, *Trans. Chem. Soc.*, 1919, **116**, 1194.

curves when no other substance is present. These curves (see Figs. 99
and 100)[1] have been prepared from data given in Tables 98 and 99. From

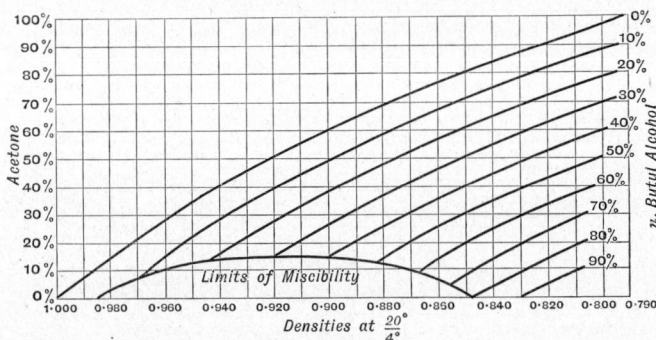

FIG. 99.

Fig. 99 the density of all possible mixtures of acetone, *n*-butyl alcohol,
and water may be ascertained.

TABLE 98

Table of Densities and Contractions of Mixtures of Acetone and Water.

Acetone per cent.	$D_{20°}^{20°}$ found.	$D_{4°}^{20°}$ found.	$D_{4°}^{20°}$ calc.	Vol. 1 gram found.	Vol. 1 gram calc.	Con- tractions per cent.
100	0·79231	0·79091	0·79091	1·26437	1·26437	...
94·98	0·80832	0·80689	0·79923	1·2393	1·2512	0·951
89·58	0·82367	0·82221	0·80841	1·2162	1·2370	1·681
79·92	0·85171	0·85020	0·82535	1·1762	1·2116	2·922
71·10	0·87370	0·87215	0·84154	1·1466	1·1883	3·509
57·46	0·90629	0·90469	0·86753	1·1054	1·1527	4·103
50·03	0·92220	0·92057	0·88250	1·0863	1·13315	4·134
37·49	0·94667	0·94499	0·90893	1·0581	1·1002	3·827
29·62	0·95949	0·95779	0·92635	1·0441	1·0795	3·279
19·31	0·97480	0·97307	0·95012	1·0277	1·0525	2·356
9·93	0·98707	0·98532	0·97290	1·0149	1·02785	1·260

The methods (other than by density determination) available for
the estimation of the water are limited to the use of a reagent which
reacts or combines with the water alone without action on the other two
constituents. Such a reagent as calcium carbide or an anhydrous salt
might be used. A much simpler and more accurate procedure is to
estimate the acetone by one of the methods available after suitably
diluting with water. Of these, a modification of the method suggested
by Messinger [2] or by Denige [3] is convenient, and gives good results.

Details of the method adopted are as follows :—

Dilute the sample with distilled water so that 10 c.c. diluted solution
contains not more than 0·005 gram acetone. Take 10 c.c. of the dilute
solution, add 5 c.c. 20 per cent soda solution, and run in from a burette

[1] Reilly and Ralph, *loc. cit.*   [2] *Ber.*, 1888, **21**, 3366.
[3] *Comp. rend.*, 1898, **127**, 963.

25 c.c. N/10 iodine solution. Shake at intervals for 15 minutes, keeping temperature at 15° C. Then add 5 c.c. 20 per cent HCl (of same strength as the 20 per cent NaOH), and titrate the liberated iodine with N/10 thiosulphate, using starch as indicator.

Let x = excess of N/10 iodine in c.c.

Then $(25 - x)\ 0 \cdot 967$ = number of milligrams acetone in 10 c.c. of dilute sample.

TABLE 99

Densities and Contractions of Mixtures of n-Butyl Alcohol and Water.

n-Butyl Alcohol per cent.	$D\dfrac{20°}{20°}$ found.	$D\dfrac{20°}{4°}$ found.	$D\dfrac{20°}{4°}$ calc.	Vol. of 1 gram found.	Vol. of 1 gram calc.	Contraction per cent.
100	0·81097	0·80953	0·80953	1·2353	1·2353	...
98·93	0·81318	0·81174	...	...	...	...
97·89	0·81538	0·81394	0·81277	1·2286	1·2304	0·143
96·96	0·81731	0·81586	...	...	...	...
95·97	0·81935	0·81790	...	...	...	...
95·06	0·82108	0·81962	0·81716	1·2201	1·2238	0·302
93·98	0·82234	0·82088	...	...	...	...
93·02	0·82513	0·82367	0·82036	1·2141	1·2190	0·402
91·97	0·82689	0·82543	...	...	...	...
90·96	0·82883	0·82736	...	...	...	...
89·96	0·83066	0·82919	0·82522	1·2060	1·2118	0·479
89·26	0·83216	0·83068	...	...	...	...
88·06	0·83436	0·83288	...	...	...	...
83·03	0·84345	0·84196	0·83633	1·1877	1·1957	0·666
80·64	0·84777	0·84627	...	...	...	...
79·94	0·84917	0·84770	0·84140	1·1797	1·1885	0·740
7·90	0·98862	0·98687	0·98018	1·0133	1·0202	0·676
7·32	0·98946	0·98771	0·98164	1·0124	1·0188	0·628
7·06	0·98968	0·98793	...	...	...	...
6·11	0·99111	0·98936	0·98421	1·0108	1·0160	0·511
5·05	0·99244	0·99068	...	...	...	...
3·95	0·99382	0·99202	0·98913	1·00805	1·0110	0·291
3·05	0·99532	0·99356	...	...	...	...
2·27	0·99651	0·99474	...	...	...	...
2·00	0·99678	0·99502	0·99364	1·0050	1·0064	0·139
1·61	0·99742	0·99566	...	...	...	...
1·04	0·99830	0·99653	...	...	...	...
0·61	0·99888	0·99711	0·99681	1·0029	1·0032	0·030

The estimation of the n-butyl alcohol requires a longer time, and may be carried out by a modification of the method of Verley and Bolsing for hydroxyl estimation.[1]

To a known weight of the " n-butyl alcohol, acetone, and water " mixture anhydrous sodium sulphate was added in proportion to the water present (approximately estimated from the density), and the mixture extracted several times with xylene. The hydrocarbon extract was made up to a known volume. A measured amount was heated gently on a sand-bath, with an excess of a pyridine solution of acetic anhydride, contained in a large flask fitted with a reflux condenser. Two hours' heating is usually sufficient to complete the esterification.

[1] *Ber.*, 1901, **34**, 3354.

TABLE 100

Mixtures of *n*-Butyl Alcohol, Acetone, and Water.

Acetone.	Water.	*n*-Butyl Alcohol.	Density $\frac{20°}{4°}$.
9·93	90·07	...	0·98532
9·44	85·63	4·93	0·97903
8·92	80·95	10·13	0·97087
19·31	80·69	...	0·97307
18·33	76·61	5·06	0·96668
17·36	72·54	10·10	0·95837
16·38	68·44	15·18	0·94947
15·40	64·36	20·24	0·93976
29·62	70·38	...	0·95779
26·59	63·19	10·22	0·94255
24·01	57·05	18·94	0·92763
21·12	50·18	28·70	0·91155
37·49	62·51	...	0·94499
33·69	56·17	10·14	0·93079
29·86	49·80	20·34	0·91421
26·11	43·53	30·36	0·89916
22·37	37·30	40·33	0·88498
50·03	49·97	...	0·92057
45·01	44·95	10·04	0·90754
39·84	39·79	20·37	0·89434
35·06	35·01	29·93	0·88264
30·02	29·98	40·00	0·87080
28·23	22·20	49·57	0·86012
57·46	42·54	...	0·90469
51·53	38·15	10·32	0·89338
45·85	33·95	20·20	0·88211
40·06	29·68	30·26	0·87153
34·41	25·46	40·13	0·86157
22·92	16·95	60·13	0·84291
71·10	28·90	...	0·87219
63·56	25·83	10·61	0·86467
56·80	23·09	20·11	0·85980
49·89	20·22	29·89	0·85059
42·60	17·31	40·09	0·84366
35·50	14·43	50·07	0·83741
79·92	20·08	...	0·85020
63·81	16·02	20·17	0·84026
47·51	12·19	40·30	0·83107
31·95	8·02	60·03	0·82300
15·97	4·01	80·02	0·81559
89·58	10·42	...	0·82222
80·55	9·37	10·08	0·82045
71·75	8·35	19·90	0·81880
62·72	7·30	29·98	0·81705
54·11	6·30	39·59	0·81559
44·81	5·21	49·98	0·81423
35·85	4·17	59·98	0·81321
18·01	2·09	79·90	0·81111
95·10	4·90	...	0·80690
30·86	...	69·14	0·80360
50·00	...	50·00	0·79976
70·47	...	29·53	0·79637

The excess of acetic anhydride remaining was estimated in the usual way, and the n-butyl alcohol content ascertained.

In the case of mixtures of acetone, n-butyl alcohol, and water, the density of the solution, together with a determination of one of the constituents by the methods, already mentioned, gives sufficient data to calculate the composition of the mixture. For example, a mixture of acetone, n-butyl alcohol, and water was found to have a density of 0·8842 at 20°/4°. The acetone content on analysis proved to be 27·9 per cent. Now all mixtures having a density 0·8842 lie upon a curve between the heavily drawn lines marked 0·880 and 0·900 in the triangular diagram, Fig. 100. Similarly all mixtures containing 27·9 per cent acetone will lie upon a line parallel to the base of diagram. The intersection of these two lines will give on inspection the percentages of alcohol and water in the mixture. In this example there were found to be 35·6 per cent and 36·5 per cent respectively. It will prove most convenient in reading off the percentage composition of n-butyl alcohol and water to turn the triangular diagram into such a position that the substance to be read forms the apex of the triangle, this being the 100· per cent point.

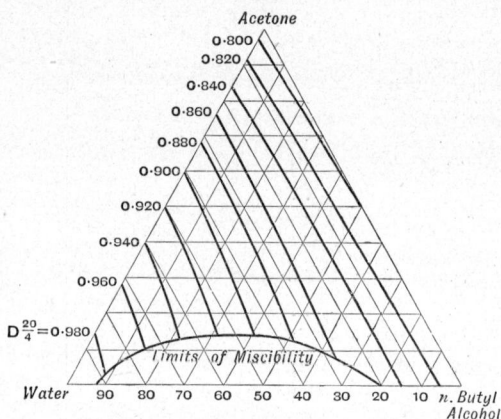

FIG. 100.

TABLE 101

Saturated Solutions of Acetone, n-Butyl Alcohol, and Water at 20°.

n-Butyl alcohol.	Acetone.	Water.	$D\frac{20°}{4°}$.
7·90	...	92·10	·9869
12·00	9·26	78·74	·9670
18·64	11·62	69·74	·9484
24·68	12·65	62·67	·934
28·15	12·95	58·90	·9260
36·91	13·42	49·67	·9071
47·02	13·10	39·88	·8874
53·86	11·65	34·49	·8764
63·68	8·28	28·04	·8633
79·94	...	20·06	·8477

Table 100 gives the density of a wide range of mixtures of acetone, n-butyl alcohol, and water, while Table 101 gives the composition of saturated solutions of these three substances.

DISTILLATION OF ALCOHOL
ON THE MANUFACTURING SCALE

The Hon. F. R. HENLEY, M.A., F.I.C.

AND

JOSEPH REILLY, M.A., D.Sc., F.R.C.Sc.I., F.I.C.

CHEMIST-IN-CHARGE AT THE ROYAL NAVAL CORDITE FACTORY, HOLTON HEATH, DORSET

CHAPTER XXVI

DISTILLATION OF MIXTURES OF ETHYL ALCOHOL AND WATER
(THEORETICAL)

A VERY large proportion of the ethyl alcohol now produced is prepared by the fermentation of sugar derived from grapes, beet, molasses, grain, or wood. The design of the apparatus used for distilling depends on the nature of the fermented liquids and the quality of the finished products required. From the distiller's point of view the substances contained in the liquids to be distilled may be divided into three classes—(1) the volatile products of the fermentation, from 1 to 12 per cent of ethyl alcohol; small quantities of the higher alcohols, amounting to 0·05 to 0·7 per cent of the raw spirit, principally iso-amyl alcohol and d-amyl alcohol, and smaller amounts of propyl alcohol; aldehydes, acids, esters, and furfurol; (2) large quantities of water; (3) the solid residue derived from the raw materials used and from the yeast. The production of alcohol, freed as far as possible by distillation from water and the by-products of fermentation, is carried out in two distinct stages :—

(1) *Distillation proper, i.e.* the extraction of the volatile from the non-volatile constituents of the fermented wash by boiling in a suitable vessel.

(2) *Rectification.*—The separation in as pure a state as possible, by means of fractionation, of ethyl alcohol and the other volatile products of fermentation.

Before describing the practical methods adopted for obtaining strong alcohol on the commercial scale it will be well to review the available knowledge of the principles underlying the distillation of—

1. Mixtures of ethyl alcohol and water.
2. Mixtures of water and ethyl alcohol with the other volatile products of fermentation.

On this knowledge is based the design and control of the stills.

Distillation of Mixtures of Ethyl Alcohol and Water.—In general, when a mixture of ethyl alcohol and water is boiled, the vapour produced contains a higher proportion of alcohol than the original mixture. There is no simple rule connecting the alcoholic content of liquid and distillate in every case, but a great deal of research has been done to establish these relations for a large number of different mixtures and to determine their boiling points.

The results obtained by Sorel (*Distillation et rectification industrielle, 1899*) and Lord Rayleigh [1] are shown on Table 102 and Fig. 101.

TABLE 102

Section 1

Boiling point °C	Sorel Alcohol in liquid Weight %	Sorel Alcohol in vapour Weight %	Lord Rayleigh Alcohol in liquid Weight %	Lord Rayleigh Alcohol in vapour Weight %
84·7	31	63·79	25·86	68·03
84·8	30	63·44		
85·0	29	63·10		
85·2	28	62·72		
85·7	27	62·30		
86·2	26	62·08		
86·4	25	61·75		
86·7	24	61·44		
87·0	23	61·12		
87·4	22	60·80		
87·7	21	60·54		
87·9	20	60·14		
88·25	19	59·75		
	18	59·3		
	17	58·82		
	16	58·39		
	15	57·50		
90·6	14	56·47	9·88	51·45
	13	55·17		
	12	53·36		
	11·3	52		
92·6	10	48·61		
93·29	9	46·13		
93·8	8	43·66		
	7·28	40		
	6·3	38		
95·8	5·0	33·49	3·98	31·59
	4·0	29·54		
	3·0	24·8		
	2·0	17·5	1·97	17·5
98·2	1·5	13·8		
	1·0	9·52		
	0·5	4·96		
	0·105	1·0		

Section 2

Boiling point °C	Sorel Alcohol in liquid Weight %	Sorel Alcohol in vapour Weight %	Lord Rayleigh Alcohol in liquid Weight %	Lord Rayleigh Alcohol in vapour Weight %
80·65	64	76·50		
80·75	63	76·08		
80·95	62	75·60		
81·1	61	75·1		
81·2	60	74·61		
81·4	59	74·19		
81·6	58	73·76	58	78·59
81·7	57	73·76		
81·9	56	72·85		
82	55	72·56		
82·28	54	72·13		
82·6	53	71·78		
82·75	52	71·38		
82·95	51	71		
83·1	50	70·63		
83·3	49	70·25		
83·4	48	69·88	45·62	74·12
83·5	47	69·50		
83·7	46	69·13		
83·85	45	68·76		
84·15	44	68·38		
84·3	43	68		
	42	67·67		
	41	67·29		
	40	66·94		
	39	66·61		
	38	66·36		
	37	65·87		
	36	65·43		
	35	65·04		
	34	64·74		
	33	64·42		
	32	64·12		

Section 3

Boiling point °C	Sorel Alcohol in liquid Weight %	Sorel Alcohol in vapour Weight %	Lord Rayleigh Alcohol in liquid Weight %	Lord Rayleigh Alcohol in vapour Weight %
	94	94·61	99·234	99·239
	93·51	94	95·55	95·45
	93	93·41		
	92·79	93	92·41	92·84
	92	92·37		
	91	91·6		
	90	91		
78·4	89	90·38	85·94	88·49
	88	89·76		
79·3	87	89	82·21	86·22
79·4	86	88·27		
79·55	85	87·65		
79·7	84	87·06	77·39	84·14
79·75	83	86·49		
79·95	82	85·7		
	81	85·3		
	80	84·8		
80·2	79	84·3		
80·3	78	83·7		
80·4	77	83·2		
	76	82·59		
	75	82·08		
80·5	74	81·45	66·06	79·76
80·6	73	80·90		
	72	80·38		
	71	79·9		
	70	79·36		
	69	79		
	68	78·42		
	67	77·93		
	66	77·45		
	65	76·98		

It will be observed that the ratio of alcohol in the vapour to alcohol in the liquid is very high when mixtures of low alcohol content are boiled, as is shown by the steepness of the curve on the left of the figure. But as the alcohol content of the liquid rises the curve becomes nearly horizontal, showing that there is not much more alcohol in the vapour than in the liquid boiled.

[1] Rayleigh, *Phil. Mag.*, 1902 [6], **4**, S, 521.

At a point where the aqueous alcoholic liquid contains a particular weight of alcohol, the vapour evolved has the same composition as the liquid. According to Young and Fortey[1] this proportion of alcohol is 95·57 per cent by weight. Wade and Merriman[2] give approximately the same figure, *i.e.* 95·59 per cent. A mixture of this composition boils at a slightly lower temperature than any other mixture (see Table 103). According to Noyes and Warfel[3] there is a lowering of

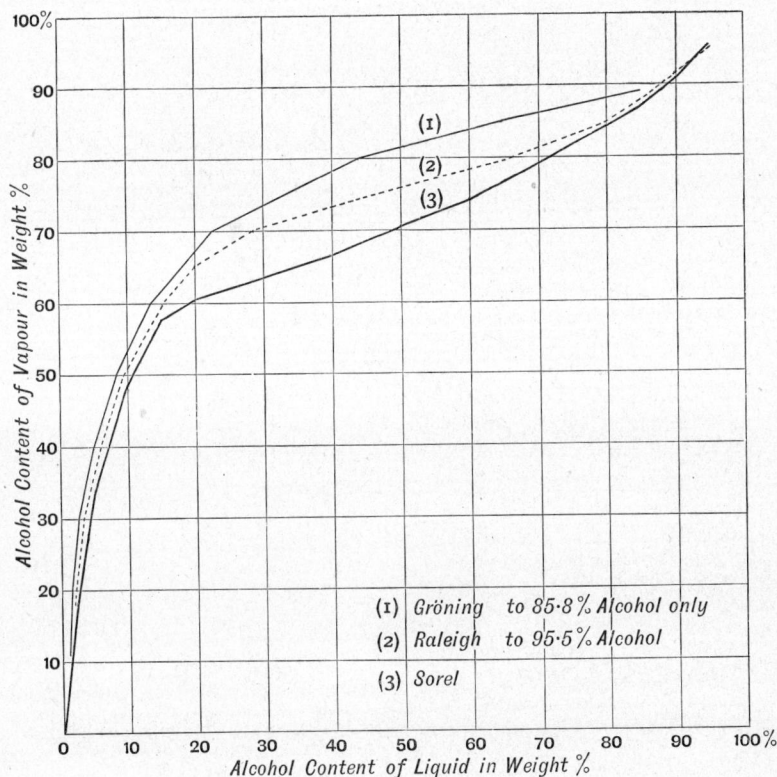

FIG. 101.

the boiling point of pure ethyl alcohol by 0·126°, while according to Merriman[4] the lowering is 0·15°.

By distillation alone, alcohol stronger than 95·57 per cent by weight cannot be prepared. Table 102 cannot be used directly for the calculation of the number of plates required in a rectifying column to produce spirit of any required concentration from a given mixture. For instance, if the liquid to be distilled contains 10 per cent of alcohol, the vapour evolved by this contains 48·61 per cent (Sorel) of alcohol. This, if completely condensed, will evolve a vapour containing 69 to 70 per cent of alcohol and so on.

[1] *Trans. Chem. Soc.*, 1902, **81**, 717.   [2] *Ibid.*, 1911, **99**, 997.
[3] *Amer. Chem. Soc.*, 1901, **23**, 463.   [4] *Trans. Chem. Soc.*, 1913, **103**, 628.

TABLE 103

Noyes and Warfel's Table.

Alcoholic strength. Per cent by weight.	Boiling point.	Alcoholic strength. Per cent by weight.	Boiling point.
100	78·300	85	78·645
99	78·243	80	79·050
98	78·205	75	79·505
97	78·181	65	80·438
96	78·174	55	81·77
95	78·177	48	82·43
94	78·195	35	83·87
93	78·227	26	85·41
92	78·259	20	87·32
91	78·270	10	91·80
90	78·323	0	100·00

In practice it is not possible on economic grounds to condense the whole of the vapour supplied to each plate and then re-evaporate this. What actually happens is that only a portion of the vapour is condensed. The liquid so produced will evolve a vapour less rich in alcohol than in the former case. Concentration of the product as it rises from plate to plate will proceed more slowly than in the ideal case. As a larger and larger proportion of the original vapour evolved is condensed the nearer will the conditions approach the ideal.

When any mixture of ethyl alcohol and water is boiled, unless the percentage of alcohol is equal to or greater than that in the mixture of constant boiling point—which would never occur in practice, starting from weak spirit—the alcohol content of the liquid progressively falls and with it the alcohol content of the vapour evolved at any moment.

TABLE 104

Boiling point. ° C.	Weight of residue.	Alcohol content. Weight per cent.	Boiling point. ° C.	Weight of residue.	Alcohol content. Weight per cent.	Boiling point. ° C.	Weight of residue.	Alcohol content. Weight per cent.
79·1	1000	85·8	81·9	56·6	50·2	86·6	27·6	21·3
79·2	536	83·2	82·1	53·3	48·3	87·1	26·7	19·6
79·4	355	80·7	82·4	50·2	46·3	87·7	24·9	17·9
79·6	264	78·3	82·6	47·6	44·4	88·3	25·7	16·3
79·7	208	75·9	82·8	45·2	42·5	89·0	24	14·6
79·9	172	73·6	83	43·1	40·7	89·7	23·2	13
80·1	146	71·3	83·3	41·1	38·8	90·6	22·3	11·3
80·3	127	69·1	83·5	39·2	37	91·5	21·5	9·7
80·5	111	66·8	83·8	37·6	35·2	92·6	20·7	8·1
80·6	99·3	64·6	84·1	36	33·4	93·9	19·9	6·4
80·8	89·9	62·5	84·4	34·6	31·6	95·2	19·0	4·8
81	82·0	60·4	84·7	33·3	29·9	96·6	18·1	3·2
81·2	75·3	58·3	85	32	28·1	98·2	17·1	1·6
81·3	69·5	56·2	85·3	30·8	26·4	100	14·5	0·0
81·5	64·7	54·2	85·7	29·7	24·7			
81·7	60·4	52·2	86·2	28·6	23			

E. Donitz[1] has prepared Table 104, which shows the amount and composition of the distillate and liquid at any point in the course of the distillation of a mixture of alcohol and water. This table is based on Gröning's figures for the distillation of alcohol-water mixtures (Fig. 101). It should be noted that the figures given by different experimenters differ to a considerable extent.

By the use of this table the following problems can be solved :—

1. The total weight and composition of the original mixture and the composition of the residue being known, to find the amount and composition of the distillate.

Example :—

Weight of original liquid $=1$ kilogram.
Composition of original liquid, 50 per cent by volume.
Composition of the residue, 10 per cent by volume.
50 per cent by volume $=42\cdot5$ per cent by weight.
10 per cent by volume $=8\cdot1$ per cent by weight.

See Table :

From $45\cdot2$ kg. at $42\cdot5$ per cent remain $20\cdot7$ kg. at $8\cdot1$ per cent.

$\therefore$ from 1 kg. at $42\cdot5$ per cent remains $0\cdot458$ kg. at $8\cdot1$ per cent.

$\therefore$ $1 - \cdot458 = 0\cdot542$ kg. has been distilled over, and the alcoholic

strength of the distillate $= \dfrac{1 \times 42\cdot5 - 0\cdot458 \times 8\cdot1}{0\cdot542} = 71\cdot5$ per cent by weight.

2. What proportion must be distilled from mixtures of various strengths before the whole of the alcohol is contained in the distillate, and what will be the strength of the distillate ?

Example :—

1000 kilos. liquid containing 14 per cent by volume of alcohol,
14 per cent by volume $=11\cdot3$ per cent by weight.

From Table 3 :

$22\cdot3$ kilos. must be evaporated down to $14\cdot5$ kilos.

$\therefore$ 1000 kilos. must be evaporated down to $650\cdot2$ kilos.

Thus $349\cdot2$ kg. are distilled off.

The alcoholic strength of the distillate $=32\cdot3$ per cent by weight.

3. If the distillate obtained in (2) be redistilled until all the alcohol has passed over, and this process be repeated, what will be the alcoholic content of the successive distillates ? By repeating the calculation explained in (2) for successive distillations it can be found how many distillations are required to raise the concentration of dilute alcohol to any required degree.

A mixture of alcohol and water containing $11\cdot3$ per cent of alcohol must be distilled five times before the alcohol content of the distillate is 83 per cent by weight.

Strength of 1st distillate, $32\cdot3$ per cent by weight.
,,   2nd   ,,   55   ,,
,,   3rd   ,, $70\cdot3$,,
,,   4th   ,, $78\cdot5$,,
,,   5th   ,, $83\cdot0$,,

---

[1] Maercker-Delbrück, *Spiritus-Fabrikation*, 1908, p. 694.

These calculations show the results obtained when condensation in the upper part of the still is prevented as far as possible. They are not applicable to stills with rectifying columns.

The distillation of mixtures of ethyl alcohol and water has been dealt with by E. Hausbrand, *Rektifizier- und Destillier-Apparate—1916*. A theoretical basis is laid down for the proper design of the various parts of the still, and tables are given showing the alcoholic strength of vapour and liquid over any plate in the rectifying and wash columns under various working conditions.[1]

The Rectifying Column (see Figs. 104, 121, 124, 126, 127).—The plates of the column should ensure the even distribution of the rising vapour to the liquid on the plate. The vapour rising on to a plate should be completely condensed in the liquid and an entirely fresh lot of vapour evolved. To assist the complete mixing of vapour and liquid on the plate the condensed liquid flowing down from above must be evenly distributed all over the plate. Drops of liquid must not be carried upwards, from one plate to another, by the rising vapour.

Composition and Amount of the Condensate returned to the Still.—If a very large proportion of the vapour evolved in the still be condensed, the composition of the condensate will approximate to that of the vapour evolved. If a very small proportion of the vapour be condensed the composition of the condensate will approximate to that of the original liquid in the still. The residual uncondensed vapour will contain a higher percentage of alcohol in the first than in the second case. If a large proportion of the vapour evolved in the still be condensed, the alcoholic strength of the vapour as it rises from plate to plate in the column will increase rapidly, and a comparatively small number of plates will be required to produce alcohol of high concentration. But the large amount of condensate returned to the still entails a correspondingly high heat consumption in the still to produce a definite weight of finished product.

If, on the other hand, only a small proportion of the vapour be condensed the alcoholic strength of the residual uncondensed vapour will only rise slowly as it passes from plate to plate up the column, and a large number of plates will be required to produce alcohol of the same strength as before. But the amount of condensate returned to the still being smaller than in the first case the heat consumption in the still will be less for the same output of finished product.

In practice the plates should be as numerous as is reasonably possible so as to reduce the fuel expenses.

Condenser or Dephlegmator (see Fig. 116).—The function of the condenser is (1) to produce the amount of condensate required for the efficient working of the plates of the column. (2) It may also to some extent increase the alcoholic strength of the vapour passing on to uncondensed to the cooler. This effect is only produced to any marked extent when the alcoholic strength of the vapour entering the condenser is low. To attain this object as far as possible the vapour should pass as slowly

[1] Cf. L. Gay; *Chim. et Ind.*, 1920, **4**, 178-188.

as possible through the condenser. The temperature difference between cooling liquid and vapour should be as small as possible. The vapour should rise and the cooling liquid fall in the condenser. A large proportion of the vapour must be condensed. The condensate should be kept in as small drops as possible, so as to ensure as perfect contact as possible between vapour and condensed liquid.

Some of these requirements are fulfilled in the best German dephlegmators, which can really be better described as rectifiers and condensers combined. They enable the column to be built with fewer plates, but their construction is complicated, and it is questionable whether a rather higher rectifying column with ordinary tubular condenser is not preferable on account of simplicity of construction.

The condenser also serves to preheat the wash to be distilled, which replaces the cooling water in many continuous stills.

Use of Wash to replace Water as Cooling Liquid in the Condenser.—If the alcoholic strength of the wash is not above 9 per cent the wash alone will suffice to provide the necessary condensate for the rectifying column when the alcoholic content of the finished product does not exceed 90 to 92 per cent by volume (85·7 to 88·3 per cent by weight). But if it is required to produce spirit containing 94·6 per cent alcohol by weight, the wash will only be sufficient to provide the necessary condensate provided its alcoholic content is not above 3 per cent.

Position of the Condenser.—If the condenser be placed above the top of the rectifying column, the volume of the condensate on the upper plates is large and gradually decreases as it flows downwards. But if several condensers be arranged at intervals down the column the upper plates receive little condensate, though the total flow returned to the still be the same in both cases.

This latter arrangement is not so efficient as the former and necessitates a larger number of plates in the column than would be required to produce the same result with the condenser above the top of the rectifying column.

Should the Rectifying Column be lagged ?—If the column be made of metal a considerable loss of heat may take place by radiation and air currents. When the wash is preheated in the condenser this means a complete loss of a considerable amount of heat. In any case the loss of heat which will take place from all points on the outside of the column will lead to a certain amount of condensation of the vapour. As previously stated, it is preferable to supply the whole volume of condensate to the top plate of the column. Condensation produced by cooling the surface of the column produces a similar effect to several condensers arranged at intervals down the column, and should therefore be avoided.

The loss of heat from a square meter of surface per hour for each degree (centigrade) difference of temperature between the surface and surrounding air is approximately :—

Copper	. 5 to 6·5 calories.
Wrought iron	. 8 „ 9 „
Cast iron	. 9 „ 10 „

The following is an example of what a high percentage of the heat supplied to a still may be lost in this way :—

	Litres.	Litres.
Volume of wash dealt with per hour . .	1000	10,000
	Calories.	Calories
For a yield of 100 litres of spirit the loss of heat		
would be for unlagged columns . .	3000	600
For well-lagged columns	600	120

Total heat required to produce 100 litres of spirit amounts to from 9000 to 12,000 calories.

The loss is much lower in the larger still, but even here it amounts to at least 5 per cent of the total heat supplied, when the column is not lagged. With the columns lagged the loss is reduced to 1 per cent of the total heat supplied.

Discontinuous Stills.—*The minimum possible volume of condensate and amount of heat required during the progress of distillation.*—Table 105 shows that to produce 1 kilo of alcohol containing 94·61 per cent (by weight), more condensate must be produced and more heat supplied to the still the lower the alcoholic strength of the liquid to be distilled. But the amount of heat required does not steadily decrease as the alcoholic strength of the liquid to be distilled is increased. A minimum is reached when the alcoholic strength of the liquid to be distilled is about 26 per cent by weight. Above this strength the amount of heat required gradually rises again.

Supposing that the still be charged with alcohol of 90 per cent strength the amount of condensate and heat required per kilo of finished product will gradually fall until the alcoholic strength of the liquid left in the still has reached about 26 per cent by weight. After this point a steady rise takes place until towards the end of the distillation the amounts of condensate and heat required are very great. The condenser must, of course, be capable of providing the maximum amount of condensate required if an even output of finished product is required throughout the distillation. In the same conditions the source of heat must be capable of supplying the maximum heat requirements.

Number of Plates required in the Rectifying Column.—As already stated, this depends on the strength of the liquid to be distilled, and on the strength of the finished product and the amount of heat abstracted by the condenser. Tables 106 and 107 give the theoretical optimum results obtained under various working conditions.

Continuous Stills.—The Wash Column.—In this type of still the liquid to be distilled is supplied continuously at a steady rate to a wash column, the construction of which is similar to that of the rectifying column (see Figs. 104, 121, 125, 127). The wash flows on to the top plate

TABLE 105

Theoretical minimum amount of heat (in great calories) to be abstracted in the condenser from vapour entering it from a rectifying column, in order to produce 1 kilo of alcohol 94·61 per cent by weight, or 85·76 per cent by weight, from liquid in the still, the alcoholic strength of which varies from 0·052 per cent to 92·79 per cent by weight. The liquids treated are assumed to be already at their boiling points.

Alcoholic content of liquid in still. Weight per cent.	Heat units to be abstracted in condenser per 1 kilo alcohol 94·61 per cent by weight produced.	Heat units to be abstracted in condenser per 1 kilo alcohol 85·76 per cent by weight produced.	Alcoholic content of liquid in still. Weight per cent.	Heat units to be abstracted in condenser per 1 kilo alcohol 94·61 per cent by weight produced.	Heat units to be abstracted in condenser per 1 kilo alcohol 85·76 per cent by weight produced.
92·79	819		8·95	717	641
92·29	1048		8·21	775	701
91·38	1022		7·66	837	765
88·33	870		7·28	907	836
85·69	731		6·30	977	913
82·14	661		5·76	1,070	1,003
78·49	577	103	5·13	1,167	1,091
74·89	542	140	4·54	1,259	1,190
71·2	530	236	4·19	1,400	1,320
67·12	514	262	3·50	1,531	1,460
62·99	494	284	3·09	1,695	1,626
58·58	489	307	2·88	1,888	1,821
53·56	457	315	2·60	2,108	2,038
48·31	455	322	2·31	2,380	2,320
43·00	452	327	2·07	2,690	2,646
37·07	443	337	1·79	3,116	3,064
31·16	434	337	1·52	3,635	3,598
25·97	430	335	1·35	4,364	4,298
19·91	435	346	1·07	5,365	5,329
15·57	448	374	0·84	6,865	6,799
13·63	478	407	0·63	9,450	9,370
12·37	512	442	0·42	14,490	14,420
11·30	559	488	0·21	29,690	29,620
10·63	607	536	0·105	59,950	59,800
9·66	658	587	0·052	120,000	119,900

TABLE 106

Alcohol content (weight per cent) of liquid and vapour on each plate of a rectifying column, when heat removed in the condenser is equivalent to 5000, 6000, 8000, 15,000 great calories per 10 kilos of alcohol, produced at 85·76 per cent by weight.

Number of plate from the top.	Liquid. Per cent alcohol by weight.	Vapour. Per cent alcohol by weight.	Liquid. Per cent alcohol by weight.	Vapour. Per cent alcohol by weight.	Liquid. Per cent alcohol by weight.	Vapour. Per cent alcohol by weight.	Liquid. Per cent alcohol by weight.	Vapour. Per cent alcohol by weight.
	5000 Calories.		6000 Calories.		8000 Calories.		15,000 Calories.	
	82	85·76	82	85·76	82	85·76	82	85·76
1	77·1	83·4	77·1	83·30	77·06	83·15	76·1	82·8
2	73·0	80·7	72·9	80·70	71·0	79·5	67·2	78·0
3	67·5	78·2	66·1	77·65	60·0	74·59	50·6	70·9
4	61·39	75·2	57·7	73·6	46·4	69·5	19·0	59·6
5	55·18	72·2	45·34	68·9	21·33	60·6	6·15	37·2
6	45·60	69·10	28·12	62·7	10·85	50·5	3·71	28·98
7	33·70	64·70	12·97	55·01	7·68	42·7	3·20	27·85
8	18·80	59·78	10·09	48·87				
9	12·50	54·25						
10	11·3	51·90						

of the column, and thence downward to the outlet, and leaves the column almost completely freed from alcohol.   Live steam is the source of heat. It is supplied at the base of the column and causes the liquid on each plate to boil.   The function of the wash column is to free the wash

TABLE 107

Alcohol content (weight per cent) of liquid and vapour on each plate of a rectifying column, when heat removed in the condenser is equivalent to 16,000 and 30,000 great calories per 10 kilos alcohol, produced at 94·61 per cent by weight.

Number of plate from top.	16,000 calories.		30,000 calories.		Number of plate from top.	16,000 calories.		30,000 calories.	
	Liquid.	Vapour.	Liquid.	Vapour.		Liquid.	Vapour.	Liquid.	Vapour.
	93·77	94·61	93·77	94·61	16	90·54	91·65	50·18	70·70
1	93·75	94·00	93·75	94·00	17	90·20	91·06	13·00	55·30
2	93·56	93·90	93·52	93·77	18	89·81	90·90	3·20	25·20
3	93·33	93·70	93·27	93·62	19	89·41	90·50	1·99	17·72
4	93·28	93·63	92·24	92·74	20	88·32	90·00		
5	93·18	93·54	92·00	92·52	21	87·20	89·20		
6	93·15	93·49	91·00	91·58	22	86·20	88·40		
7	92·90	93·35	90·80	91·46	23	85·20	87·40		
8	92·73	93·16	90·20	91·10	24	83·20	86·60		
9	92·45	92·97	89·75	90·65	25	80·15	84·90		
10	92·25	92·75	88·50	90·19	26	77·50	83·40		
11	92·08	92·64	87·10	89·10	27	71·80	80·20		
12	91·43	92·45	85·00	87·73	28	62·45	75·70		
13	91·35	92·00	81·90	85·85	29	43·75	68·30		
14	91·24	91·91	77·00	83·07	30	12·57	54·40		
15	91·00	91·74	68·00	78·62	31	4·09	33·02		

TABLE 108

Wash Column.

Theoretical minimum amount of heat (in great calories) to be supplied to wash column to produce 100 kilos of outflow water free from alcohol, from wash the alcoholic content of which varies from 15 to 0·5 per cent by weight.

Alcohol content. Weight per cent.		Calories.	Alcohol content. Weight per cent.		Calories.
Liquid.	Vapour.		Liquid.	Vapour.	
15	57·50	12,320	7	39·54	8,600
14	56·47	11,610	6	37·00	8,100
13	55·17	11,000	5	33·49	7,600
12	53·36	10,025	4	29·54	6,600
11	51·00	10,000	3	24·80	6,360
10	48·61	9,800	2	17·50	6,050
9	46·13	9,650	1	9·52	5,750
8	43·66	9,500	0·5	4·96	5,600

from alcohol as completely as possible.   The heat required to do this increases as the alcoholic strength of the wash to be distilled increases (Table 108).

Table 109 shows the results obtained on each plate of a wash column when the alcoholic strength of the wash and heat consumption vary, but outflow of water from the base of the column remains constant.

For instance, if the original wash contains 8·45 per cent of alcohol, seven plates will be required if the heat consumption is 20,000 calories, but about seventeen plates are necessary if the heat consumption is only 10,000 calories.

TABLE 109

Wash Column.

Alcohol content (weight per cent) of liquid and vapour on each plate of a wash column when 10,000, 12,000, 20,000, 50,000, and 125,000 great calories of heat are supplied at the base of the column, per 100 kilos of water leaving the column.

Number of plate from the bottom.	10,000 calories.		12,000 calories		20,000 calories.		50,000 calories.		125,000 calories	
	Alcohol per cent by weight in liquid.	Alcohol per cent by weight in vapour.	Alcohol per cent by weight in liquid.	Alcohol per cent by weight in vapour.	Alcohol per cent by weight in liquid.	Alcohol per cent by weight in vapour.	Alcohol per cent by weight in liquid.	Alcohol per cent by weight in vapour.	Alcohol per cent by weight in liquid.	Alcohol per cent by weight in vapour.
21	10·2	49·0								
20	10·07	48·84								
19	9·78	48·0								
18	9·11	46·4								
17	8·14	44·84	14·2	56·7						
16	7·25	41·01	13·7	56						
15	6·54	38·5	12·7	54·7						
14	5·22	34·2	11·15	51·3						
13	3·82	28·74	9·26	46·8						
12	2·74	21·84	7·20	41·01						
11	1·624	16·24	4·94	33·0						
10	0·990	9·90	3·057	24·0	22·5	60·6				
9	0·561	5·61	1·73	15·57	21	60·4	41·24	67·41	60·4	74·8
8	0·349	3·49	1·09	10·0	15·1	57·5	41·1	67·33	60·0	74·5
7	0·221	2·21	·593	5·98	8·45	44·6	40	67·00	58·5	73·75
6	0·141	1·41	·325	3·25	3·71	27·5	35·6	65·3	54·3	72·25
5	0·090	0·90	·178	1·78	1·47	13·0	19·1	59·8	39·7	66·85
4	0·0583	0·583	0·1	1·0	0·53	5·3	5·28	35·5	22·8	55·00
3	0·0371	0·371	0·059	0·59	0·195	1·95	1·125	10·68	3·44	25·00
2	0·024	0·241	·033	0·33	0·0721	0·721	0·233	2·33	0·485	3·85
1	0·0156	0·156	·018	0·18	0·0268	0·268	0·048	0·48	0·069	0·693
	0·01	0·10	0·01	0·10	0·01	0·10	0·01	0·10	0·01	0·10

The amount of heat given in Table 105 represents not only that required to free the wash from alcohol but also to effect the concentration of the spirit in the rectifying column.

This total is made up in the following way :—

(1) Latent heat of vaporisation carried off to the cooler by the finished product.

(2) If the wash is preheated in the condenser it will be supplied to the top plate of the wash column at a temperature above that at which it entered the condenser, but below its boiling point. It cannot be heated to its boiling point in the condenser, for the temperature of the strong alcoholic vapour in the condenser is only about 78° C. to 85° C.,

and it is impossible to heat the wash to a temperature of more than 8 to 10 degrees below that of the vapour in the condenser. Heat must therefore be supplied in the wash column to raise the wash to its boiling point on the top plate of the wash column.

(3) To produce the necessary condensate in the rectifying column heat is abstracted in the condenser. This amount of heat must have originally been supplied in the wash column, and is represented by the amount of heat taken up by the wash as it is preheated in the condenser.

These three items together account for the greater part of the heat requirements of the still, and together represent the total given in Table 105.

But there are three other small amounts of heat which must be added to make up the total heat requirements of the still.

(a) The heat required to raise the temperature of the condensate from that at which it leaves the condenser to that existing on the top plate of the wash column.

(b) The heat required to raise the temperature of the wash from its boiling point on the top plate of the wash column to its temperature as it leaves the base of the wash column (about 102° C.). In calculating the amount of steam to supply the total requirements of the still, the temperature of the spent wash as it leaves the wash column must be taken into account. This temperature is about 102° C. From this it follows that 1 kilo of saturated steam will yield to the liquid in the column $637 - 102 = 535$ calories.

(c) Replacement of heat unavoidably lost by radiation, etc., from the still.

To determine the number of plates in the rectifying and wash columns and capacity of the condenser required to produce spirit of a certain concentration, the following procedure may be adopted. It is first necessary to determine the alcoholic strength of the liquid and vapour on the top plate of the wash column. This will depend upon the alcoholic strength of the wash and on the temperature to which the wash has been preheated in the condenser.

Table 110 gives the necessary information for wash containing 1, 5, 7, and 10 per cent of alcohol by weight. It will be noted that the higher the temperature of the wash the lower the alcoholic strength of the liquid and vapour on the top plate of the wash column.

Having found the alcoholic strength on the top plate of the wash column, the number of plates in the rectifying column and capacity of the condenser required to raise the alcoholic strength to the required extent can be found from Table 106 or 107.

The number of plates required in the wash column is found in Table 107, which also gives the heat requirements of the still as already explained.

A concrete example will make clear how the information respecting wash and rectifying columns given in the tables can be combined so as to show the most economical working arrangement theoretically possible.

Example.—From 112·8 kilos of wash containing 10 per cent alcohol (by weight), temperature 20° C., to produce spirit containing 85·76 per cent of alcohol by weight, the wash being preheated to 70° C. in the condenser.

TABLE 110

Alcoholic strength (weight per cent) of liquid and vapour on the top plate of a wash column when the alcoholic strength and temperature of the wash varies.

Alcoholic content of wash.	0°	10°	20°	30°	50°	70°	90°	Degrees centigrade by which the wash from the condenser is below its boiling point on the top plate of the column.		
1	9·52	11·2	13·8	17·5	27·5	38·5	50·2	Alcoholic content of vapour.		
	1·0	1·25	1·5	2·0	2·0	6·45	10·7	,,	,,	liquid.
5	33·49	35·5	38·1	41·5	48·61	55·17	58·15	,,	,,	vapour.
	5	5·6	6·33	7·5	10	13	15·75	,	,,	liquid.
7	39·54	42·6	45·5	48·81	54·8	58·1	58·9	,,	,,	vapour.
	7	7·80	8·7	10·05	12·65	15·6	17·5	,,	,,	liquid.
10	48·61	51·5	54·2	55·6	58·39	59·4	60·3	,,	,,	vapour
	10	11·15	12·3	13·55	16	18·1	20·2	,,	,,	liquid.

In these circumstances the strength of the liquid and vapour on the top plate of the wash column will be 12·3 and 54·2 per cent by weight respectively (see Table 110).

From Table 109 it is seen that to produce 100 kilos water practically free from alcohol as outflow from the wash column when the strength of liquid on the top plate is 12·7 per cent the minimum heat consumption is 12,000 calories, and the number of plates required in the wash column is 15.

Table 106 shows that with a minimum abstraction of 5000 calories in the condenser per 10 kilos of finished product of the required strength (85·76 per cent by weight) the column will require 8 plates.

Let us now attempt to assess the heat requirements of the still in the given case by another method.

Let us assume as the basis of calculation that 100 kilos of spent wash practically free from alcohol leave the boiling column. This quantity of liquid (assuming that the amount of steam used for heating purposes, and condensed in the wash, is left out of account) would be produced from 112·8 kilos of the original wash, which would also yield 13·16 kilos of finished spirit (containing 85·76 per cent of alcohol by weight).

Let the specific heat of the wash be 1·01, the latent heat of vaporisation of water and alcohol being 544 and 205 respectively.

Then the total heat requirements of the still will be made up, as already stated, of the following items (see page 277) :—

(1) Heat of vaporisation of finished product leaving the top of the still. Total weight of finished product 13·16 kilos containing 11·28 kilos of alcohol and 1·88 kilos of water.

This heat therefore amounts to—

$$(11·28 \times 205) + (1·88 \times 544) = 3335·1 \text{ calories.}$$

(2) Heat required to raise the wash from its temperature when preheated to its boiling point on the top plate of the wash column. From Table 102 the

boiling point of the liquid on the top plate of the wash column is 90° C.   The
wash is preheated to 70° C.   The amount of heat required here is :—

$$112 \cdot 8 \times 1 \cdot 01 \ (90 - 70) = 2278 \cdot 5 \text{ calories.}$$

(3) The amount of heat taken up in the wash in the condenser :—

$$112 \cdot 8 \times 1 \cdot 01 \ (70 - 20) = 5696 \cdot 4 \text{ calories.}$$

To these must be added the amount of heat required for the three
small additional requirements referred to on p. 278.   This amounts to
884 calories.   For the calculation of this see Hausbrand, *loc. cit.* p. 92.

These four items together give a total of 12,194 calories as the total
heat requirements in the still.

If these figures be compared with those given on Tables 106 and 109,
it will be seen that the calculated total heat requirement of the still is
nearly equal to that given in the table.

In comparing the heat abstracted in the condenser as given in
Table 106 with that found by calculation, it must be remembered that
the former is for an output of 10 kilos of spirit and that the amount
obtained is 13·16 kilos.

By reference to Table 105 it is seen that for a production of 1 kilo of
alcohol the greatest abstraction of heat in the condenser for the interval
between spirit of 12·3 per cent and 85·76 per cent by weight is 442
calories.   To produce 13·16 kilos of such spirit the amount required
here will therefore be at least 5816 calories, which is only very slightly
greater than the figure obtained in the calculation, *i.e.* 5696 calories.

The wash here has been alone sufficient practically to supply the
necessary condensate to the rectifying column.   Had it been required
to produce a finished product containing 94 per cent of alcohol from the
same wash a considerable amount of cold water in addition to the wash
would have been required in the condenser.

It must be remembered that the figures in the tables referred to
represent the minimum theoretical heat requirements.   In practice it
would therefore be necessary to use a certain amount of cooling water
in addition to the wash (cf. p. 273).

CHAPTER XXVII

DISTILLATION OF MIXTURES OF WATER AND ETHYL ALCOHOL, WITH THE OTHER VOLATILE PRODUCTS OF FERMENTATION

THE problem of how to separate the ethyl alcohol from the water has already been dealt with. The rectification or removal of the volatile by-products remains to be considered. The majority of these substances are produced in the fermentation, but some of them are formed by chemical changes taking place during the process of distillation. To the latter class belong certain esters produced by the interaction of acids and alcohols present in the fermented wash. The elimination of these acids in the early stages of rectification will reduce the amount of esters produced.

The volatile by-products include the following substances : [1]—

Aldehydes.	Acetal.
n-Propyl alcohol.	Glycerine.
Iso-propyl alcohol.	Fatty acids, including formic acid.
n-Butyl alcohol.	
Iso-butyl alcohol.	Ethyl acetate and formate.
Iso-amyl alcohol.	Ethyl butyrate.
d-Amyl alcohol.	Ethyl iso-valerate.
Hexyl and heptyl alcohols.	Iso-amyl acetate.
Furfurol bases.	Iso-amyl iso-valerate.
Iso-butyl glycol.	Terpenes.

The by-products can be roughly divided into two classes : (1) " head products," which tend to pass over into the distillate more readily than ethyl alcohol, and (2) " tail products," which do not do so. But no hard-and-fast distinction can be drawn, as some of the by-products pass from class 1 to class 2 as the composition of the liquid to be distilled varies.

The head products are more readily removed from dilute alcoholic solution (fermented wash) than from concentrated alcoholic solutions which have not received a preliminary purification. In the initial stages of the heating of the fermented wash the liquor is practically subjected to a process of distillation in a current of steam. Owing to the relatively small proportion of impurities present, the first fractions

---

[1] Harden, *Alcoholic Fermentation*, 1914, p. 85 ; Maercker Delbrück, *Spiritus-Fabrikation*, 1908, p. 761 ; Windisch, *Arb. Kais. Gesund.*, 1892, **8**, 228.

of the distillate will contain the greater proportion of the impurities (fusel oil) contained in the original fermented wash. The distillation constants of the ester impurities are also high, and these bodies will also come over in the first fraction.

DISTILLATION OF MIXTURES OF ALCOHOL AND WATER

In the distillation of mixtures of liquids four groups are usually considered.

(*a*) Non-miscible liquids.
(*b*) Partly miscible liquids.
(*c*) Closely related liquids—miscible in all proportions.
(*d*) Liquids not closely related but miscible in all proportions.

As the mixture of ethyl alcohol and water will come in group *d*, the remaining groups (*a*, *b*, and *c*) will only be briefly referred to here—their more detailed study has been considered earlier in this volume.

With non-miscible liquids (group *a*) the boiling point of each individual liquid in the mixture depends on the partial pressure of its own vapour and generally is not influenced by the other substances present. Since the constituents of these mixtures may be considered as independent of each other the laws of distillation are simple and the theoretical calculated results obtained from the known vapour pressures agree very closely with those found experimentally. These laws have been developed in the main by Naumann,[1] Pierre and Puchot,[2] Brown.[3]

In the working up of the by-products from fermentation liquors sometimes two layer mixtures are formed—*e.g.* some higher alcohols and water. These mixtures would be of the type included under group *b*.

In mixtures of liquids included in group *c*, the relation between the vapour pressure and molecular composition may be represented by a straight line, and for such a mixture as chlorobenzene and bromobenzene the relation holds accurately.[4] The relation between the composition of the liquid mixture and that of the vapour evolved from it according to Brown's formula may be expressed as follows :—

$$\frac{m'_A}{m'_B} = \frac{m_A}{m_B}\frac{P}{P_B},$$

where m'_A and m'_B are the relative masses of the two substances in the vapour, m_A and m_B their relative masses in the liquid mixture, and P_A and P_B the vapour pressures of the pure substances at the boiling point of the mixture. Substituting a constant *c* for the ratio P_A/P_B, a better agreement was obtained between the calculated and observed results. As shown earlier in this volume, Brown's experimental results do not agree well with this formula, and the experimental evidence avail-

---

[1] *Ber.*, 1879, **10**, 1421, 1877.   [2] *Compt. rend.*, **73**, 599, **74**, 224.
[3] *Trans. Chem. Soc.*, 1879, **37**, 547.
[4] Cf. Brown, *Trans. Chem. Soc.*, 1880, **35**, 541, **37**, 49 ; 1881, **39**, 304 and 517 ; Young and Fortey, *Trans. Chem. Soc.*, 1903, **81**, 768, 902, and **83**, 45, etc.

able points to the conclusion that the formula is only applicable to these liquid mixtures of which the relation $P = \text{M}P_A + (1 - \text{M})P_B$ holds good, P being the vapour pressure of the mixture and P_A and P_B the vapour pressure of the two pure substances at the same temperature, and M the molar fraction of the substance A.[1]

Duhem and Margules independently suggested the formula

$$\frac{d \log P_A}{d \log \text{M}} = \frac{d \log P_B}{d \log (1 - \text{M})}$$

for the relation between the molar composition of the liquid mixture and the partial pressures P_A and P_B of the components in the vapour, and the formulae of Lehfeldt and of Zawidski are based on this equation. Doubt has been thrown on the graphic method of Margules (which relies upon the measurement of the slope of the total pressure curve at its two ends) by the subsequent work of Rosanoff and co-workers.[2] These latter have developed a general law which may be expressed as

$$\frac{dP}{d\text{M}} = (1/K) \log \left[P_A (1 - \text{M})/P_B\text{M} \right],$$

where $1/K = (P_A - P_B)/(\log P_A - \log P_B)$.

The alcohols may be considered as being formed from water by the substitution of an alkyl group for one of the hydrogen atoms. They may also be considered as hydroxyl derivatives of the saturated paraffins. From both of these aspects the subject has been studied by Young[3] in his investigation of the relationship of the alcohols to water on the one hand and to the paraffins on the other hand. In general the influence of the OH group diminishes and that of the alkyl group increases as we pass from water up the series of alcohols. The vapour pressures of mixtures of methyl alcohol and water are always intermediate between those of the pure components, and the curve representing the relation between the vapour pressure and the molecular percentage of methyl alcohol deviates only slightly from a straight line. The maximum differences between the pressures represented by the curve and a straight line joining the two ends of the curve is 43 mm. on a certain scale. On the same scale the maximum difference for mixtures of ethyl alcohol and water mixtures is 315 mm.[4] For further study of this relationship of the alcohols see earlier sections in this book. The behaviour of methyl alcohol and water mixtures appears to place these mixtures in group c rather than group d, but with ethyl alcohol and water mixtures there is a closer connection with group d. With the higher soluble alcohols, such as the butyl alcohols or the amyl alcohols, the connection with group d is very complete within the limits of solubility, and if the formula used for non-miscible or closely related miscible liquids is applied large deviations from the calculated results are obtained.

[1] Cf. Young, *Proc. Royal Dublin Soc.*, 1920, **15**, 47.
[2] *J. Amer. Chem. Soc.*, 1911–1920.
[3] *Trans. Chem. Soc.*, 1902, **81**, 707.
[4] Konowalow, *Wied. Ann.*, 1881, **14**, 34.

The study of the vapour pressures of mixtures of group d was undertaken early in the nineteenth century by Gay-Lussac (1815), Boit (1816), Magnus (1836), Regnault (1853), and later by Berthelot, Young, and others. It is, however, to Duclaux[1] that we are indebted for the first co-relation of the earlier work on the study of the distillation of dilute solutions of certain substances (such as alcohols or fatty acids) and for the establishment of a relation between the amounts of the constituents in the mixture to be distilled and their amounts in the vapours.

Distillation of Dilute Aqueous Solutions of Substances Volatile in Steam

Duclaux originally studied the rates of distillation from the analytical standpoint. He distilled a known volume of a dilute solution of the volatile acid or alcohol and collected the distillate in several equal fractions. By expressing the amount of acid or alcohol in each fraction as a percentage of the total amount distilled, he was able to obtain a series of constants which served to identify the particular acid or alcohol. For mixtures of two acids or two alcohols each substance approximately follows its own law of distillation. The same line of reasoning applies in the case of either alcohols or acids, and as the distillation of the volatile acids was first studied a typical example of the distillation of dilute acid solution will be considered.

A solution containing at the most 1-2 grams of acid is made up to 110 c.c. and distilled in a flask of 200-300 c.c. capacity, using an ordinary Liebig condenser. Ten successive fractions of 10 c.c. each are collected. Each fraction is titrated in turn, so that the acid in all the 10 fractions can be determined. The titration of each fraction is determined as a percentage of the total titration for 100 c.c. of distillate.

Suppose in the distillation of a solution of a volatile acid the following values are obtained for titration of each fraction :—

			Total.		Percentage.
1.	4·1	c.c.	4·1	c.c.	15·5
2.	3·5	,,	7·6	,,	28·8
3.	3·1	,,	10·7	,,	40·6
4.	2·85	,,	13·55	,,	51·3
5.	2·65	,,	16·2	,,	61·4
6.	2·45	,.	18·65	,,	70·6
7.	2·2	,,	20·85	,,	79·0
8.	2·1	,,	22·95	,,	86·9
9.	1·75	,,	24·7	,,	93·6
10.	1·7	,,	26·4	,,	100·0

Total titration ⎫
for 100 c.c. ⎭ 26·4 ,,

On this principle tables have been constructed showing the percentage of acid in all fractions, calculated on the amount distilled in 100 c.c. of distillate.

The following table gives these values obtained by Duclaux :—

---

[1] *Ann. de phys. et chim.*, 1878, and later.

Distillate.	Acetic acid.	Propionic acid.	n-Butyric acid.
20 c.c.	15·2	24·0	33·6
30 ,,	23·4	35·3	47·5
40 ,,	32·0	46·2	60·0
50 ,,	40·9	56·8	70·6
60 ,,	50·5	66·7	79·5
70 ,,	60·9	76·2	86·5
80 ,,	71·9	85·0	92·5
90 ,,	84·4	93·0	97·0

For mixtures of two acids, Duclaux states that each acid follows its own laws of distillation. This is not strictly true, but the error involved is generally less than the experimental error.

It follows that if a mixture containing two volatile acids is distilled, the distillation values will lie between those of the two acids, and will be a measure of the composition of the mixture. This can be seen in the following example. A mixture of 1·25 molecular proportion of butyric acid and 1 of acetic gave the following results of distillation by Duclaux's method :—

	Percentage.
30 c.c.	36·7
40 ,,	47·6
50 ,,	57·4
60 ,,	66·7
70 ,,	75·2
80 ,,	82·4

For 30 c.c. of distillate the corresponding values for acetic and butyric acids are 23·4 and 47·5 respectively. If the acid is originally supposed to be butyric acid to which increasing amounts of acetic acid have been added, the difference between the value for butyric acid figures and the corresponding experimentally derived value will be a measure of the acetic acid in the mixture. Thus the molecular proportion of acetic acid present in the solution is—

$$\frac{47 \cdot 5 - 36 \cdot 7}{47 \cdot 5 - 23 \cdot 4} = 0 \cdot 448.$$

Similarly for fraction 40 c.c. the molecular proportion of acetic acid present

$$= \frac{60 \cdot 0 - 47 \cdot 6}{60 \cdot 0 - 32 \cdot 0} = 0 \cdot 443 ;$$

for 60 c.c.

$$\frac{79 \cdot 5 - 66 \cdot 7}{79 \cdot 5 - 50 \cdot 5} = 0 \cdot 442.$$

The method of calculation can be expressed by the general formula—

$$\frac{C_1 - C_3}{C_1 - C_2} = \text{mol. ratio of lower acid in the mixture} ;$$

where $C_1 =$ distillation constant of acid 1 having higher distillation constant,

$C_2 =$ distillation constant of acid 2 having lower distillation constant,

$C_3 =$ distillation constant of mixture of 1 and 2.

Similarly the molecular ratio of the acid of high molecular weight is given by general expression—

$$\frac{C_3 - C_2}{C_1 - C_2} = \text{mol. ratio of acid of high molecular weight.}$$

For mixtures of unknown acids, the analysis is somewhat tedious. Calculations must be made for trial mixtures until a certain mixture gives constant results for the proportions of acids present. The calculations are simplified by constructing tables showing the results which would be expected by mixtures of known composition.

While the original method outlined by Duclaux did not give concordant results with different workers, a large amount of experimental work has been carried out to remedy this defect. It has been shown that loss of heat from the exposed flask and still-head, by convection currents and radiation, leads to irregular results. Various arrangements for preventing this have been used. It is advisable to keep the exposed flask at no higher temperature than that of the vapour, otherwise errors will be introduced by splashing and complete evaporation of drops of solution. The best conditions for preventing condensation are obtained by surrounding the flask with a steam jacket.

In the case of acids the acidity of the distillate may be determined by simple titration, but with alcohols the method of estimation is not so direct. The "drop method" used by Duclaux[1] depends on the fact that the surface tension of different mixtures of alcohols and water tends to diminish with the increased percentage of alcohol.

The Coefficients of Duclaux, Sorel, and Barbet.—The original formula of Duclaux may be expressed thus :—

$$\frac{da}{db} = c\frac{a}{a + b}.$$

where a and b represent the percentage by volume of alcohol and water respectively in the original liquid, and da and db the percentages of alcohol and water respectively in the vapour. The above relation may be represented by a hyperbola. The value of c with dilute solutions of various alcohols varies from 10·9 for methyl alcohol to 50 for amyl alcohol and 61 for capryl alcohol. The coefficient c increases with the molecular weight of the alcohol, and the higher alcohols pass over on distillation of their aqueous solutions more readily than do the lower alcohols.

In Chapter XXVI. reference is made to Gröning's figures for the distribution of alcohol-water mixtures. It was from a study of these figures that Sorel was led to his investigation on the distillation of alcohol and water mixtures.[2] He used a copper retort of 5 litres capacity submerged in a bath of glycerine for the heating of 4 litres of liquid. The distillate was collected in about 40 equal fractions, and the alcohol content in each fraction estimated by a density determina-

[1] *Ann. Inst. Pasteur*, 1895, **9**, 575.
[2] *Comptes rendus*, 1892.

tion. If the volumes are plotted as abscissae and the concentrations of the remaining liquid as ordinates, we have what Sorel calls the curve of purification of the original liquid. If V is the volume remaining at any particular moment, a the Gay-Lussac concentration of the liquid, i.e. the percentage by volume of the alcohol in the mixture, and U the Gay-Lussac concentration of the liquid distilled, i.e. the percentage by volume of alcohol in the fraction distilled ; then at any moment

$$Va = (V - dV)(a - da) + dVU,$$

or
$$U = a + \frac{da}{dV}.$$

The value of U is obtained from the curve of purification by finding the angular coefficient of the tangent.

Mariller[1] makes special use of a constant K, usually called the coefficient of solubility, or, better, the coefficient of enrichment, which is given by the following equation :—

$$K = \frac{\text{percentage by weight of alcohol in vapours}}{\text{percentage by weight of alcohol in liquid}},$$

or in above notation—

$$= \frac{da}{a} = \frac{100c}{ac + 100}.$$

With dilute alcohol solution (1 per cent) the enrichment coefficient K on Sorel's calculations is 9·9 for ethyl alcohol at 1 per cent. concentration, and this figure gradually diminishes as a increases. With $a = 50$, $K = 1 \cdot 5$; between $a = 50$ and $a = 95 \cdot 47$, K falls from 1·3 to 1. The corresponding values of K from Duclaux's and Gröning's figures are nearly equal, but somewhat higher figures are obtained from Sorel's data.

Although Duclaux mainly dealt with the problem of dilute alcoholic distillation from the analytical standpoint, both Sorel and Barbet (especially the latter) were concerned with the industrial problem of the distillation of fermentation liquors. If S is the weight of the impurities present in 1 kilogram of the mixed vapours from fermentation liquors, and " s " the weight of impurities in 1 kilogram of the original liquor, then according to Sorel—

$$S = K_1 s + K_2 s^2 + K_3 s^3 + \ldots$$

If " s " is considered to be small then $S = K_1 s$.

The distillation of mixtures of ethyl alcohol, higher alcohols, various esters, aldehydes, etc., and water over a wide range of alcohol concentrations has been made by Barbet (*La Rectification et les colonnes rectificatrices*, 1895, p. 46).[2]

---

[1] *Le Bulletin de l'Association des Chimistes*, 1911, 473-490, and *La Distillation fractionnée*, p. 23.

[2] See also Sorel, *Comptes rendus*, 1894, **118**, 1213 ; Sorel, *La Rectification de l'alcool*, 1894, pp. 18-33 ; Sorel, *Société d'encouragement, Bulletin de mai*, 1891, pp. 226, 240.

The ratio

$$\frac{\text{percentage of `` impurities '' in ethyl alcohol in distillate}}{\text{percentage of `` impurities '' in ethyl alcohol in liquid}}$$

he designated by K', and it is known as the coefficient of purification.

If $k' = \dfrac{\text{percentage by wt. of impurities (\textit{e.g.} amyl alcohol, etc.) in vapour}}{\text{percentage by wt. of impurities (amyl alcohol, etc.) in liquid}}$,

then
$$K' = \frac{k'}{K}.$$

In other words, K' indicates how far distillation will remove this impurity from the alcohol apart altogether from the quantity of water present. This " coefficient of purification " is of a more practical value than the coefficient of enrichment. The experimental methods of Barbet are also more closely related to the methods employed in the distillery than are those of Sorel.

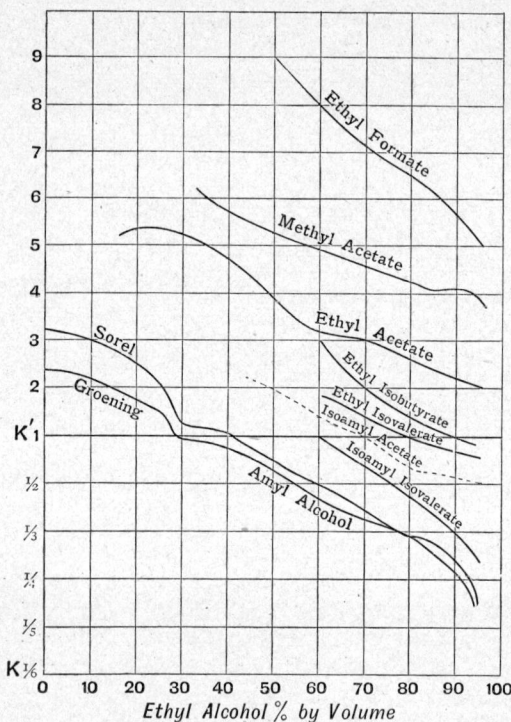

FIG. 102.

Fig. 102 gives the value of K' for various impurities found in fermented liquid. At the point where $K'=1$ there is a change of direction in the course of the purification. So long as $K'>1$ the alcoholic vapour is richer in the impurity than the alcohol in the original liquid. When $K'=1$ the concentration of impurity is the same in liquid and vapour. Finally, when $K'<1$ the impurity tends to become concentrated in the liquid.

It should be noted here that the amount of impurity present in the original liquid as examined by Sorel never exceeded 2 per cent. In practice the concentration of some of the impurities considerably exceeds 2 per cent at certain stages of the rectification. Their behaviour will depend largely on their vapour pressures under the varying conditions. In the case of such an impurity as amyl alcohol the formation of a mixture of maximum vapour pressure with water will greatly influence the result. In practice there is always more than one impurity present, and this will to some extent alter the values of K'. In all the cases shown the value of K' falls with rising alcoholic content of the liquid. This is due partly to the fact that, as the alcoholic strength rises, the temperature of ebullition of the liquid falls, and with it the vapour pressure of the impurity. Another factor affecting the result in the same direction is the great solubility of all the impurities in alcohol and the very slight solubility of most of them in water. With the exception of propyl alcohol and aldehydic bodies, the greater part of the impurities are insoluble or only partially soluble in water.

The fact that the tail products are more readily removed from dilute alcoholic solutions than from more concentrated solutions is utilised in at least one type of modern still. According to the patent of E. Guillaume [1] the alcoholic liquors to be purified are diluted if necessary in such a way that the products considered as tail products, e.g. amyl alcohol, may behave during the distillation like head products relative to the mixture of ethyl alcohol and water. These liquors are then fractionally distilled in a continuous distilling column of sufficient power and number of plates to enable a large part of the products hitherto considered as tail products to be removed at the same time as the so-called head products, so that to the bottom of this distilling column there comes only a practically pure mixture of water and ethyl alcohol.

DISTILLATION OF AQUEOUS ALCOHOL SOLUTION AT CONSTANT VOLUME

The work of Duclaux, Sorel, and Barbet deals with the distillation of alcoholic solutions at varying volumes. A brief consideration will be given to the distillation of aqueous solutions of volatile substances at constant volume. [2]

As a result of the distillation of aqueous phenol solutions at constant volume, Naumann and Müller [3] came to the conclusion that under constant conditions of temperature and pressure, the amount of substance distilling was proportional to the amount of substance in the flask. They deduced values for the ratio of the amount of substance in the flask to the amount distilling in each fraction, and also for the ratio of the titration or value of a fraction to the one preceding it. Stein [4] distilled aqueous solutions of certain volatile substances in a current of steam, and observed certain regularities in the distillation. To express the rates of distillation, values were calculated from the formula

$$\frac{1}{v} \log \frac{a}{a-x}$$

[1] Eng. Pat. 5794/1902.
[2] Cf. Reilly and Hickinbottom, *Proc. Royal Dublin Soc.*, XV. 37, 514.
[3] *Berichte*, 1901, **34**, 224.   [4] *J. pr. Chem.*, 1913, **88**, 83.

where v = volume distilled, a = amount of substance originally present, x = amount of substance distilled.

According to Nernst's law of distribution

$$\frac{\text{concentration in vapour phase}}{\text{concentration in liquid phase}} = \text{a constant.}$$

The proportion of constituents in the distillate are assumed to be the same as the ratio in the vapour phase in a state of equilibrium. This assumption is probably not strictly accurate, but approaches accuracy if the distillation is carried out slowly and regularly.

Let x = the amount of alcohol in distillate after volume v has been distilled,

$\quad a$ = initial amount of alcohol in distillation vessel,

$\quad \rho$ = density of water vapour in distillation vessel,

$\quad \sigma$ = weight of water per unit volume of distillate,

then if dx is the quantity of alcohol coming over in a quantity dv of distillate,

$$\frac{\rho \delta x}{\sigma \delta v} = \text{concentration in vapour phase approximately,}$$

$$\frac{a - x}{V} = \text{concentration in liquid phase approximately,}$$

where V = constant volume of liquid in a flask, then—

$$\frac{\rho dx}{\sigma \delta v} = + k \frac{a - x}{V},$$

$$\frac{\delta x}{\delta v} = + \frac{k\sigma}{\rho V}(a - x) = + \lambda(a - x),$$

which gives $\qquad x = a(1 - e^{-\lambda v}),$

or $\qquad a - x = a e^{-\lambda v},$

$$\frac{a}{a - x} = e^{\lambda v},$$

$$\lambda = \frac{1}{v} \log_e \frac{a}{a - x},$$

writing $\qquad A = \frac{1}{v} \log_{10} \frac{a}{a - x}.$

Then A is a constant, assuming $\dfrac{\sigma}{\rho}$ is constant.

It may be readily seen that the coefficient K used on p. 287 is related to the distillation constant (A) given above as follows :—

$$K = \frac{\text{concentration in vapour phase} \times \sigma}{\text{concentration in liquid phase} \times \rho} = \frac{k\sigma}{\rho} = \lambda V = A \times V \times 2 \cdot 3026.$$

The distillation constant (A) varies to some extent with the concentration of the alcohol (cf. distillation constants in the distillation of 0·8 per cent and 4·8 per cent ethyl alcohol solutions). The values for A can strictly be compared only with the values of K for dilute solutions. Variations in the temperature of the aqueous alcoholic solution may partly account for the alteration in the distillation constant.

Another disturbing effect may be due to association of the alcohol. Murray [1] has adduced evidence in favour of the view that the molecules are associated in the liquid state but are not generally associated in aqueous solution. If there still exist in the solution some associated molecules at the concentration employed, there will be a continual change in the state of aggregation as the distillation proceeds. Under these conditions Nernst's law of distribution will only hold approximately. The available evidence, however, mainly favours the view that if there is any combination between the alcohol and water it must be very slight.[2]

The above treatment only deals with the case of a single alcohol in solution. The theory of mixtures may be considered as follows :—Let a and b be the amounts of each alcohol present in the distillation vessel initially and x_n and y_n the amounts distilled over in n-fractions.

We get $\qquad a - x_n = ae^{-\lambda_1 nf}$ for first alcohol,

$\qquad\qquad\quad b - y_n = be^{-\lambda_2 nf}$ for second alcohol,

or $\qquad\quad a + b - (x_n + y_n) = ae^{-\lambda_1 nf} + be^{-\lambda_2 nf}.$

The fraction of original alcohols left—

$$1 - \frac{x_n + y_n}{a +} = \frac{a}{a+b}e^{-\lambda_1 nf} + \frac{b}{a+b}e^{-\lambda_2 nf},$$

write $\qquad\qquad \dfrac{a}{a+b} = m \qquad \dfrac{b}{a+b} = n.$

m and n represent the ratio in which alcohols were present initially, and we have $m + n = 1,$

$$\frac{x_n + y_n}{a + b} = \frac{\text{percentage of alcohol distilled over}}{100} = \frac{P_n}{100}.$$

We have then

$$1 - \frac{P_n}{100} = me^{-\lambda_1 nf} + ne^{-\lambda_2 nf},$$

writing $\qquad\qquad e^{-\lambda_1 f} = c_1 \text{ and } e^{-\lambda_2 f} = c_2,$

$$1 - \frac{P_n}{100} = mc_1{}^n + nc_2{}^n$$

if we write p'_n = per cent of 1st alcohol coming over in n-fractions,

,,   ,, P''_n =   ,,   ,,   2nd   ,,   ,,   ,,

$$1 - \frac{p'_n}{100} = c_1{}^n \qquad\qquad 1 - \frac{p''_n}{100} = c_2{}^n,$$

$$1 - \frac{P_n}{100} = m\left(1 - \frac{p'_n}{100}\right) + n\left(1 - \frac{p''_n}{100}\right),$$

or $\qquad\qquad P_n = mp'_n + np''_n.$

When a mixture of alcohols in dilute solution is distilled, each alcohol

[1] *Amer. Chem. Journ.*, 1903, **30**, 193.
[2] Reilly and Hickinbottom, *Proc. Roy. Dublin Soc.*, 1921.

will distil at a rate independent of the other alcohols present but depending on its own concentration.

The distillation constant (A) of several of the lower alcohols have been determined at constant volumes.[1] The experimental results are given in Table 111.

It may be seen that the distillation constants of the alcohols increase in an approximately regular manner with an increase in the molecular weight. The influence of other impurities such as aldehyde, esters, etc., is not considered as the quantity of these present is much less than that of the higher alcohols. Their distillation constants, however, may be calculated from the ceofficient given in Table 111.

TABLE 111

METHYL ALCOHOL 200 c.c. of the solution contained 9·0 grams methyl alcohol.		
Weight of distillate.	Percentage of alcohol collected in distillate.	$\frac{1}{v} \log \frac{a}{a-x}$.
10·14 grams	23·2	0·0113
20·01 ,,	40·5	0·0113
29·97 ,,	55·1	0·0116
50·55 ,,	72·8	0·0112
ETHYL ALCOHOL 200 c.c. solution contained 1·62 grams ethyl alcohol.		
10·52 grams	41·7	0·0222
21·17 ,.	64·4	0·0211
31·38 ,,	77·3	0·0205
41·74 ,,	84·0	0·0191
52·22 ,,	88·4	0·0179
200 c.c. of solution contained 9·62 grams ethyl alcohol.		
9·71 grams	32·1	0·0173
19·51 ,,	55·1	0·0178
29·65 ,,	72·2	0·0187
39·36 ,,	82·2	0·0190
49·31 ,,	88·1	0·0187
59·29 ,,	91·0	0·0176
Alcohol.	Concentration in grams of solute per 100 c.c. of solution.	
n-propyl alcohol	5·0	0·026
n-butyl ,,	9·7	0·030
iso-butyl ,,	6·9	0·047
sec-butyl ,,	5·6	0·050
iso-amyl ,,	4·3	0·053

The value $\dfrac{1}{v} \log \dfrac{a}{a-x}$, as will be seen from p. 290, is equal to $\dfrac{k\sigma}{\rho V}$ and

[1] Reilly and Hickinbottom, *Proc. Royal Dublin Soc.*, 1921.

if $\dfrac{\sigma}{\rho}$ is assumed to be constant $\dfrac{1}{v}\log\dfrac{a}{a-x}$ is inversely proportional to the initial volume of liquid in distilling vessel. This value varies also to a small extent with the concentration of the liquid distilled and, therefore, in specifying distillation constants of the above type it is necessary to specify the initial volume of liquid distilled and also its concentration.

As will be seen from Table 111 the distillation constant (A) for ethyl alcohol is lower than the corresponding constants for the higher alcohols. In the case of fermented wash the higher alcohols present (fusel oil) will correspond mainly to a mixture of iso-amyl alcohol with smaller quantities of iso-butyl and other alcohols. The total quantity of fusel oil present is small compared with the ethyl alcohol content (1 : 20). It is evident that in the distillation of fermented wash the fusel oil will tend to concentrate in the first fractions of the distillate. From the value of the distillation constants a table can be made for any known mixture of ethyl alcohol and fusel oil, which will show the extent of the separation of the alcohols in the various fractions of the distillate. The presence of other impurities such as esters will influence these results.

Removal of By-products by Rectification.—As the means adopted for the removal of volatile by-products in discontinuous rectification differ from those employed in continuous rectification the two cases will be considered separately.

Discontinuous Rectification.—The still is charged with raw spirit at about 38 per cent by weight, for it is more difficult to eliminate the impurities from very concentrated alcohol.

The raw spirit contains organic acids which react with the various alcohols present to produce esters. This action is more pronounced in discontinuous rectification, for in this case the acids remain in contact with concentrated alcohol for a longer time than when the rectification is continuous. It is therefore desirable to neutralise the acids by the addition of caustic soda.

The raw spirit must, however, not be completely neutralised for then amines may be evolved which will give an unpleasant smell to the alcohol.

Behaviour of various Impurities during the Rectification.—Impurities for which K' is very high, such as ethyl formate, methyl acetate, ethyl acetate (see Fig. 102), will tend to pass over in the first runnings.

For a reason which will be explained later the concentration of the ethyl alcohol on the plates of the rectifying column is allowed to attain its maximum before the collection of the " first runnings " is started.

In these circumstances impurities such as those we are considering are easily eliminated from the still, but their separation from the concentrated alcohol on the upper plates of the column is much more difficult. The result is that the distillate will continue to be contaminated by such impurities for a considerable time. The same difficulties occur but in a more aggravated form with such substances as ethyl iso-butyrate. For this substance $K'=1$ when the alcoholic strength is 89·7 per cent by weight, so that this impurity will tend to collect near the plate containing alcohol of this concentration. This point is dangerously near the level at which the finished product is run off, so that the continuous contamination of the distillate is very probable.

The contamination of the finished spirit by head products is prevented to a considerable extent by running off the ethyl alcohol from a point a few plates below the top of the column. The head products mixed with ethyl alcohol pass to the top of the column and enter the condenser. The major part of this mixture is condensed and returned to the top plate of the column, but a small portion rich in head products passes uncondensed through the condenser and is collected separately. By this process, which was first employed by Barbet, and called by him " Pasteurisation," the ethyl alcohol obtained is freed from head products to a very considerable extent.

Impurities such as amyl alcohol, for which K' is low, are retained in the still or lower parts of the column as long as possible. Completely to prevent amyl alcohol from passing over into the distillate, it is necessary to maintain a very high concentration of ethyl alcohol on the upper plates of the column, and the alcoholic content of the distillate should be not less than 93·8 per cent by weight. For this purpose a very large number of plates are required in the column.

As amyl alcohol is the principal volatile impurity it is important to retain it effectively from the start of the rectification. The alcoholic strength on the plates is therefore raised to a high point before the first runnings are collected, so as to prevent the amyl alcohol from mounting to the upper part of the column and contaminating the finished product.

The amyl alcohol will be present in greatest concentration near the point in the column at which the ethyl alcohol strength is 33·4 per cent by weight, and $K'=1$. Towards the end of the rectification, as the amount of ethyl alcohol remaining in the still decreases, the point at which the alcoholic strength is 33·4 per cent by weight will gradually rise in the column and the distillate will become more and more contaminated with amyl alcohol.

The danger of amyl alcohol passing over into the distillate increases as the rectification proceeds ; for as the alcoholic strength of the liquid in the still decreases the proportion of the total amyl alcohol contained in the column increases owing to the rise of K' with the fall of the alcoholic strength in the still.

Impurities such as iso-amyl acetate, iso-butyl alcohol, and propyl alcohol behave in a similar manner to amyl alcohol, but as they tend to collect at points in the column where the alcoholic strength is 62·5 to 73·5 per cent by weight, they are more difficult to retain than amyl alcohol.

" The distillation of a mixture of three constituents and the continuous separation of these in different phases have recently been investigated by Gay.[1] Formulae have been worked out for calculating

---

[1] *Chim. et Ind.*, 1920, **4**, 735-748 ; *J. Chem. Soc.*, A. 1921, 120, ii. 85.

the composition of the different phases, and rules established concerning the minimum and maximum heat compatible with the correct operation of the fractionating and rectifying column.

Continuous Rectification. — The raw spirit to be rectified is supplied to the still with alcoholic content of 33·4 per cent to 38 per cent by weight. In this case it is not found necessary to neutralise the liquid before rectification, as the acids are largely removed in the preliminary purification, and therefore do not remain for long periods in contact with the concentrated alcohol.

The raw spirit is supplied at a point near the top of a column similar in principle to the wash column of a continuous still, steam being admitted near the base.

The head products tend to pass away from the upper part of the column mixed with some ethyl alcohol. To free them as far as possible from the latter, a few plates are placed above the feed plate.

The distillate leaving the top of the column is rich in head products for which $K' > 1$. The raw alcohol flows down the column, being progressively freed from head products, and finally leaves the base of the column with a greatly reduced content of head products, but with its content of ethyl alcohol and higher alcohols almost unchanged.

This preliminary purification is in general only efficient for the removal of substances for which K' is considerably over 1 at the greatest alcoholic strength which exists in the purifying column.

The less volatile by-products are best removed in the final rectifying column.

The partially purified alcohol is then supplied to a point near the base of a rectifying column, in which it is caused to boil by steam introduced below the supply plate. The distribution of the various constituents of the mixtures takes place as in the column of a discontinuous rectifying still.

But in this case the composition of vapour and liquid at any point in the column remains constant throughout the rectification. This is very important, as the various constituents can be run off continuously from the points in the column at which they accumulate. In the discontinuous still this is not generally possible, or is very much more difficult to control, as these points are continually changing as the rectification proceeds.

The remaining head products (mixed with a certain proportion of ethyl alcohol) pass continuously uncondensed through the condenser at the top of the column. The ethyl alcohol is run off from a plate somewhat lower down.

Impurities such as iso-amyl acetate, iso-butyl alcohol, and propyl alcohol are continuously run off from the point in the column where the alcoholic strength is about 68 to 73·5 per cent by weight. Amyl alcohol is run off from a point at which the alcoholic strength is 33·4 per cent by weight.

The withdrawal of these two classes of by-products is not begun until they have had time to accumulate to a considerable extent. This is generally not the case until the still has been running for about two days.   But it is not advisable to allow them to accumulate beyond a certain point for fear of contaminating the finished spirit.   For this reason the mixtures withdrawn always contain much ethyl alcohol and have to be fractionated separately, or, if the impurity is only slightly soluble in water, they may be purified by washing with water and decantation.

A comparison of the two methods of rectification just described shows that the fixed conditions of the continuous process combined with the preliminary purification of the dilute spirit renders the removal of the volatile by-products much easier and more efficient—the head products being largely removed under the most favourable conditions, and the higher alcohols and other less volatile impurities being continuously prevented from mounting to the point in the column from which the finished spirit is withdrawn.

CHAPTER XXVIII

Discontinuous Distillation. — **The Pot Still** (Fig. 103). — In the earliest types of plant the two processes of distillation and rectification were carried out in the same vessel, the latter process being very incompletely effected. A certain volume of the fermented wash was boiled in a large kettle and the vapour so produced led directly to a condenser without passing through any rectifying apparatus. The distillate was then again distilled, and this process had to be repeated several times before strong alcohol was produced. To remove aldehydes and higher alcohols the distillate from the final distillation was collected in three fractions. The first, called "foreshots," contained, besides some ethyl alcohol, most of the aldehydes and esters. The middle fraction contained most of the ethyl alcohol, and the third (feints) contained a large percentage of the higher alcohols.

The plant was simple, but the process was very uneconomical both in time and materials, and the finished alcohol still contained a considerable quantity of aldehydes and higher alcohols In the manufacture of whisky, brandy, and rum

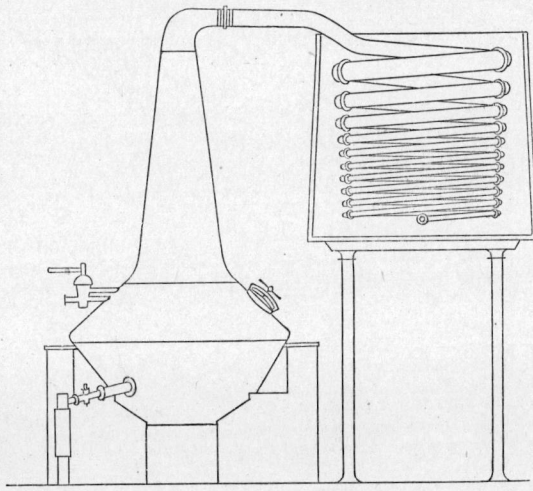

FIG. 103.

the presence of these by-products is to some extent desirable, as they produce characteristic tastes and odours. For this reason the distillation of these beverages is still sometimes carried out in the above manner in the so-called "Pot Still." Fig. 103 shows a fireheated pot still. In the earliest form of pot still the neck was not so high as that shown in the figure. The object of lengthening the neck was to prevent the liquid passing over into the condenser, as much frothing takes place in the still, and to produce an increased rectifying action.   Many pot stills now in use in Scotland are fitted with " purifiers," which consist of circular vessels cooled by water, placed between the neck and condenser ; the condensate produced by them flows back to the still.   In some cases the " purifier " consists of a form of multitubular water-cooled dephlegmator.   In Ireland the pot stills are much larger, up to 20,000 gallons capacity.   The pipe connecting the top of the still with the condenser, called the " Lyne Arm," is from 30 to 40 feet long and is surrounded by a trough through which passes water, the rate of flow of which can be regulated.   A pipe from the end of the " lyne arm " furthest from the still serves to return the liquid condensed in the lyne arm to the still.   Similar return pipes are sometimes fitted between the cooling coil and the still.   The rate of flow of liquid through these return pipes can be controlled by taps.   The point at which the lyne arm enters the condenser is frequently 35 to 40 feet above the top of the still.   It is obvious that rectification of the distillate must take place to a considerable extent in such a lyne arm.

In Scotland the process of distillation is carried out in two stages. The wash, to which soap is often added, principally to prevent frothing, is distilled in the wash still.   One fraction only is collected. The distillation is stopped when the hydrometer shows that the distillate no longer contains alcohol.   The residue in the still is run to waste.

The distillate is transferred to a second still, called the " low wines " still, and redistilled.   In this case three fractions are collected, (1) Foreshots, (2) Whisky, (3) Feints.   The collection of the last fraction ceases when the distillate no longer contains alcohol as shown by the hydrometer.   The residue in the still is run to waste.

The point at which to begin collecting a new fraction can be judged from the alcoholic strength of the distillate and from the presence or absence of turbidity on adding water to a sample of the distillate.   But the decision depends largely on the experience and judgment of the distiller.   The first and third fractions are mixed together and added to a subsequent charge in the low wines still, or they may preferably be treated separately.   The strength of the middle fraction is about 60 per cent by weight.   In Ireland three stills are generally used and the distillate is collected in a larger number of fractions.   The strength of the whisky fraction in this case is about 80 per cent by weight.[1]

The alcoholic strength of the middle or whisky fraction will have a marked effect on its content of impurities.   As already stated certain volatile impurities tend to accumulate at a point in the still at which the alcoholic content is from 68 to 73·5 per cent by weight.   If the distillate be only raised to a strength of 60 per cent a large part of these substances will pass over into the distillate.   But at a strength of slightly over 80 they will be condensed to a considerable extent in the lyne arm, and returned to the still.

[1] P. Schidrowitz, *J.I.B.*, 1906, p. 496.

The coefficients of purification of the other impurities vary with the content of ethyl alcohol in the mixture. So that the amount of these in the whisky will also depend to some extent on the alcoholic strength at which the foreshots and whisky fractions are collected.

To prevent amyl alcohol from passing over too freely in the whisky fraction, the alcoholic content of the distillate is raised to a high point before any distillate is collected. When this is the case the separation of the aldehyde and esters will be more difficult. These substances will not be completely eliminated at the end of the foreshots fraction and will contaminate the whisky fraction to some extent.

The character of the whisky produced will therefore depend upon the amount and nature of the impurities present in the raw spirit and on the methods of controlling the distillation.

Discontinuous Rectifying Stills. — The alcohol produced by the stills so far described is not sufficiently pure for many purposes. To produce highly concentrated alcohol containing only traces of volatile impurities a still with a very efficient fractionating column is required. Fig. 104 diagrammatically shows one of these stills.

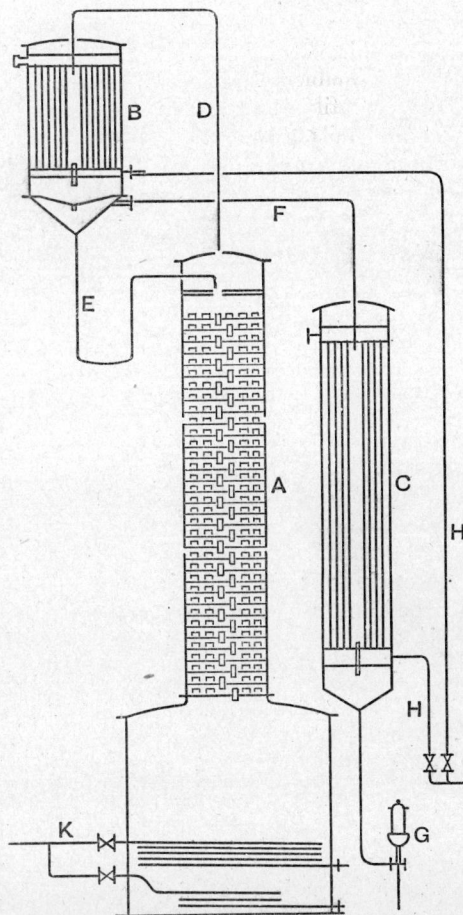

FIG. 104.

A. Rectifying column.
B. Dephlegmator.
C. Condenser for finished spirit.
D. Vapour pipe.
E. Reflux pipe from dephlegmator to column.
F. Pipe carrying vapour uncondensed in the dephlegmator to the condenser.
G. Test glass.
H. Cold water pipe.
K. Steam pipes.

The various types of plates in these columns are described under continuous rectification. The plate with bells is most frequently employed.

Tests of the efficiency of the various forms of plates have been carried out by Barbet and others. It is difficult to make a just comparison between them as varying conditions of working so greatly affect the results obtained. The plates with improved type of bells and directed flow appear to give as good results as any.

The number of plates in the column is very considerable. Table 112, taken from Sorel, *Rectification de l'alcool*, 1895, p. 55, shows the alcoholic content in volume per cent of the liquid on the plates of a discontinuous rectifying column. It will be noticed that there are a very large number of plates on which the alcoholic strength is over 95 per cent by volume.

The object of this, as already explained, is to keep amyl alcohol (and other by-products which behave similarly) from passing over with the distillate until as late a stage of the distillation as possible.

The details of the stills and columns vary greatly in practice, as do also the number of fractions collected and the degree of purity obtained.

TABLE 112

Alcoholic content of liquid on plates of a rectifying column at the middle of the distillation.

Number of plate from bottom.	Alcohol. Volume per cent.	Number of plate from bottom.	Alcohol. Volume per cent.
49	96·3	19	94
44	95·9	14	93
39	95·6	9	92
34	95·3	4	87·5
29	95	Still	33·6
24	94·6		

Continuous Distillation.

The simple pot still does not economically deal with large volumes of wash. Stills were therefore designed which could be continuously supplied with wash.

Fig. 105 gives a diagrammatic representation of one of the earliest of these stills, introduced by Pistorius.

The wash is run from the pipe a into the heater (economiser) A, thence by the pipe bb into B, and finally into C through the pipe c; A and B are then filled with wash. The still C is heated either by a furnace, as shown in the diagram, or by steam, the wash being constantly stirred by a chain. The weak alcoholic vapours pass through the pipe D into the second heater B, where they condense and heat the wash to the boiling point, the waste heat from the furnace being also utilised. The stronger vapours from B now pass by the pipe EE' into the outer chamber of A (where more weak alcohol is condensed and flows back by the pipe d into B) then up through the narrow passage FF into the dephlegmator G, shown on an enlarged scale below. After passing through two or more of these dephlegmators the strong alcoholic vapour reaches the condenser by the pipe H. The plan adopted here of passing the vapour evolved in one still into liquid in a second, led to the gradual evolution of the modern type of plate column which effects the concentration of the spirit by exactly the same principle differently applied.

One of the first efficient continuous stills with plate columns to be employed in England was that designed by Æneas Coffey in 1831, and stills of this type are still largely used to this day (see Fig. 121). The wash is distilled in one column called the analyser and the vapour so produced is led to a second column called the rectifier, in which the majority of the water and fermentation by-products are removed. This arrangement has formed the basis for most of the continuous stills

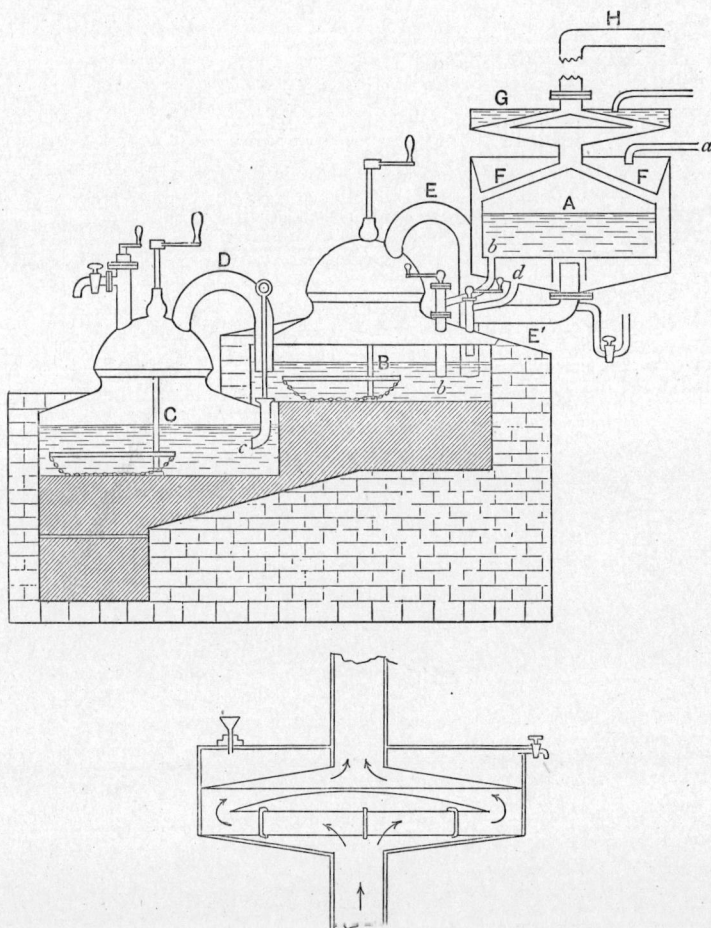

FIG. 105.

used at the present day, all of which have some form of analysing or boiling column combined with a rectifying column. Before describing some modern stills an account will be given of the essential parts of various types of distilling or boiling and rectifying columns.

Distilling or Wash Columns.—The continuous boiling of the wash is carried out in a variety of ways, all of which have this in common, that the wash flows down some kind of column and meets a rising current of steam. By this means the wash is freed from the volatile products of fermentation, and flows away from the base of the column as " spent wash."

The methods adopted fall under two heads :—

(1) A. Plate columns.
    B. Sloping columns.
(2)   Full columns.

Plate columns aim at bringing every part of the wash to be distilled into intimate contact with the vapour by causing it to flow over a number of plates in a shallow stream through which the vapour is forced to bubble. They are more efficient in carrying this out than the full column, but are more difficult to regulate when using thick wash. The sloping column is an attempt to unite the advantages of both types when distilling thick wash. The wash supplied to the boiling column is generally already heated to a fairly high temperature by heat exchange with the spent wash leaving the base of the wash column, or by its use in place of cold water in the condenser or rectifying column. A considerable saving of fuel is thus effected.

1. Various Types of Plate Columns.—Perforated Plates.—In

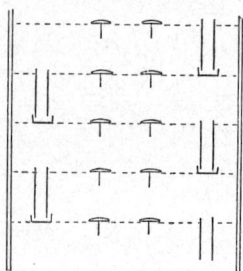

FIG. 106.

Coffey's still the boiling column or analyser consists of a series of copper plates arranged horizontally at intervals of 9 inches (Fig. 106). (In all plate columns it is most important to have the plates absolutely level so as to maintain an equal depth of liquid all over the plate.) The plates are perforated with holes arranged 1 inch apart. The number of plates is about 20 to 24. Through each plate (see Fig. 106) passes a copper pipe, the upper end of which stands about 1 inch above the perforated plate. The lower end of the pipe dips into a small pan placed on the plate below. Every plate is provided with simple safety valves (shown in the figure) to prevent excess of pressure being set up between any pair of plates owing to blocking up of perforations or sudden increase of steam supply to the boiling column. The action of these plates can be followed by reference to the illustration of Coffey's still (Fig. 121).

The wash, previously heated by passage (in the wash pipe) through the rectifying column, flows on to the top plate of the wash column, and passes by a zig-zag course from plate to plate by the drop or overflow pipes, a depth of about 1 inch of the liquid being maintained on each plate. The wash in its descent meets steam which is introduced at the base of the column. The vapour produced rises through the perforations and bubbles through the liquid on the plates. The pressure of steam is sufficient to prevent the liquid flowing down through the perforations. But if the necessary pressure is not maintained the liquid on the plates will flow down through the perforations and the charge of the column be lost ; this disadvantage is avoided with the type of plates described subsequently, which do not empty themselves if by accident the pressure is allowed to fall too low. The liquid to be distilled is brought into very intimate contact with the vapour, and a very efficient separation of the volatile products of fermentation is produced. The spent wash which leaves the base of the column by a syphon pipe is practically free from alcohol. The dimensions of the syphon determine the maximum pressure in the column. This type of plate works well with thin and moderately thick wash, but would tend to become choked by very thick wash such as is used in some foreign distilleries. The chief objection to these plates is that the perforations are gradually enlarged by the passage of the vapours, and when they become inconveniently large the whole plate has to be replaced, which necessitates the complete dismantling of the boiling column.

Plates with Bells.—In another form of boiling column the liquid flows from plate to plate by means of reflux tubes similar to those in Coffey's still. A slightly greater depth of liquid is maintained on each plate, but not in this case by the pressure of the ascending vapours. The perforations in the plates are replaced by a comparatively small number of larger holes, each of which is covered with the arrangement shown diagrammatically in Fig. 107. The vapour passes through the central pipe A (Fig. 107) (the upper rim of which is above the surface of liquid on the plate) and is forced to bubble through the liquid on the plate by the bell B. The depth of liquid is determined by the height to which the top of the drop or overflow pipe C rises above the plate. The depth of liquid through which the vapour bubbles is determined by the difference of level between the rim of the bell B

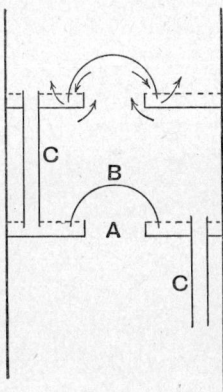

FIG. 107.

and the top of the overflow pipe C. Fig. 108 shows the arrangement of bells on the plate. Fig. 109 represents a form of bell designed by

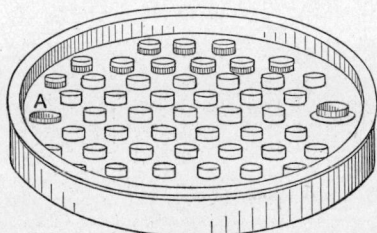

FIG. 108.

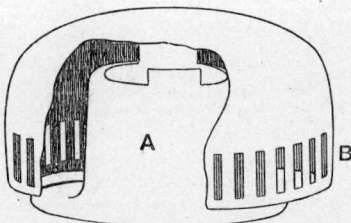

FIG. 109.

Barbet. The slits on the side tend to break up the vapour as it enters the liquid. Fig. 110 shows a bell designed for use with thick wash; the conical form is intended to prevent the deposit of solid matter from the wash. The efficiency of this form of plate depends on the proper arrangement of these holes and bells over the surface of the plates. They must be arranged so as to cause the wash to flow evenly all over the plate.

When properly designed this type of plate does not get blocked even by thick wash, and has the advantage over the perforated plates that the bells can be replaced one by one without dismantling the column when the slit becomes worn out. The liquid remains on the plate if by chance the pressure falls too low.

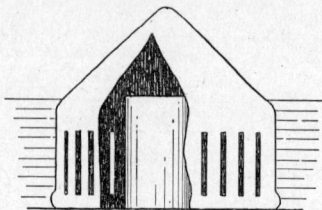

FIG. 110.

Sloping Columns.—Fig. 111 shows the Guillaume Inclined Column.

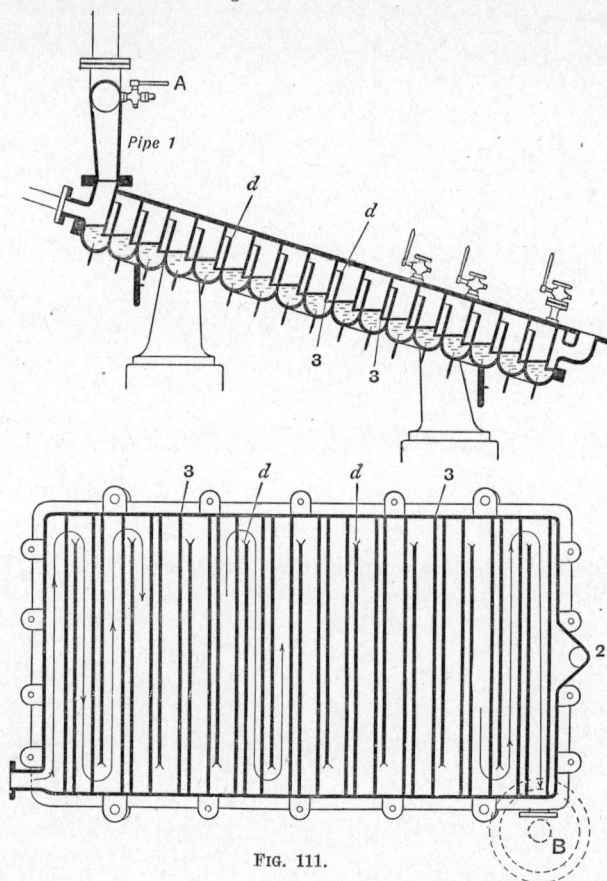

FIG. 111.

This apparatus is constructed of an upper and lower part which can be easily separated for cleaning purposes. The lower half is fitted with upstanding baffle plates d, which alternately stop short of either side.

The wash which enters at the top by pipe 1 follows a zigzag course down the column as shown by the arrow. The rate of flow is regulated by a valve at A, and the spent wash leaves the column by B. The upper half is provided with depending baffle plates (3) extending across the whole width of the column.

Steam enters the base of the column by pipe 2. On its way up the column the steam must pass under each of the depending baffle plates (3) attached to the upper half of the column. In so doing it bubbles through the down-flowing wash, carrying with it all the volatile products of the fermentation.

This column is said to work very well with thick wash, and a more even flow of the liquid is obtained than in the ordinary plate column. This is important, as it means more complete removal of the volatile products. Blocking of any part of the passage is not likely to occur ; it has also the advantage of being more compact and less lofty than some of the other forms of boiling columns. The top of this column being so low enables the condensate from the base of the rectifying column to flow back into it by gravity. With a high boiling column, this can only be attained by having an unduly high building.

2. The Full Column.—The " full column " is completely filled with wash. The wash which enters at the top and flows away at the base is made to circulate by a large number of baffle plates arranged in the column as shown in Fig. 127, D. Steam enters at the base. It is not so efficient as the plate type of column, as it is impossible to effect so thorough a mixture of the vapour and liquid, but it is not so liable to blockage when thick wash is being used.

Preheaters.—Before entering the boiling column the fermented wash is subjected to a preliminary heating in some form of heat exchanger. The wash may either be used as a substitute for cold water in the condenser, dephlegmator, or rectifying column, as in Coffey's still, or a special vessel may be constructed in which a heat exchange takes place between the hot spent wash and the cold wash to be distilled. The former method has the advantage of compactness and economy of design. No extra apparatus is required, but it may not be practicable when very thick wash is used. It should be noted here that the arrangement adopted in Coffey's still for preheating the wash is not the ideal one from the point of view of economy of rectification.

Theoretically the best arrangement is to have one dephlegmator at the top of the rectifying column so that every plate of the rectifier has the full benefit of the maximum volume of the condensate. If part of the condensate is produced at points lower down the column the efficiency of the rectifier is diminished (see p. 273). However, the most vigorous condensation must take place at the top where the wash is cold. When the wash has become fairly hot the amount of condensation of head products and ethyl alcohol must be greatly diminished.

For thick wash the preheater must be specially designed to prevent

x

the deposit of solid matter from the wash and to facilitate cleaning and the removal of obstructions should any be formed. Whatever system of distillation be adopted, a large amount of heat is always wasted unless it be utilised for some purpose external to the distillation.

Rectification.—In the earliest types of still the concentration of the alcohol was effected by repeated fractionation. In a discontinuous apparatus this was a slow and wasteful process. The distillation and rectification are now in the majority of cases carried out by continuous processes. But the final rectification is frequently carried out in a discontinuous still. In some systems the continuous distillation is independent of the continuous rectification; in others the two processes are combined and carried out simultaneously. In the former the vapour produced by distillation is condensed and the liquid so produced is fed to an independent rectifying plant. In the latter the alcoholic vapour produced by distillation is led directly into a rectifying apparatus. The latter is evidently more economical of heat. The process of rectification is essentially the same in both cases. The number of plates required in the rectifying column to obtain spirit of a certain strength depends on (1) the alcoholic content of the wash, (2) the strength of the alcoholic vapour supplied by the boiling column, (3) the amount of liquid refluxed by the dephlegmator. *Ceteris paribus*, the larger the number of plates the less the amount of dephlegmation required and the smaller the consumption of steam by the still.

Rectifying Columns.—The various methods adopted, though they differ greatly in details, all make use of one or more fractionating

columns, from which the fractions to be collected are drawn off simultaneously. The columns used may be classified, as in the case of the wash columns, under the following types:

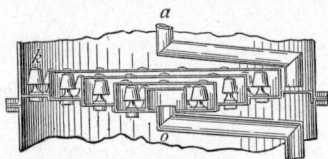

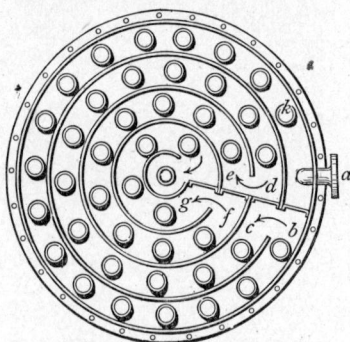

1. Plate Columns.
2. Sloping Columns.
3. Film Columns.
4. Spraying Columns.

1. Plate Columns.—The two types already described under wash columns are frequently employed for rectifying columns.

Both types give good results, but the plates with perforations (Fig. 106) have the disadvantage that the vapour rising through the perforations tends to carry small drops of liquid with it on to the plate

FIG. 112.

above. This objection is overcome by the construction shown in

Fig. 107. In this case the vapour passes horizontally into the liquid on the plate, and splashing up of liquid on to the plate above is reduced.

In some cases the flow of liquid on the plate is directed by baffles so as to ensure an even distribution and mixture with the rising vapour. To this class of plate belong those designed by Egrot, Fig. 112. Tables 106 and 107 show the number of plates required under various working conditions.

2. Sloping Column.—The sloping column already described may also be used for rectifying purposes.

3. Film Columns.—Ilges has made use of towers packed with glass balls as rectifying columns. As pointed out by Tungay,[1] the nature and arrangement of materials used for packing such towers is of importance. The packing should be composed of units of regular form. There should be a large free space compared with the total volume of the column filled. The packing should not be acted on by the vapours in contact with it and should not be brittle. Direct free passage of the vapour should be avoided. Continual change of velocity of the vapours should be produced so as to effect thorough mixing.

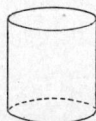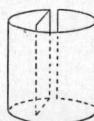

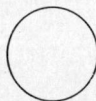

FIG. 113.   Fig. 113a.

These objects are well attained by Raschig rings (see Fig. 113). These consist of thin sheet-iron rings, 1 inch in diameter and 1 inch long. They are arranged promiscuously, and give a scrubbing surface of 200 square metres per cubic metre of space filled. The free space is also large, amounting to about 90 per cent of the total space filled. Lessing's rings (Fig. 113a) give an even larger scrubbing surface.

4. Spraying Columns.—A form of spraying column has been designed by Egrot, E.P. 3561, 1903, and is shown in Fig. 114.

Another type of spraying column has been invented by Dr. K. Kubierschky, E.P. 15300; 1913. Shown in Fig. 115.

The column is divided into separate chambers by strainers 30, 31, which allow the trickling liquid to pass down, but do not allow the gases to rise. The sieves 30, 31 are arranged alternately. The sieves 30 carry axially disposed tubes 32, which lead from the lower strainer 31, namely the base of the lower compartment, to the top of the compartment above, that is to say, to the next sieve 31 above.

The sieves 30 are arranged in annular sleeves 33, which allow an annular channel 36 to be mounted between themselves and the casing

[1] *Chemical Age,* 21.6.19, p. 11.

of the column apparatus, so that said annular channel leads from one sieve 30, that is, from the base of one compartment, to the next sieve, and consequently to the top of the next compartment. Steam and vapours entering the lowest tube 32 pass, owing to the arrangement of the central tube 32 and the sleeves 33, in the direction indicated by the arrows. It will be obvious that in the several compartments the vapours will descend with the downward trickling liquid and will be forced on the bottom of each com-

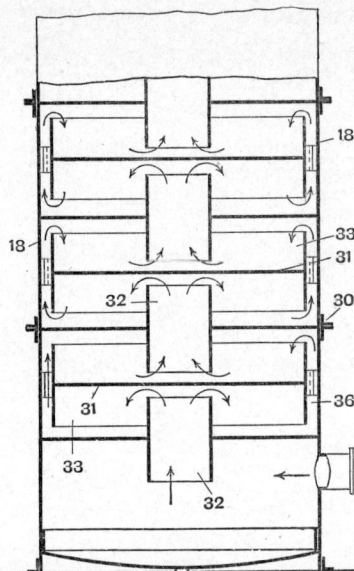

FIG. 114.

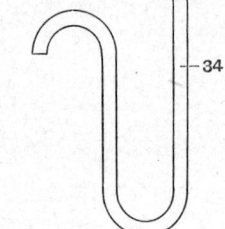

FIG. 115.

partment towards the column casing by vapours following. The vapours then ascend through the annular channel to the top of the next compartment, to then again pass down in the same direction as the trickling liquid to the base of the compartment and thence through the centrally arranged pipe 32 to the top of the next compartment.

It is stated by S. J. Tungay,[1] that this type of still was used by Messrs. E. Merck of Darmstadt for the distillation of alcohol, and that good results were obtained.

The vapour is compelled to pass from a point low down in each chamber to the chamber above. Vapour rich in alcohol and having a higher density than the steam is supposed to collect in the lower part of each chamber. By drawing the vapour

[1] *Chemical Age*, 21.6.19, p. 11.

from the lowest possible point, a more rapid concentration of the alcoholic vapours in the column is said to be effected. But with the turbulent motion in each section, differences of density would probably have very little influence.

Dephlegmation or Partial Condensation of the Vapour.— The fractionating columns used for the purification of alcohol on the manufacturing scale are so large that they cannot depend, as many laboratory columns do, on the cooling effect of the surrounding air to provide the necessary condensed liquid on the plates. The necessary condensation is generally provided either, as in Coffey's still, by causing the cold fermented wash to flow in a pipe running in a zigzag course over every plate of the rectifying column, or, as in most of the French

and German stills, by means of a dephlegmator placed so that the liquid condensed flows back to the top of the rectifying column. Theoretically the latter arrangement, from the point of view of fractionation, is the most efficient and economical (see p. 273).

The construction of and effect produced by the dephlegmator have already been considered (see p. 272).

If the fractionating column is efficient, the form most frequently employed consists of a simple multitubular condenser. The cooling liquid passes through the tubes and the vapour between them. The condensate is returned to the top plate of the column and the uncondensed vapour passes to the cooler.

If it be desired to reduce the height of the column, some form of dephlegmator is employed which exercises a certain rectifying action on the vapour entering it from the column. Fig. 116 represents

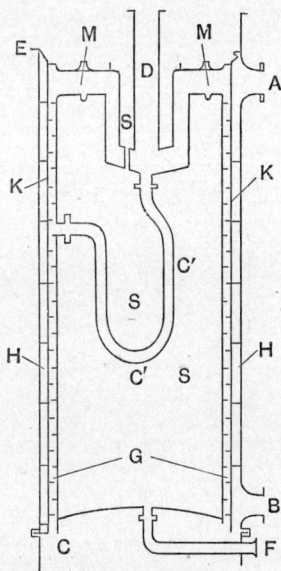

FIG. 116.

Verchow's dephlegmator. The wash to be distilled enters at A and follows a spiral course down the annular space H. It flows away by B to the wash column.

The vapour from the column enters at C and passes up the annular space G, where it is partially condensed owing to the presence of the cold wash on the other side of the plate K.

The uncondensed vapour passes into the vessel M, which is entirely surrounded by water, which also completely fills the space S, S, S. Part of the vapour entering M is condensed and returns to G by the pipe C'; the uncondensed vapour passes out of the dephlegmator by D and is conveyed by means of a pipe to a condenser.

Control of Continuous Stills.— To obtain final products of constant composition from continuous stills it is most important to keep a very steady flow of steam and wash to the still and of finished products from the rectifiers.

Control of Wash Supply to the Still.—The wash flows from a reservoir to the regulator shown on Fig. 117. This works by means of a float A, which rises with the level of wash in the regulator and automatically closes the valve v when a certain level of liquid is reached in the regulator. When the level of the liquid falls, the valve falls open again by its own weight.

Regulation of Steam Supply to the Still.—The steam supply to the still is automatically regulated as the pressure near the base of the rectifying column rises or falls. This is effected by the

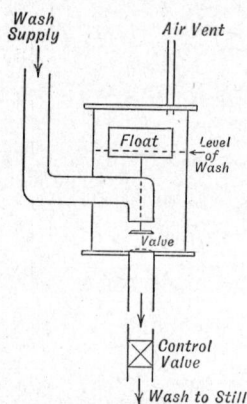

FIG. 117.

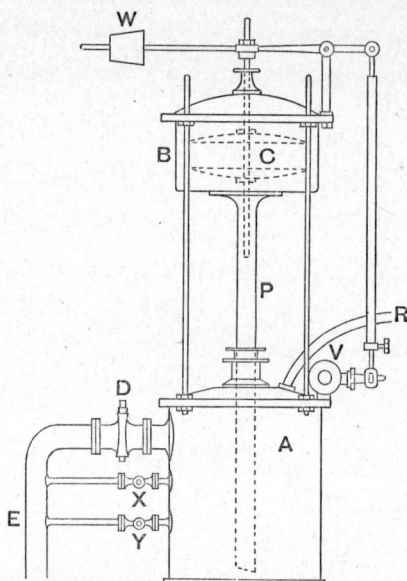

FIG. 118.

apparatus shown in Fig. 118. The chamber A is filled with water to the level of the outlet cock D.

The pipe R is connected to the base of the rectifying column, so that whatever pressure exists there will be transmitted to the chamber A. Increasing pressure forces the water in A up the pipe P into the chamber B, causing the copper float C to rise and gradually close the steam valve v which controls the steam supply to the boiling column. Decreasing pressure allows water to flow down again into the chamber A, the float to fall and the steam valve to open. The weight w ensures the falling back of the float when the water sinks. Valve D is set so as to adjust the amount of liquid in A. The overflow containing some alcohol from the vapour entering at R is returned by F to the boiling column. The working pressure can be varied by opening X or Y.

Regulation of Rate of Outflow of Finished Products from the Rectifying Column.—Fig. 119 represents a test glass which will measure the rate of flow of the finished product and enable the flow to be regulated to the desired rate.

The glass vessel F contains a known volume of liquid measured from the point A to the point B.

To test the rate of flow the vessel is emptied to A by opening the cock D. The latter is then closed and the time taken for the liquid to reach B is noted. The process is repeated and the flow is adjusted by the cock C until the desired rate is obtained. The cock D is then partly opened. When the level of liquid in F is steady, this level is noted on the scale E and maintained constant.

Fig. 120 shows an arrangement devised by Barbet and often employed on French stills. The even running of the still is controlled by the means already described.

The vapour leaving the top of the column A passes into the condenser E and is partially con-

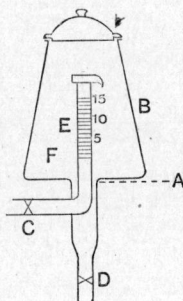

FIG. 119.

densed. The condensate is returned to the top plate of the column at B.

The uncondensed vapour passes on to F, in which it is completely condensed.

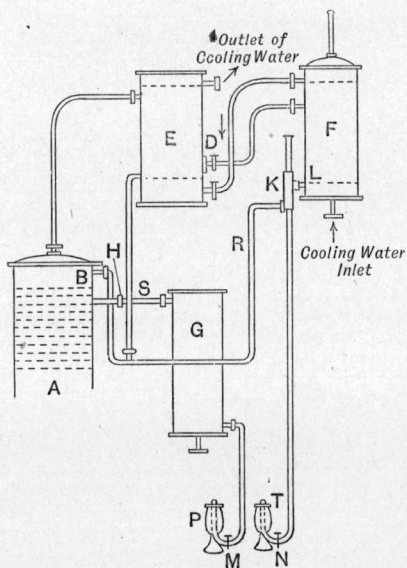

FIG. 120.

Finished spirit is run off from about the third plate from the top of the column, and is cooled in G. The rate of flow is regulated at the test glass P.

The liquid condensed in F flows to the test glass T. The rate of flow at T is fixed in a certain relation to that at P. In practice the flow at P is often fixed at 95 per cent, and the flow at T at 5 per cent of the total flow from both points. Should the condenser E fail to condense enough of the vapour passing through it, owing to irregularity of supply of the cooling water, the excess of uncondensed vapour will be condensed in F, which is always supplied with an excess of cooling water, and will be returned by pipe R, together with the condensate from E, to the top plate of the column.

The liquid drawn off at T is rich in head products. It has been found that the alcohol drawn off from the third plate from the top of the column contains a much smaller proportion of head products than alcohol collected from the top of the column ; and this is not surprising, for the condensers E and F together supply the whole of the condensate for the still. This liquid, which is often equivalent to 6 or 7 times the volume of finished spirit run off through P, flows on to the top plate of the column, where violent ebullition takes place. By this means the head products tend to be driven to the highest point in the still away from the point at which the finished spirit is drawn off. This process is called pasteurisation.

By the use of two condensers in series the temperature of the major part of the condensate returned to the top of the column is more easily kept steady, by maintaining a steady temperature in condenser E. The second condenser F ensures the complete condensation of the vapour entering it, but the low temperature of this small part of the condensate does not greatly affect the average temperature of the condensate returned to the column. Were only one condenser used it would be necessary to cool the whole condensate to a lower temperature so as to ensure complete condensation. Any irregularity of supply of condenser water would have a much more harmful effect in upsetting the smooth working of the still.

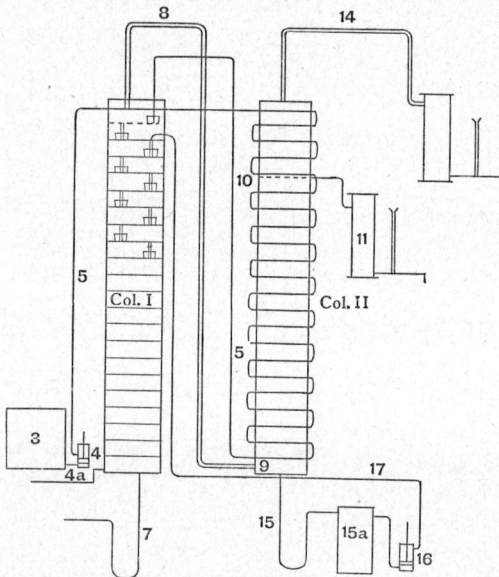

FIG. 121.

The specific gravity of the distillate and the temperatures registered at the base and top of the rectifying column are also useful checks in the proper running of the still.

Coffey's Still.— Description and method of working.[1]—Wash is pumped from a reservoir 3, Fig 121, up the pipe 5, and passes down the zigzag pipe 5 (Col. II.), where it is heated by the ascending vapour ; then up the pipe 5 into the highest section of the analyser, column I. It then descends from section to section through the drop pipes, and finally escapes by a trapped pipe 7, at a temperature of about 102° C.

Steam is passed into the analyser by the pipe 4a, and causes the wash to boil, so that by the time it has reached the bottom it is free from alcohol. The ascending vapours pass through the perforations in the plates and bubble through the liquid on them, part of the aqueous vapour being thus condensed, and the descending wash heated.

[1] J. A. Nettleton, *Manufacture of Spirit.* 1893, pp. 133-145.

On reaching the top of the column the concentrated alcoholic vapour passes through pipe 8 into the base of the rectifier at 9, and then ascends through perforated plates similar to those in the analyser ; the ascending vapour does not, however, meet wash in the rectifier, but spirit formed by partial condensation of the vapour. As the vapour ascends the rectifier it becomes richer in ethyl alcohol, and at a certain level in the rectifier an unperforated copper plate (" the spirit plate " or Dumb plate) 10 is fixed.   In this plate, Fig. 122, a wide pipe or pipes

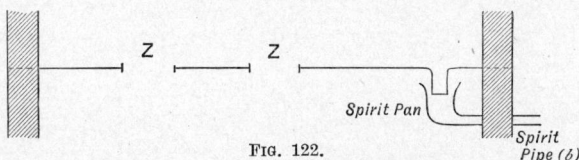

FIG. 122.

z are fixed which stand up about 1 inch above the plate ; through these pipes the vapour passes. Any liquid produced by the condensation of vapour in the column above this level flows from the spirit plate into the spirit pan which feeds the spirit pipe, from whence it is run to a condenser 11 (Fig. 121). As the level of the liquid rises in the spirit pan it overflows on to the perforated plates below. The flow of the finished spirit is controlled by a cock on the spirit pipe b, Fig. 122. The spirit so produced may contain as much as 94 per cent by weight of alcohol. It may also contain most of the aldehyde and esters and traces of the higher alcohols originally present in the fermented wash. The more volatile by-products may, however, be made to pass together with carbon dioxide away through pipe 14, Fig. 121. They may then be recovered by means of a condenser and gas scrubber. If very pure alcohol is required, the condensate from the top of the rectifying column must not be returned to the analyser, but must be collected and treated separately. The higher alcohols should also be drawn off from the point in the rectifying column at which they tend to accumulate. But this is seldom done in England. The condensate from the base of the rectifying column flows by pipe 15 to the hot feints receiver 15a, from which it is returned by the pump 16, through pipe 17, to a point near the top of the analyser. The usual way of removing the fusel oil is as follows. When shutting down the still the fusel oil drops with the charge from the plates of

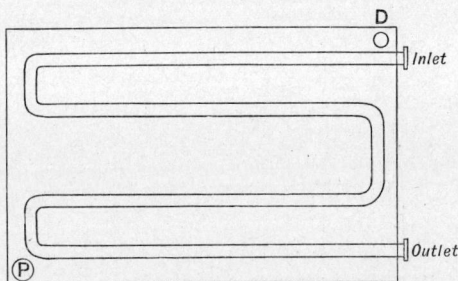

FIG. 123.—The inlet is connected by a bend to the corresponding pipe on the plate above. The outlet to the corresponding pipe on the plate below.

the rectifying column into the hot feints receiver. Subsequently water is added to the feints, from which the oil separates on cooling.

Plates.—The plates in both analyser and rectifier are perforated as already described. The arrangement of the wash pipes over the plates in the rectifier is shown in Fig. 123.

Starting and Control of Coffey's Still.—To start the still, steam is turned on so as gradually to heat up first the analyser and then the rectifier. When the latter is warm enough, as shown by the temperature at the level of the spirit plate, the wash pump is started. After a time the rectifying column becomes charged with enough alcohol to begin running off. When running smoothly the steam pressure at the base of the analyser and at the upper part of the vapour pipe are kept as steady as possible. The actual pressures vary with the size and proportions of the still. Samples are continuously taken from—

1. Vapour at the base of the analyser to test the exhaustion of wash.
2. The vapour pipe from analyser to rectifier.
3. The spirit plate.

The running of the still is controlled by keeping the temperature correct at the spirit plate, and by the indications of the pressure already referred to, and by the strength of the samples.

French Stills.—In France the rectification of alcohol has been brought to a very high pitch of perfection. Innumerable modifications in design of the stills have been introduced with the object of purifying the final product.

Two types of distilling plant will be described. A large number of the stills in use are based on the same principles as those to be described, though the differences in detail are very numerous. For further details see Mariller,[1] and Messrs. Egrot and Grangé's trade catalogue.

Continuous Distillation Direct Rectification Plant.—Fig. 124 is a diagram of a still of this type. The wash to be distilled flows from A to the preheater B, where its temperature is raised by heat exchange with the spent wash. The preheated wash enters the wash column C at H. From the rectifying column D, the pasteurised alcohol (see p. 311) is drawn off near the top, passing to F, and the head products pass to M. The amyl alcohol is removed at G.

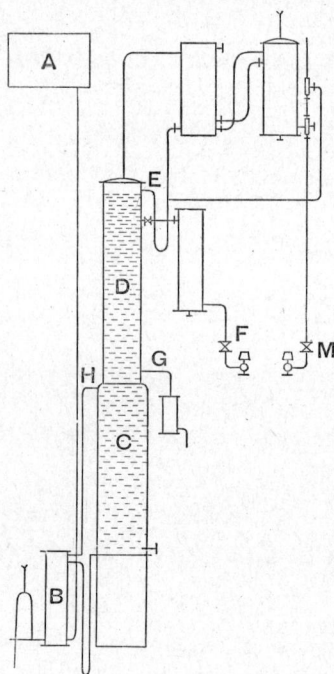

FIG. 124.

Sometimes the wash column is placed by the side of the rectifying column. The procedure followed, and the results obtained, are very similar to those of the Coffey still.

[1] *La Distillation fractionnée*, 1917.

In some cases small subsidiary columns are provided to free the amyl alcohol and head products from ethyl alcohol. The latter is in this case returned to the main column.

As already stated, the removal of the head products is more difficult to effect from concentrated than from dilute aqueous alcohol. If a more complete removal of these impurities is required, a purification of the alcohol before concentration in the rectifying column must be effected. The distilling plant now to be described is designed for this purpose.

Continuous Distillation and Indirect Rectifying Plant.—
The wash is first distilled in the still shown in Fig. 125.

The wash to be distilled flows from the reservoir A through the pre-heater B, in which its temperature is raised by a heat exchange with the spent wash. It enters the column at G and flows down, meeting steam which is admitted at H. The plates above G exercise a small rectifying effect.

The vapour from the top of

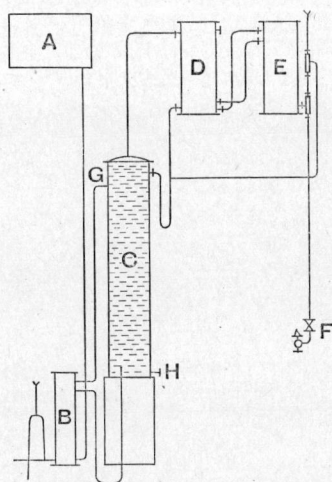

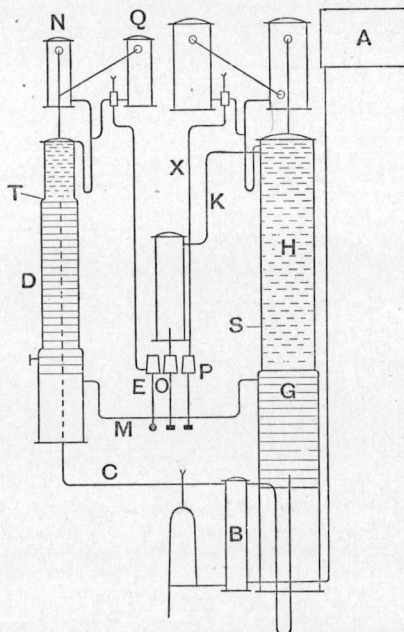

FIG. 125.   FIG. 126.

the column passes to the condensers D and E, which supply the necessary condensate to the column. The distillate flows down to F.

By this means spirit containing about 40 per cent of alcohol is obtained, which is then continuously rectified in a second still, Fig. 126.

The spirit prepared in the previous operation flows from A to the preheater B, in which its temperature is raised by heat exchange with the exhausted liquid leaving the base of G. It then passes by the pipe c to a point T in the upper part of column D. At this point it is discharged on to a plate of the column, down which it flows, being subjected to a gentle ebullition. By this means it is gradually freed from head products which, together with some alcohol, tend to accumulate at the top of the column. The plates above T exercise a certain fractionating action by means of which the head products are freed to a considerable extent from alcohol and pass after partial condensation in N and Q to the test glass E.

The spirit, partially freed from head products but still containing most of the impurities of the amyl alcohol class, leaves the base of the column with a slightly reduced alcoholic content, owing to the unavoidable loss of alcohol mixed with the head products at E. It passes by pipe M to the distilling column G. The vapour evolved in G is rectified in H in the manner already described. The finished alcohol is run off at K, and the oils from some point S near the base of the column. A further quantity of head products are run off by X to the test glass P. In some cases this part of the distillate is returned to the appropriate point in the column D, so as to avoid collecting two separate lots of head products.

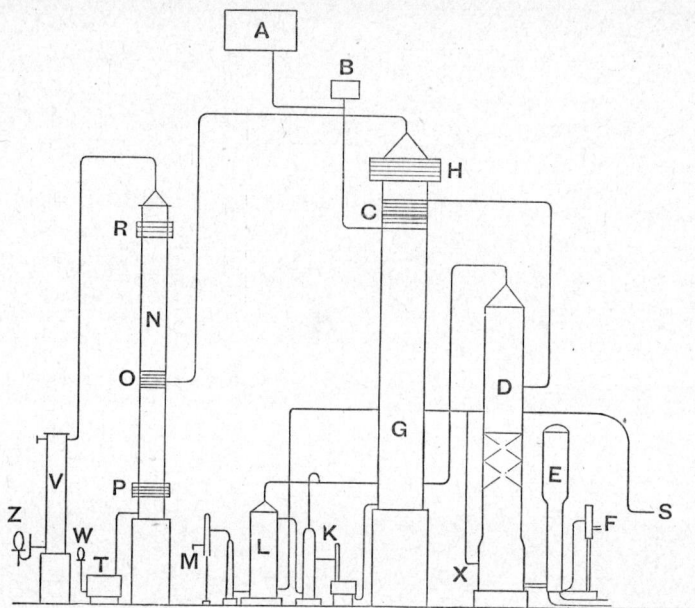

FIG. 127.

German Still.—Fig. 127 represents the Ilges continuous distilling and rectifying plant.

The wash to be distilled flows from the reservoir A and regulator B through the tubular dephlegmator C to the wash column D. The wash flows down D, meeting steam which is introduced at X, and its rate of outflow is regulated by E, and its freedom from alcohol tested at F. The vapour from D enters the base of the column G, which is provided with an additional water-cooled dephlegmator H. The column G is filled with porcelain balls. The liquid condensed at the base of G flows through the oil separator K to a small column L down which it flows, and leaves the base of this column free from alcohol.   The vapour evolved in the column L is returned to G.

Samples of the liquid leaving the base of L are tested at M to see that they are free from alcohol.

The vapour leaving the top of G consists of very strong alcohol freed from oils but still containing aldehyde and other head products. The removal of these is effected in the column N, which is packed with porcelain balls.   The vapour on entering the column N is heated by the steam coil O.   The head products tend to accumulate at the top of the column (near which a water-cooled dephlegmator is placed to supply the condensate necessary for the proper working of the column) and pass over to the condenser V and test glass Z, having been freed as far as possible from alcohol.   The condensate in the column flows downwards and is again submitted to a partial evaporation by the steam coil P. The liquid leaving the base of the column consists of purified alcohol. It is cooled in the condenser T.   The rate of flow is regulated by the test glass W.

FRACTIONATION OF FUSEL OIL DISTILLATE

A technical distillate from fusel oil, on fractionation on a laboratory scale with a Young 12 pear-bulb still-head, gave indications that it is capable of being resolved into fairly pure fractions of the individual compounds.   One litre of it was therefore distilled from a glass flask fitted with a plain glass tube as still-head, 4 cm. in diameter by 140 cm. high, which was completely filled with Lessing contact rings made of copper ; approximately 4000 $\frac{1}{4}$-inch rings were employed.

TABLE 113

Fractionation of Fusel Oil Distillate with Lessing Contact Ring Still-head.

Fraction No.	Temperature. ° C. corr.	Percentage by vol. Total.	Percentage by vol. Fraction.	Specific gravity.	Refractive index η_{D}^{20}
I.	78·0-87·4	9·5	9·5	0·858	1·3714
II.	87·4-88·6	19·5	10·0	0·870	1·3745
III.	88·6-89·0	24·6	5·1	0·872	1·3750
IV.	89·0-89·5	46·9	top layer 20·1 water 2·2	0·861	1·3800
V.	89·5-107·2	50·6	3·7	0·814	1·3900
VI.	107·2-117·0	58·2	7·6	0·808	1·3921
VII.	117·0-130·0	62·3	4·1	0·814	1·4005
VIII.	130·0-131·0	69·8	7·5	0·815	1·4029
IX.	131·0-131·4	92·4	22·6	0·814	1·4034
X.	steam distillate	95·5	3·1	0·832	
XI.	residue in flask	98·6	3·1	0·822	
	loss	(100·0)	1·4		

With a rate of distillation of 5 c.c. per minute, the results shown in Table 113 were obtained by Dr. Lessing, who kindly supplied the curves. The volume distilled was observed for every rise of 0·2°.   The fractions taken off were tested for specific gravity and refractive index and these results are also recorded.

The most striking feature of the curve is the right-angle formed after 46·9 per cent is distilled over.   The much sharper separation of the azeotropic isobutyl alcohol-water mixture from the anhydrous isobutyl alcohol than that of the isobutyl alcohol from the amyl alcohol is also very marked.   Chemical evidence shows that the residue remaining in the flask and still-head at this point is absolutely anhydrous.

The first three fractions are apparently aqueous mixtures of propyl and isobutyl alcohol, with possibly traces of ethyl alcohol in the fore-

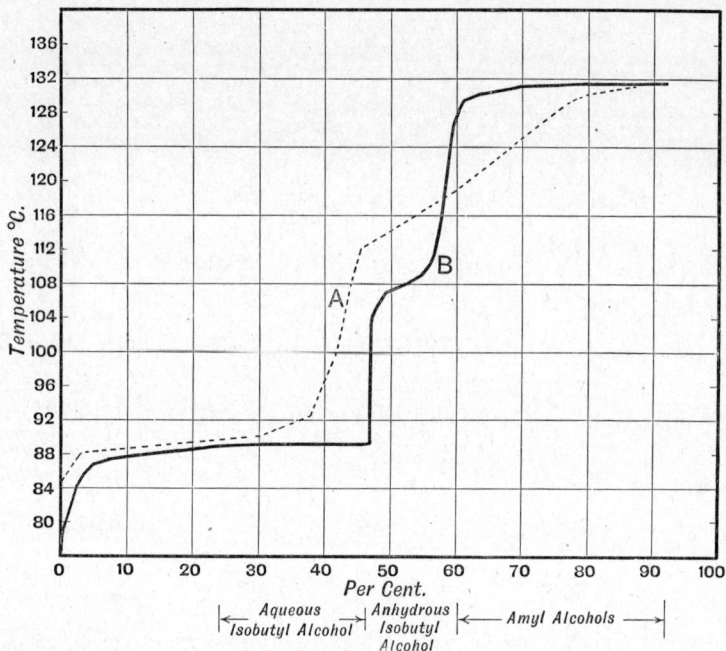

FIG. 128.

runnings.   The appearance of undissolved water in the distillate was observed repeatedly at 89·0° with considerable sharpness.   The fraction then coming over consists of saturated isobutyl alcohol forming a top layer over water.   As soon as the last drop of water is removed, the remainder of the isobutyl alcohol distils over in anhydrous condition and leads rapidly to the isoamyl fraction.

The residue in the flask at the end of the distillation amounted to 3·1 per cent, and a similar quantity was recovered by steam distilling the still-head.

It should be noted that the results obtained with the 12 pear-bulb still-head are not directly comparable with those of the ring still-head, as the former was only 45 cm. high with a maximum diameter of the bulbs of 3·5 cm.

FRACTIONAL DISTILLATION
AS APPLIED IN THE PETROLEUM INDUSTRY

JAMES KEWLEY, M.A., F.I.C.

MEMBER OF COUNCIL OF THE INSTITUTION OF PETROLEUM TECHNOLOGISTS

CHAPTER XXIX

INTRODUCTION

FROM time immemorial petroleum products, in the first instances crude oils and natural bitumens, have been employed in the service of man, but the real development of the industry, which has now assumed such large proportions, dates from the application of distillation methods to the working up of crude oils, and distillation, in one form or another, still affords the most practical method of splitting up crudes into their commercial products. Although many and great modifications of the original process have been developed, it can hardly be claimed that really accurate or intensive fractionation as yet finds any extensive application in the petroleum industry. This is not surprising, seeing that (1) the industry is now only in its infancy, only a limited number of products, generally boiling over considerable ranges of temperature, being in demand ; and that (2) crude oils are very complex mixtures of hydrocarbons, the boiling points of which, in many cases, lie close together, so that the extraction of any one of them by distillation alone becomes an exceedingly difficult and quite impracticable operation.

Crude petroleums are found in many countries, occurring in various geological formations, the most prolific deposits being generally of tertiary age. A discussion of their origin is beyond the scope of this volume. They differ widely in character, ranging from very light, slightly coloured, naturally filtered oils such as have been found in Western Canada, Russia, and elsewhere, to very heavy asphaltic oils containing little or no volatile fractions, such as occur in Mexico, and even to natural asphalts such as those of Trinidad and Venezuela and the natural waxes or ozokerites. Some of the light crude oils of Sumatra contain as much as 40 per cent of benzine and 50 per cent of kerosene, whereas many heavy crudes contain no benzine or kerosene fractions whatever.

Crude oils differ not only in the relative proportions of the various commercial products which they contain, but also in the chemical composition of these products. Certain crudes, e.g. those of the Eastern fields of the United States, are composed primarily of hydrocarbons of the paraffin series ; certain Russian crudes and others are composed largely of hydrocarbons of the naphthene or alicyclic series ; certain others consist of paraffins and naphthenes in varying proportions in admixture with hydrocarbons of the aromatic series, those of

East Borneo being conspicuous in this respect. Crude oils show much variation also in respect of the content of sulphur compounds. These are generally regarded as impurities, but, in the higher boiling fractions and residues especially, they must be regarded as intrinsic components of the crude. While the chemical composition of the lighter components of many crudes has been fairly well worked out, it is safe to say that little is so far known of the chemical nature of the components of the higher boiling fractions, and practically nothing of that of the residues and of the heavy asphaltic crudes.

Of recent years fractional distillation has been to some extent supplemented by fractional condensation, a process in which the requisite fractions are distilled off from the crude oil *en bloc*, and separated during the condensing process. This method possesses certain technical advantages, the chief being the smaller and more compact plant required for any given throughput of crude oil and the lower operating costs. It finds its chief application so far in the distillation or topping of crude oils containing only small percentages of volatile fractions.

Further, of recent years, owing to the steadily increasing demand for light petroleum fractions, systems of destructive distillation (so-called " cracking processes ") have been introduced. This method of treatment is, however, beyond the scope of this work.

The distillation of petroleum may be conveniently considered under two main headings, viz. periodic and continuous methods. Periodic methods are the older and have been largely replaced in modern refineries by continuous methods, which offer many advantages, although for certain purposes, *e.g.* the manufacture of special spirits of narrow boiling point ranges, of lubricating oil and of asphalts, the periodic methods are still largely in vogue. There is little doubt, however, that in course of time, except for such operations as distilling down to coke, continuous methods will eventually be almost entirely adopted. As the early forms of distilling apparatus were naturally crude and inefficient, and are now merely of historical interest, little need be said of them.

In consequence of the great differences in character displayed by various crude oils, different methods of treatment are called for, but as this work deals with the processes of distillation involved, rather than with their application to various types of oil, detailed descriptions of methods of working up of different crudes would be out of place. A few remarks on this subject, however, may not be inappropriate.

Crude oils may be roughly subdivided into three classes, viz. asphalt base oils, paraffin (wax) base oils, and mixed base oils, each of which may be divided into light and heavy crudes. Several types of crude may occur in the same field, but at different geological levels. In general, however, any one field produces a particular type of crude, *e.g.* the crudes of the famous Russian fields on the west shore of the Caspian Sea are of the light asphalt base type, containing little or no paraffin wax ; those of Burmah are of the light paraffin wax type ; while those of Mexico belong chiefly to the heavy asphalt type.

The method of working up of crude oils depends to a large extent on their nature, and on the nature of the commercial products which they will yield. Thus certain crudes are used direct as liquid fuel in their natural state, others require the removal of a small percentage of light fractions, in order to raise the flash-point of the residue to liquid fuel standard, while others must be subjected to a fairly complete distillation and subsequent redistillations, in order to separate off the relatively small proportion of residues they contain, and split up the distillates into their commercial products.

A method of working up of an asphalt-base crude oil is set forth in the diagram.

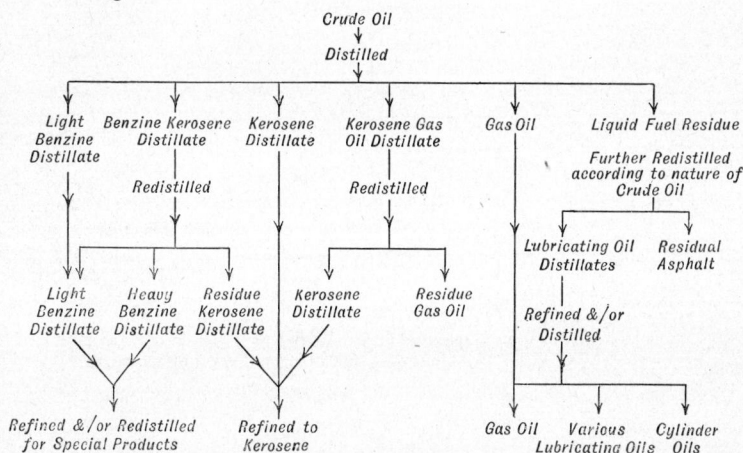

Diagram illustrating method of working up of an asphalt-base crude oil.

The working up of paraffin wax base crude oils is more complicated, owing to the filtration and working methods involved in the separation and purification of the wax.

For diagrams illustrating the working up of various typical crudes reference may be made to Engler-Höfer, *Das Erdöl*, vol. iii. ; Bacon and Hamor, *The American Petroleum Industry*, vol. ii., etc.

REFERENCES

For general information on the petroleum industry reference may be made to :—

ABRAHAM: *Asphalts and Allied Substances.*
BACON AND HAMOR: *The American Petroleum Industry.*
CAMPBELL: *Petroleum Refining.*
ENGLER-HÖFER: *Das Erdöl.*
REDWOOD: *A Treatise on Petroleum.*

CHAPTER XXX

PERIODIC distillation methods are still largely applied to :—the distillation of crude oil, the redistillation of benzine (spirit, petrol, gasoline, naphtha, etc.) and kerosene fractions, the manufacture of lubricating oils, and of asphalts, paraffin wax, and petroleum coke. In nearly all of these cases continuous distillation methods may be, and often are, applied, but particularly in those cases where a residue of definite properties, e.g. lubricating oil or asphalt, is required, periodic methods afford easier, though more costly, operation.

Periodic oil distillation may be carried out (1) at ordinary pressures without the introduction of live steam ; (2) at ordinary pressures with the introduction of live steam, or (3) with live steam under vacuum. The first is the method used for distilling certain crudes when large yields of benzine and kerosene are required, and when the residue is not required for cylinder oil manufacture. The second is the method in general use, and the third is as yet only employed occasionally, chiefly for the manufacture of lubricating oils. Distillation without steam and under increased pressure is also employed in certain " cracking " processes, the treatment of which lies outside the scope of this work.

When the distillation is carried out at ordinary pressures without the introduction of live steam, a certain amount of cracking or decomposition takes place, which is accompanied by an increased yield of the lower boiling fractions. The quality, however, of the distillates is thereby somewhat impaired, the colour is not so good, and a more intensive refining is necessary in order to produce good marketable products. The quality of the residue is also impaired from the point of view of manufacture of lubricating oils, the viscosity and flash-point both being reduced.

Distillation with live steam is the method usually employed, especially when decomposition of the oil is to be avoided, as in the case of lubricating oils and asphalts especially. The steam is usually admitted through perforated pipes placed near the bottom of the still. It assists not only in the agitation of the oil, thereby preventing long contact with the hot bottom of the still, but also in the actual distillation, in that the distillates come off at relatively lower temperatures than is the case when steam is not used. The decomposition of the higher boiling fractions is thus arrested, so that the nature of these fractions may be considerably varied by varying the quantity of steam. As a general rule, limited steam supply produces distillates of low viscosity and flash-point, of poor colour, and, in the case of paraffin wax, of low melting point and good crystallising quality ; ample steam supply produces distillates of high viscosity, high flash-point, and, in the case of wax, high melting point, but poor crystallising quality (owing partly to the more viscous nature of the oil associated with the wax in the distillate).

The difference between "dry" and "steam" distillates is well illustrated by the following :—

A certain Mexican crude was distilled in both ways. With steam, 0·9 per cent of light distillate of ·848 sp. gr. was obtained, whereas without steam, 9·0 per cent benzine of sp. gr. ·742 was obtained, and a further 10·8 per cent of distillate of sp. gr. ·816. This is a particularly striking example, as the crude oil in question was evidently very liable to "crack."

Vacuum distillation of heavier fractions, in conjunction with ample use of steam, is sometimes employed for lubricating oil manufacture, very fine distillates of high viscosity, high flash-point, and good colour being so obtained. This method of distillation is, however, rarely employed for any other purpose.

As the plant and methods employed for dry distillation and steam distillation are very similar, the description of the latter method may be taken to apply to both. In the case of dry distillation the stills are often unlagged to allow of a certain amount of condensation and dropping back of distillate into the hot oil, but in the case of steam distillation the stills are always well insulated to prevent loss of heat.

" Periodic " Crude Oil Distillation.—As crude oils are so complex in character and contain components of such widely varying characters as, for example, light motor spirit and paraffin wax, efficient fractionating apparatus is not usually fitted to a periodic still which has to yield such a series of distillates from every run.

Modern periodic crude oil stills are usually cylindrical in shape (sometimes of egg-shaped or elliptical cross-section) of capacities up to 150 tons or more. The bottom is, when possible, composed of one plate. Large stills are often fitted with internal fire tubes like those of a Lancashire boiler. The still is provided with the following fittings (*vide* Fig. 129) :—A cast steel draw-off pipe (1) fitted with internal and external valves ; a steam spider line (2) perforated, lying just above the bottom of the still for the purpose of admitting live steam ; two or more manholes (3) ; a safety valve (4) ; gauge glasses (5) ; thermometer well (6) ; inlet pipe (7) ; dome and vapour pipe (8) ; and a pressure and vacuum gauge (9). The still is usually set on brickwork, so that the furnace gases after passing along the bottom may be returned along the sides, or diverted direct to the flue when the still contains small quantities of oil.

The capacity of such a still for crude oil varies, naturally, with the percentage distilled off. As a general rule it may be taken that 1 square metre of heating surface will give $1\frac{1}{2}$ tons of distillate per 24 hours, or 1 square foot will give about 1 barrel of distillate in the same time.

It is important that the vapour pipe be sufficiently large to allow of the vapours getting freely away.

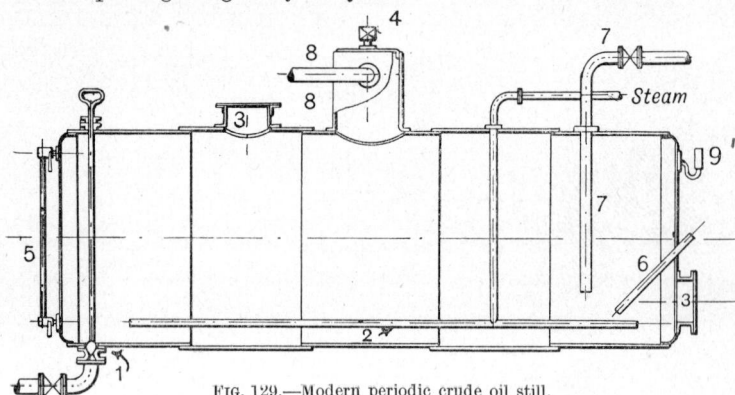

FIG. 129.—Modern periodic crude oil still.

Such a still may be used for handling any grade of dry crude. In the case of heavy crudes, however, containing water in emulsion, difficulties will be met owing to foaming over. Such crudes are best handled by one of the types of continuous topping or dehydrating plants, described later (*vide* p. 345).

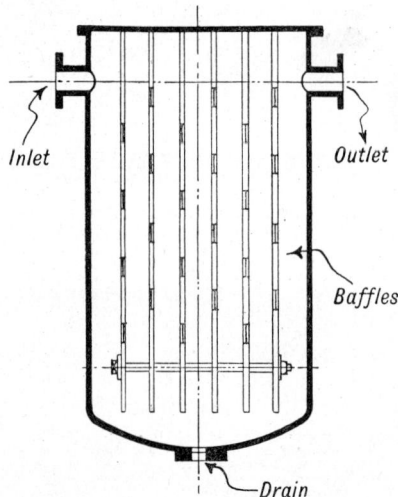

FIG. 130.—Oil arrester.

The vapour pipe is often fitted with an oil arrester (Fig. 130), which is merely a cylindrical vessel fitted with perforated plates, rather like a water trap in a steam line, for the purpose of retaining any spray of crude oil carried over mechanically. Owing to the reduced velocity in the arrester, this spray settles down and is returned by a trapped pipe to the still. Such an oil arrester considerably improves the colour of the distillates obtained, and reduces to some extent the amount of redistillation necessary.

In distilling a crude oil the temperature is brought slowly up to about 130° C., and at this point live steam is usually admitted, in small quantities at first, the distillation being thereby much facilitated. By the use of live steam the distillates come off at temperatures lower by 50° C. or more than would otherwise be the case. The steam should be superheated to about the same temperature as that of the oil in the still, and the amount of steam used should be gradually increased as the distillation proceeds. When the higher lubricating or wax fractions are being given off, the rate of distillation is increased and more steam is given, so that the volume of water in the mixed oil and water condensate may be equal to that of the oil, and even, in the case of the last distillates taken off when reducing down to asphalt, three or four times greater.

The residue from such a steam distillation may be a liquid fuel, if the distillation has been stopped after removal of the kerosene fractions, or earlier ; a steam refined cylinder stock if lubricating oils and wax have been distilled off ; a waxy or asphaltic residue ; or an asphalt to a particular specification, depending on the nature of the crude oil worked and the products desired.

In periodic working the distillation is often interrupted owing to the small volume of residue left, this being then pumped out to be transferred to other stills for completion of the process. This means a waste of time and heat, which can be obviated by continuous working methods. In the case of dry distillation the process is sometimes carried on further in special coking stills until the residue is completely coked, the coke being removed as soon as the temperature is low enough to allow men to enter the still. Such distillation to coke is very severe on still bottoms, so that an ordinary cylindrical still does not stand many runs before patching is necessary. In some refineries, therefore, cast-iron stills with hemispherical bottoms have been used for carrying out this distillation down to coke, the removal of the coke being sometimes facilitated by an iron anchor or grapnel placed in the still, which becomes embedded in the coke, which can be lifted out *en bloc* when the top of the still is removed.

The " cutting " or dividing of the fractions as they distil over into separate portions depends entirely on the crude oil worked and on the products required. The method can, therefore, only be indicated in a general way. The first runnings will probably be good enough to be put into light benzine distillate (for motor spirit). The laboratory distillation of the fractions will determine this in the first instance, and, after some experience, the specific gravity of the distillate will give a good working indication of the point at which a change must be made. The next distillates will go into benzine distillate for redistillation, as they will contain a percentage of heavier distillates which will eventually find their way, after redistillation, into kerosene. As soon as the flash-point or distillation test allows, the distillate will be run into kerosene distillate according to requirements. When the distillate becomes too highly coloured, or when it contains higher boiling fractions than are permissible in a kerosene (up to 300° C.), then it may be run into a fraction for further redistillation, the distillate from which will go into kerosene and the residue to solar oil or liquid fuel. In practice the cuts are made according to the specific gravity of the distillate, the properties corresponding to various specific gravities having been previously determined in the laboratory. Thermometer control is also largely adopted, especially in continuous plants.

The reducing of lubricating oils and the manufacture of asphalt need not be here described, as the distillation is merely carried out in order to obtain a residue of properties conforming to certain requirements, with minor regard to the properties of the distillates. The ample supply and uniform distribution of the live steam plays an important part in such distillations.

Periodic stills are often fitted with some form of simple so-called dephlegmator (really an air-cooled condenser) the condensate from which is either separately cooled or returned to the still according to circumstances (Fig. 131). Such dephlegmators are usually simply cylindrical air-cooled vessels, in some cases fitted with baffle plates and perhaps with water coils so that the amount of distillate condensing therein may be varied.

By the use of one or more fractional condensers, much subsequent

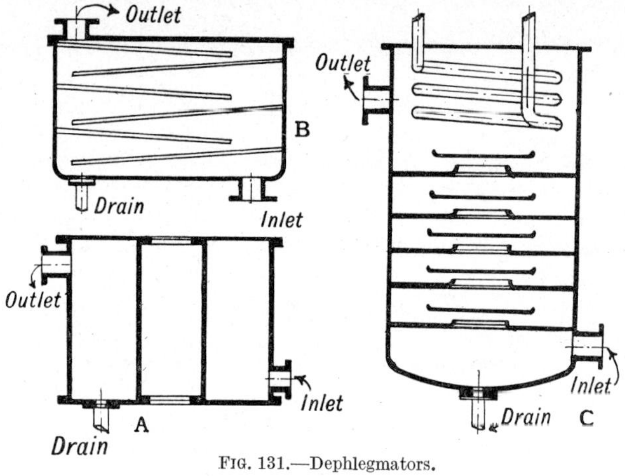

FIG. 131.—Dephlegmators.

redistillation may be avoided, e.g. when the still is producing a heavy distillate towards the end of the kerosene cut, some gas oil may be eliminated which would otherwise require to be separated by redistillation from that particular fraction. As an example of the separation which may be expected with such simple fractional condensers, the following example may be given from actual practice, from a still fitted with two such dephlegmators of the type, Fig. 131 C, with water cooling, in series.

Three samples drawn simultaneously yielded the following results :—

1. Distillate from bottom of Dephlegmator I. Sp. gr. 0·840 Flash 80° C.
2. ,, ,, ,, ,, ,, II. ,, ,, 0·816 ,, 43° C.
3. ,, ,, main condenser . . ,, ,, 0·795 ,, low.

The distillate from Dephlegmator I. was yellow in colour, while the other two were colourless. From this it is evident that the use of these simple fractional condensers avoids much redistillation. In the above case, for example, fraction 1 was run into kerosene-gas oil distillate for redistillation, fraction 2 was run into kerosene distillate, and fraction 3 into kerosene-benzine distillate for redistillation. Without these condensers the whole of the distillate would have been of bad colour and low flash and would have required redistillation *in toto*. Further examples will be found under " Continuous Distillation."

In place of the fractional condensers above described simple atmospheric condensers consisting merely of a series of horizontal or vertical pipes, each fitted with a drain from which the distillate is drawn off, are sometimes employed. As the vapours pass through the series of pipes, they condense, giving a series of distillates which become successively lighter and more volatile as the vapours pass along the condenser.

Various forms of condensers are used dependent to some extent on the nature of the crude oil to be distilled. Many crudes give off sulphuretted hydrogen during distillation, *e.g.* those of Mexico, Egypt, Persia, and those of certain American fields, and some crudes which contain chlorides give off hydrochloric acid. For use with such crudes condenser pipes of cast iron stand longer than those of wrought iron or steel.

Water is still generally adopted as the cooling medium, though this necessarily entails the loss of the heat given out by the vapours. To obviate this, crude oil-cooled condensers (heat exchangers or distillate preheaters) have been in many cases adopted. Details of these are given in the later chapters. For high boiling distillates such as paraffin wax and lubricating oils air-cooled condensers are often used, and in some modern plants direct contact spray condensers find application. The form of condenser still in most general use is the old type in which the vapours traverse a coil of pipes, arranged either as one continuous coil or as several coils in parallel, immersed in water. These pipes may be made of decreasing diameter downwards. The cooling surface provided should be ample, especially in the case of condensers for light distillates, in order to avoid loss of the more valuable lighter vapours by imperfect condensation. From one to three square feet of condensing surface per gallon of distillate per hour is the usual practice.

The quantity of water required for efficient cooling may be easily calculated if the temperature of the incoming vapour is known. For this purpose the specific heat of petroleum oils may be taken as roughly 0·5. The latent heats of petroleum oils are low, ranging between 60 and 80 (centigrade) calories. When the nature of the crude permits, economies may be effected by using condensers of the tubular type. In this form of condenser the water circulates through the tubes which are of mild steel. As the walls of these tubes are much thinner than those of the cast-iron tubes, the heat transmission is much better, though the liability to corrosion is greater and the life of the condenser is shorter. Such condensers, however, may be easily retubed.

A vertical condenser as depicted in Fig. 132, about 1 metre in diameter and 5 metres high with 150×3 cm. tubes, would have a cooling surface of about 80 square metres.

Simple forms of condensers or rather condensate coolers may be made of water-jacketed pipes, the ordinary types of pipes in general use in a refinery being available for the purpose.

In the case of high boiling fractions, lubricating oil, paraffin wax, etc., air-cooled condensers may be used. Such condensers are usually made of ordinary piping arranged either vertically or horizontally, the end of each pair of pipes being connected by bends. From each pair of pipes a separate distillate fraction may be drawn off.

The principle of direct contact is also sometimes made use of, jets of water being sprayed into cast-iron pipes or vessels containing the vapours to be condensed, the resulting mixture of oils and water being subsequently separated by settling. Such condensers are very simple in construction and are particularly useful in cases where the distillates contain acid vapours or sulphuretted hydrogen, as these are to some extent, if not completely, removed and washed out by the condensing water, the subsequent chemical treatment necessary being thereby somewhat cheapened.

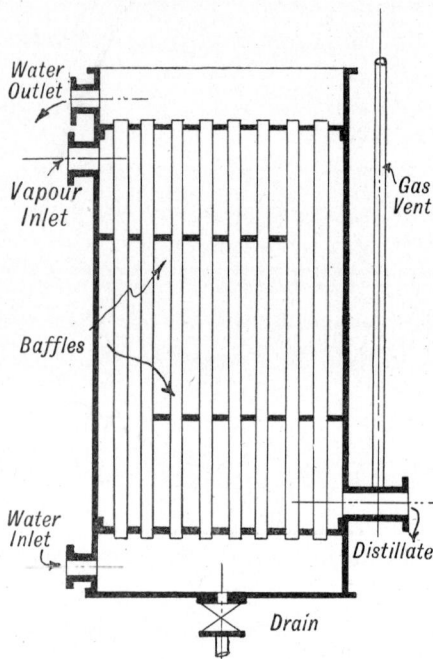

FIG. 132.—Vertical condenser.

Periodic Redistillation.—Certain of the products obtained from the crude oil distillation require redistillation, *e.g.* a portion of the distillate which requires sharply separating into benzine and kerosene, and a portion which requires separating into kerosene and gas or solar oil. The redistillation of the latter calls for no special comment, being usually carried out in a crude oil still fitted with simple dephlegmators as above described. The redistillation of a mixture of benzine and kerosene, and the working up of benzine into products of limited boiling point ranges, requires, however, more accurate fractionation. This is usually conducted in a so-called steam still, fitted with much more efficient fractionating apparatus according to requirements.

The earlier types of redistillation plant, many examples of which are still in use, consisted merely of large stills, often of the cheese-box type, near the bottoms of which were laid series of perforated steam pipes. The distillation was conducted merely by the blowing in of live steam. Large quantities of water were thereby condensed and drawn off from time to time. Such stills were, in fact, simply crude oil stills minus the fire heating arrangements. They have now been to a

large extent replaced by more efficient forms, as more efficient fractionation and a sharper cut between kerosene and benzine is required.

Redistillation or steam stills differ from crude oil stills in that they
are fitted with nests of steam coils, connected to steam traps. Each
of these coils should preferably be in one piece in order to avoid joints

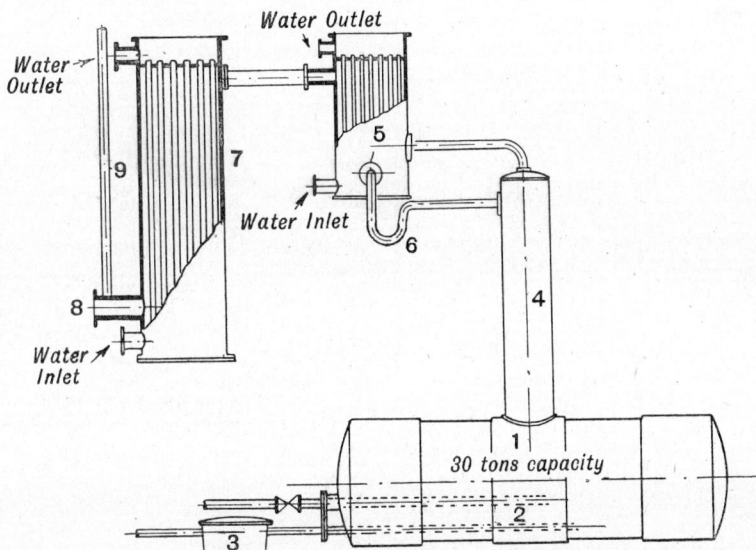

FIG. 133.—Petroleum redistillation and fractionating still.

inside the still. They should be of drawn steel piping about $1\frac{1}{2}$ inches
to 2 inches diameter, and should be placed near the bottom of the still.
Steam at pressure up to ten atmospheres is generally used, which
enables a temperature of about 160° C. to be obtained in the still.
Such stills are often set in brickwork so that direct firing may be used
to supplement the steam heating.

Fractionating columns of all types,
such as simple columns fitted with baffle
plates, highly efficient bubbling columns
of the Heckmann type and columns of
the absorption tower type, are fitted to
steam redistillation stills.

The following description of an actual
petroleum redistillation and fractionating
still may be taken as typical. The plant
(Fig. 133) consists of a still of about 30
tons capacity (1) fitted with three or four
heating coils of 2 inches diameter (2)
having a total heating surface of about
600 square feet. Each coil is connected
to a separate steam trap (3). On top of the still is placed a
Heckmann column (4) (shown in detail in Fig. 134), 18 feet long and

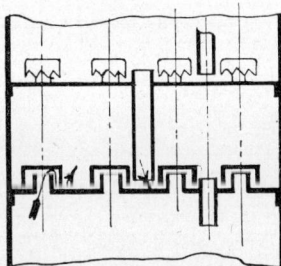

FIG. 134.—Heckmann column.

$4\frac{1}{2}$ feet in diameter, fitted with 23 plates, each plate being fitted with 46 cups. The vapours escaping from the top of the column pass into the dephlegmator (5) which is practically a reflux condenser. This is 9 feet high and 3 feet in diameter, and has a cooling surface of 320 square feet. The water supply to this dephlegmator is controlled, so that the quantity of heavy distillate returned to the column by the trapped pipe (6) may be varied according to requirements. The vapours which pass the dephlegmator go on to the tubular condenser (7), 20 feet long, 3 feet in diameter, of cooling surface 900 square feet, whence the condensate passes away by pipe (8) fitted with gas vent (9).

Such a still can distil at the rate of from 500 gallons an hour or less according to the amount of fractionation required. The degree of fractionation is controlled by varying the water supply to the dephlegmator and by the rate of distillation. For relatively high boiling fractions the supply of water to the dephlegmator may be heated. Table 114 gives some actual results obtained by the above described plant.

TABLE 114

ENGLER DISTILLATION

Sample.	Sp. gr. of distillate.	Initial boiling point,° C.	Boiling to 60° C	60°– 70.	70°– 80.	80°– 90.	90°– 100.	100°– 110.	110°– 120.	Final boiling point,° C.
1	·701	27	50	26	22	...	...	...	...	75
2	·710	32	40	29	30	...	...	...	...	80
3	·732	40	6	49	38	6	...	...	...	83
4	·749	46	1	30	62	6	...	...	...	86
5	·755	54	...	8	79	12	...	...	...	89
6	·760	65	...	...	76	18	5	...	...	91
7	·759	67	...	...	43	48	8	...	...	99
8	·760	74	...	...	1	69	29	...	...	99
9	·764	78	...	...	...	24	74	1	...	101
10	·766	78	...	...	...	11	84	4	...	106
11	·770	82	...	...	...	2	87	10	...	104
12	·773	87	...	...	...	...	75	24	...	105
13	·783	91	...	...	...	...	24	75	...	106
14	·788	92	...	...	...	...	3	96	...	107
15	·797	93	...	...	...	...	2	97	...	108
16	·803	95	...	...	...	...	...	99	...	110
17	·807	100	...	...	...	...	...	99	...	110
18	·819	101	...	...	...	...	...	99	...	111
19	·820	101	...	...	...	...	...	99	...	110
20	·829	102	...	...	...	...	...	90	...	113
21	·826	102	...	...	...	...	...	89	...	113
22	·805	102	...	...	...	...	...	45	54	119
23	·776	105	...	...	...	...	...	5	85	124
24	·774	110	...	...	...	...	...	...	87	130

The spirit distilled in this case was a petroleum spirit rich in aromatic hydrocarbons, the object being to obtain a fraction into which the toluene was largely concentrated. The fractionation was, therefore, closely controlled only over the range in which the toluene would be found, i.e. fractions 13 to 22. The original spirit boiled between 47°

and 170° C. in an Engler flask under normal conditions, and had a specific gravity of ·785 at 15° C. The distillation was carried out at first at 300 gallons per hour, the rate being reduced to 200 gallons during the periods of careful fractionation. The samples were taken hourly and the distillation figures given are those obtained by distilling in an Engler flask under normal conditions. The table clearly shows the effect of the intensive fractionation. Fraction number 20 contained as much as 70 per cent of toluene, the original spirit containing only 14.

Table 115 gives some idea as to what takes place in the fractionating column. The same benzine as in the previous case was used. Five samples of vapour were drawn simultaneously from five points at different levels in the column, condensed and analysed, with the following results :—'

Temperature of benzine in still, 117° C.

Temperature of water entering dephlegmator, 29° C., leaving, 81° C.

TABLE 115

ENGLER DISTILLATION

Samples from cloumn.	Initial boiling point,° C.	Boiling to 105° C.	° 105–110.	° 110–115.	° 115–120.	120– 125° C.	Final boiling point,° C.
Top 1	98	48	46	5	...	...	112
2	100	37	54	8	...	...	114
3	101	14	68	13	4	...	117
4	102	4	66	21	4	4	124
Bottom 5	103	trace	32	38	16	7	134

Other forms of fractionating columns, e.g. ring columns of various types, find application in the petroleum industry too, but, as much more intensive fractionating is employed in the coal-tar industry, the reader is referred to the section dealing with that subject for details of such columns.

In general, petroleum products are not required to conform to such rigid specifications as are demanded for pure toluene, benzine, etc., made from coal-tar. This is only natural as petroleum spirits are such complex mixtures of various hydrocarbons. It would not be commercially practicable to isolate pure hydrocarbons by distillation, nor, indeed, would it be possible in many cases owing to the formation of constant boiling point mixtures. Petroleum spirits of relatively narrow boiling point ranges, e.g. 90° to 110° C., 100° to 120° C., etc., are used as extraction spirits and cleaning spirits. The preparation of these by an efficient fractionating column presents no difficulty. The ordinary benzines or motor spirits have, however, much larger ranges of boiling point, e.g. from 35° C. up to 200° C., and so require little fractionating except in so far as it is necessary to obtain a sharp cut between the benzine and kerosene fractions.

CHAPTER XXXI

CONTINUOUS DISTILLATION UNDER ATMOSPHERIC PRESSURE

CONTINUOUS distillation was first introduced on a working scale in Persia as far back as 1883. It is now very extensively applied in all modern refineries for the treating, not only of crude oils, but also of distillates, and for the making of lubricating oils and asphalt. This method can, in fact, be applied to practically all the operations ordinarily carried out in periodic stills, with the exception of those cases in which the residue is a solid material.

The continuous system possesses several advantages over the periodic, the chief of which are :—(a) increased throughput for plant of given size, with consequent lower depreciation charges ; (b) much less wear and tear on the plant, owing to uniform temperature conditions, with consequent lower maintenance charges ; (c) lower operating wages and working losses, and (d) lower fuel consumption. The comparatively recent introduction of tubular stills has made this system of distillation even more general.

The earlier types of continuous batteries consist of simple stills of the kind described in the previous chapter, fitted merely with domes, vapour pipes, and condensers. The stills are arranged in series, as many as twelve or more being sometimes employed. They are usually all of the same dimensions, but as the volume of oil which has to pass through the later stills of the series is naturally relatively smaller, these are often constructed of smaller dimensions. They are set in cascade fashion with a fall of 9 inches or thereabouts between each still. In order to allow for expansion of the brickwork they are usually arranged in groups of three or four. They resemble the periodic stills previously described except in so far as regards the arrangement of the feed pipes. These are arranged so that the oil flows from one still to the next, throughout the series, the level of the oil in any one still remaining constant. By-pass valves A (Fig. 137) are fitted so that any one still can be cut out for repair without affecting the working of the bench as a whole. Expansion pieces B are usually fitted between each still. The inlet pipe C usually extends to the far end of the still, terminating near the bottom of the still. Both inlet and outlet pipes should be fitted with open tee-pieces at the end, or, if bends are used, then these must be provided with small holes on the top to act as gas vents, as otherwise vapour locks may form and interfere with the free running

FIG. 135.—A complete petroleum distillation plant.

of the system. In this connection, too, it should be noted that all valves on the feed line should be of the gate type, as globe valves seem to give opportunities for vapour locks in the line.

The main feed line should be of ample diameter, not less than 6 inches, better 8 inches or 10 inches, depending on the capacity of the bench, as the fall between successive stills is only a few inches. The stills are kept about $\frac{5}{8}$ths to $\frac{2}{3}$rds full, and the oil should flow through regularly, each still being kept at a certain definite temperature. The working temperatures will depend naturally on the number of units in the bench, on the nature of the oil, and on the products being distilled off. The conditions should be adjusted so that the work is

Fig. 136.—A continuous distillation bench.

distributed regularly among the stills. If this is not done, then irregular running will result, owing to different pressures obtaining in various stills consequent on different rates of condensation in the coolers. One square metre of still heating surface may be taken to be capable of giving one ton of distillate per twenty-four hours.

The outflowing residue is usually allowed to flow through a heat exchanger in which it gives up its heat to the inflowing crude oil. These heat exchangers may be of various types. A usual form consists of a high vertical tank (Fig. 138) fitted with a nest of residue coils, provided with a float and gauge glass F for showing the level of the crude oil, with a water draw-off A, and with an outlet for the crude E at such a level that it will gravitate to the bench of stills. By keeping the oil at a constant level in this preheater, a steady flow of crude to the bench is assured.

In the case of crude oils which yield high percentages of residue, the crude oil in the preheater becomes sufficiently warm to allow distillation to begin. The preheaters are, therefore, often fitted with a vapour

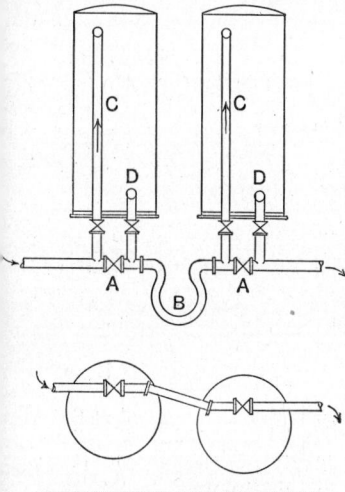

Diagrammatic arrangement of
Feed Pipes for Continuous Stills

A. *Bye-pass Valves*   C. *Inlet Pipes*
B. *Expansion Bend*   D. *Outlet Pipes*

FIG. 137.

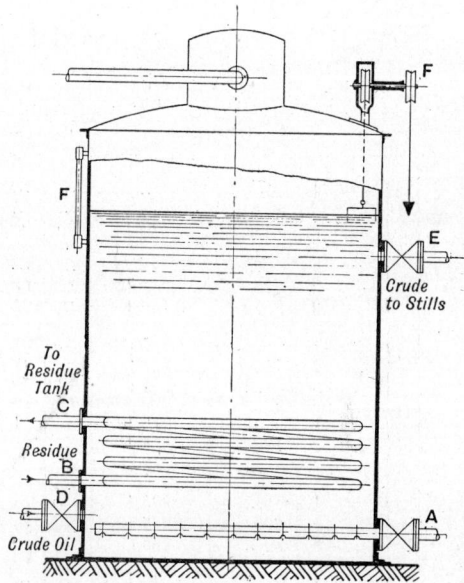

FIG. 138.—Crude oil preheater or heat exchanger.

dome, pipe and condenser, and function as a still. They should be fitted also with safety valves. If a crude which contains any water is being handled, care should be taken that the temperature in the

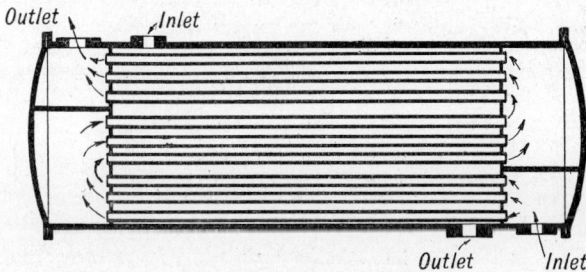

FIG. 139.—Tubular heat exchanger.

preheater never reaches 100° C., or otherwise the contents may foam over through the vapour pipe or safety valve.

Tubular heat exchangers are also commonly employed. These are constructed on the lines of a tubular condenser, baffle plates being usually placed in the end chambers in order to direct the flow of residue through the tubes (Fig. 139).

z

Each of the stills of the bench is, in the simple types of plant, coupled direct to its condenser. As each still is kept at a constant temperature, it yields a distillate of constant composition. A simple

TABLE 116

Still No.		Temp. of oil in still.	Sp. gr. at 15° C. of distillate.	Composition of distillate.					Flash point of distillate.
				Boiling up to					
				160.	150.	200.	250.	300° C.	
1	Cond. 1	97° C.	...	...	...	...	...	...	...
	,, 2	...	...	...	...	...	...	...	...
	,, 3	...	·730 B	30	90	all	...	...	ord. temp.
2	,, 1	...	...	...	...	...	...	...	...
	,, 2	125	·765 B	12	63	95	all	...	ord. temp.
	,, 3	...	·750 B	18	75	all	...	...	,, ,,
3	,, 1	...	·787 C	...	32	80	all	...	ord. temp.
	,, 2	155	·777 C	...	48	88	all	...	,, ,,
	,, 3	...	·765 C	2	59	92	all	...	,, ,,
4	,, 1	...	·812 K	...	1	50	78	all	40° C.
	,, 2	180	·808 K	...	...	...	...	...	30
	,, 3	...	·785 C	...	28	90	all	...	ord. temp.
5	,, 1	...	·825 K	...	...	...	...	...	45° C.
	,, 2	200	·816 K	...	...	...	...	...	43
	,, 3	...	·795 C	...	5	74	97	all	30
6	,, 1	...	·840 K	...	...	...	...	...	80
	,, 2	225	·833 K	...	...	...	...	...	56
	,, 3	...	·817 K	...	...	...	...	...	32
7	,, 1	...	·854 S	...	...	...	7	63	95
	,, 2	250	·845 M	...	...	...	...	...	81
	,, 3	...	·828 K	...	...	...	...	...	41
8	,, 1	...	·867 S	con	tains	wax,	solid	at −8° C.	121
	,, 2	275	·860 S	...	...	...	...	...	111
	,, 3	...	·847 M	...	...	...	...	...	30
9	,, 1	...	·873 S	con	tains	wax,	solid	at −8° C.	132
	,, 2	305	·866 S	,,	,,	,,	,,	−17° C.	113
	,, 3	...	·859 S	...	...	...	...	...	30
10	,, 1	...	·880 S	con	tains	wax,	solid	at +2° C.	141
	,, 2	325° C.	·876 S	,,	,,	,,	,,	−7° C.	127° C.
	,, 3	...	·870	show	s evi	denc	e of	cracking	ord. temp.

B, fractions running into Benzine distillate.
C, ,, ,, ,, Benzine-kerosene distillate for redistillation.
K, ,, ,, ,, Kerosene distillate.
M, ,, ,, ,, Solar-kerosene for redistillation.
S, ,, ,, ,, Solar oil.

battery of 5 stills working continuously on a crude oil of about ·850 specific gravity taking off benzine and kerosene distillates would be operated in this way. The crude oil, after leaving the heat exchangers, would have a temperature of perhaps 90° to 100° C., but this would depend, naturally, on the percentage of residue left after distilling off the distillates. The percentage of residue may vary between very broad limits from as much as 95 per cent down to as little as 10 per cent, depending on the nature of the crude oil. The temperatures in the stills would be kept at about 130°, 170°, 220°, 260°, 300° C., the temperature of the vapours being 20° or 30° less. The distillates coming off would have specific gravity of about ·750, ·785, ·810, ·830, ·850.

The five distillates so obtained would be by no means sharply cut. That from the first still might be used as a straight run benzine (after chemical treatment), but would probably require redistillation for conversion into a better grade. The distillates from the second and third stills would require redistilling to split them into kerosene and benzine fractions ; that from the fourth still would be straight kerosene distillate, while that from the fifth would require redistillation to separate off its higher boiling constituents. A detailed description of the working of a simple battery of four stills is given by Wadsworth in the *U.S. Bureau of Mines Bulletin*, No. 162. Such a simple system of stills is, however, capable of considerable modification, the efficiency being thereby much improved.

The vapour pipe of each still may be and often is fitted with traps of the type, already described (*vide* p. 326), which arrest and return to the still any fine particles of crude mechanically carried over in the form of spray. It may be further fitted with one or more dephlegmators or fractional condensers, so that the vapour from any one still may be subdivided into two or three fractions, which are separately cooled and collected. By this simple device much otherwise necessary redistillation may be avoided.

Table 116 is given to exemplify the operation of a battery of 10 stills so fitted, working up a rather heavy crude from which about 25 per cent of distillate was being taken off.

It will be noticed that still 7 is yielding three distinct types of distillate, and several of the others two. This indicates how the arrangement of fractional condensers on the vapour line from a still reduces to some extent the redistillation necessary.

A further example of the distillation of a lighter crude worked in a similar bench of six continuous stills is here given (Table 117).

The efficacy of the fractional condensers is here again well exemplified. In the case of still 3, for example, distillate (1) might go into the kerosene fraction, (2) into the benzine-kerosene fraction for redistillation, and (3) into the benzine fraction. The dephlegmators or fractional condensers used in both the cases above cited were those of type C, Fig. 131.

There is apparently no reason why the efficiency of such a continuous bench of stills could not be considerably improved by the introduction of more efficient fractional condensers.

Further examples of the use of such dephlegmators will be found in Chapter XXXII.

As one of the chief objects to be aimed at is low fuel consumption, it will doubtless occur to the reader that much heat might be saved by utilising not only the heat of the residue, but also that of the distillate vapours, condensing those by crude oil instead of by water. This principle is indeed applied in many modern continuous benches.

TABLE 117

Still No.		Sp. gr. of distillate at 15° C.	Engler distillation test of distillates showing percentages boiling up to ° C.									
			75.	100.	125.	150.	175.	200.	225.	250.	275.	300.
1	Cond. 1	·725	...	...	...	...	...	...	...	...	...	...
	,, 2	·720	...	...	...	...	...	...	...	...	...	...
	,, 3	·705	...	...	...	...	...	...	...	...	...	...
2	,, 1	·730	2	25	55	76	88	94	all	...	...	...
	,, 2	·725	6	30	61	70	90	95	all	...	...	...
	,, 3	·710	8	38	68	84	92	all	...	...	...	...
3	,, 1	·774	...	...	3	16	35	55	72	90	95	all
	,, 2	·769	...	...	9	28	55	77	90	95	all	...
	,, 3	·750	...	2	31	61	84	95	all	...	...	...
4	,, 1	·807	...	...	...	...	3·5	16·5	37	69	80	93
	,, 2	·801	...	...	...	1	2	23	53	78	92	all
	,, 3	·785	...	...	4	24	49	73	88	95	all	...
5	,, 1	·836	...	...	...	...	...	...	...	5	21	52
	,, 2	·831	...	...	...	...	...	...	5	18	47	80
	,, 3	·812	...	...	...	...	...	7	24	57	80	95
6	,, 1	·856	...	...	...	...	...	...	...	...	...	...
	,, 2	·849	...	...	...	...	...	...	...	...	...	...
	,, 3	·836	...	...	...	...	...	...	...	...	...	...

Each still, or pair of stills, is fitted with a distillate-crude oil heat exchanger or distillate preheater. This is constructed somewhat like a steam-heated benzine still. The form is that of a normal still (Fig. 140) with inlet and outlet pipes A, B, dome and vapour pipe C, draw-off pipe D, and the usual fittings. Two or more nests of pipes are fitted with their ends opening into vapour boxes at the end of the still. The vapours enter at inlet pipe E, and after passing through the two nests of tubes emerge partially or wholly condensed at F. The condensate is led to a cooler and the vapours are further condensed in a separate condenser. The vapours given off from the distillate preheater are condensed in the usual way. Each of these distillate preheaters is thus virtually a crude oil still fired without any extra fuel consumption. The path of the crude oil through such a system is diagrammatically represented in Fig. 141 : A-F, crude oil stills ; G, crude oil residue heat exchanger ; HIJ, distillate-crude oil preheaters (functioning as condensers for distillates from main stills and as preliminary crude oil stills) ; K-P, coolers for distillates from main stills ; Q-S, coolers for distillates from distillate heat exchangers ; T, cooler for distillate from crude oil residue heat exchanger.

The crude oil enters the system at the heat exchanger G, where it

is heated by the residue flowing away from the last still ; it then flows
through the distillate heat exchangers, where it is
partly distilled by the heat of the vapours from
the main stills, these being thereby partly condensed. It then flows on through the continuous
stills in the usual way. The portions of the
distillate vapours from the main stills which are
condensed in the distillate heat exchangers are
separately cooled, the uncondensed portions being
separately condensed and cooled. Each of the five
main stills in such a bench gives, therefore, two
fractions, (a) the fraction condensed in the dis-

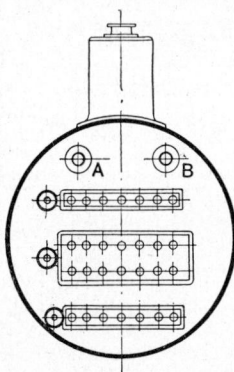

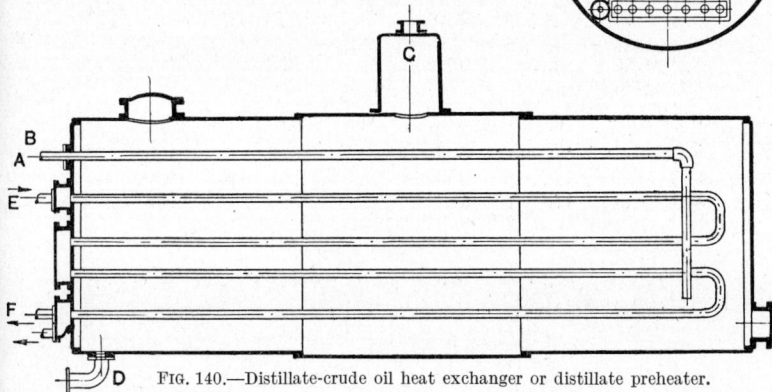

FIG. 140.—Distillate-crude oil heat exchanger or distillate preheater.

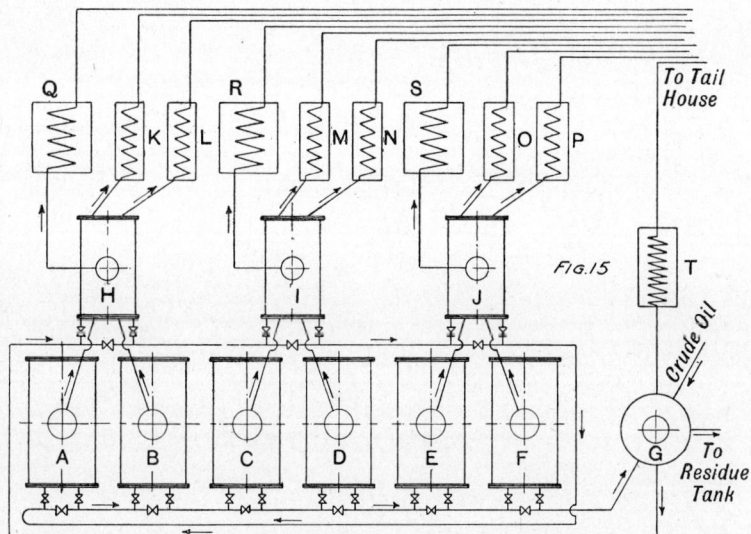

FIG. 141.—Path of crude oil through bench of stills fitted with distillate preheaters.

tillate heat exchanger, (b) the fraction which escapes condensation

there, but which is condensed by water, and in addition each distillate
heat exchanger gives a separate fraction.   Such a bench, therefore, of
5 stills and 3 distillate heat exchangers would yield at least 13 distillates.

With such an arrangement a crude oil yielding over 70 per cent
of distillates can be distilled with a fuel consumption (oil fuel) of below
1·5 per cent.   The difference in character of the distillates which are
condensed and which escape condensation in the distillate preheaters is
shown by the following analyses :—

Still No.		Sp. gr. at 15° C.	Flash point.	Percentage boiling up to					
				100° C.	150° C.	200° C.	240° C.	280° C.	300° C.
2	D1	·760	...	6	86	...	...	...	...
	D2	·773	42° F.	...	62	93	...	...	...
5	D1	·783	62	...	51	95	...	...	...
	D2	·802	104	...	...	57	89	...	...
8	D1	·825	120	...	...	16	70	95	...
	D2	·843	170	...	...	...	15	74	90

D1 in each case represents the vapour which passes on uncondensed,
and D2, the portion condensed in the distillate preheater.

CHAPTER XXXII

Of recent years the usual type of continuous distilling bench has been largely replaced by tubular stills or retorts. Such plants were originally designed, and still find their chief application, for the working up of heavy crude oils, from which a relatively small percentage must be distilled or " topped " off in order to raise the flash-point of the residue to the point demanded for liquid fuel. Such plants are usually called " topping plants." They find a further application for the dehydrating of crude oils containing emulsions. There is no reason why such plants could not be used for the working up of crudes containing large percentages of benzines and kerosenes, and their use for such purposes is indeed being rapidly extended. Many types of such plants have been designed and are in operation, the general principle being much the same in most cases. They always operate on the continuous system, and for the space occupied and capital outlay have a large capacity.

In simple plants the crude oil to be topped or dehydrated is pumped through residue-crude heat exchangers on its way to the retorts. These latter usually consist of steel piping set in brickwork, arranged for heating by gas or liquid fuel. Evaporation largely takes place in these retorts, so that the issuing oil is more or less in the state of foam. This is led into a vessel, which may really be regarded as a vapour-liquid separator, which may or may not be externally heated by the waste gases from the furnaces. The vapours are led off from the top of this vessel and the residue from the bottom.

These plants fundamentally differ from continuous still plants, in that the complete evaporation of the necessary fractions is carried out in one operation, so that the vapours leave the residue *en bloc* instead of successively. As a result of this, the residue is in equilibrium with a mixed vapour which may contain light fractions, whereas the residue from a bench of continuous stills is in equilibrium with the relatively heavy vapour from the last still only. The residue from a topping plant is, therefore, likely to have a somewhat lower flash-point ; or, to put it in another way, in order to get a residue of the same flash-point in both cases, a slightly higher percentage would require to be distilled off in the case of the topping plant.

Such a simple system as above described finds application specially for (*a*) dehydrating crude oils containing emulsion and water. The great difficulties met with in distilling such oils in ordinary stills are the risk of the contents of the still boiling over, and the slowness with which the distillation must be effected. These difficulties are eliminated in the tubular still type, as the contents of the retorts are practically boiling and frothing over all the time. (b) For topping off small quantities of distillate in order to raise the flash-point of the crude a few degrees, so that it may comply with liquid fuel specifications, (c) for the manufacture of asphalt ; the advantage in this case being that the oil is exposed to the high temperatures for a few minutes only.

The advantages of the system are (a) initial low capital expenditure for a given throughput, (b) lower fuel consumption, (c) simplicity of operation.

The disadvantages are (a) the obtaining of one composite distillate, which requires further fractionation. This disadvantage has, however, been overcome to a large extent by the modification of the condensing arrangements, so as to condense the distillate in fractions, and by the introduction of redistillation plant, so incorporated with the main plant that most of the redistillation is effected by the heat of the residue. This is particularly easily effected when the percentage of distillate is small and that of the residue consequently large. (b) The necessity for closing down the whole, or at least half, of the plant in order to replace any one damaged retort tube.

This system has been developed to a large extent in America, but many such plants may now be found in operation in other parts of the world. Very full details of the operation of the topping plants in America are given in the *Bulletin* No. 162 of the U.S. Bureau of Mines, to which the writer is indebted for much information.

The earlier forms of plant were of very simple construction, consisting merely of :—

(1) A battery of retorts or stills A (Fig. 142),
(2) A separating chamber B,
(3) Condensers C,
and (4) Residue-crude heat exchangers D.

One of the first plants designed to top crude oil by means other than the conventional stills, was erected in 1908 for dealing with Santa Maria oils. This process was patented by H. S. Burroughs.[1] This plant consisted of a series of 12-inch pipes mounted in sets one above the other. The oil flowed through the pipes downwards, being heated by superheated steam circulating through coils placed in the 12-inch pipes. The vapours were taken off from the 12-inch pipes to condensers in the ordinary way.

A more successful type is that known as the Brown-Pickering (Fig. 143). This is composed of a rectangular still A, fitted with cross baffles, which is set in an inclined position over a furnace in which are set 3-inch pipes arranged as a coil B. The hot gases from this furnace pass under an ordinary boiler C, which is used as a preheater. The crude oil, after passing through the heat exchanger D, goes through the boiler heated by waste gases and then through the pipe retorts, eventu-

[1] Amer. Pat. 998837, July 1911.

ally finding its way into the still. It follows a zigzag course through
the still, determined by the baffle plates, giving off its vapour and
passing off as residue through the heat exchangers to the residue

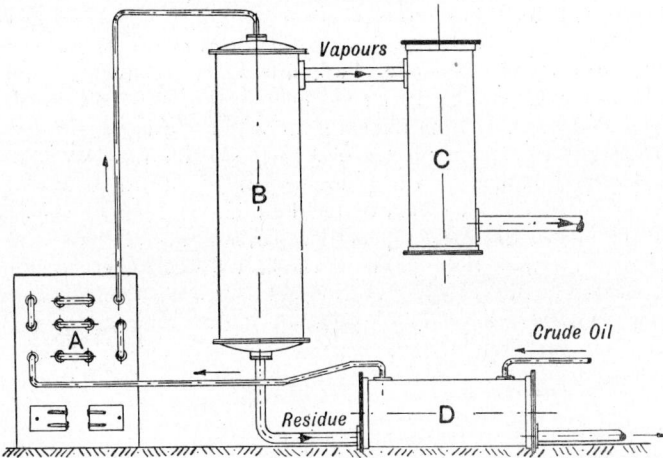

Fig. 142.—Simple dehydrating or "topping" plant.

tank. The vapours, after passing through a separating chamber E
(dephlegmator or fractional condenser) go on into the condenser F.
Several plants of this very simple type are working satisfactorily in
California.

Several plants of the Bell design are also operating successfully in

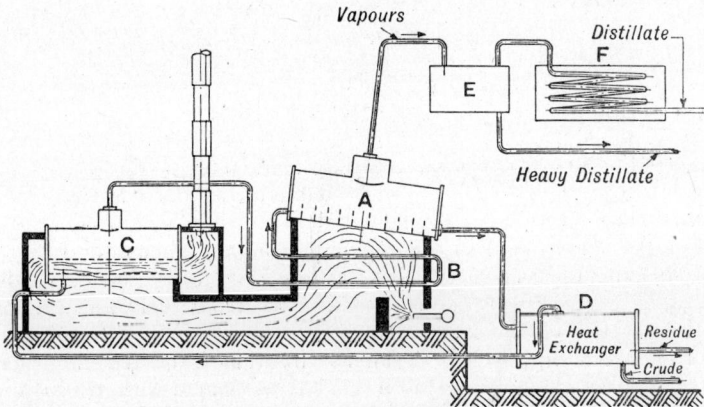

Fig. 143.—Brown-Pickering plant.

California. These plants consist of batteries of retorts of 4-inch pipes
arranged as continuous coils. The crude oil after passing first through
the heat exchangers, where it is heated by residue in the usual way,
flows through these retorts and thence to a separating tower, where the
vapours are separated from the residue.

Each battery of retorts (Fig. 144) consists of a suitable brick setting which carries three rows, one above the other, of thirty-six 4-inch wrought-iron or steel solid drawn pipes, 20 feet long, placed at right angles to the direction of the fire gases. The pipes in each row are connected end to end by return bends so as to form a continuous coil.

The furnace is fitted with Dutch ovens so that the flames may not impinge directly on the coils. Two further sets of five rows of four 4-inch pipes, 20 feet long, similarly connected to form a continuous coil, are placed in the flues from the main furnace to act as preheaters.

The separating tower consists of a vertical cylindrical still, 6 feet in diameter, 25 feet high. The residue is drawn off through a 6-inch pipe, placed about 4 feet above the bottom. A 12-inch pipe placed about 14 inches from the top acts as vapour take-off, and a 4-inch pipe, 44 inches from the top, allows the heated frothing crude to enter. A vertical partition divides the tower into two equal compartments, from a point just above the oil inlet, to a point about 20 feet above

Outlet   Inlet

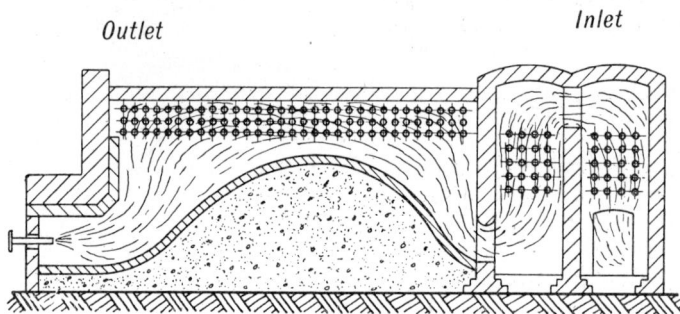

FIG. 144.—Bell plant.

the bottom. The condensers and heat exchangers may be of the usual type.

Such a plant produces only one distillate, which naturally requires redistillation in order to separate it into marketable products.

Such a plant can handle wet oils containing as much as 20 per cent of water or more, oils which could only be handled in ordinary stills with very great trouble indeed. Such simple topping plants are designed primarily for the removal of light fractions from heavy crudes in order to raise the flash-point of the residue to liquid fuel specifications. Although they present many advantages these are set off to some extent by several drawbacks. As already mentioned, the distillate comes off *en bloc* as the vapours are intimately associated with the residue in the separating towers. The fractionation, therefore, is relatively poor. Residue from such a topping plant run at the same temperature as that of stills, on the same oil, flashes at about 10° F. lower temperature.

This disadvantage can, however, be largely overcome by the introduction of fractional condensation, as is applied in the modern plants working on the Trumble system. Such a system of fractional condensation has long been in use in connection with lubricating oil distillation plants, simple pipe air condensers being usually employed. A description of a modern plant working on this system is here given (Fig. 145).

The plant consists of one battery of two heaters, each containing 6 horizontal rows of 12 solid drawn 4-inch pipes, 18 feet 9 inches long, their ends being connected by steel flanged return bends, so as to make one continuous coil. These pipes are set in a brick furnace where they can be heated by the combustion of liquid fuel. The return bends outside the wall are well insulated and are, in modern plants, further enclosed in a chamber fitted with steel doors.

The outside effective heating surface of these pipe retorts is 2650 square feet. The crude oil is heated in 6 horizontal tubular heat exchangers of 556 square feet heating surface by hot condensate, and in 4 similar heat exchangers of 1600 square feet by hot residue.

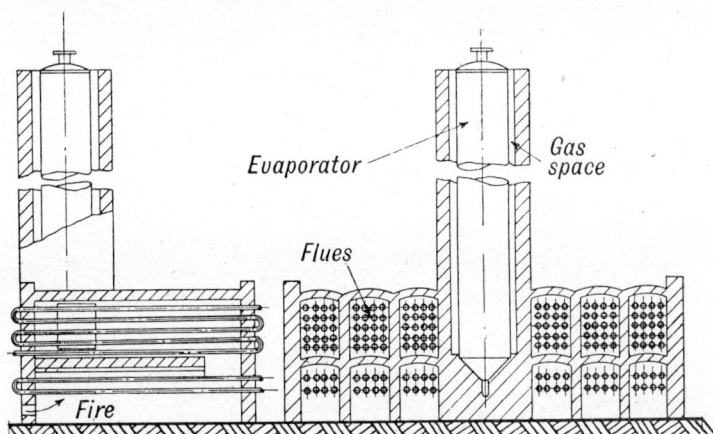

FIG. 145.—A form of Trumble plant.

The heated oil from the retorts flows to the top of a single separating tower or evaporator. In the Trumble system this evaporator is enclosed in a brickwork stack and is heated by the waste flue gases circulating in the annular space between the brickwork and the evaporator. The evaporator (Fig. 146) is a vertical cylindrical vessel, 6 feet in diameter by 25 feet high, with dished ends. At the top is a 6-inch pipe for admission of the heated crude, and at the bottom an 8 inch pipe for outgoing residue. The heating surface is 471 square feet. Inside the evaporator a central vapour-collecting 12-inch pipe supports several umbrella-shaped baffles, the object of which is to divert the flow of crude oil to the sides of the evaporator. Immediately below the umbrellas are perforations in the central 12-inch pipe to allow of the exit of the vapours which are taken off by one or more pipes to a vapour header outside the evaporator. The bottom of the evaporator is provided with a perforated steam coil, through which live steam can be blown to assist in the evaporation. An oil catcher is placed in the vapour line to return any mechanically carried spray back to the crude. The vapours, after leaving the evaporator, pass through the dephlegmators arranged in series. Each of these (Fig. 147) is 16×6 feet. They are fitted internally with a number of horizontal circular baffle plates of saucer shape. Half of these are fitted closely to the inside surface of the vessel, having circular holes in their centres to permit of the passage of the vapours. The other half are placed alternately with these and are of smaller diameter without a central hole, so that the vapours pass through the annular space between the plates and the inner surface of the vessel. At the top of the dephlegmator is a water circulat-

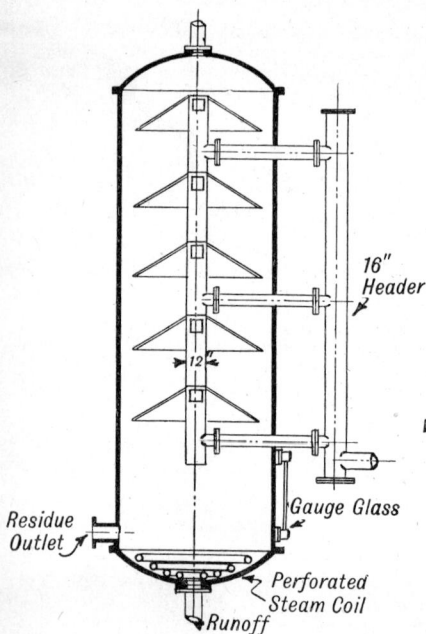

FIG. 146.—Trumble evaporator.

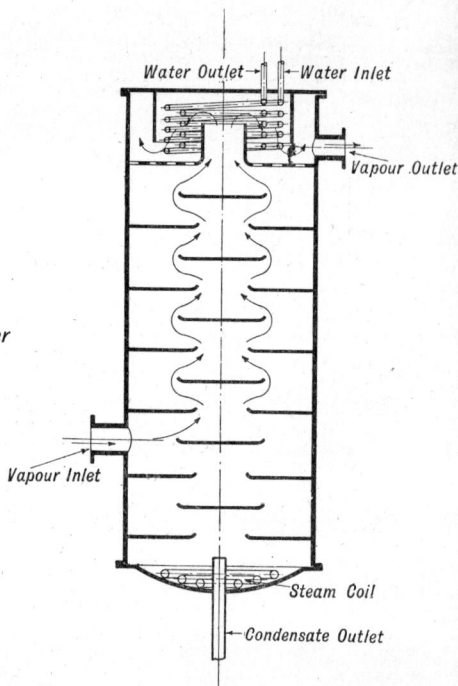

FIG. 147.—A type of dephlegmator for Trumble plant.

ing coil. At the bottom of each dephlegmator is a perforated steam coil. The condensate draw-off pipe taken from the bottom of the dephlegmator is trapped by a syphon pipe, 4 or 5 feet long, to prevent the blowing out of vapours. The area of the baffle plates is 400 square feet, of the outer radiating surface of the dephlegmator 378 square feet, and of the cooling surface of the water coils 256 square feet.

In this way the crude vapours are fractionally condensed into 7 fractions, one flowing from the bottom of each dephlegmator, and one from the vapour outlet of dephlegmator 6. These condensates are dealt with according to circumstances and the nature of the crude. They may be led off as a ready-product, being cooled by the crude-oil condensate heat exchanger mentioned above, or they may be treated as an intermediate product and be subjected to further redistillation in the separators or in another plant. The quality of these dephlegmator condensates may be to some extent controlled and varied by the regulation of the steam and water supplied to each dephlegmator.

For example, the condensate from dephlegmator 1 may be a gas oil of high flash-point, that from 3 or 4 may be a kerosene distillate, that from 6 a benzine distillate, those from the others may require redistillation.

The separators consist of steel rectangular boxes, 18×6 feet $\times 40$ inches high, with a manifold at either end connecting to a series of heating pipes which run through the box. The box is divided into six portions by longitudinal partitions about $\frac{2}{3}$rds of the height of the box, each compartment containing six heating pipes, heating surface 3 to 4 square feet, connected to the outside manifold. The distillates to be redistilled enter at one side of the apparatus flowing through holes in the partitions at successive levels. Distillation takes place, which may be assisted by live steam, the vapours passing off by a vapour outlet from the top of the box, the residual portion (separator bottoms) flowing out at the far end of the box. Hot residue or distillate flowing through the heater pipes supplies the necessary heat.

Condensers of the usual tubular type working on the counter current principle are employed, and jacketed

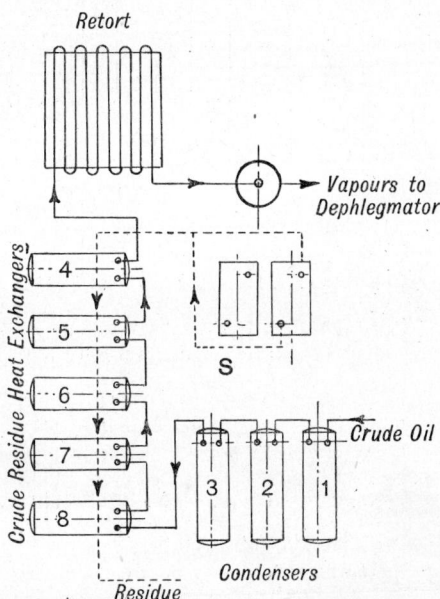

FIG. 148.—Path of crude through Trumble plant.

line coolers are used to cool the condensed distillates on their way to the tailhouse. The flow of oil through the plant is depicted on the accompanying flow sheet (Fig 148).

The crude is first pumped through three large condensers 1, 2, 3, where it is heated by vapours and hot distillates, thence through the heat exchangers 4, 5, 6, 7, 8, where it is heated by residue from the evaporator. It then enters the bench of retorts, where it passes through the pipes in an upward direction and on into the top of the evaporator column. In the evaporator the vapours are disengaged and find exit through the side vapour take-off under the umbrellas, while the residue flows out at the bottom through the residue crude heat exchangers 4, 5, 6, 7, 8. A part of the residue may be diverted through the heating tubes of several of the separators, so that its heat may be utilised for redistilling the distillates condensed in the dephlegmators. One or more of the dephlegmators may be arranged to work on the vapours from a separator. The actual arrangement of the plant and the flow of the crude, residue, and distillates depend on the nature of the crude and on the manner in which it is to be worked. A general idea only can be here given (Fig. 149); the actual system used must be evolved on the spot. A plant such as described above would have a daily capacity of about 900 tons when distilling off 20 per cent from the crude oil.

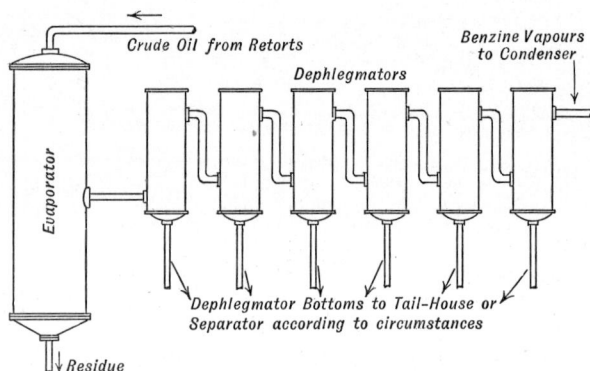

FIG. 149.—Evaporator and dephlegmators of Trumble plant.

As an indication of the manner of operating the plant and of the results obtained, the following data may be given :—

Crude Oil distilled, sp. gr at 15° C.   0·916 (15 per cent being distilled off).
Temperature of crude in evaporator 220° C.
Temperature of vapours in dephlegmators—
No. 1 . . . 185° C.   No. 2 . . . 160° C.   No. 3 . . . 150° C.
No. 4 . . . 120° C.   No. 5 . . . 110° C.   No. 6 . . . 100° C.

The condensates from these dephlegmators had the following properties :—

Condensate.	Sp. gr.	Flash point.	Percentage in Engler flask boiling up to			
			100° C.	150° C.	200° C.	250° C.
D 1	·860	140° F.	...	...	9	40
D 2	·849	130° F.	...	...	15	65
D 3	·828	95° F.	...	7	47	92
D 4	·815	90° F.	...	14	80	98
D 5	·800	...	...	30	92	100
D 6	·787	...	...	45	96	100
D 6 issuing vapour	·750	...	29	90	100	...
Residue	·927	150° F.	...	1	11	15

The following data give some idea as to the nature of the fractionation effected by the separators :—

Fraction.	Sp. gr.	Percentage in Engler flask boiling up to						
		100° C.	125° C.	150° C.	175° C.	200° C.	225° C.	250° C.
Vapour from separator	·798	...	4	24	75	94	100	...
Residue from separator	·835	...	...	...	1·5	44	82	96

One of the great advantages of a modern Trumble plant is the low working loss. This is a result of the compactness and continuity of operation of the plant.

Products requiring redistillation are not stored, allowed to cool and again reheated, but are redistilled as they are made, so that the system embodies the functions of a continuous bench and redistillation stills all in one. In the working up of a crude yielding $2\frac{1}{2}$ per cent of benzine, and 16 per cent of other distillates, a working loss of only 0·8 per cent of the crude oil was obtained.

A further advantage is the relatively low fuel consumption consequent on the efficient manner in which all possible heat content of vapours and residue is utilised, e.g. in working up a crude which yields 20 per cent of distillates only 1·1 per cent of oil fuel was used directly in the plant. This represents an over-all efficiency for the plant of 57 per cent, a figure which exceeds that of a plant of conventional stills, though with similar arrangements for fully utilising the waste heat of the flue gases, residues, and distillates there is no reason why the thermal efficiency of a bench of stills should not be as high.

Topping or distilling plants of this type are decidedly cheaper as regards both capital outlay and operating costs than stills of the conventional type. Under the present unsettled conditions, however, with constantly varying costs of fuel, material, and labour, it is impossible to give any figures of value as to the operating charges of any plant.

REFERENCES

J. M. WADSWORTH : "Removal of the Lighter Hydrocarbons from Petroleum by Continuous Distillation." U.S. Bureau of Mines. *Bulletin* No. 162, 1919.

A. F. L. BELL : "Important Topping Plants of California." *Transactions Amer. Institute of Mining Engineers*, Sept. 1915.

J. M. WADSWORTH : "Construction and Operation of Toppers of Bell Design." *Nat. Pet. News*, Feb. 4, 1920, p. 33.

CHAPTER XXXIII

DISTILLATION UNDER REDUCED PRESSURE

VACUUM distillation in the petroleum industry has not up to the present received the attention it deserves. Plants for the distillation of lubricating oils are often operated under vacuum, but this is seldom sufficiently high to give the best results, in many cases, indeed, being so low as to have little marked effect on the distillation.

The advantages of distilling under vacuum are :—

The fractions are distilled off at relatively lower temperatures, cracking or decomposition being thus largely avoided. The distillates obtained are of better colour and of higher flash-point ; the residues are also of better quality, as they have not been cracked and so contain little or no free carbon. From wax base oils, for example, a better yield, not only of the heavier lubricating oils but also of the higher grade waxes, may be obtained. Good distillates suitable for concentrating to cylinder oils may thus be obtained. The difference in character between wax and asphalt oils, as far as yield of lubricating oils is concerned, largely disappears under high vacuum distillation.[1] Vacuum distillation plants are naturally more costly as regards capital outlay and operating wages. This is, however, counterbalanced by the fact that the products obtained are of better quality and require less subsequent chemical treatment. The system finds application, in the petroleum industry, chiefly for the manufacture of lubricating oils ; but there is no reason why it should not be applied to other operations as well, the treatment of which lies outside the scope of this work.

Vacuum distillation may be applied either to periodic or continuous distillation. The simpler forms of plant consist of a still of the ordinary type, structurally strengthened to withstand the external pressure, with dephlegmators, coolers, and receiving boxes, all under vacuum. The system is connected to the top of a barometric condenser and exhausted to the required vacuum by an air pump. Many forms of plant, differing in detail but similar in principle, have been devised, e.g. that of Henderson, designed as far back as 1883 (Eng. Pat. 5401 of 1883 and 17332 of 1889), those of Lennard (Eng. Pat. 944 of 1892) (applied in the coal-tar industry), of Zaloziecki, Palmer, Wanklyn and Cooper (Eng. Pat. 4097 of 1893) and others, all of which operate at relatively low vacuums. Fig. 150 illustrates the main features of a simple vacuum plant.

[1] L. Singer, *Petroleum*, Berlin, **10**, 605.

The still is of the ordinary type, internally strengthened to withstand the external pressure. It is fitted with perforated pipes in the usual way, as the distillation is invariably conducted with the aid of steam.

The vapour pipe, which is always of large diameter, is connected to several domes on the still. It leads to dephlegmators, or atmospheric condensers, where the heavier fractions are condensed. The first air-cooled condenser often takes the form of a number of large diameter horizontal pipes. The condensate from these air coolers flows through a cooler into a receiving tank connected to the air pump. The receiving tanks are in duplicate, so that that which receives the distillate can always be kept under vacuum, while the other is being pumped out. Several air-cooled condensers may be employed, and also water-cooled condensers, the condensates from which run off through coolers to receiving tanks under vacuum. The vapours containing the lightest fractions and the steam are finally condensed in a spray condenser, elevated and connected with a water pipe terminating in a water

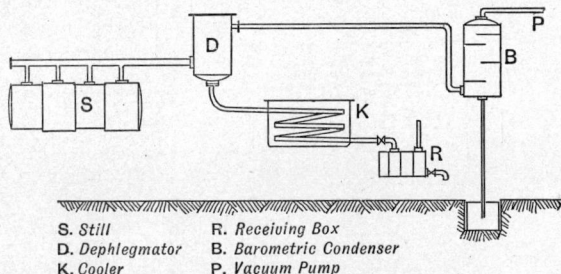

S. Still   R. Receiving Box
D. Dephlegmator   B. Barometric Condenser
K. Cooler   P. Vacuum Pump

FIG. 150.—Simple vacuum distillation plant.

seal below, of such a height that it functions as a barometer tube, the vacuum in this barometric condenser being maintained by an air pump. An apparatus designed for working at high vacuum is that of Steinschneider (U.S. Pat. 981953 of 1911). One of the chief objections to the previously described schemes is that the receivers for the distillates are under vacuum, an inconvenient arrangement, as they are not under complete observation and control. Further, if evacuation takes place *via* the distillate receiver, the lighter vapours are retained in contact with the distillates, whereas they should be removed as quickly as possible. This could be avoided by placing each receiver tank at the bottom of an oil barometer, but this would necessitate either building the plant very high, or else much excavating, both of which are expensive. Steinschneider avoids these difficulties in the following way (Fig. 151). The distillate vapours pass from the still s through dephlegmators or air condensers D to the cooler C and on to the elevated barometric condenser B. The distillates condensed in the dephlegmators D flow away through the coolers E to the receivers R, which may be fitted with floats for regulating the level of the liquid contained in them.

2 A

The fractions collected in these receivers are pumped out by low level pumps, each fraction having its own pump, into the distributing box T. The non-condensable gases are sucked out of the elevated condenser B (which stands at the top of a barometric column) by the air pump F.

This arrangement enables the distillation to be carried out under high vacuum without its being necessary to make the heights of the discharge pipes G correspond to the vacuum. By employing these pumps P, it is possible to make the fall of the discharge pipes G such that the column of liquid in this pipe requires to have merely that height which this pump is able to maintain when evacuating.

There would appear to be no advantage in applying vacuum distillation to the distilling off of the volatile fractions from crude petroleum, as the temperatures at which the distillates come off are low and as extra condensers or scrubbers at atmospheric pressure would be required to retain the lighter vapours.

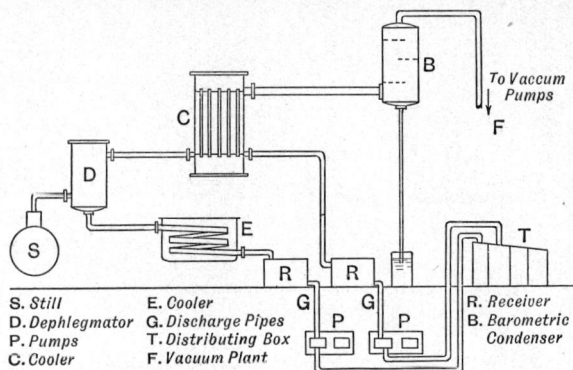

FIG. 151.—Steinschneider's vacuum distillation plant.

S. Still   E. Cooler   R. Receiver
D. Dephlegmator   G. Discharge Pipes   B. Barometric
P. Pumps   T. Distributing Box   Condenser
C. Cooler   F. Vacuum Plant

In this connection a few words concerning distillation and condensation at higher pressures may not be out of place. Distillation at high temperatures and under higher pressure is the principle involved in many patented processes for the cracking of petroleum. By such methods a partial decomposition of the heavier into the more valuable and lighter fractions is obtained. Such decomposition is, under these conditions, accompanied by increased losses owing to the formation of hydrogen and uncondensable gases, and the quality of the residue is also impaired owing to the formation of coke. Excessive coke formation is, indeed, one of the chief drawbacks of such systems. Systems of condensation under pressure are much in use for extracting further very volatile fractions from petroleum distillates, especially from the natural gas so often given off at the well heads, as natural gas may be regarded as the distillate from crude petroleum which comes off at the ordinary temperature. The volatile fractions so condensed are largely used for blending purposes in the United States and elsewhere, being usually termed " casinghead gasoline." In place of compression plants, absorption plants, using a gas oil or other relatively high boiling distillate as absorbing agent, are frequently employed.   De Brey and the Bataafsche Petroleum Mij. (*vide* Eng. Pat. 123522 of 1919) have patented a plant for the distillation and rectification of such volatile liquids under pressure.   The discussion of cracking processes, however, lies outside the scope of this volume.

CHAPTER XXXIV

VARIOUS DISTILLATION METHODS

In addition to the fundamental methods of distillation described in the preceding chapters, attention must be drawn to various methods not at present in general use, some of which possess features rendering them specially applicable in certain cases.

In the case of the Merrill process (Pat. U.S.A., 1918) oil circulating in pipes is employed as the heating agent. It is claimed by the inventor that the process is particularly valuable for replacing the usual steam heating. In the case of steam stills, for example, hot oil, previously heated in a well-designed tubular furnace, is circulated through the heating pipes of an ordinary steam still.

The advantages claimed are the following :—

(*a*) The heating medium is at ordinary atmospheric pressure, leakages thus being minimised.

(*b*) Ease of control.

(*c*) Ample " Thermic Head," *i.e.* the temperature of the heating medium is well above that of the oil to be heated.

(*d*) Low loss of heat, as the heating oil is returned to the furnace, whereas in the case of condensed steam much of the heat is necessarily lost even if the hot water is returned to the boiler plant.

The system is already largely used in the United States for melting asphalt, greases, waxes, etc., and in a few cases for distillation.

The method appears to offer many advantages over steam heating, and should doubtless find extensive application in the petroleum industry.

A difficulty which is frequently met with in the distillation of certain crude oils is that of the formation of coke on the still bottoms or on the internal walls of tubular retorts. This occurs particularly when the oil is heated to high temperatures, as is the case in cracking plants. Many devices have been proposed for overcoming this difficulty ; for example, Wells and Wells (Am. Pat. 1296244 of 1919) patented a process in which the oil to be heated lies on a bath of molten metal, an alloy of suitable melting point being chosen. As no coke adheres to the surface of the molten metal, but merely floats on the top and can be removed, any burning of the still bottom is avoided.

A similar patent has been granted to Coast (U.S. Pat. 1345134 of 1920), who suggests covering the bottom of an ordinary still with a melted alloy. The still is provided with an arrangement by which the molten metal can be circulated through the oil to be heated.

A very obvious method of overcoming this difficulty is that of applying the well-known principle of heating by direct contact with gases, the transference of heat through still bottoms being thus avoided. This method is well known and largely used in other industries, *e.g.* for the concentration of sulphuric acid.

The idea of utilising gases as direct heating agents has indeed been adopted. Wells and Wells in the above-mentioned patent also claim the passing of a permanent gas into the molten metal bath, and thence into the oil to be distilled. Day and Day (Eng. Pat. 119440 of 1918) claim distillation by introducing into a body of oil gases composed of products of combustion mixed with cracked vapours, these vapours being introduced by a perforated pipe on the bottom of a still.

As far back as 1862, Trachsel and Clayton (Eng. Pat. 2966 of 1862) patented arrangements for distilling liquids by the passage through them of hot gases. In 1862 Broonan (Eng. Pat. 3037 of 1863) distilled liquid bituminous substances by direct application of heated gases, specifying products of combustion among others.

Wells (Fr. Pat. 379521 of 1907) adopted a similar idea in utilising the volatile products of petroleum in place of direct steam in the distillation of the heavier fractions. The same idea was utilised in the plant of Alexieef, which was operated for several years in Baku, without, however, any degree of success.

It seems strange that this method, which apparently offers certain decided advantages, has so far never been successfully developed in the petroleum industry.

Scores of patents have been granted for various types of distillation plant and fractionating apparatus. Only those methods which have so far found general application in the industry, or which seem to present prospects of future application, have been described. Intensive fractional distillation finds as yet little application in the petroleum industry, possibly a future demand for pure light products may some day necessitate it.

FRACTIONAL DISTILLATION IN THE COAL TAR INDUSTRY

T. HOWARD BUTLER, Ph.D., M.Sc., F.I.C.

MANAGING DIRECTOR OF WILLIAM BUTLER & CO. (BRISTOL), LTD., TAR, ROSIN, AND
PETROLEUM DISTILLERS

CHAPTER XXXV

INTRODUCTION

Early History.—Few industries are more dependent on distillation than that of coal tar. It was in consequence of a discovery by Bethel in 1838 that it was realised that valuable products could be obtained from coal tar by submitting it to distillation. Since that date the industry has thrived, and every year brings out discoveries of further ingredients in tar, and further uses for the bodies that have for a long time been known to exist in it.

Bethel's discovery was that an oil suitable for the effective preserving of timber could be obtained by the distillation of tar. The first distillation of tar actually dates back much further than this, although the value of the process was not then realised.

Henry Haskins took out a patent in 1746, but probably used wood tar, as gas tar was very little known so early as this. Again, Longstaffe and Daleton aver that they erected the first distillery in Leith in 1822, for the purpose of producing an oil that was used as a waterproofing medium. It is also asserted that a plant was erected in the neighbourhood of Manchester in 1834 for manufacturing an oil that would dissolve the residual pitch to make a black varnish.

Nature of Coal Tar.—Coal tar is essentially the thick viscous liquid which is obtained by the carbonisation of coal, and is a by-product in the ordinary manufacture of coal gas. It should be differentiated from all the other forms of tar.

When coal is submitted to dry distillation in horizontal or vertical retorts, the products of carbonisation are gas, ammoniacal liquor, and tar. The tar comes off in the form of a fog suspended in the gas, from which it is condensed or scrubbed out by suitable means, when the resulting products contain varying proportions of ammoniacal liquor and tar.

Tar is a thick black liquid, and is a very complex mixture of many widely different substances. Its composition varies enormously according to the mode of its formation in the gas or coke works. Some of the principal factors governing its composition are the following :—

1. **The Nature of the Coal.**

2. **The Temperature of Carbonisation.**—This is probably the most

361

important factor of all, as tar obtained from low temperature carbonisation differs widely from that from high. The former is by far the most valuable to the tar distiller, as it contains more naphthas and phenol, and less pitch and free carbon. But the production of gas is much lower, and as the tar is a by-product it is not a commercial proposition to utilise this method.

3. The Kind of Retort.—A horizontal retort gives an entirely different tar from that afforded by a vertical retort. The tar from the former contains larger quantities of aromatic hydrocarbons, more phenol, and more naphthalene, but less phenol homologues. The vertical retort tends to produce more paraffins, less pitch, and less free carbon.

4. The Shape of the Retort.—The length of time the hot gases evolved from the coal remain in contact with the walls of the retort has a great influence on the nature of the tar produced and depends largely upon the size and shape of the retort.

5. The Depth of the Coal Charge in the Retort.—The depth of the charge in a vertical retort is naturally much greater than in a horizontal one. Again, similar horizontal retorts worked under identical conditions, except that the depth of the coal charge is varied, will produce tar of quite different composition.

There are many other details in gas or coke manufacture which may affect the composition of tar, but they need not be enumerated here; and it should always be borne in mind that the coal is carbonised for the manufacture of gas or coke, and consequently the quality of the tar has too frequently to be left to chance.

Different Kinds of Tar

The field of the tar distiller is a very wide one, and although generally confined to coal tar and coke oven tar, he frequently has to deal with other forms of tar. They may be very briefly mentioned :—

1. Coke Oven Tar.—It was not until the discovery of Knab, Hauport and Carves in the late 'fifties that it was realised what enormous waste was going on in the making of metallurgical coke for the iron industry in the old beehive oven. In this form of oven all products went to waste, as the coal was carbonised purely for the production of coke. Even after many of the new ovens had been erected on the Continent there was much prejudice against them in this country, as it was thought that an inferior coke was obtained if the plant was enclosed so as to collect the tar and ammonia.

In 1887 further progress was made by Franz Brunck of Dortmund, who introduced the recovery of benzole from coke oven products, a step of extreme importance to all industries connected with coke ovens and benzole.

It is not necessary to go further into this question of by-product coke ovens beyond mentioning the names of some of the most important ones in use at the present time, such as—

> The Coppee Oven.
> The Otto-Hoffmann Oven.
> The Otto-Hilgenstock Oven.
> The Simon-Carves Oven.
> The Hussener Oven.
> The Semet-Solvay Oven.

The quality of coke oven tar is, like that of coal tar, very variable indeed, but is similar enough to enable the distiller to mix the tars and work them up together. Generally speaking, however, coke oven tar is less viscous, has a lower specific gravity, contains less naphtha and phenols, although often as much or even more total tar acids, and it leaves less pitch.

2. Blast Furnace Tar.—The combustible material used in blast furnaces for the manufacture of pig iron is either coke or coal. The most modern plants use coal, but the earlier forms only coke, which naturally does not produce much volatile material such as ammonia and tar. On the other hand, the amount of these produced when coal is used is very large indeed, and in some cases the quantity is 12 to 13 times as great as that obtained from the same kind of coal when heated in ordinary gas retorts. The tar comes off similarly in a fog suspended in the gas, but is in a physically different form, which makes it even more difficult to separate from the gas. Many plants and processes have been devised and patented to separate the tar, but it is hardly necessary to mention them here. The gas is thus purified and is used for raising steam, distilling the tar, etc., and a well-equipped works need use hardly any coal outside that fed into the furnaces.

The composition of blast furnace tar is very different from that of gas tar. It contains very small quantities of the lower aromatic hydrocarbons, large quantities of bases and phenols, some solid paraffins, and a small quantity of pitch. The phenols are the higher homologues, and in this fact lies the value of the tar, for blast furnace creosote is used largely for obtaining disinfectants having high Rideal-Walker coefficients, whereas gas tar creosote is unsuitable for this purpose, as it is not so rich in the higher homologues of phenol.

Blast furnace tar is distilled and treated in a similar way to coal tar, but, owing to the difference in its composition, cannot be efficiently worked up with it.

3. Carburetted Water-gas Tar.—Water gas is obtained by passing steam through white-hot coke or anthracite, and is formed according to the following reactions :—

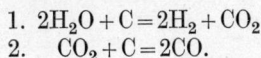

> 1. $2H_2O + C = 2H_2 + CO_2$
> 2. $CO_2 + C = 2CO$.

The mixture of carbon monoxide and hydrogen thus formed is useful for heating, but not for lighting, as it burns with an almost non-luminous blue flame, hence the term " Blue Water Gas." It is usual to " carburett " the water gas with crude petroleum hydrocarbons by passing the two together through red-hot retorts. This carburetted water gas gives up on scrubbing a certain amount of tar, which unfortunately forms an emulsion with water, which is very difficult to separate. Long standing has very little effect, and the most efficacious remedy is centrifuging.

This tar is of little value, as it contains only small quantities of naphthas, no phenols, but chiefly high-boiling hydrocarbons, and it leaves little pitch. In fact, it is not greatly changed from its original composition as it entered the retorts to serve as the carburettor.

It can be mixed with coal tar and distilled with it, provided the proportion is not large.

4. Producer Gas Tar.—This is very viscous and contains a large quantity of water which is exceedingly difficult to separate. When distilled, it produces no light oils, but paraffins, high-boiling tar acids, and a large percentage of pitch.

5. Oil-gas Tar.—Reference need only be made to this, as very little is made. Oil gas is formed by passing aliphatic hydrocarbons, generally crude petroleum, through red-hot tubes. Part of the oil escapes decomposition and part is decomposed, leaving a residue of carbon, the result being the oil-gas tar.

6. Wood Tars.—These tars are produced in almost every country of the world which contains big forests, and where charcoal burning is practised. They are entirely different from coal tar and are rarely distilled.

COMPOSITION OF COAL TAR

The composition of coal tar, as has already been stated, varies so enormously, according to its mode of production, that it is useless to attempt to give any definite analysis. It can be said, however, that it generally consists of—

Water.—Properly ammoniacal liquor, the quantity of which is generally about 5 per cent, although it may reach even 40 per cent. A tar with this amount of water is an emulsion, is separated with great difficulty, but has the appearance of a good tar free from water.

Naphthas.—The most important of these are benzene, toluene, and the three isomers of xylene. Amount 1-5 per cent.

Light Oils.—Containing aromatic hydrocarbons, phenol, cresol and other homologues of phenol. Amount 5-10 per cent.

Creosote Oil.—Containing also high aromatic hydrocarbons, high phenol homologues and naphthalene. Amount 15-25 per cent.

Yellow Oil.—Containing still higher aromatic hydrocarbons and anthracenes. Amount 5-10 per cent.

Pitch.—Amount 50-65 per cent.

The Tar Industry

The magnitude of the tar-distilling industry in this country can be judged by reference to the following figures, showing the amounts of tar produced and distilled. It seems impossible to obtain complete statistics of any pre-war period, as reliable data were not then kept, and figures relating to these periods can only be estimated roughly. From the Annual Reports on alkali and other works, by the Chief Inspector, are obtained the following data :—

Tar distilled in	1916.	1917.	1918.	1919.
Gas and Coke Oven Works	1,420,867	1,526,209	1,510,065	1,402,987
Other Works . .	138,552	126,966	131,325	106,056
Total in Tons .	1,559,419	1,653,175	1,641,390	1,509,043

From the Reports for 1916 and 1917 it would appear as though the tar had been obtained from gas works only, but there is no doubt that the figures given should refer to both kinds of works. The Ministry of Munitions collated complete returns of tar produced and distilled during the war and have kindly given the writer the following summaries :—

COAL TAR AND COKE OVENS TAR DISTILLED, IN TONS

1913	1,320,000
1914 1915 Average	1,376,397
1916	
1917	1,574,863
Sept. 1917 to Aug. 1918	1,417,919
,, ,, (actual production) . .	1,523,401

For the years prior to Sept. 1917 the total production of tar may be taken as approximately 100,000 tons annually more than that shown as distilled.

For the year Sept. 1917 to Aug. 1918 the sources of tar produced were :—

Gas Works and Produce Plants	949,397
Coke Ovens	558,561
Water-gas Plants	15,443
Total . . .	1,523,401

Against these figures estimates of the tar produced have been made,

366 DISTILLATION PRINCIPLES AND PROCESSES CH. XXXV

and Dr. Beilby, in an address to the Society of Chemical Industry, 1900, gives such an estimate for the year 1899 as follows :—

TAR PRODUCED, IN TONS

Gas Works	650,000
Blast Furnaces	150,000
Coke Oven	62,000

An interesting estimate of the tar produced by all countries of the world for the year 1901 appears in *Chemical Discovery and Invention in the Twentieth Century*, by Tilden (p. 305) as follows :—

	Tons.
United Kingdom	908,000
Germany	590,000
United States (including water-gas tar)	272,400
France	190,680
Belgium, Holland, Sweden, Norway, and Denmark	272,400
Austria, Russia, Spain, and other European countries	199,760
All other countries	227,000
Total	2,660,240

From this it is evident that the industry in the United Kingdom is more important than in any other country in the world, and, comparing the total of these figures with the tar produced during the most important period of the war, one sees that the United Kingdom produced then very nearly as much as the rest of the world at the beginning of the twentieth century.

CHAPTER XXXVI

Dehydration of Tar.—One of the greatest difficulties a tar distiller is faced with is the elimination of water, which, as already stated, is present in varying proportions. The tar is generally allowed to settle in large wells or tanks, from the bottom of which it is pumped into the stills. Its water content is thus reduced to anything below 5 per cent. Some tars cannot be even partially freed from water in this way, as an emulsion forms, and no separation takes place even after long standing. These high water tars generally contain a large percentage of free carbon, and the difficulty experienced with any individual tar in separating water by mere settling may almost be taken as a measure of its free carbon content. The cost of distilling out only a few parts per hundred of water is very considerable ; and in some cases it is as high as that of distilling off the remaining 30-40 per cent of volatile products.

Dehydration of tar may be carried out by mechanical means, such as are mentioned in the Patents of J. and R. Dempster (B.P. 3245, 1882), Kunuth (Ger. P. 15255), and many others. These mechanical separators are of little value except where small quantities have to be handled.

Centrifuging may also be employed for separating tar and liquor, and the difference in specific gravity makes this process very feasible. The mixture is fed into the machine at about 50° C., and the tar, being heavier, rapidly goes to the periphery of the machine, the liquor remaining in the interior. The two liquids are drawn off by pipes inserted into the revolving mass at suitable depths. Tar so dehydrated contains less than 1 per cent of water. This method can be very effectively employed with water-gas tar and is the only mechanical one that can be used for this particular emulsified mixture.

Chemical methods of separation have been tried, but none of them are of any practical importance. In one such method the watery tar is treated with chromic and sulphuric acids, when the heat evolved during the oxidation distils off without frothing the water and naphtha.

The universal practice is to heat the aqueous tar either by means of live steam, steam coils, or fire. Live steam has in the past been used fairly largely for the primary distillation of tar, but is now rarely employed except in a few instances in Scotland. In these cases, of course, super-heated steam is used, and together with the water some of the more volatile naphthas are distilled off, leaving a thick tar useful for road work, varnish, roofing felt, etc.

- A plan most commonly adopted now for the separation of water is to charge the tar still up to a given height, and warm the contents to nearly 100° C., when the water commences to boil. At this stage the firing is discontinued, and, if necessary, the liquid is allowed to settle for · a short time. It is found that nearly all the water now has separated, forming a layer above the tar. A small cock is fitted in the side of the still at the correct height, so that most of the water can be drawn off. Still better results can be obtained if a swing pipe is fitted with a raising and lowering rod connected to it and projecting out of the top of the still through a gland. It is found that, by this means, the majority of the water can be separated so that the amount that has to be distilled is small.

It is not necessary to say more here about the dehydration of tar, as the subject will be discussed later under the distillation proper. Confusion should be avoided in the use of the word dehydration, as it is very loosely employed in connection with tar. There are many plants existing which were erected for the manufacture of the so-called dehydrated tar. In effect these plants were intended for taking the water out of tar, but, generally speaking, are used for distilling out the water together with some naphthas and light oils that come over with the water, their primary object being the production of a tar suitable for road-spraying.

INTERMITTENT METHODS OF DISTILLATION

Description of Stills.—The plant used for distilling tar is the ordinary form of pot still. The variations in construction are legion and are dependent on the individual ideas of the distiller, although generally they follow certain broad principles. The best material to use is wrought iron, although mild steel is used in some cases with success. Cast iron would be an excellent medium from the point of view of its power of resistance to corrosive action, but it offers too great mechanical difficulties. For instance, it would be very difficult to cast a still large enough and perfect enough to be free from blow holes ; the capital cost would be too high, and, owing to the high temperature (300° C.) involved, there would be a liability for the metal to crack. When once this happened the still would have to be scrapped, as it is impossible to patch or repair cast iron satisfactorily.

Ordinary Tar Stills.—Fig. 152 gives a diagram of an ordinary tar still. It will be noticed that no dimensions are given, as these depend so largely, again, on the individual distiller. Some prefer to work their stills over a period of several days ; others, on the other hand, like to have a small still and work it off in a day. Generally speaking, however, 15-20 tons is the usual capacity, although stills to take 10-12 tons are used entirely in some large works. On the Continent, however, 30-40 and even 50-ton stills are quite usual. A is the charging pipe for the raw tar, B the sides or shell of the still, generally of ½-inch thickness. Some makers recommend that the shell should be made of mild steel, although the remainder is of wrought iron. This may be satisfactory in some respects, but at the join, which is a channel iron, where the crown or dome C is riveted to it, there must be severe strains, which are accentuated when the parts of the still are made of different metal, for the whole of the bottom of the still is subject to great changes of temperature.   The bottom of the still is of about ⅝-inch thickness, and is in this concave form for many reasons. There is a great saving of fuel, as it forms a large heating surface. Again, the shape helps to take up expansions and contractions on heating and cooling between charges. The run-off pipe D can also be connected at the lowest point of the still, and yet be out of contact with the fire heat.

The run-off pipe is made of cast iron, and the cock is usually

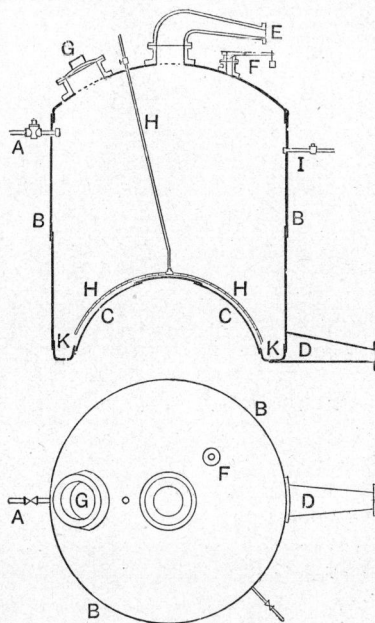

FIG. 152.—Ordinary Tar Still.

A, charging pipe ; B, sides or shell; C, crown or dome ; D, run-off pipe ; E, vapour pipe ; F, safety valve ; G, manhole ; H, steam pipe ; I, water run-off pipe ; K, channel iron.

fixed outside the brick work. The channel iron K is made in the same thickness of metal as the crown. The one drawback to the concave bottom is that there is difficulty in cleaning out the still efficiently, and, unless precautions are taken, workmen will not see that the channel iron is properly chipped round the rows of rivets on each side of it.   The top of the still, which is made of ⅝-inch plate, should not be high-pitched, but as low as convenient. It has connected with it the cast-iron vapour pipe E, safety valve F, cast-iron manlid G, and steam pipe H. This steam pipe is connected to the bottom of the still, where it is spread out in some sort of star formation, each branch pipe being perforated.   In this way, as soon as the steam is turned on, the whole of the tar in the bottom of the still is brought into contact with small jets of live steam. The small connection and cock I is used for drawing off the water after the tar has been warmed as already described.

A thermometer pocket is frequently fitted either in the top of the still or in the vapour pipe, although it is more usual either to measure the fractions collected from the tar, or to estimate their composition from their specific gravity.

The manhole should be large enough to admit a man easily, as the still has to be frequently cleaned out, for the life of a still is largely dependent upon this. It only needs one or two charges to be worked on a dirty still for dirt and foreign matter to get burnt on the bottom, when the iron assumes a brittle crystalline form and will very shortly crack. This means either that a new plate or sometimes a new bottom must be put in.

Great care must be taken in the setting of the still, as its efficiency largely depends on this. The usual plan is to rest the channel iron on an annular ring built several courses high and sprung from the floor of the combustion chamber. Opposite to the fire bars and door are one or two openings which lead into the flues which circle round, above and outside the annular ring to a convenient height up the sides of the still. Some makers prefer to protect the bottom of the still by a curtain arch which is simply a body of brickwork (covering, but not necessarily touching, the bottom of the still) through which there is an arch leading from the fire to the flue inlets in the annular ring.

Still with Convex Bottom.—A still which has been tried, and is in use in a large works to-day, is similar in all respects to the one just described except that it has a convex instead of a concave bottom. The heating surface is obviously the same, and it is more easily and effectively cleaned out. Its great drawback is that the draw-off pipe for the pitch must be connected in the centre of the bottom, and the junction is therefore subjected to the greatest heat during the distillation. The engineering problem how to overcome the consequent strains set up must be a difficult one.

Horizontal Still.—The form of still just described is, as already mentioned, the usual one, but there are a few works that prefer a horizontal still, which, in most cases, has a similar dome-shaped bottom. There seems no distinct advantage in the horizontal still, but rather a disadvantage, as the fuel consumption is higher, in many cases even twice as high.

Tubular Heating.—Many patents have been brought out for heating the tar in a still with tubes inserted through which the flue gases pass. In principle they are exactly similar to the ordinary water-tube steam boiler. This method of heating has given fairly good results, but it has been applied more successfully to continuous stills, details of which appear later.

Vapour Pipe.—The vapour pipe is best made of cast iron and should be large enough to allow free flow of vapour. There is a tendency for manufacturers to make this pipe too small so that there is a danger of the vapour being throttled.

Condenser.—The best metal for the coil is cast iron, as wrought iron corrodes too easily. Lead is sometimes used, but it is quite unnecessary to incur the high initial expense entailed by the employment of this metal.

Sight Box.—The outlet of the coil from the condenser goes into a sight box or seal which allows all the gases to pass away by a gas pipe. This is very important, as the gases evolved just when the still begins to run, and that liberated when the anthracene oil is coming off, are very poisonous, and it is advisable to pass them through the gas pipe to an iron oxide purifier. The chief poisonous ingredient, hydrogen sulphide, is acted on by the oxide, and the purified gas is usually carried round and fed into the fire of the still or into some convenient fire.

It is customary in a works where there are several stills in a row to pass the gas from each still into a common gas pipe leading to the purifier. Where this is the practice great care should be taken, when cleaning out any individual still, to disconnect the gas pipe of that still; for if at the same time other stills are working, the gas can easily work back into the one under repair and so cause danger to the men inside. Several fatal accidents have occurred in this manner.

Receivers.—The condensed vapour from the sight box is run into receivers. There are many simple devices by which the various fractions are made to run into their respective store tanks.

Pitch Cooler.—The running-off pipe for the pitch is connected directly to the pitch cooler, which is placed as near to the still as possible. It must be remembered that when the distillation is complete the residual pitch has a temperature approximating to 300° C. The pitch is allowed to remain in the still for some hours until its temperature is considerably lowered, so that, when "tapped out," too much strain on the still shall not be caused by sudden cooling and contraction. The further strain on the metal of the still which would be caused by immediately charging it again with cold tar should be avoided. Even at the lower temperature it is not safe to expose the pitch freely to air, as it would be liable to catch fire.

For this reason an intermediate vessel or pitch cooler is used. It is either an iron boiler or sometimes a "pitch-house" made of iron work. The pitch is cooled in this vessel until it is only fluid enough to run into the cooling bays, where it remains until quite cold and sufficiently workable to send out.

Heat Interchangers or Economisers.—Most plants have some form of heat interchangers or economisers. They work on the principle of utilising the heat in the vapours leaving the still and that in the pitch to warm up the raw tar for the next charge.

These arrangements have been very thoroughly worked out and enough heat may be recovered to distil off all the water together with some naphthas from the raw tar while in the economiser, so that the actual still charge is hot enough to come on to the run soon after charging, and is free from water.

Lay-out of Still.—Fig. 153 shows the lay-out of such a still; and it will be noticed that in this case there are two economisers: the first is fired in the same way as the still, in order to ensure the removal of all the water ; the second is simply a condenser for the vapours from No. 1 with the coil surrounded with raw tar.

Frothing.—A good many arrangements have been devised and patented with the object of preventing the tar from frothing and boiling over when distillation commences. The underlying idea is to cool the top portion of the still or the vapours in the top of the still. They are none of them very successful and, indeed, are little used, continuous dehydration giving better results, as will be shown later.

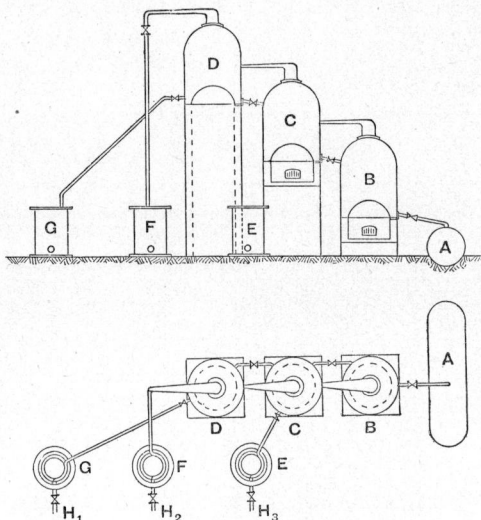

FIG. 153.—Lay-out of Still.

A, pitch cooler ; B, tar still ; C, No. 1 economiser ; D, No. 2 economiser ; E, heavy oil condenser ; F, naphtha condenser ; G, light oil condenser ; H₁, H₂, H₃, outlets to receivers.

CONTINUOUS METHODS OF DISTILLATION

Numerous attempts have been made to distil tar continuously, and although to-day the great difficulties that have to be faced are overcome, many distillers prefer the old intermittent methods. Continuous distillation is expensive, and can only pay where the still can be kept going continually, or when very few stoppages are required. This, then, is the great difficulty, as the tar available in most districts is limited, and very often the quantity differs greatly at different seasons of the year.

Some of the earlier experiments are very ingenious, and may be usefully mentioned.

Mason's Still.—Mason's still (Ger. Pat. 66097) consists of a raw-tar tank from which the tar is fed into the still, which consists of a series of horizontal and vertical pipes. Each horizontal pipe has a vapour pipe and condenser attached, and as many of these are used as fractions are required.

The tar flows through the zigzag still pipes, which are heated with fire from the bottom one upwards. The lightest fraction comes off the top one, and the heaviest off the bottom, and the last horizontal pipe emits the pitch to the cooler. The difficulty of regulating the firing of such a still must be tremendous and must condemn the plant.

Ray's Still.—Ray of Turn (Fr. Pat. 348267) elaborates Mason's still by using a series of retorts in zigzag formation and maintains that good results are obtainable.

Pfropfe's Still.—Pfropfe's still (Ger. Pat. 55025, 1890) consists of a series of semi-cylindrical tanks, each separately fired and regulated, and each having its own vapour pipe and condenser. The principle has been more carefully carried out in the Hird Still, which is described later.

FRACTIONAL CONDENSATION

Kohn's Still.—Walter Kohn, of Lubeck, has invented a most ingenious arrangement which works on the principle of fractional condensation. The tar is pumped under pressure through a series of heating coils from which it is ejected through a nozzle into the still. The nozzle baffles the flow of tar and breaks it up into a fine spray, from which the residual pitch falls to the bottom of the still and thence flows to the pitch cooler, whilst all the volatile vapours pass on into a series of condensers. These are supplied with coils kept at different temperatures, and so the required fractions are obtained. The Lennard still, which is more fully described later, is similar in general outline.

The Wilton Still.—The Wilton continuous still, as manufactured by the Chemical Engineering & Wilton's Patent Furnace Company, is shown in Fig. 154, and is there represented with plant for taking off two fractions. The makers maintain that it can be used for distilling down to pitch, and the distillate can be split up into as many fractions as required.

The tar in tank A is pumped up by the pump B, through the pipes C into the heat interchanger or economiser D. In this it takes up heat from the dehydrated tar or pitch according to the use the plant is being put to. It then traverses the pipe C_1, and enters the furnace S, which is heated by an ordinary fire of coke or breeze and consists of a series of cast-iron coils superimposed one above the other. It leaves the furnace through the pipes E or E_1, at a temperature and pressure predetermined and varied by the rate and by the temperature of the furnace. It is then allowed to expand suddenly under atmospheric pressure in the vapour box F, whence all the volatile materials pass over through the vapour pipe G, whilst the pitch or dehydrated tar, as the case may be, overflows into the economiser D, and thence to the pitch cooler or storage tank. The vapours from G are cooled by water in the condenser tank H, so that the heaviest of them are condensed and received in I, whilst the more volatile pass over through pipe K into another condenser L, and so on.

The fractions can thus be varied by using as many condensing units as required. The temperature and pressure of working, as already stated, are dependent upon what is required, and must be determined for the individual tar to be distilled ; but those recommended by the makers for dehydrated tar are about 130-140° C. and 30-40 lb. respectively.

It is doubtful whether the plant, as described, is very suitable for distilling tar to pitch, as the fractional condensation of the distillates is likely to give inconsistent results.

Improved Wilton Still.—Probably a better plan, which the makers adopt, is to combine this plant with the ordinary pot still.

The drawing in Fig. 155 is self-explanatory after the descriptions already given. The plant can be run almost continuously, as the dehydrating portion is working the whole time and filling the one pot still while another is being worked off, and the third is cooling before tapping the pitch. If the operations are properly adjusted the third can be tapped and be ready for charging as soon as the first still is full and ready to work.

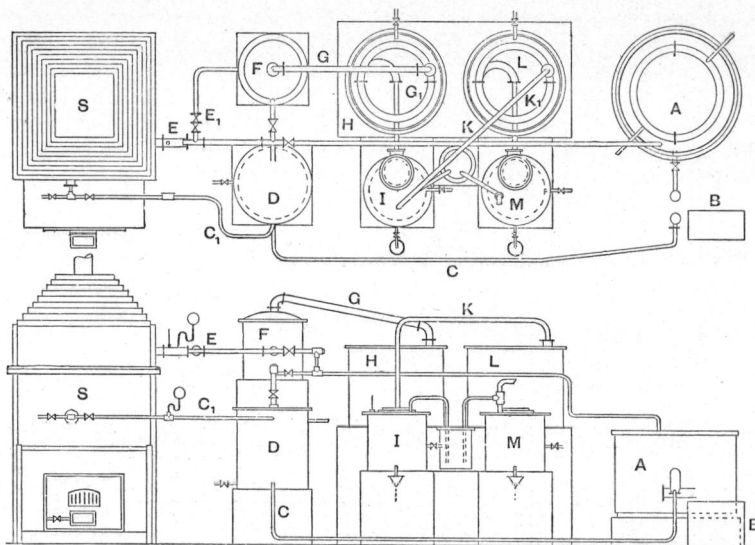

Fig. 154.—The Wilton Still.

A, raw tar tank; B, pump; C, C₁, tar delivery; D, dehydrated tar tank and economiser; E, E₁, heated tar pipes; F, expansion box; G, G₁, vapour pipe and coils; H, first condensing tank; I, first condensate receiver; K, K₁, second vapour pipe and coils; L, second condensing tank; M, second condensate receiver; S, dehydrating still furnace.

It is obvious that great economies can be effected by this means, both in saving of fuel and—a most important point—in wear and tear on the pot stills, as they are never subjected to strains set up by sudden cooling through charging cold tar into hot stills.

Lennard's Still.—The continuous still of F. Lennard, of Forbes, Abbot & Lennard, Ltd., is worthy of description, as it is based entirely on the principle of fractional condensation and is capable of dealing with large quantities of tar.

The still is shown in Figs. 156 and 157. The heating portion proper consists of a number of cast-iron pipes running to and fro across the width of a brick-work oven, and forming a long coil which is heated by means of producer gas or oil fuel. One end of the coil is connected to the feed pump P, and the other to the pitch scrubber D. The coil is divided into two sections; the front one is known as the "front bath" A, and the back one as the economiser B. The only difference in the two is that the economiser coil is directly heated by the flue gas from the still furnace and steam boiler, while the other is protected by brick-work.

The heated tar, passing into the pitch scrubber D, meets in its descent a jet of steam which sets free all the vapours, which pass on through the connecting vapour pipe into the anthracene condenser E. The pitch falls to the bottom of the scrubber and eventually to the cooler and pitch beds. The pipe extending from D to E connects the top of the pitch scrubber with the condensing coils of the specially constructed anthracene condenser. The lower chamber of this condenser, from which the anthracene is collected, is connected with the creosote condenser F, by a pipe through which the vapours pass forward.

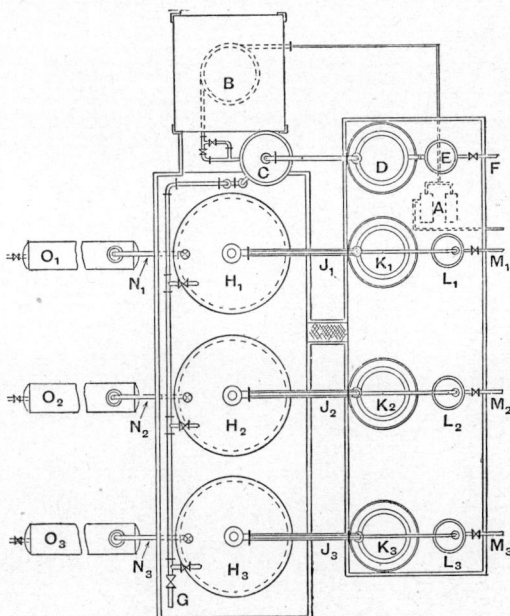

FIG. 155.—Improved Wilton Still.

A, tar pump ; B, dehydrating still furnace ; C, vapour box ; D, naphtha and water condenser ; E, naphtha and water receiver ; F, naphtha and water outlet to storage ; G, dehydrated tar pipe ; H_1, H_2, H_3, pot stills ; J_1, J_2, J_3, pot still vapour pipes ; K_1, K_2, K_3, pot still condensers ; L_1, L_2, L_3, pot still receivers ; M_1, M_2, M_3, pot still outlets to storage ; N_1, N_2, N_3, pitch tapping pipes ; O_1, O_2, O_3, pitch coolers.

The creosote condenser is built similarly to E, and has also a cooler T, the coil of which extends to the receiving tanks. Again the lower chamber of the creosote condenser communicates with the light oil condenser G.

The condensing medium used in E and F is the raw tar, which, after exchanging heat in the process, overflows from the top of both condensers into the top of the tar scrubber H. It is there subjected to the action of raw super-heated steam, which drives off all the water and some naphtha, which are condensed in J. The dehydrated tar flows into the hot tar boilers 1 and 2, and is thence pumped by the pumps P (one always kept in reserve) into the still coil A and B.

The receiving tanks are provided with an exhaust pipe, connecting to the vacuum pump Y, the exhaust gases passing through the vapour absorbers R.

The working of the still is carried out by feeding the tar through the

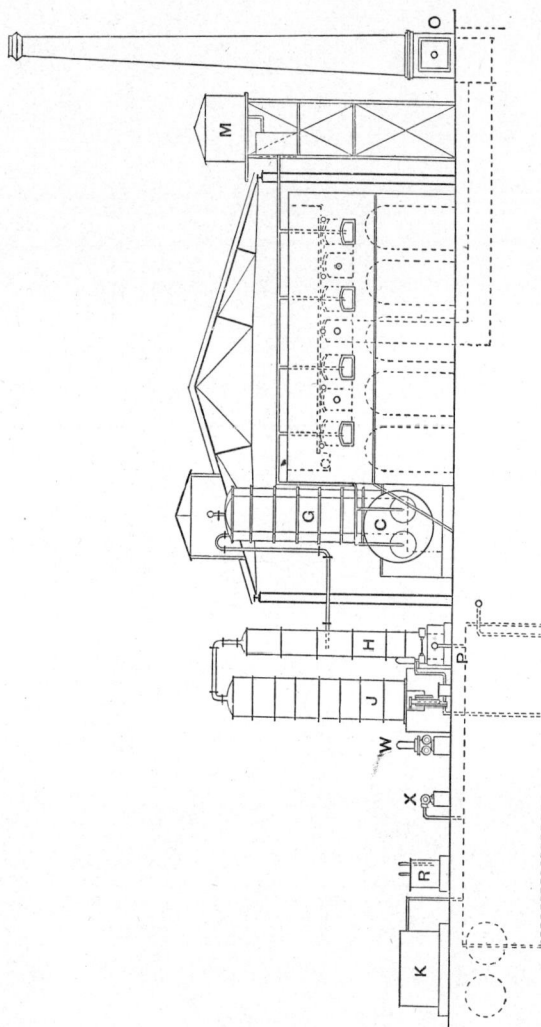

FIG. 156.—Lennard's Still.

C, steam boiler; G, light oil condenser; H, tar scrubber; J, naphtha condenser; K, hot tar boiler vapour condenser; M, oil fuel tanks; O, chimney stack; P, hot tar pump; R, vapour absorber; W, raw tar pump; X, vacuum pump.

still coils, in which it is heated to the final temperature of about 280° C. It is under pressure when it leaves the coil, and suddenly expands on entering the pitch scrubber. There by aid of live steam all the volatile constituents pass over into the anthracene condenser, while the pitch falls to the bottom. Enough cold raw tar is passed through the anthracene condenser to condense only the anthracene oil fraction, while the remaining vapours pass over into the creosote condenser, and similarly a creosote and a light oil fraction are separated.

One of the secrets of success of the plant is that good heat interchange is applied, and the tar is dehydrated in the tar scrubber before it falls into the hot tar boilers which are the supply tanks for the pumps to feed the still coil.

Better adjustment and regulation of the fractions have been obtained by slightly varying the above arrangements. The change can be effected by reversing the operations in the anthracene and

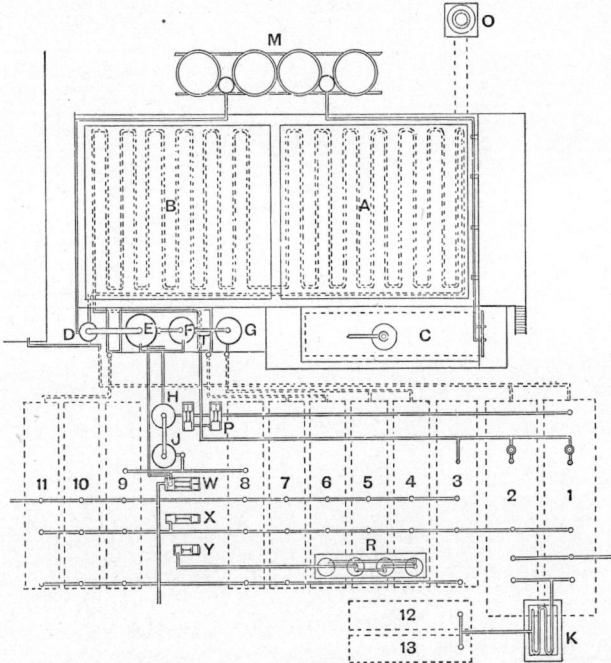

FIG. 157.—Plan of Lennard's Still.

A, front batch of still coil ; B, economiser batch of still coil ; C, steam boiler ; D, pitch scrubber ; E, anthracene oil condenser ; F, creosote oil condenser ; G, light oil condenser ; H, tar scrubber ; J, naphtha condenser ; K, hot tar boiler vapour condenser ; M, oil fuel tanks ; O, chimney stack ; P, hot tar pumps ; R, vapour absorber ; T, water condenser ; W, raw tar pump ; X, Y, vacuum pumps ; 1, 2, hot tar boilers ; 3, spare receiver ; 4, 5, light oil receivers ; 6, 7, creosote oil receivers ; 8, 9, naphtha receivers ; 10, 11, anthracene oil receivers ; 12, 13, receivers for vapour off K.

creosote condensers. That is to say, the vapours are allowed to pass into the top of the column and to cool to the desired degree by letting the cold raw tar circulate in a coil in the condensers.

METHOD OF SUCCESSIVE STILLS

The Hird Still.—The Hird Still is manufactured by W. C. Holmes & Co., Ltd. It is worked on an entirely different principle, and consists of a series of small stills, each one responsible for distilling off one fraction. Here again, as many fractions as desired can be obtained according to the number of units supplied in the plant (see Figs. 158 and 159).

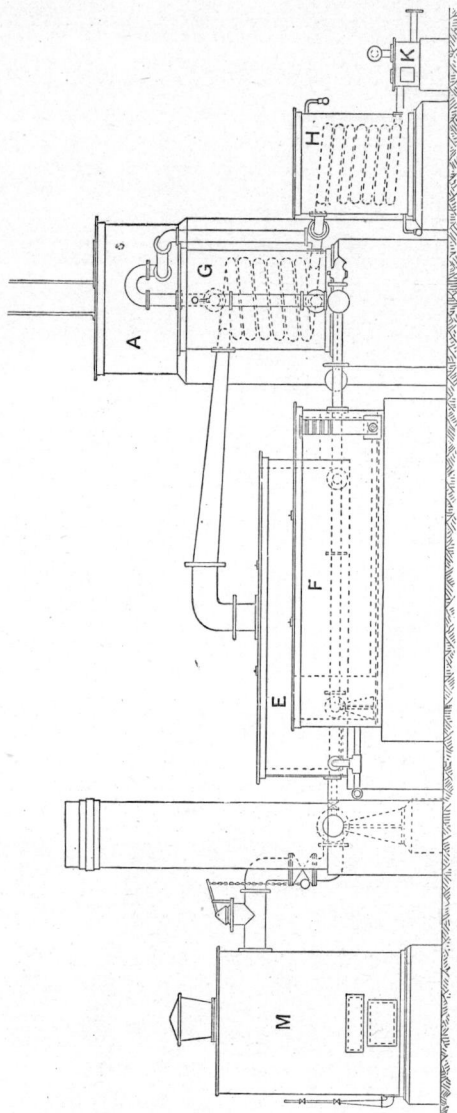

FIG. 158.—The Hird Still.

A, tar regulation tank ; E, No. 4 still ; F, pitch cooler ; G, combined heater and cooler ; H, coil condenser ; K, sight box ; M, gas producer.

The tar flows from the regulating tank A through all the vapour condensers G, and thus takes up enough heat to drive off most of the water and some naphthas which are condensed in the cooler J. It is further heated by passing through a coil in the pitch cooler F, whence it flows to No. 1 still B. This and the other stills are of the same construction and consist of cast-iron tanks heated by means of fire tubes running longitudinally and close to the bottom of the still. The firing can be carried out by oil or gas and is done by means of the burners N.

The height of the tar in the stills is only 12-16 inches, and thus relatively small quantities of tar are in the whole plant at any one time. The contents of No. 1 still B are heated sufficiently to drive off only the first fraction, namely the light oil, when the tar overflows into No. 2 still C, where it is subjected to greater heat and another fraction is distilled off. So the process is continued until the tar has passed through

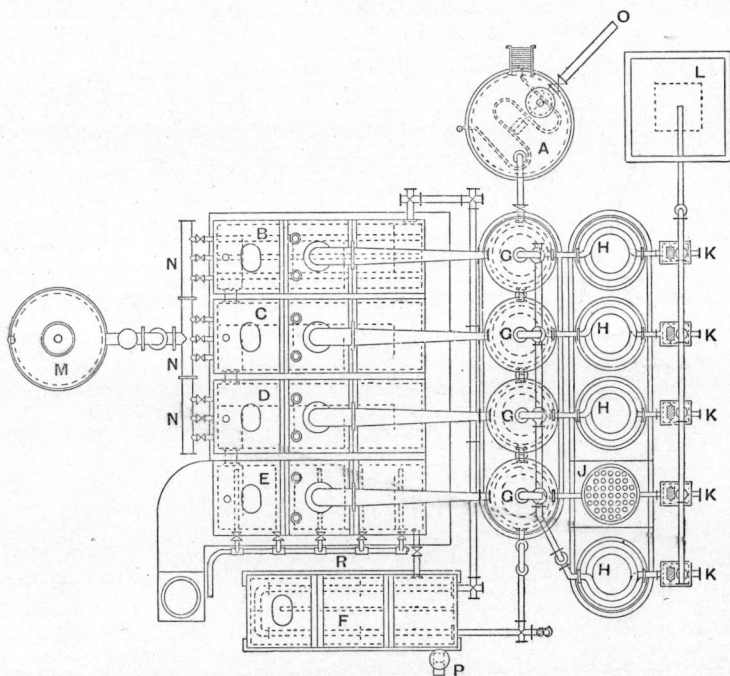

FIG. 159.—Plan of the Hird Still.

A, tar regulation tank ; B, No. 1 still ; C, No. 2 still ; D, No. 3 still ; E, No. 4 still ; F, pitch cooler ; G, combined heater and cooler ; H, coil condenser ; J, water tube condenser ; K, sight box ; L, oxide purifier ; M, gas producer ; N, gas burners ; O, tar inlet ; P, pitch outlet ; R, steam pipes.

the required number of stills. There are usually four, which give off naphtha, light oil, creosote oil, and anthracene oil, respectively. No. 4 still is not, like the others, provided with burners, but with perforated steam pipes which allow live steam to be blown into the partially distilled tar, this being the usual method of getting off the remaining anthracene oil.

The pitch overflows from No. 4 still into the pitch cooler F, and eventually into the pitch beds. The vapours from the stills are partially condensed in the heat interchangers, and finally in the water condensers H, and received in the sight boxes K.

The working of the still is very simple and is very easily and regularly adjusted, so that with little attention regular and constant fractions can be obtained.

In the early types of this plant difficulty arose with the joints between the fire tubes and the sides of the still, as leaks were frequent and corrosion heavy. This difficulty has been surmounted by expanding tubes into steel end plates and bolting the whole on to the flanges of the cast-iron casing, as the more modern plants are supplied with cast-iron instead of wrought-iron stills.

A COMPARISON BETWEEN THE INTERMITTENT AND THE CONTINUOUS DISTILLATION PLANTS

Some of the advantages of the continuous over the intermittent still may be enumerated, viz. :—

1. Generally a reduction of fuel consumption.

2. Less loss through bad manipulation of the plant.

A continuous plant when once well started can be fairly easily handled with small chance of anything going wrong. On the other hand a pot still requires a very experienced man to watch it, and even then frothing and " bolting " or boiling over are not infrequent.

3. Wear and tear by corrosion and overheating is not so heavy, as the coils or stills, which have to be subjected to the greatest amount of heat in the distillation, are made of cast iron. Also the overheating of a cast-iron coil does far less damage than that of the bottom of a pot still, especially if it is a little dirty and cleaning out has been neglected.

4. Saving of labour in cleaning out.

On the other hand some of the disadvantages of a continuous still as compared with an intermittent still may be mentioned, viz. :—

1. It is claimed that some continuous stills require less labour than periodic plant of equivalent size. This is very doubtful, for a row or battery of pot stills run in proper rotation can be worked with very little labour and expense indeed.

2. The continuity of the continuous still is in many cases a drawback. In works where the tar producer is also the distiller, as in the case of many coke oven works, it is obvious that he can estimate fairly accurately how much tar he will have to deal with, and again, his production will be more or less consistent, so that he can erect his continuous plant to meet his needs, and keep it continually running.

On the other hand the tar distiller who does not produce his own tar cannot estimate the quantity he may expect to obtain from the various gas works, as their make is a seasonal one and is also dependent upon climatic conditions. Unless a continuous plant can be kept more or less consistently running, the charges mount up enormously, and if such a plant has to be frequently started and stopped it is doomed from a financial point of view.

3. Most continuous stills do not produce such a clear-cut fractionation of the products as does the pot still.

Corrosion of Tar Stills.—This is a trouble which has a very great bearing on the whole industry of tar distilling, for the wear and tear on stills, and particularly those of the pot still type, by corrosion is a very great financial factor.

The problem has been studied by many investigators, but no satisfactory solution of the difficulty has been found. The corrosion is caused chiefly through the tar being mixed with liquor, and if distillation could be carried out with water-free tar it would be almost eradicated.

Warnes and Davey [1] give a summary of the work done on this subject and maintain that the corrosion is caused rather through chemical or mechanical than electrolytic action. It seems certain that dissociation of ammonium chloride, ammonium sulphide, ammonium hydrosulphide, and ammonium cyanide takes place at certain temperatures, and that the dissociation products act upon the iron. The action is most pronounced at parts of the still that have been subjected to mechanical strains in manufacture, and parts that are continually in contact with hot vapours rather than hot liquids. For instance, stills that have had the rivet holes punched instead of drilled show the most marked corrosion round the rivet holes, and it is very noticeable around a manlid mounting or charging pipe. Again, the sides of the still are most corroded at the height where the tar and liquor are in violent ebullition at the beginning of the distillation when the water is coming off.

Warnes and Davey state that the process of corrosion goes on at a greater rate during the latter portion of the distilling operation, principally during the period when steam is used to assist in the distillation.

Whatever the causes may be, electrolytic action undoubtedly plays a certain part in the process, and great care should be taken in the selection of suitable plates for the still, whether they be of mild steel or iron. It should be seen that there is no sign of lamination, crystalline structure, or blisters. Indeed, in bending the plates metal hammers should not be allowed but only wooden mallets.

FRACTIONS

The question of the number of fractions, size of fractions, etc., that are to be taken off when tar is distilled is dependent on so many factors that it is only possible to give generalisations. For instance, the kind of tar to be dealt with obviously plays a big part, while the kind of plant that is available to deal with the fractions is another, but the market value of the products is probably the most important. For if the price of any individual product should be too low for its extraction to be profitable, the fractions would probably have to be entirely altered. Generally speaking, however, they are the following :—

No. 1 Fraction.—In this is contained the ammoniacal liquor, and naphthas, which are mixtures of benzene, toluene, xylenes, and pyridine.

[1] Warnes and Davey, *J.S.C.I.*, 1910, **29**, 657.

The boiling-point range is from 80° C. to about 140° C., and the specific gravity ·870 to ·950. The quantity of water is of course dependent upon the amount in the original tar, and whether it has been partially taken out before distillation. It separates easily from the naphtha, and is drawn off from the bottom, and sent direct to the ammonia works.

Great care must be taken in getting off the first fraction, as frothing is very prevalent, particularly in a tar with a high free carbon content. The point when this danger is passed can be easily noticed by the noise that is heard inside the still, known as the "rattles." When nearly all the water is off, globules of water condense on the inside of the top of the still and occasionally fall back into the hot liquid below : they are immediately turned into vapour again with almost explosive force, with the resulting rattling noise. The same phenomenon can be observed in distilling a liquid like benzole containing traces of water in the laboratory.

No. 2 Fraction.—This is known as the light oil fraction, and boils from about 140° C. to 200° C., and has a specific gravity of about ·950 to 1·000. It contains the higher hydrocarbons of the benzene series such as mesitylene, cumenes, some naphthalene, also phenol, and higher homologues of pyridine. Many distillers do not separate this fraction, but mix No. 1 and 2 together ; on the other hand, in districts such as Lancashire and Yorkshire where the tar is rich in these valuable light products, it pays to collect the two fractions.

No. 3 Fraction.—This fraction is collected purely to obtain the phenol in as concentrated a state as possible, and is consequently called the carbolic oil or middle oil fraction. It boils between 200° C. and 240° C., has a specific gravity of 1·000 to 1·025, and contains phenol, cresols and higher hydroxy acids, much naphthalene and creosote hydrocarbons. In the distillation great care must be taken to see that the condenser water is quite hot, so that crystallisation of the naphthalene shall not take place in the coils. The cold water should be turned off in the middle of No. 2 fraction, and if the cooling water does not get warm quickly enough, steam should be turned into the condenser.

Safety valves are supplied to all stills, and these should be looked to before each distillation, to see that they are working freely, as explosions have occurred in tar works through the jamming of the valves.

This carbolic oil fraction is not always separated, as the acid in some cases does not pay to extract, and in others the tar contains too small a quantity. Again, it is sometimes found more economical to re-distil the creosote fraction, as described later.

No. 4 Fraction.—This, known as the creosote oil fraction, is the largest of all, and contains naphthalene and heavy oils, which are aromatic hydrocarbons with a high carbon and hydrogen content, and cresols and other phenol homologues. The boiling point is about 240° C. to 280° C., and specific gravity 1·025 to 1·065.

No. 5 Fraction.—This fraction is marked by its distinctive colour, and is consequently called the green oil, yellow oil, or anthracene oil fraction. Its specific gravity is 1·065 to 1·100, and boiling point from 280° C. upwards to the end of the distillation. It contains still higher aromatic hydrocarbons, anthracene, phenanthrene, also carbazol, etc.

Numerous attempts have been made to largely increase the number of fractions taken off the tar with the idea of better isolating the products. All these have failed, as the distillates obtained are no purer, so many complex azeotropic mixtures being formed. Again, nothing is saved, as many of the fractions have to be mixed together again for treatment in subsequent processes.

CHAPTER XXXVII

FURTHER DISTILLATION AND RECTIFICATION OF PRIMARY FRACTIONS OF TAR

Methods of Separation.—The usual methods of separation of substances from the complex mixtures contained in the fractions are employed. The solids such as naphthalene and anthracene are crystallised out and freed from the other liquids. The acids such as phenol are extracted by washing with alkali, generally caustic soda, and the acid is separated out from the sodium salt by treatment with mineral acid. The bases such as pyridine are subjected to the reverse treatment, as they are converted into the sulphate by the action of sulphuric acid, and subsequently neutralised with alkali, generally ammonia. Lastly, the remaining bodies, which are mostly hydrocarbons, are separated by further distillation.

The methods of further separation and treatment of the primary fractions may be mentioned, but it must be borne in mind that many of the products to be extracted, and therefore also the processes employed, are common to several of the fractions. For instance, phenol and tar acids generally are common to all the fractions after the crude naphtha, and under certain market conditions every fraction has to be subjected to the treatment of washing out with soda. Only one typical case of each separation will therefore be described.

Nos. 1 and 2 Fractions.—These will be left for the next chapter, as the separations can only be carried out by fractional distillation, and the whole of the next chapter is devoted to this subject.

No. 3 Fraction—Naphthalene.—The oil contains fairly large quantities of naphthalene, which is sometimes allowed to cool out first, but generally after extraction of the acids, as the solubility of naphthalene in tar acids is greater than in the creosote hydrocarbons. Better separation of the naphthalene is therefore obtained if the acids are washed out first.

Phenol and Cresols.—The crude oils are washed in washers or wrought-iron tanks of about 2000 to 5000 gallons capacity. The tanks are fitted with good agitating gear, as described in the next chapter, also they are either steam-jacketed or contain steam coils, in order to keep the oils hot while the washing is carried out, otherwise crystallisation

384

of the naphthalene would cause trouble. Caustic soda of a strength of about 16° to 22° Tw. is the alkali used, and, after settling, the sodium phenate and cresolate, commonly known as the carbolate and cresolate, are drawn off from the bottom.

Separation of Phenol from Cresols.—The separation of the phenol from the cresols or higher homologues is carried out direct in the washing process, and is based on the fact that phenol is more acidic than its homologues. The method is to wash the light oil first of all with the so-called crude carbolate solution. This is the middle wash out of three and is a saturated solution of sodium phenate and cresolate. The action of bringing this into contact with fresh oil is to decompose the cresolate in the crude with the phenol contained in the oil, so that an interchange of phenol from oil to aqueous solution is obtained. Thus by regulating the quantity of crude carbolate used in the wash, sodium phenate fairly free from sodium cresolate can be obtained. The light oil, now containing relatively small quantities of the phenol and large quantities of the cresols, is washed with clean caustic soda and the resulting solution is the crude carbolate for the next charge. The oils are now fairly free from phenol and contain chiefly cresols and homologues which are extracted by washing with the required quantity of clean caustic soda.

In order to obtain the acids from the solutions some mineral acid must be used. The old method was simply to neutralise with sulphuric or hydrochloric acid, in an open vessel. The method most in use at the present time is to treat the solutions in upright tanks or gassing cylinders with carbon dioxide. These tanks are run in series, so that any gas not absorbed in the first is caught up by the next. The carbon dioxide is usually obtained from a lime kiln, for, besides the gas, limestone produces lime which is required in the process. After the solutions are neutralised by the carbon dioxide, the tar acids can be easily separated from the top, and the sodium carbonate at the bottom is used to react with the lime from the kiln to make caustic soda for the washing of further quantities of light oils. The process is a continuous one and works in a cycle, viz. :—

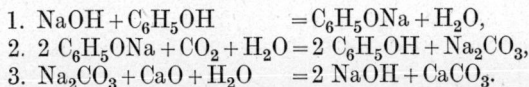

1. $NaOH + C_6H_5OH$ $\quad = C_6H_5ONa + H_2O,$
2. $2\ C_6H_5ONa + CO_2 + H_2O = 2\ C_6H_5OH + Na_2CO_3,$
3. $Na_2CO_3 + CaO + H_2O$ $\quad = 2\ NaOH + CaCO_3.$

Washing with a cream of lime instead of caustic soda can be employed, and has been worked out successfully by some distillers.

Again, washing the light oil with sodium carbonate solution and freshly burnt lime will effect the same result, the action taking place at quite low temperature, viz. :—

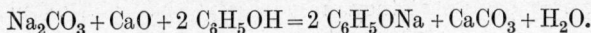

$$Na_2CO_3 + CaO + 2\ C_6H_5OH = 2\ C_6H_5ONa + CaCO_3 + H_2O.$$

Both of these methods present mechanical difficulties, as objectionable emulsions are formed with consequent loss of phenol.

2 C

REFINING OF THE TAR ACIDS

Phenol.—The acids as separated above are in the form of dark-brown oily liquids, and contain 12 to 15 per cent of water. The phenol is prepared to the specification known as 50's, 60's, or 70's crude carbolic acid. The names are derived from the result obtained by submitting the acids to the "Lowe" test, and do not denote the quantity of absolute phenol in the crude substance, although the figures, as a matter of fact, do approximate to these values. So that a 60's crude carbolic acid may be taken to contain about 60 per cent absolute phenol, 15 per cent water, and 25 per cent cresols and homologues.

Distillation and Crystallisation.—The refining of the phenol and its separation from the cresols is carried out entirely by fractional distillation and fractional crystallisation. We may take 60's crude carbolic acid as a good example, as this is the usual form in which the material is prepared from the light oils.

The mixture is first distilled in an ordinary iron pot still with the object of separating the water, cresols, and pitchy matter from the phenol. The first fraction, coming over below 180° C., consists of water, neutral oils, and some phenol. The second fraction is collected between 180° and 205° and contains mostly phenol. The third fraction, 205° to 220°, consists of cresols with a small quantity of phenol. The residue contains practically no phenol and may be collected from successive charges and reworked for the yield of cresols or pale cresylic acid or, as it is known to the trade, "Liquid Carbolic Acid."

The fraction collected between 180° C. and 205° is run into small pans holding 25 to 50 gallons, and allowed to crystallise. These pans are arranged so that cold water can be circulated round them, or in some instances the pans are kept in a refrigerating house. After crystallisation has taken place the liquid is drained off and mixed with the fraction from 205° to 220°, which is again distilled. The crystals of phenol are dried by means of a centrifugal machine, when they show a fusing point of about 30° C. These crystals are melted and subjected to a further distillation and a further crystallisation, as above, when the fusing point should reach about 35° C. To obtain 42° crystals the same process must be repeated again. It will be seen that the process is a long and tedious one and that many fractions are obtained, but by careful observation of their boiling points many of them can be mixed together and reworked.

The distilling apparatus is sometimes supplied with a fractionating column, but often not. There can be no doubt that a column adds to the efficiency if it is of a good type ; but there is no doubt that distillers do not pay enough attention to the columns they use, and consequently are apt to conclude that they are unnecessary, while the fault really lies in the construction of the column itself.

In cases where very pure phenol is required it may be advisable to redistil once more, and to use either zinc or silver still-head pipe and condenser coil, as it has been found that iron, copper, lead, and aluminium are unsuitable.

Cresols.—These are recovered from the liquor from the fractional crystallisation of phenol, and also in the crude form when washed out of the tar oil with caustic soda. In the second case they are mixed with higher homologues such as the xylenols and cumenols. The cresols are present as a mixture of the three isomers in proportions varying according to the source of the tar, but averaging approximately—

> Ortho-cresol 40 per cent (b.p. 191°).
> Meta-cresol 35 per cent (b.p. 203°).
> Para-cresol 25 per cent (b.p. 202°).

It will be seen by the boiling points that the separation of the three by distillation is impracticable, although the ortho can be isolated from the other two by this means. Many methods of separation have been attempted and they are all based on the principle of making crystallisable salts or sulphonation or nitration products, separating these, and reconverting to the original cresol. But the best results are obtained if the mixture is first fractionated, when most of the ortho can be separated, and the residue can then be treated for the separation of the remaining two isomers.

Raschig's Still for Phenol and Cresols.—Probably the firm that, prior to the war, made more pure phenol and cresols than any other in the world was Raschig of Ludwigshafen. Raschig describes [1] an apparatus for the preparation of carbolic oils. It is shown in Fig. 160, and is simple in structure and principle, except that the distillation is carried out *in vacuo*, but the distillate collected under atmospheric pressure. Raschig maintains that to distil tar acids and

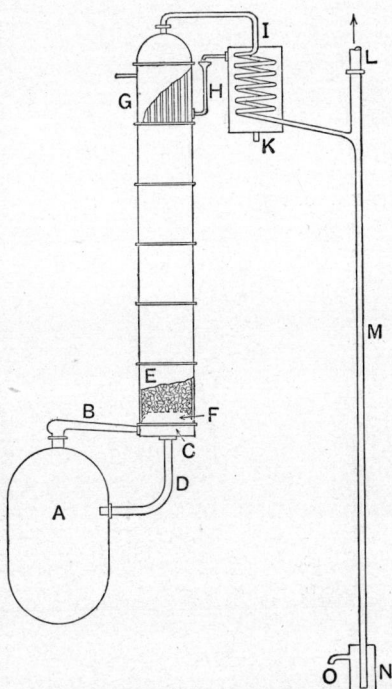

FIG. 160.—Raschig's Still for Phenol and Cresols.

A is the still ; B is the wide tube which carries the vapour to the column ; C is the lower part of the column with back flow tube ; D leads the condensed oil back into the still ; E is the column which is filled with Raschig rings, which are carried on a perforated plate F at bottom ; G is the dephlegmator, which can also be fitted as a worm condenser ; H shows the stream of condensing water ; K is the condenser with the worm (I) ; L is the connection with the vacuum pump ; M is the fall tube for the distillate ; it must be 39 feet long from L to a point O where the distillate runs out from the top of the receiver N.

tar oils containing acids *in vacuo* is distinctly advantageous, as

[1] Raschig, *Zeitschrift f. angte. Ch.*, 1915, **84**, 28.

decomposition, that would be likely to occur under atmospheric conditions, is avoided. Also, by using some such plant as the above, the necessity of observing the course of the distillation is obviated, as the exact flow from the still can be seen and also average samples of the distillate at any time can be accurately taken.

The column is 45 feet in height, and is packed with Raschig rings, which are simply sheet iron ferrules about 1 inch long by 1 inch diameter. The apparatus is simple and ingenious, but could probably be greatly improved by the use of a good seal and bubbling column instead of the packed column, for the reasons given in the next chapter.

The Separation of Naphthalene

To proceed with the separation of the products in Fraction 3.

Crystallisation.—Fraction No. 3.—It is now necessary to crystallise out the naphthalene. The fraction has been washed practically free of all tar acids, so that the oil, on cooling, will more readily throw out the naphthalene. The quantity that will crystallise is, of course, dependent on the amount present and the saturation of the oil. The solubility very rapidly decreases as the temperature is lowered, and the oil is therefore cooled as much as possible. Ordinary atmospheric conditions are usually suitable and due advantage is taken of winter temperatures. It is sometimes advisable to use refrigerating plant where special creosote oils free from naphthalene are desired. After the oil is cooled it is run off from the crystals, which are dug out into conveyors which carry them to the drying house.

The drying is done in several ways, the crudest and simplest method being to allow the crystals to remain in a heap for the oil to drain out, when they still contain 10 to 15 per cent of oil. To free the salt more completely from oil, pressing or centrifuging is employed. The latter is the more usual method of dealing with the naphthalene, and a product is obtained containing only a small percentage of oil. On the other hand, if quite dry crystals are required, horizontal presses supplied with steam jackets must be used. By either of these two methods naphthalene is produced which has a melting point of 77° to 78° C. (pure naphthalene melts at 79·7° C.).

Pure Naphthalene

Washing.—Pure naphthalene is prepared and marketed in various forms, such as " crystal " or " flake," or moulded forms like " balls," " candles," " moth-block," etc. In every case it is necessary to wash the crude naphthalene with concentrated sulphuric acid, water, caustic soda, and finally with water. The washing is carried out in cast-iron or lead-lined washers supplied with the usual form of agitating gear, and steam-jacketed or provided with steam coils to keep the naphthalene in a molten state.

Redistillation.—To obtain crystal naphthalene the washed crude has to be redistilled. In this case very little fractionation is required, so that a column to the still is not necessary. The distillate is run into pans, and, when cold, dug out and packed ready for the market. Care should be taken in this distillation to see that no blocking takes place in the condenser coils ; the condenser water must therefore be kept hot enough to prevent solidification.

Sublimation.—The flake naphthalene is more difficult to manufacture, as the washed substance has to be sublimed from a pan or some open vessel into large chambers. Fig. 161 shows a useful plant for making this product. The plant is arranged in such a way that it can be worked continuously, the sublimed product being raked out of the subliming chambers from time to time through the trap-doors. The heating medium shown in the diagram consists of steam coils, but in some cases fire heat is found more suitable. The sides of the chamber slant inwards and are made of light material such as galvanised iron on wooden beams. By gently tapping the sides from the outside the

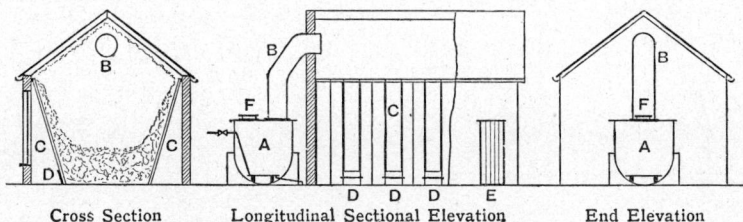

Cross Section   Longitudinal Sectional Elevation   End Elevation

FIG. 161.—Naphthalene Sublimation Plant.

A, naphthalene pan with steam coil ; B, uptake to subliming house ; C, slanting settling sides ; D, extracting doors ; E, entrance to extracting chamber ; F, charging manhole.

naphthalene will slide to the bottom where the trap-doors are fixed. The process is naturally a very slow one, but the output is very considerably increased by working the plant continuously, and the very unpleasant task of cleaning the chamber out (which is necessary in an intermittent plant) is avoided. The walls of the chamber can be made of any suitable material that will not chip or leave any foreign material in the naphthalene. An ordinary brick building with the inside plastered with a good smooth cement plaster meets the requirements. Precautions should be taken against fire and explosion, for although naphthalene is not inflammable at ordinary temperatures, a mixture of it in a finely divided state with air may certainly be explosive. For this reason no light should be brought near the building, nor should workers be allowed inside the house unless wearing rubber shoes. No iron implements like shovels should be allowed.

The candles, balls, and other moulded forms are generally made from the crystals, but may, if necessary, be cast from the flake naphthalene. In either case the process is simple, and consists of melting the pure material, running it into the desired moulds, and pressing it out when cold and solid.

Fraction No. 4.—This contains chiefly creosote oil, but, as mentioned in the last chapter, it is sometimes advisable to start collecting this fraction a little earlier so that Fraction No. 3 is then richer in tar acids. In this case the fourth fraction can with advantage be redistilled, to collect a small first runnings fraction again containing a fairly high percentage of tar acids and all the phenol. By this means slight expense is entailed by the additional distillation, but it is saved by having a cleaner oil to wash and less bulk of oil to treat with caustic soda.

The distillation is carried out in iron stills similar in all respects to the tar pot stills, with a fractionating column attached. Here again the type and efficiency of the column employed are of great importance.

The redistillation of the creosote is in no way different from the other distillations, and needs no further description. It presents, however, one slight difficulty, in that there is always a certain amount of trouble in distilling off the water. Arrangements are generally made for sweating this off direct down the still-head pipe, and passing the vapours into the column after the water is removed. Difficulties will be experienced if it is attempted to work the whole distillation through the column. The creosote fraction off the tar frequently has a specific gravity nearly approaching that of water, so that troublesome emulsions are formed and the water is difficult to separate. Quite frequently as much as 5 per cent of water has to be distilled off.

Fraction No. 5.—It is not usual to redistil this fraction except in a few instances where a special heavy yellow oil is required. In these cases the ordinary pot tar still is used, and the amount taken off is regulated to suit the specification of the required oil.

STEAM DISTILLATION

All the hydrocarbons occurring in the distillates from tar may, if desired, be distilled by means of steam. The use of steam for distilling the anthracene oil has already been mentioned, and this is generally the only fraction that is dealt with in this way. No particular economy or advantage is gained by using steam except in cases where decomposition is liable to commence or where high temperatures would otherwise have to be dealt with.

The effect of using steam is simply to lower the temperature at which the mixture distils, whereby decomposition is either prevented or hindered. Also by keeping down the temperature in this way wear and tear on the stills is reduced.

The steam distillation is carried out in the ordinary way, and the only precautions that are necessary are to see that the steam is dry, and that it is introduced into the contents in several places, so that intimate admixture of steam and oil and good agitation are obtained. The external heat on the stills should also be sufficient to avoid condensation of steam. Steam distillation can be used with great success for the prevention of "bolting." For instance, when, by careless firing, the tar in an ordinary distillation bolts or boils over, the trouble can often be quickly overcome by letting in live steam through the perforated pipes as shown in Fig. 152.   The result is a rapid reduction of the temperature of the still contents, when the distillation again proceeds in a normal manner.

Anthracene.—The separation of the anthracene is similar in many respects to that of the naphthalene.   The oil is cooled and the solid anthracene collected in filter presses.   The filter cake is further dried by hot pressing or centrifuging, and the resulting dark green semi-crystalline mass contains about 40 per cent of pure anthracene.

CHAPTER XXXVIII

FRACTIONAL SEPARATION OF THE NAPHTHAS AND LIGHT OILS

Naphthas and Light Oils.—As has been explained previously, the first distillates from the crude tar are the naphthas and light oils which come over from 80° C. to 200° C. They consist chiefly of the hydrocarbons of the benzene series ranging from benzene itself to the mesitylenes. The separation of these is carried out entirely by fractional distillation, and, owing to the similarity in chemical composition and, in some cases, the slight difference in boiling point, the problem is by no means a simple one. Very efficient apparatus is required, and probably there is no instance where, technically, more attention has to be paid to the type of plant installed.

Products required.—The products that the distiller has to prepare in a pure state are—benzene (b.p. 80·4° C.), toluene (b.p. 110·6° C.), and sometimes ortho-xylene (b.p. 141·5° C.), meta-xylene (b.p. 139° C.), and para-xylene (b.p. 138·5° C.). Generally speaking, however, the last three (if they are required at all) are made and sold as a mixture of the three isomers with a boiling-point range of 2 to 5°. Commercially pure benzene, which is used for making nitro-benzene, synthetic phenol, and other intermediates, must boil within a range of 0·5° C. and contain only small quantities of thiophene and other impurities. Pure toluene, for the manufacture of trinitro-toluene and dye intermediates, must also pass a similar specification.

Prior to the Great War neither of these products were prepared to any great extent in this country, and it was only through the urgent need of them for the manufacture of high explosives that distillers realised the importance of efficient apparatus. The quantities of benzene and toluene produced during the war were far in excess of the demands of the dye trade in normal peace times, so that there is now no further need to separate them to any large extent into the pure hydrocarbons, but they can be sold as a mixture for motor fuel. The xylenes are occasionally refined into a mixture boiling within a few degrees, where there is a demand for the production of intermediates, but generally they are sold in the form of solvent naphtha distilling 90 per cent at 160° C., a product which is a mixture of toluene, xylenes, mesitylenes, and other homologues.[1]

[1] "By distilling 90 per cent at 160° C." is meant that, on distillation, 90 per cent of the naphtha comes over between 120° C. and 160° C.

First Distillation.—The crude naphtha, or light oils whether separate or mixed, is first of all distilled in an ordinary fire still. This is similar in construction to the pot tar still already described, except that it is not generally so large and is fitted with a good fractionating column. It is found that a still of 2000 to 3000 gallon capacity is large enough for most purposes, as one of this size can usually be worked off in a day. Here again, economy can be effected by interposing a heat interchanger as condenser, in which the coil carries the vapour from the column, and the cooling medium is the naphtha for the next charge in the still. The column is of the type of any of those described later, except that, owing to the higher working temperatures, no dephlegmator is used. The object of this first fire distillation is to make a primary split of the distillate into fractions suitable for subsequent distillations, some of which can be carried out in a steam still and some in a fire still.

Fractions obtained.—The fractions generally obtained are three,

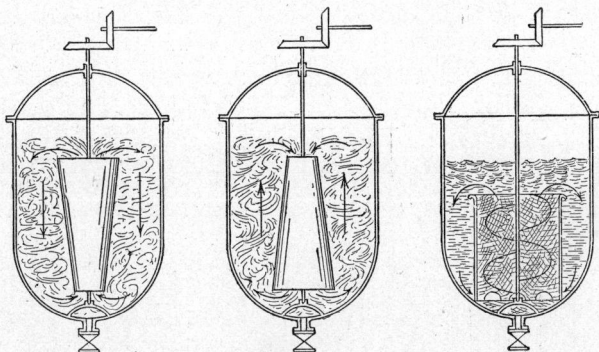

FIG. 162.—Washers.

namely, crude benzole, crude solvent, and crude heavy naphtha. These, when redistilled, should have the following approximate boiling-point ranges : in the case of crude benzole 60 per cent should come over below 120° C. ; in the case of crude solvent 60 per cent from 120° to 160° C. ; and in the case of crude heavy naphtha 60 per cent from 160° to 200° C. A few experiments with the plant in use will readily show the temperatures at which the changes from one fraction to another are to be made. The residues from this primary distillation are composed of creosote and naphthalene and are consequently treated in the manner already described (p. 388).

Washing the Naphthas.—Before the various fractions can be further distilled it is necessary to eliminate from them the numerous impurities they contain. The purification is carried out by violently agitating the spirit with sulphuric acid and afterwards caustic soda.

Washers.—Various forms of washers or mixers are used, some of which are shown in Fig. 162. They are made either of wrought iron lined with lead, or of cast iron, and are of greater capacity than that of the stills in use, so that sufficient room is left in the washer to agitate a full still charge. Of the two metals, cast iron, in the experience of the author, has been found to be the more economical, as in practice very little action takes place between this metal and the acid. Many kinds of agitators have been devised, from the simplest air-blowing form to those illustrated, which are probably the best. A very efficient type is an elaboration of the cone agitator where two cones, one at the top and one at the bottom, are fitted on the same spindle. The washer should be so designed that the bottom is well dished or conical, in order that the last traces of acid and alkali can be drawn out without loss of naphtha.

Some distillers prefer that two washers should be employed, one in which to conduct the acid wash, and the other the caustic soda and the final washes. If the acid and soda washes are conducted in one washer, there is a great risk of some of the resinous matter, produced by the action of the acid, adhering to the sides of the washer and suffering decomposition by the caustic soda, the products then being dissolved by the spirit.

Distillates obtained from naphthas, rendered impure in this way, possess a peculiarly unpleasant odour and rapidly go off colour.

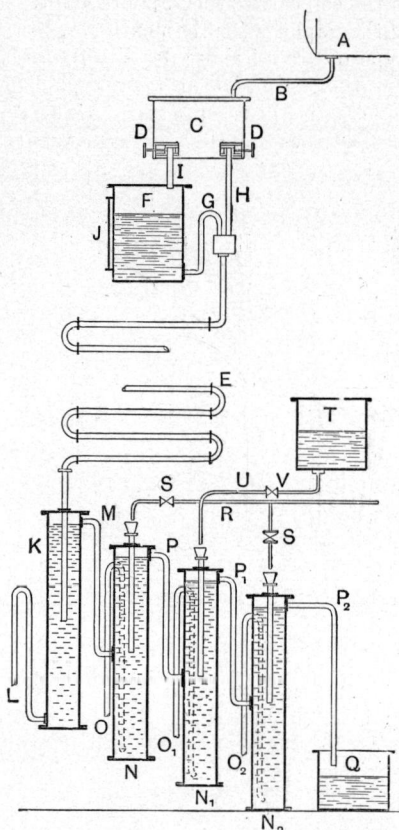

FIG. 163.—Continuous Washing Plant.

Continuous Washing.—Several attempts have been made to make the washing of naphtha or crude benzole a continuous one, and the arrangement of Blyth and Miles (Eng. Pat. 123839, 1918), as manufactured by Newton Chambers, Ltd., may be described. The invention is not only applicable to the washing of naphthas, but may be used for the purification of petroleums or the extraction of tar acids from creosotes by means of soda.

As shown in Fig. 163, A is a storage tank connected by a pipe B to a regulating tank C, provided with two outlets controlled by valves D, or similar devices, one of which controls the flow of benzole to the upper end of a coiled or convoluted pipe E, the other to a tank F, containing sulphuric acid or other reagent. The lower part of this tank F is connected by an overflow pipe G to the upper end of the pipe E. If now the valves D are opened, benzole will flow from the tank C directly into the pipe E, through the pipe H ; it will also flow through the pipe I to the tank F, and accumulate on the surface of the reagent contained therein, whereby an amount of the reagent corresponding to the weight of benzole thus admitted will be forced through the overflow pipe G into the pipe E. Thus the proportions of benzole and sulphuric acid (or other reagent) supplied to the pipe E can be determined by suitable adjustment of the valves D. If desired, the tank F may be duplicated or divided into separate compartments, so that when one tank or compartment is emptied of sulphuric acid or other reagents the other may continue in use whilst the first is recharged with the reagent. The tank F is provided with a gauge glass J.

Having passed through the coil E, in which the acid or other reagent effects the purification of the benzole flowing along in contact with it, these materials enter a receiver K, from the lower end of which the spent reagent is drawn off through an overflow pipe L. The benzole passes from the upper part of this receiver K, through a pipe M to a washing vessel N, which it enters at a point below the level of the water or purifying reagent contained therein. The foul water or spent reagent is drawn off from the bottom of this washing vessel through an overflow pipe O, and the benzole is drawn off from the top of this vessel through a pipe P. The overflow pipe O extends upwards from the bottom of the vessel N, to a sufficient height above the inlet of the pipe M. The benzole passes successively through a number of such washing vessels N_1, N_2, provided with overflow pipes O_1, O_2, and benzole pipes P_1, P_2, and is finally delivered to a purified benzole receiving tank Q.

In the arrangement illustrated for the treatment of benzole the washing vessels N, N_2, are supplied with water from a pipe R, in quantities regulated by valves S, and the vessel N_1 is supplied with caustic soda solution from a tank T, through a pipe U, under the control of a valve V.

With this apparatus, the regulating valves D having been once adjusted, it is only necessary to maintain the supply of unpurified benzole to the tank C, the supply of sulphuric acid to the tank F, the supply of caustic soda solution to the vessel N_1, and of water to the vessels N, N_2, in order to obtain a continuous flow of purified benzole into the receiving tank Q.

If desired, the material treated may be passed in succession through more than one proportioning system such as that comprising the tank F, and coils E, and any number of washing vessels such as the vessel N may be provided connected one to the other in succession.

Impurities to be removed.—The naphtha impurities that have to be extracted are phenolic bodies, pyridine bases, and numerous other complex bodies chiefly containing sulphur. The first two are valuable and therefore have to be saved.

The naphthas, before they are sent to the refinery, should have had all the phenols extracted from them (p. 384), but generally traces are still present. If the quantity is greater than about 2 per cent, further washing with soda is necessary, otherwise difficulty will arise in extracting the bases with acid. This is possibly due to the acid forming sulphonation products of the phenols which in some way prevent the solution of the pyridine in the acid.

In most cases, however, the naphtha may be directly agitated with dilute acid (sp. gr. 1·300) which extracts the pyridine bases in the form of pyridine sulphate.

The next agitation is with concentrated acid to remove the impurities, and generally this has to be repeated several times until the required purity is obtained. The quantity of acid used varies of course with the quality of the crude naphtha and the purity of the finished article that is required, but may range from 2 to 10 per cent of the spirit. Slight economy in washing may be obtained by using the pyrites brown acid (B.O.V. 140° Tw.) in the primary washings and finishing off with (C.O.V. 168° Tw.).

The alkali wash, which is generally preceded by a washing of water to remove the majority of the acid, is caustic soda of the usual 16° to 20° Tw. strength ; and sufficient is used to ensure the complete extraction of all phenolic bodies and the neutralisation of traces of acid that may still be present. Finally, the naphtha is agitated several times with water, so that it may be quite neutral before it is run into the still. The first water wash after the acid and soda should be only slightly agitated, and even so objectionable emulsions are liable to be formed. In such cases it is advisable simply to allow the water to trickle down through the naphtha.

The dilute acid wash, as has been mentioned, extracts the pyridine bases, and the recovery of these is generally carried out in conjunction with the manufacture of ammonium sulphate. The ammoniacal liquor obtained from the dehydration of tar is boiled in good contact stills or columns with excess of lime, whereby ammonia is set free. This is passed into the pyridine acid or sulphate until the solution is neutralised. After well settling, the ammonium sulphate solution is run off and concentrated until the salt crystallises. The crude dark brown pyridine is refined by two or sometimes three distillations, and dehydrated by means of lump caustic soda. The refined bases have to pass a rigid specification and are sold for denaturing alcohol. The agitation with concentrated acid produces a dark brown or red viscous mass which is known as " acid " or " vitriol " tar. This product has been the subject of innumerable researches and patents, but little of practical importance has yet been discovered. Even as a waste product there is great difficulty in disposing of it. If deposited into a river it kills all fish life, and if burnt it produces so much sulphurous fumes that generally factory inspectors object. One of the best methods of dealing with it is to boil it up with the water washes to extract all free acid, the solution then being diluted to the correct strength for subsequent washing of naphtha for pyridine. The residue, after cooling, is a pitchy

substance which floats on the top and can be disposed of without
further treatment by burying it in a tip.

Sources of Benzene and its Homologues.—The chief source of
benzene, toluene, and solvent naphtha is not coal tar but coke oven gas
and coal gas, from which these substances are extracted by washing
with various solvents such as tar or special creosotes. Some idea of
the proportion of crude benzole obtained from the various sources can
be obtained from the following interesting statistics :—

Production of Crude 65's Benzole from Coke Oven Plants

1913 (estimated)	18,920,000 gallons.
1914 (actual)	21,877,000 ,,
1915 ,, ...	25,148,000 ,,
1916 ,, ...	31,081,000 ,,
1917 ,, ...	33,552,000 ,,
1918 ,, ...	32,162,000 ,,

Figures compiled by the Ministry of Munitions for the crude
benzole produced by gas companies during the latter part of the war
period are as follows :—

1915 (September to end)	91,533 gallons.
1916	3,828,555 ,,
1917	6,694,340 ,,
1918	7,146,115 ,,

No figures are obtainable for the crude benzole produced by the tar
distilleries, that is to say, directly from the distillation of tar, but it is
estimated that these distilleries make approximately 4,000,000 gallons
per annum, and this quantity does not vary very much from year to
year. It will be seen that the height of production was reached in the
years 1917 and 1918, when about 44,000,000 and 43,000,000 gallons
respectively were produced from all three sources. Since this time the
production has rapidly dwindled down until in 1920 the estimated
production from all sources was not much more than one half of this
amount. The reduction in output is chiefly from the coke ovens and
gas companies, as the extraction of the benzole at the present time is
not a paying proposition.

M. Grebel [1] gives the following interesting figures about the production of crude benzole in France :—

Production in 1918

Coke ovens	2,100,000 gallons.
Gasworks	3,250,000 ,,
Tar distilleries	25,000 ,,
	5,375,000 ,,

Production in 1913 from coke ovens was 2,600,000 gallons. With

[1] H. Grebel, *Gas Journal*, 1920, **149**, 364.

proper legislations and enterprise he foretells that the annual productions could be increased to—

Gasworks	7,000,000	gallons.
Coke ovens	17,000,000	,,
	24,000,000	,,

Fractionation of Naphthas

Plant required. — The plant required for fractionating the naphthas may be conveniently described in four separate parts as follows :—

1. The Still, or heating medium, in which the liquid is converted into vapour.

2. The Column, or scrubbing tower, in which the vapour and condensed liquid are brought into intimate contact.

3. The Dephlegmator, analyser or constant temperature still-head, in which the vapour is partially condensed and passed back into the column.

4. The Condenser and Receiver, where the desired vapour is finally condensed and received.

1. The Still

It is not necessary to further describe the fire still which is used for the primary distillation of the naphthas, and again for the preparation of the refined heavy naphthas. The plant is simple and has been fully discussed earlier.

The steam still is simple in construction and can be made of either wrought or cast iron, and of any size from 2000 to 10,000 gallons capacity according to requirements. The shape of the still is usually a horizontal cylinder, although any convenient second-hand iron vessel will suit. This latter is often used, as the wear and tear on it is very small.

The coils, which must be of sufficient length to give the required heating capacity, are made of cast iron in sections, or preferably of wrought iron in one piece so that only two joints are required, one at the entrance and one at the exit. The cost of installing these wrought-iron coils is greater, but considerable economy is gained by saving heat, and also the chance of leaks is thereby reduced.

Copper coils should not be used, for although they are recommended by some makers, they do not work out economically. Their initial cost is high, and for some unknown reason they rapidly corrode and wear thin.

The slightest leakage of steam into the still is fatal at stages in the distillation when the very intensive fractionation is being carried out. In a large still several sets of coils are used so that more economical regulation of temperature can be effected. The outlet of the coil is connected with a steam trap so that constant pressure and temperature are maintained in the coil. The still is also supplied with a raw steam inlet at the bottom which enables the solvent naphtha to be brought off by steam, as the temperature required for this is too high for the steam boiler pressure found in most works. There is a charging pipe from the washer. The end of the still has a large manlid, generally the whole width of the still, which permits of easy cleaning and also allows the one-piece steam coil to be taken in and out. The column either rests directly on the still, the vapour passing into it through the large outlet hole, or is separated from it, the vapour being then conveyed through a wide connecting pipe.

2. The Column

This part of the plant is by far the most important of the whole apparatus, and too much attention cannot be given to it. Its function is to attain intimate contact between the vapour and condensed liquid.

It has been shown by Rosanoff and Bacon [1] that the washing or scrubbing in a column is quite apart and distinct from the cooling attained in a dephlegmator or analyser and is far more important. So that although dephlegmation is an important adjunct to a fractionating apparatus it is essential that the column should be of sound construction before it is added.

The evolution of the present-day column, so far as coal tar distillation is concerned, is an interesting one and may be described in order of employment as (a) Packed Column, (b) Perforated Plate Column, (c) Cup and Seal Column.

(a) **Packed Columns.**—The earliest types of columns consisted of a cylinder of wrought or cast iron in which are promiscuously thrown bricks, stones, or any material that will break up the vapours and so cause intimate contact between them and the condensed liquid ; such a crude type of plant was patented by B. Hoff (Eng. Pat. 3468, 1889).

Raschig's Ring Still.—F. Raschig (Eng. Pat. 6288, 1914 ; and Ger. Pats. 286122, 1915 ; 292622, 1916 ; 297379 and 298131, 1917) further developed the idea of packing columns by employing his rings ; to which reference has been made in a previous chapter. These rings are pieces of thin-walled piping, the length and diameter of which are equal. They are thrown into the column and allowed to arrange themselves promiscuously just as they fall into position. A very uniform distribution is thus obtained which does not tend to assist the formation of particular channels for the vapours. The rings should be of such a size (25 mm.) that not less than 55,000 go to a cubic metre. The surface exposed per cubic metre of volume with such a filling amounts to about 300 square metres. Further, as little as 10 to 20 per cent only of the volume is actually occupied by the material of the rings. The column, moreover, offers little resistance to the vapours, e.g. a column 10 metres high and of 1 square metre cross-section will pass 500 cubic metres of vapour per hour under a pressure of only 2 mm of water.

---

[1] Rosanoff and Bacon, *J. Amer. Chem. Soc.*, 1915, **37**, 301.

One of the chief points claimed in Raschig's Patent is the irregular arrangement of the rings in the column, whereby both vapour and liquid in their passage are compelled to change their direction so frequently that their intimate and frequently repeated contact is assured.

FIG. 164.—Raschig Rings.

Other Packed Stills.—Modified forms of packing rings have been devised by Goodwin (Eng. Pat. 110260, 1917), Lessing (Eng. Pat.

FIG. 165.—Lessing Rings.

139880, 1920), and Prym (Ger. Pats. 317166 and 317167). Somewhat similar hollow balls and cellular bodies were described by Guttmann (Eng. Pats. 14774, 1896 ; and 4407, 1907).

Goodwin's rings consist of "units constructed or shaped in the fashion of two hollow truncated circular cones, united at their smaller diameter, at which part both the interior and the exterior junction points are rounded and smoothed off, leaving a unit substantially the same external shape and internal section as a dice-throwing box, but, of course, open at both top and bottom " (Fig. 166).

Such rings are filled into a fractionating column either regularly or irregularly. The vapours in passing through this filling are constantly throttled and allowed to expand, their velocities being thereby alternately increased and decreased with the resultant formation of eddies which ensure efficient contact between vapour and liquid.

Lessing's and Prym's rings differ from others in having " a more or less central partition," in consequence of which they offer a larger surface without materially increased obstruction to the passage of the vapour and liquid " (Fig. 166).

Packed columns are convenient in some respects in that they are cheap and easy to erect and the rings are very light. They fail, however, in one particular as they do not regularly bring vapour and liquid together. It stands to reason that where the packing is promiscuous the ascending vapour will go by the passage giving the least

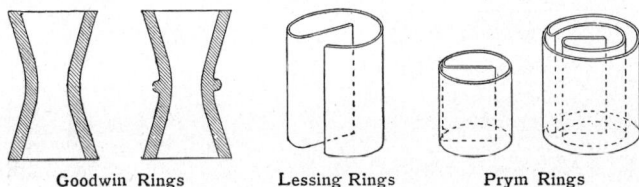

Goodwin Rings   Lessing Rings   Prym Rings

FIG. 166.

resistance. Again, the descending liquid will come down the channel of least resistance. It may easily happen that these two passages are not the same, so that regular and intimate contact is not assured. This so-called " channelling " is more marked when the column has been used for some time and is dirty, and chiefly in this fact lies the alleged improvement in using the various kinds of rings described instead of such articles as brickbats.

Packed columns can well be used for crude distillation, but the tendency to channelling makes them of little use for intensive fractionation compared, at any rate, with the more modern cup and seal column.

(b) **Perforated Plate Columns.** — This column (Fig. 167) is simple in construction and is composed of a cylinder of the required height containing horizontal partitions throughout the whole length. The partitions are bored over the whole surface with small holes. The number and size of holes are variable and made according to the rate at which the still is required to be worked. They must be of small enough area to keep a certain quantity of liquid on each tray so that the ascending vapour continuously bubbles through this liquor. The earlier inventors of this class of columns were Coffey and Savalle.

In the actual working of a perforated plate column difficulties arise,

2 D

for if the still is run faster than the rate for which the holes are designed, the sections begin to fill up ; while on the other hand, if it is worked too slowly the whole column drains down, leaving no liquid on the plates. This difficulty was partially got over by inserting overflow pipes in each

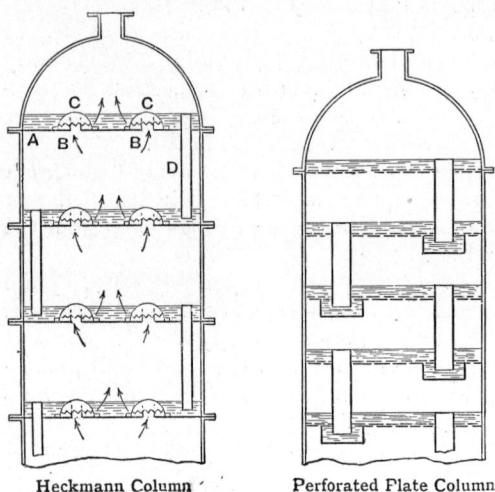

tray, the upper end of each pipe protruding some few inches above the level of the plate. Each tray was supplied with one overflow pipe which dipped into the liquid on the tray below it, this tray having its overflow on the opposite side of the column. Thus the returning liquid had to traverse a zigzag course on its way down the column. D Savalle et Fils, of Paris,[1] erected such an apparatus and obtained good results.

Heckmann Column   Perforated Plate Column

FIG. 167.

A, fixed plate ; B, ascending vapour pipe ; C, hood, dome, or cup ; D, return pipe.

Many distillers, as seen later, still prefer the perforated plate type of column, and maintain that it gives as good results as the cup and seal column. In the opinion and experience of the author they are not so good in practice. To carry out efficiently a separation by fractionation it is necessary to alter the rate of working frequently, and with any kind of perforated plate column it is impossible to slow the still down to a lower speed than the area of the holes will allow, otherwise the plates will drain. The only method of overcoming this difficulty is to reduce the hole area, but the still cannot then be worked at high speed without getting an abnormal amount of pressure on the column. From this it will be seen that the flexibility of any individual column is not large, and it must be worked at a certain limited range of speed to attain the best results. In most cases it will be found that great difficulty is experienced to obtain a constant pressure in the steam boiler, so that varying temperatures are obtained in the heating coils of the still. This naturally tends to vary slightly the speed of flow of distillate and is fatal to a column, such as the one described, that is so sensitive to changes of speed. This trouble can be overcome by introducing a reducing valve before the steam enters the still, but only at a big expense of fuel. Again, if the column is erected in the open it must be covered, otherwise winds, rain and other atmospheric conditions very quickly alter the amount of condensation in the column.

[1] *Bulletin de la Société d'Encouragement*, 1876, p. 657; *Dingl. pdyt. J.* 223, p. 615.

(c) **Cup and Seal Column.** — The difficulties experienced in practice in the use of the perforated plate column induced makers to devise a different form, and Heckmann (Ger. Pat. 39557) may be said to be the first to develop the cup and seal column, which is so efficient that it has, with slight alterations, come into general use.

Heckmann Column.—Fig. 167 shows a sectional diagram of the column in which A is a fixed plate with several (according to the diameter of the column) ascending vapour pipes B. The vapour is prevented from being carried direct into the next section by means of the inverted hood, dome or cup C. The cups, which have serrated edges, are partially immersed in the condensed liquid, the depth of this liquid on the plate being regulated by the height of reflux pipe D.

The vapour when it strikes the dome C is forced down and through the liquid on the tray, and so an actual scrubbing action between liquid and vapour is attained. The return pipes D are placed on opposite sides of the column in each tray and dip into the liquid in the tray immediately below. Thus the returning liquid flows across each tray, and the ascending vapour can only rise through the pipes B.

Egrot's plant was devised on similar lines to that of Heckmann, and was used very successfully in France in the 'nineties of last century, and was then considered a great improvement on Savalle's perforated plate column. Many patents have been brought out since, making slight variations on the cup and seal column. For instance, Metallwerke Nehien A.-G. (Ger. Pat. 294781, 1914) introduced baffles on each plate of the column to ensure the ascending vapour taking a more zigzag course through the liquid on the plate. The improvements effected are not, however, very noteworthy.

The practical advantages of the cup and seal column are that the column is not so sensitive to irregularity of working caused by variation in atmospheric conditions and by change of steam pressure. It can also be worked at almost any speed with efficiency, although, of course, as in all cases of distillation, the slower the speed within certain limits the better the fractionation. The inverted cup plays an important part in the column, as there is no doubt that the vapour, as it bubbles through the liquid on a tray, mechanically carries with it a certain amount of that liquid to the next tray above. The result is that there is a continual ascension of liquid of too high a boiling point, being mechanically carried upwards with the vapour. The inverted cup lessens this action, as it allows a less violent ebullition on the trays.

3. Dephlegmation

Dephlegmation is the word usually employed at the present time to describe the action of an apparatus above the column which condenses further portions of the vapour and returns them to the column.

Dephlegmators, analysers, reflux condensers or constant temperature still-heads, as they are variously called, may all be said to perform the same function of dephlegmation. No plant can be thoroughly efficient without some form of dephlegmator, although makers have often paid too much attention to the dephlegmator and too little to the column itself. Unless the construction of the column is correct no amount of complicated dephlegmating apparatus will produce good results, but the careful combination of the two is essential.

The usual form of dephlegmator is shown in Fig. 168, in which it will be seen that the vapour passes through the nest of tubes and is partially condensed by water surrounding them.

It is found practical, in some cases, to allow the vapours to go outside the tubes and the water inside as the tubes can be more easily cleaned of lime deposit.

E. Bonnell (Fr. Pat. 395040, 1909) employs hexagonal tubes immersed in a water bath, by which he obtains greater cooling surface without unduly increasing the size of the apparatus.

C. Still (Eng. Pat. 3269, 1911) uses a horizontal dephlegmator so arranged that the vapours pass in at the top and traverse a zigzag course downwards while the water flows in the reverse direction.

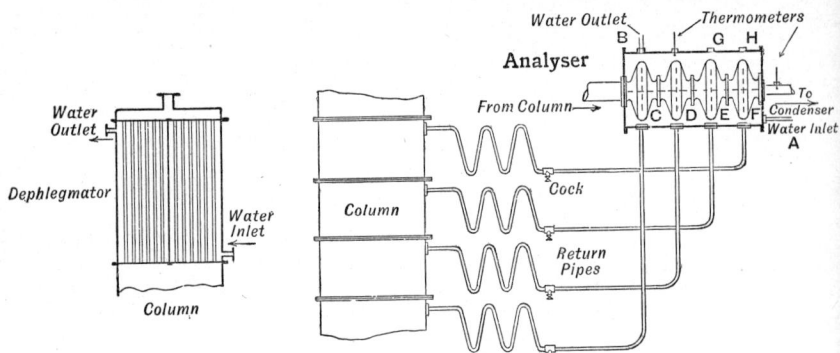

FIG. 168.—Dephlegmator or Analyser.

Other patentees of note of constant temperature still-heads and dephlegmators are—Rosanoff (Fr. Pat. 443054, 1912), Chenard (Fr. Pat. 443499, 1912 and 1914), Allen (Fr. Pat. 481134, 1916) and Aylsworth (U.S. Pat. 1250760), who regulates the temperature of his still-head by means of a thermostat.

The usual form of analyser, as shown in Fig. 168, is slightly different in principle from the usual dephlegmator in that separate condensates are obtained. Each condensate is then returned separately to a different part of the column. If the water is allowed to flow in at A and out at B, the heaviest portion of the vapour will be condensed in C, and the lightest in F. The condensate in C is, therefore, returned to a section about half-way down the column, and in D, E, F, respectively to successively higher sections in the column.

Many makers, and particularly some of the earlier ones such as Coupier, Vedle and Egrot, paid great attention to their analysers and maintained that they were the chief factor in obtaining their good results.

The author has experimented with a plant with an ordinary perpendicular tube dephlegmator and horizontal analyser, as described (p. 404), attached to a cup and seal column. The results showed that the analyser did practically no work at all and accounted for no extra separation whatever. The reason may have been that the plant was working so efficiently that before the vapours reached the analyser the separation was so complete that any further change was negligible ; at all events, this horizontal type of analyser can be of little practical value. The principle of all these forms of dephlegmator is exactly the same. Water is usually the cooling medium, although in come cases the raw material for the still charge is used in order to obtain a heat interchange. The inflow of cooling liquid is regulated so that the constant difference of temperature between cooling liquid and vapour is about 20° to 25° C. If this difference is maintained, a sufficient quantity of vapour is condensed and refluxed back into the column, thus ensuring a good descending flow of liquid through the column and a consequent intimate contact between vapour and liquid.

Combined Column and Dephlegmator.—Many attempts have been made to combine the action of a column and dephlegmator in one. Miller (Eng. Pat. 25469, 1901) interposes dephlegmators between each section of the column. Vallat (Fr. Pat. 376872, 1907) inserts horizontal water tubes in each section of the column. Lummas (U.S.A. Pat. 1216334, 1917) on the other hand surrounds the sections of the column with water coils. Wright and Atwood (U.S.A. Pat. 1278279 and 1278280) go still further, and besides surrounding the sections with coils they make arrangements to obtain, if desired, separate fractions off each. Barbet et Fils (Fr. Pat. 469979, 1913) insert water tubes in the liquid which collects on each tray in the column whereby more delicate control can be attained.

Other Stills

Perrier's Still.—Perrier (Fr. Pat. 409041, 1909) invented a most ingenious device in which he entirely did away with column and dephlegmator. Two horizontal shafts, each carrying a series of discs dipping into the condensed liquid, rotate at a high speed in opposite directions in a horizontal oval drum through which vapours from the still pass. A series of partitions divides the drum into chambers, so that the vapours pass through the chambers in turn and are subjected to the spray produced by the rotating discs.

Foucar's Still.—Foucar's " differentiator " column (Eng. Pat. 19999, 1908) is a double-walled cylinder, between the two walls of which a spiral is set. The vapours from the still pass into the space between the two walls and are there fractionated. Vapour may also be passed to the inside tube where it is refluxed back to the still. This maintains a regulating temperature on the column. The liquid descends on the spiral and is continuously brought in contact with ascending vapour. Wonderful results have been obtained with this column on a small scale, but engineering difficulties of manufacture seem to be the cause of its partial failure in works practice.

It is doubtful whether any of these ideas are of much practical value as they have not been adopted by present-day distillers.

4. Condensers and Receivers

Little need be written about these as the apparatus is so simple. Any tank containing a cast iron or drawn wrought iron coil will meet the requirements of a condenser provided it has cooling surface enough to condense all vapours completely. After leaving the condenser the liquid usually passes through a sight glass, so that the rate of flow can be noted by the operator. The receivers are arranged in duplicate and should not be too large, otherwise an unnecessary loss will be involved. For instance, if the still is just finishing to run pure benzene and a 200-gallon receiver has in it 150 gallons, it may not be possible to notice that the benzene is off until the receiver is full, so that the whole 200 gallons will have to be reworked. If the receiver had only been of 50-gallon capacity probably the expense of redistilling at least 150 gallons would have been saved. The receivers are connected directly to the storage tanks and also to a series of drums or other small tanks. When the whole charge is off, sampling and testing of these small receptacles will enable the whole distillate to be split up into the required fractions.

DESCRIPTION OF SOME OF THE MODERN PLANTS

Intermittent Types.—Hodgson Thomas Plant.—The intermittent rectifying plant of Hodgson Thomas, Manchester, as shown in Fig. 169, is one that gives good results.

The still is heated by steam passing through a nest of tubes fixed close to the bottom of the still and divided into two separate units. The column, upon which the efficiency of a still mainly depends, consists of a number of cast iron circular sections into each of which is fitted a loose plate, carrying four serrated caps. A liquid return pipe is fitted into each plate and is so arranged that a good seal is secured. The dephlegmator is carried on the top of the tower and is independent of it; it is built up of a number of tubes fitted into expansion tube

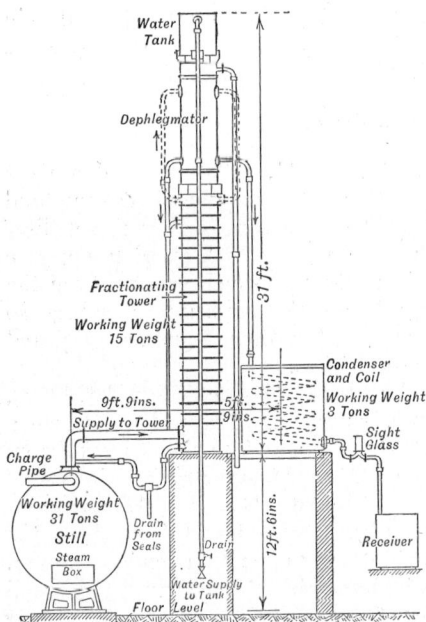

FIG. 169.—Hodgson Thomas Plant.

plates, the cooling water passing through the tubes. Baffles are arranged to break up the vapours which enter the top of the dephlegmator, and an outlet is provided for carrying the heavier fractions back to the tower. By dispensing with an analyser the initial cost of the plant is reduced and there is of course a constant economy in the use of water. The condenser is a large circular tank fitted with coil and stands, and of sufficient cooling surface to prevent waste of vapour. To the outlet of this coil is attached a special sight glass consisting of two glasses, one of which is graduated so that the rate of flow can be readily estimated, and being arranged at a convenient height the glass is visible from some distance. The receiver is made of cast iron ; it is circular in shape and fitted so that the distillate may be either run into barrels or direct into storage tanks.

The cost of the plant maintenance is said to be low, whilst its efficiency is certainly high and there is an almost entire absence of pressure when working. The plant is also very adaptable, as with slight alterations it can be used for most coal-tar products. From a charge of crude benzole (60 per cent to 65 per cent below 120° C.) the plant will produce a good standard product (95 per cent below 90° C.) in one operation, the rate of distillation varying between 120 and 150 gallons per hour according to the available steam pressure. Should pure benzole be required, the plant will give, in one distillation of a standard product at the rate of 100 gallons per hour, a distillate with a range of 0·2 to 0·5° C., the quantity of water used in the dephlegmator varying between 100 and 120 gallons per hour. All the valves are operated from the floor level, and the whole plant can be shut down by operating one valve.

Some makers still prefer the perforated plate type of column and maintain that it gives as good results as the cup type. Two such manufacturers are—The Chemical Engineering and Wilton's Patent Furnace Co., Ltd., and R. & J. Dempster, Ltd. Their columns are very much alike, so that it is only necessary to describe one of them.

The Dempster Plant.—R. & J. Dempster carry out the refining in two stages, the first to split up the crude benzole into primary fractions, the second to convert the primary fractions into pure products.

The primary fractionating still, as shown in Fig. 170, is constructed on the same principle as the refining still, without the analyser. It is used for first distillations or for the production of crude or unseparated fractions, and is also admirably adapted for blowing over crude benzole, and will produce 90's benzole from washed products. The working is the same as described for the final still, bearing in mind that there is no analyser.

The final fractionating still (Fig. 171) produces pure products and should not be used for unwashed oils. In actual practice stills of this construction have produced regularly products testing to 100 per cent with a range of less than 1° C. They are very speedy in their action and are very easily controlled by skilled operators.

The still bottom A is charged with the required amount of oil, which is vaporised by steam at high pressure contained in a series of longitudinal tubes placed in the bottom of the still. The vapour passes up the fractionating column, which is made of sufficient height to condense and separate all the heavy and intermediate fractions. From the top of the fractionating column the vapour passes into the analyser, which contains a series of water pipes controlled by cocks, where the final separation takes place, leaving vapour which, on condensation, yields a pure product. This vapour passes down to the final cooler

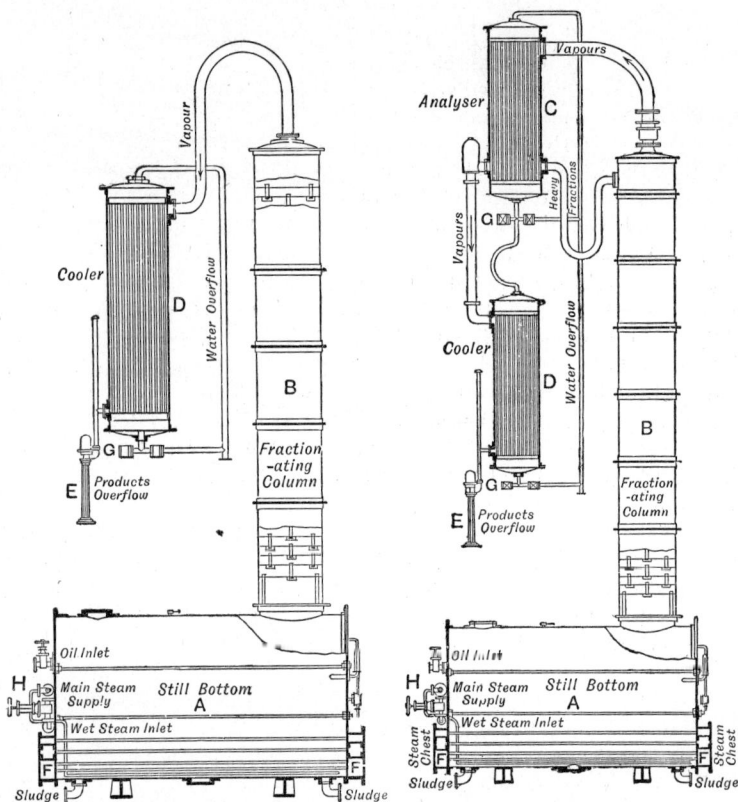

FIG. 170.
Dempster Primary Fractionating Still.

FIG. 171.
Dempster Final Fractionating Still.

and, after condensing, the refined product passes out through an ordinary overflow column E which renders the flow of oil visible to the operator. The analyser can be fed with water, either warm from the cooler below or cold from the water main, and the depth of the water in the analyser can also be controlled. The still is fitted with front and back steam chests, steam control valves, oil control valves, and the usual vacuum pressure valves. Thermometers and pressure gauges are attached to the front of the still to guide the operator in working, and thermometers are attached to the analyser where efficient control is very necessary.

It is essential that in designing the fractionating column the maker should have in view the products to be obtained, the diameter and the pitch of the perforations in the various trays being regulated accordingly.

Blair, Campbell & M'Lean Plant. — Messrs. Blair, Campbell & M'Lean, Ltd., of Glasgow, manufacture a rectifying plant which gives good results (Fig. 172). Some of the special features are the facility with which the steam heating coils may be withdrawn from the still, when any repairs are necessary, without dismantling the whole still. The column is made with plates having bells and dip pipes of extra large area to prevent blocking. The dephlegmator, the rectifier, and the condenser are constructed in such a way that repairs and replacements of tubes, when necessary, can be carried out with the minimum of trouble. The whole still is of the most substantial design and is provided with all the necessary thermometers, gauges, etc., to enable the various operations to be carried out with the greatest ease and under perfect control of the operator.

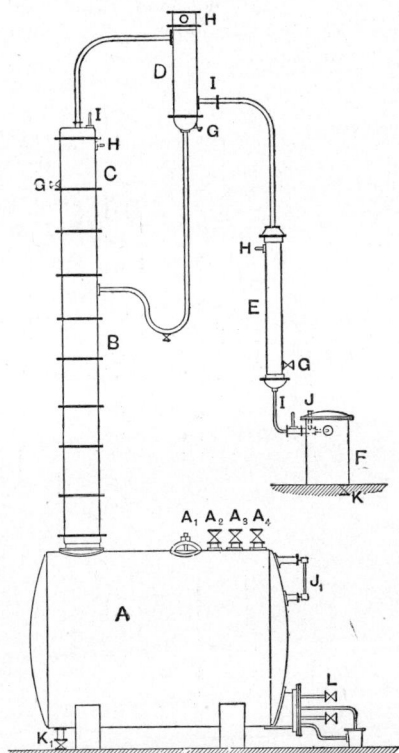

Continuous Types. — It is only within the last few years that plants of practical value for the continuous fractionation of crude benzole or naphthas have been introduced. They are economically far superior to the intermittent

FIG. 172.—Blair, Campbell & M'Lean Plant.

A, rectifying still ; A_1, manhole ; A_2, inlet valve ; A_3, safety valve ; A_4, air-cock ; B, rectifying column ; C, tubular dephlegmator ; D, tubular rectifier ; E, tubular condenser ; F, receiver ; G, water inlets ; H, water outlets ; I, thermometers ; J, observation glass ; J_1, gauge glass ; K, drain-cock ; K_1, run-off valve ; L, steam inlet.

stills, but up to now pure products have not been obtained to any large extent by means of them. Blair, Campbell & M'Lean, Ltd., make a double and triple column continuous rectifying still and maintain that it will yield pure products, but it has not been possible to get details of the plant for publication. It is no doubt similar in many respects to those described later, and with the continual improvements and the amount of research being done on this subject, one may hope to see such apparatus generally installed in the near future.

Early investigators realised the importance of continuous stills, and a good many attempts have been made to produce them. E. Sorel (Eng. Pat. 243, 1900) used such a plant which was based on fractional condensation of the steam distillates of naphtha ; for he observed that mixtures of steam and naphthalene boiled at 98° C. ; steam and xylenes at 93° C. ; steam and toluene at 80° C. ; steam and benzene at 68·5° C.

W. R. Bonsfield (Eng. Pat. 25699, 1901) endeavoured to work continuously by means of fractional condensation.

F. W. Ilges (Eng. Pat. 3302, 1909) developed the principle of using two stills continuously, each still being provided with two condensers, kept at different temperatures. Thus four fractions can be obtained. The modern plants described later are developments of this idea.

Reference may also be made to the continuous still of Barbet et Fils (Fr. Pat. 434677, 1910).

Sadler and Bellerby (Eng. Pat. 105395, 1917) were one of the first to make a continuous still of any practical importance, but their apparatus must be very delicate to control and liable to give inconsistent fractions.

Simon Carves Plant.— Simon Carves, Ltd., Manchester, and Samuel Walker and Sons, Ltd., make very similar plants, so that a description of the former will suffice. This plant is shown diagrammatically in Fig. 173. It consists essentially of a column-still provided with reheating chambers as described later, but the pro-

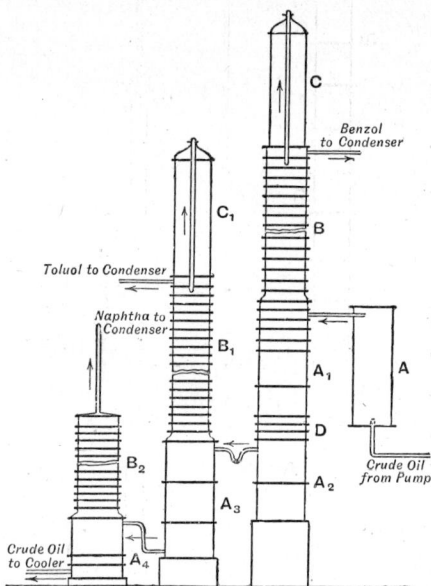

FIG. 173.—Simon Carves Plant.

A_1, A_2, A_3, A_4, steam heaters; B, B_1, B_2, fractionating trays; C, C_1, dephlegmators; D, trays.

ducts of distillation after leaving each reheating chamber are separately washed in fractionating columns and collected. Referring to the drawing—the crude benzole, which must be entirely free from water, is passed through a steam heater A, and then into the base of the fractionating column B. The benzole vapour coming off the top of the heater is delivered into the still column lower down than the point of entry of the liquid benzole. The benzole vapour therefore takes the place of the live steam jets in the usual type of still. In the base of the still, large heating chambers, A_1 and A_2, are fixed, the size of these depending of course largely upon the amount of benzole to be dealt with. These chambers are similar to those provided in the ordinary still as already described. At the head of the still is placed a large dephlegmator C, with the usual vapour connections to the condenser. The temperatures in the still, dephlegmators, etc., are so regulated that only benzole vapour is allowed to pass forward to the condenser.

The less volatile products escape at the base of the still and overflow into the second still, which is provided for the recovery of toluol. The operation of this still is similar to that of the first, but the size is naturally smaller.

The residue, naphtha and creosote, from this still overflow into the third or naphtha still, which is suitably proportioned and provided with a steam inlet at the base.

The residual creosote, when overflowing from this still, runs back to the store tank.

The fractionating column is 3 feet 6 inches in diameter, and 12 feet high, and contains 14 trays, each tray being of the usual type and having 9 bubbling bells and the usual overflow pipes. There is also a mid-feather in each tray, and the liquor flows in alternately on each side of the mid-feather.

The dephlegmator casing is 3 feet in diameter and 9 feet high, and contains 9 cast iron gilled radiating pipes of 4 inches bore, secured to tube plates at top and bottom. The gilled pipes are, therefore, 9 feet long.

The toluol column is 3 feet in diameter and 12 feet high and takes 16 trays ; the dephlegmator is of corresponding size.

A plant of this description gives good results, surpassing those obtained in the usual blow-over still, and will convert 8000 gallons, in some cases even 12,000 gallons, of crude benzole per day into 90's benzole, 90's toluol and solvent naphtha.

The Adam Still.—The Adam Patent Continuous Benzole Still is worthy of note and is shown in Fig. 174, which portrays a plant capable of distilling 10,000 gallons of crude benzole in 24 hours. It is employed in the primary separation of crude benzole into fractions, roughly representing the final products, and the elimination of the small proportion of creosote.

There are three stills, A_1, A_2 and A_3, through which the crude benzole flows in succession, and from each a fraction is obtained corresponding to the benzole, toluol and solvent naphtha final products.

The advantages usually sought for in continuous working are uniformity of operation and heat economy, and these are obtained in this plant mainly by the use of control thermometers and the utilisation of a large proportion of the heat latent in the vapours.

The crude benzole is first circulated through a tubular preheater B, where it exchanges heat with the vapour of the benzole fraction from the first still A_1 ; next, direct steam heat is applied in a special heater c, and this, broadly speaking, brings the temperature to the distilling point. The crude benzole now enters the first still, where it is heated by steam of 80 to 90 lbs. pressure. The still is a flat-bottomed one of D-shaped section, and the crude benzole follows a channel which repeatedly traverses the still longitudinally. The steam heating coil lies in the same channel, and the steam and benzole flow in reverse directions. The crude benzole on leaving the still should have parted with nearly all its benzole. That this is the case is indicated by its temperature which is registered by a thermometer in the outlet pipe.

The crude benzole is similarly treated in a second, A$_2$, and third still,

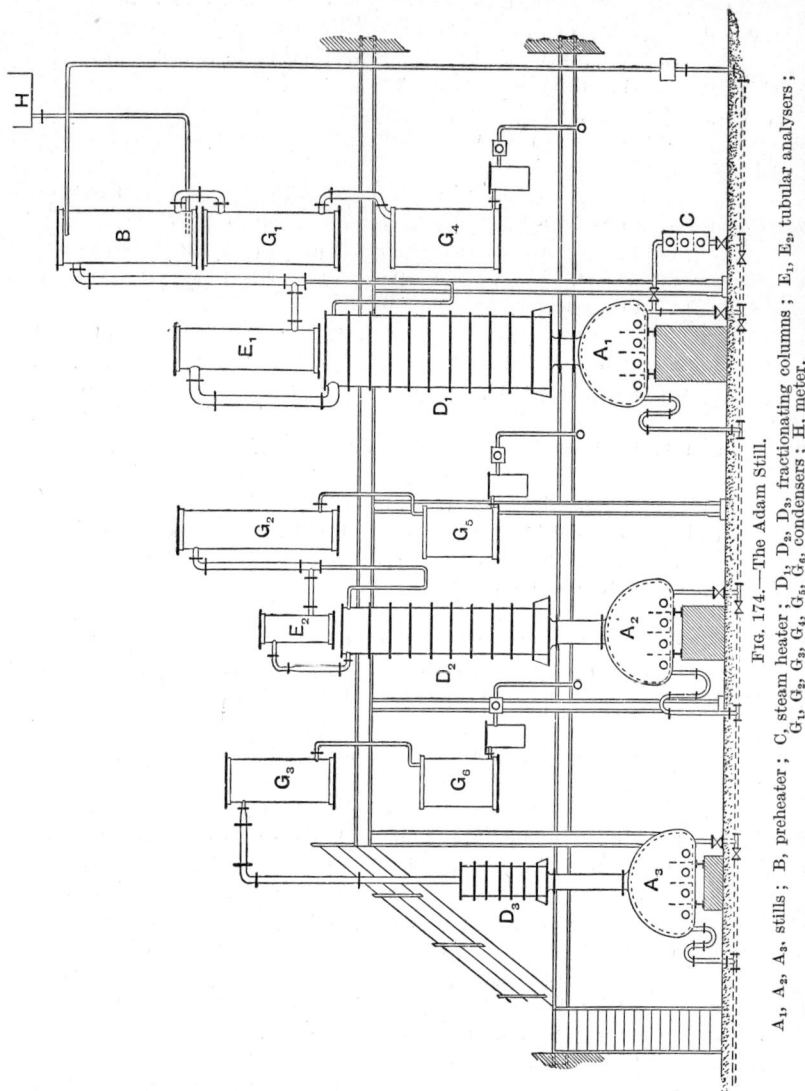

FIG. 174.—The Adam Still.

A$_1$, A$_2$, A$_3$, stills; B, preheater; C, steam heater; D$_1$, D$_2$, D$_3$, fractionating columns; E$_1$, E$_2$, tubular analysers; G$_1$, G$_2$, G$_3$, G$_4$, G$_5$, G$_6$, condensers; H, meter.

A$_3$, of the same type, and the final outflow contains no light products.

Each still is fitted with a fractionating column of dimensions corresponding to the amount of vapour released and to the dephlegmation required. The first, D$_1$, is the largest (18 feet high × 5 feet in diameter in 10 sections), the others being respectively, D$_2$ (15 feet high × 3 feet in diameter in 8 sections), D_3 (7 feet high × 2 feet in diameter in 7 sections). Each section contains an arrangement whereby the ascending vapour is compelled to pass through about 2 inches of refluxed liquid, in fine bubbles, from a perforated plate. This makes a very efficient column. It should be noted that in any section of the column the composition of the liquid and vapour remains the same, whereas in an intermittent type of still it continually alters. The liquid which flows down the column is produced by a tubular analyser or reflux condenser, E_1 and E_2 of the ordinary type (this is omitted in the last still), and the vapour then passes to the condensing system, G_1, G_2 and G_3.

As previously stated, the cold crude benzole plays a part in condensing the first or benzole fraction of the distillates. The rate of flow of crude benzole into the plant is seen on a meter H, of the Venturi type at the inlet to the preheating system, and each fraction at its outflow from the condenser passes over a V-notch calibrated to act as a meter. Further controls are provided by thermometers in the pipes carrying the vapours from the fractionating columns, and at various other points.

The products are worked up by the usual acid treatment, and are refractionated from stills of the usual type, specially designed for preparing pure products.

TREATMENT OF FRACTIONS

As the various plants required and in use for the fractionation of the naphthas have been very fully described, a survey of the method of splitting up the fractions will be of interest.

This, of course, cannot be a fixed method as it is dependent on so many factors, such as the quality of the crude material available and the kind of plant employed. However, roughly speaking, the distiller endeavours to make his fractions somewhat on the lines of those shown in Fig. 175, always bearing in mind that the fewer the fractions the less complicated the whole fractionation. Also the endeavour obviously is to get from the crude product to the finished one with as few intermediate fractions as possible. In the figure the distillations are shown as follows :—primary distillation ————, secondary — · — · —, tertiary – – – – –, final .... The figures below each fraction indicate approximately the boiling range.

1. Primary Distillation.—The crude naphtha or light oil, distilled in a fire still, produces crude benzole, crude solvent, crude heavy solvent and creosote.

2. Secondary Distillation.—The crude benzole, which should show a test of 60 to 65 per cent distilling off in a retort below 120° C., is the raw material for this distillation, and is obtained from the crude naphtha or from the washing of coke oven or coal gas with creosote. The crude is washed in the manner already described before it is run into the steam still. It can be worked in two different ways, that is, after the forerunnings containing large quantities of carbon di-sulphide have been sweated off, the charge can be run direct for pure benzene and pure toluene, or 90's benzole and 90's toluol.

In the former case the splitting of the distillate is very complicated as so many fractions have to be collected. They are, in order of collection as follows :—forerunnings, pure benzene, light crude toluol, pure toluene, heavy crude toluol, crude solvent, crude heavy solvent and residues. To obtain the pure products a very efficient plant has to be employed and, even so, the method is hardly practicable, for just before and just after the pure hydrocarbons come over, the still has to be run very slowly indeed (40 to 50 gallons per hour) and the intermediate fractions are necessarily large. It is better to sweat off some of the forerunnings and collect the fractions rapidly as 90's benzole, 90's

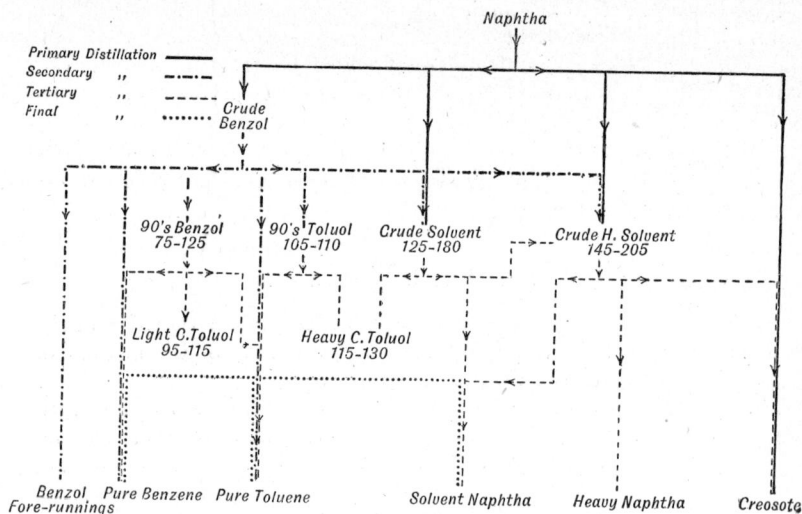

FIG. 175.

toluol, crude solvent and crude heavy solvent. The last two fractions may either be blown over by means of raw steam or, if preferred, by distilling *in vacuo*, in which case the fractionation is more marked and better separation is obtained. The residue contains mostly creosote, but, to ensure that no naphtha is wasted, it may conveniently be put back into the crude naphtha for further working.

3. Tertiary Distillation.

3. Tertiary Distillation.—The fractions should, as far as possible, be worked in rotation, so that the residue from each is the base on which a fresh charge of the next fraction is run. The 90's benzole produces a little forerunnings, pure benzene, light crude toluol and a little pure toluene.

When running from any fraction for pure products, the principle of working the still is always the same. The stage just before coming on to the pure, which can be ascertained by reference to the thermometer in the still-head pipe, is the one that has to be most carefully watched ; the still is then run very slowly, and the temperature of the water in the dephlegmator and analyser is kept very constant and well below the boiling point of the pure product. As soon as the pure commences to distil over, the rate may be very considerably increased, even up to 200 gallons per hour, and kept at that until there are indications of a change, when again the rate is decreased.

It is of very great importance to work a still at the speed which gives the best results, and this can only be ascertained by trial for each plant and for each fraction. Diligent research on this matter will enable the distiller to obtain the best possible yields of pures with the smallest quantities of intermediates.

After the pure benzene has been collected and the temperature has risen a few degrees, the charge of 90's toluol is added to the residue from the 90's benzole in the still ; the distillation is then continued, light crude toluol, pure toluene and heavy crude toluol being collected.

The distillation is again stopped, the crude solvent added to this residue and the distillation recommenced, when the first fraction to come over consists of heavy crude toluol.

The other products from this distillation, namely the solvent naphtha and crude heavy solvent, are either obtained by blowing in raw steam or by distillation *in vacuo.* The former is the cheaper way, but in the opinion of the author the latter has advantages which more than compensate for the extra cost.

The last of the tertiary distillations, that of the crude heavy solvent, must be carried out in a fire still with a column having no dephlegmator. In this case solvent naphtha and heavy naphtha are obtained. and the residue is creosote.

All the fractions before distillation are washed with sufficient amounts of acid and soda to meet the necessary requirements of purity.

4. The Quaternary Distillation.—This is the splitting of the light crude toluol into pure benzene and pure toluene, and of the residue, after addition of a new charge of heavy crude toluol, into pure toluene and solvent naphtha. In each of these cases small quantities of intermediates are formed which have to go back again to their respective fractions in the tertiary stage.

The quaternary distillation is conducted in the manner already described except that it may be generally carried out at a greater speed.

The figure shows the simplest form of separation of the various naphthas from the crude and one which would be nearly perfect from a distiller's point of view. Unfortunately, in practice, a few intermediates between those mentioned are obtained, and have to be put back to the fractions to which they most nearly agree in boiling-point range.

The control of a refinery making all these products is not a simple matter, and should be in the hands of a capable chemist who can translate the results of tests obtained in the laboratory into terms of works production.

General Practice.—The practice followed in distilling for pure benzene and toluene will necessarily vary with different distillers and with different types of plants. Generally speaking, however, the crude benzole is run and preliminary fractions separated. Although, with the present forms of column, it is possible to separate pure products direct from the crude material, the time occupied in eliminating intermediates is so large, and the washing difficulties are so important, that greater economy is obtained by running stills in pairs, one working crude and the other working intermediates for pure products.

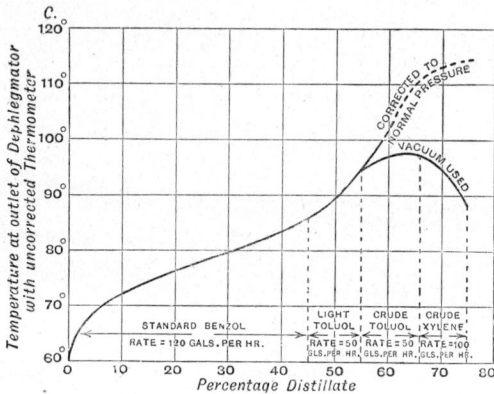

FIG. 176.
Fractionation of crude benzole. Preliminary separation.

Example of Actual Results.—The following are some results obtained by the author using a Hodgson Thomas plant (Fig. 169, p. 406). The preliminary fractions from the crude are :—(a) standard benzole, (b) light toluol, (c) crude toluene, (d) crude xylene. The distillation is represented in Fig. 176, which shows the main features. The variation in the rate of working will be noted; the rate is rapid so long as the distillate is uniform in quality, and is slow during the passage from one fraction to another. This

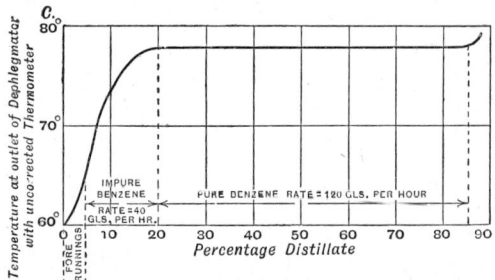

FIG. 177.—Fractionation of standard benzole.

variation in rate is more apparent in the subsequent figures. The dotted part of the curve represents the distillation converted to normal pressure, the actual distillation under reduced pressure being shown by the solid line.

Fraction (a) consists of benzene and toluene and contains 90 to 95 per cent of benzene. It is fractionated for pure benzene and a yield of 65 per cent is obtained on the first run. The results are indicated in Fig. 177, the noticeable feature being the slow rate of distillation until the temperature at the outlet of the dephlegmator reaches a constant value ; at this stage the rate can be increased and maintained until the temperature rises, when it is advisable to stop the distillation and recharge on the residue. The temperature of the dephlegmator water during this period is about 42° C., which ensures a good return to the column.

Fraction (b) consists of about 50 per cent of benzene and 50 per cent of toluene, no xylene being present. The working of this is shown in Fig. 178. The treatment of this fraction indicates, perhaps more than any other, the efficiency of the modern column. Both pure benzene and pure toluene are obtained in one distillation from the mixture. As the temperature rises after the separation of the pure benzene the rate is decreased, and is increased again when the temperature becomes constant while the pure toluene fraction is collected. The temperature of the water in the dephlegmator is maintained at about 42° C. on the benzene fraction, and allowed to rise slowly during the distillation of the intermediates until it reaches about 100° C. during the collection of the toluene.

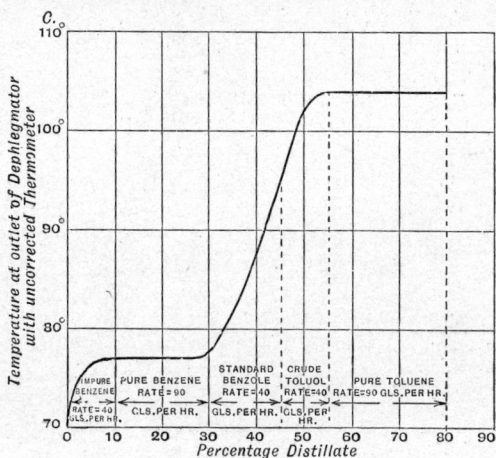

FIG. 178.—Fractionation of light toluol.

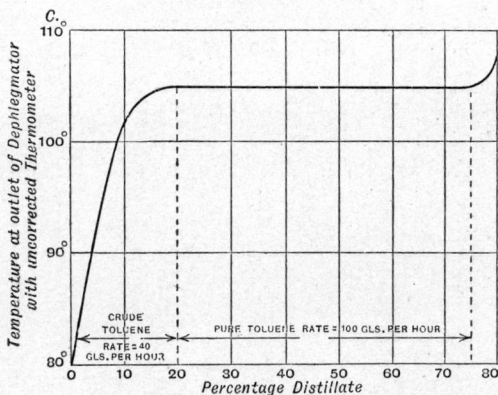

FIG. 179.—Fractionation of crude toluene.

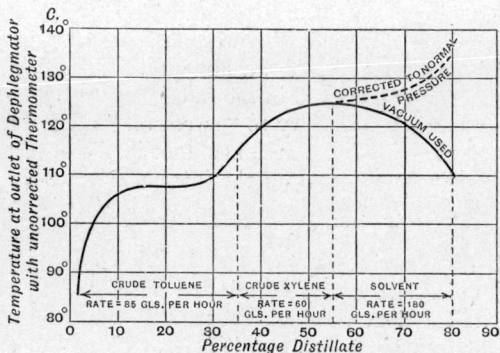

FIG. 180.—Fractionation of crude xylene.

2 E

Fraction (c) consists of benzene, toluene and xylene, the toluene content being 75 per cent. The fractionation of this is indicated in Fig. 179.

Fraction (d) consists of toluene, xylenes and higher homologues, and is the most difficult to deal with. It will be noted from the curve (see Fig. 180) that a comparatively high percentage of heavy toluol is produced and is returned for reworking. When working for solvent naphtha (a mixture of xylenes and higher homologues) the use of a vacuum or even of open steam at the end is advantageous, but should be avoided if possible when working for xylenes.

Economy in Working.—Economy in working can be obtained by observing a definite rotation. It has already been pointed out that the distillation of standard benzole is stopped when the temperature rises after the separation of pure benzene.

After successively working two or three charges in this manner the combined residues have accumulated toluene and the still may then be charged with light toluol, to be followed by crude toluol. This rotation avoids unnecessary working of intermediates, thus saving both time and steam.

The rates indicated by the curves will necessarily vary with the plant and must be determined for each plant, but they indicate what has been done and show a marked improvement on the practice which obtained previous to the war.

Another point of note is the difference between the separation of benzene from benzene-toluene mixtures, and of toluene from toluene-xylene mixtures. From experiments made by the author working on pure compounds in the laboratory, it would appear that no such difficulty should exist, that is to say, there are no properties inherent in the three hydrocarbons such as the formation of azeotropic mixtures which would cause this difficulty. The cause must therefore lie in the plant, probably in the dephlegmation, and the difficulty might possibly be overcome by use of salt solution or oil as a dephlegmating medium.

Paraffinoid Crude Benzole.—It may be of interest to record differences in the working of paraffinoid crude benzoles and those free from paraffins. When dealing with such benzoles, in addition to greater care being necessary in the washing process, involving greater loss, a further difficulty is observed in the fractionation. When working for standard benzole it is necessary to " sweat off " the forerunnings slowly in order to prevent a high paraffin content in the benzene, and to reduce correspondingly the rate on the intermediates when working toluene from crude toluene. It is very noticeable that in the case of products free from paraffin the " cuts " for pure benzene and toluene are clear and definite, whilst in the case of paraffinoid products a larger percentage of the distillate has to be rejected both at the beginning and end of the pure fraction, although the paraffin content may be as low as 0·5 per cent. For instance, Fig. 177 shows that when the temperature reaches 78° C. pure benzene can be at once separated and the separation continued until the temperature rises.   In the case of a paraffinoid product, on the other hand, as much as 10 per cent at the beginning and 5 per cent at the end has to be rejected.   The author attributes this difficulty to the formation of azeotropic mixtures of the aromatic hydrocarbons and the paraffins ; some such mixtures are known, that of benzene and normal hexane being typical (pp. 47, 51, 215).

Another feature in the working of the paraffinoid benzole is the formation of a deposit on the coil in the still, which reduces the heating capacity to a detrimental degree.   More thorough washing of the benzole does not prevent this.   The nature of the deposit cannot easily be ascertained, because it is rapidly decomposed to a highly carbonaceous compound of indefinite composition.   Frequent cleaning of the still is necessary to maintain the conditions required for successful working.

Products generally required.—It should be noted that only the usual finished products have been described and that these are not the only ones that can be obtained from the crude naphtha.   For instance, in times of peace only small quantities of pure benzene and toluene are produced and those only for the manufacture of dye intermediates. The separation is generally only carried as far as the secondary distillation, and the two fractions 90's benzole and 90's toluol are run together to form motor benzole.   Again, many special fractions frequently have to be made to meet special requirements of consumers, such as a mixture of the pure xylenes, distilling within a range say from 135° C. to 145° C.

SPECIFICATIONS OF FINISHED PRODUCTS

The specifications of the various finished products may be briefly described as follows :—

Benzole Forerunnings.—This is a by-product and the quantity obtained is kept as low as possible as it does not obtain a ready sale. It is composed chiefly of carbon di-sulphide and only the small quantity of benzole that in practice cannot be profitably extracted.   There is no definite specification for it.

Pure Benzene and Pure Toluene.—The specifications, as laid down by the Government and accepted by the dye-makers, are almost identical for the two products and are as follows :—

For Pure Benzene.—

Appearance.—To be a clear water-white liquid, free from suspended solid matter.

Specific Gravity.—The specific gravity is not to be less than 0·883 and not more than 0·887 at 15·5° C.

Boiling Point.—Must correspond approximately to 80·5° C., corrected.

Sulphuric Acid Test.—90 c.c. of benzol shaken with 10 c.c. of 90 per cent sulphuric acid for five minutes should impart only a slight colour in the acid layer.

Distillation Test—*Apparatus.*—A fractionating flask of 200 c.c. capacity, fitted with thermometer graduated to read to one-tenth of a degree Centigrade, and so adjusted that the top of the bulb is on a level with the side tube. The flask to be placed on a sheet of asbestos board having a hole one inch in diameter in the centre, in which the bottom of the flask rests, the bulb of the flask being surrounded by a cylinder of wire gauze resting on the asbestos sheet, and of such a height that the top of the cylinder is on a level with the top of the bulb.

Distillation.—100 c.c. of benzole are placed in the fractionating flask exposed to the flame (or source of heat) by the hole in the asbestos sheet, the distillation being conducted at such a rate that about 7 c.c. per minute are collected in the receiver, which is a 100 c.c. graduated cylinder.

The temperature is read when 5 c.c. and again when 95 c.c. have come over. The difference between the two readings must not be greater than 0·5° C.

For Pure Toluene.—This specification is exactly similar to the foregoing except that the limits of specific gravity are 0·868 and 0·870 and that the boiling point must correspond approximately to 110° C.

Motor Benzole.—The National Benzole Association, which protects the interests of both the manufacturer and user of motor benzole, has laid down that the most suitable benzole for internal combustion engines is one that passes the following specifications :—

1. **Specific Gravity.**— ·870 to ·885.

2. **Distillation Test (by flask).**—Benzole shall give a distillate of not less than 75 per cent to 80 per cent up to 100° C. ; of not less than 90 per cent up to 120° C. ; of not less than 100 per cent up to 125° C.

3. **Sulphur.**—The total sulphur shall not exceed 0·40 per cent.

4. **Water.**—The benzole shall be entirely free from water

5. **Colour.**—Water-white.

6. **Rectification Test.**—90 c.c. of the sample shaken with 10 c.c. of 90 per cent sulphuric acid for five minutes shall not give more than a light brown colour to the acid layer.

7. The benzole shall be entirely free from acids, alkalis and sulphuretted hydrogen.

8. The benzole shall not freeze at 25° F. below the freezing point of water.

Solvent Naphtha.—This material, used chiefly as a solvent for raw rubber, is commercially described as 90/160 or 95/160 solvent, thereby indicating that 90 per cent or 95 per cent distils off between 120° and 160° C. when tested in an ordinary retort. The liquid must be water-white, well washed and not strong smelling. It has a specific gravity of ·865 to ·875, and the flash point is 60° to 70° F. (Abel's closed test).

Heavy Naphtha.—This product has a specific gravity of ·900 to ·910 and flash point of 95° to 105° F. It should be water-white, well washed and fractionated well enough to ensure the presence of only small quantities of naphthalene. Its distilling points are generally 90 per cent off up to 190° C., and when this is the case it is consequently known as 90/190 naphtha. It is used chiefly as a paint-vehicle, particularly for ships' anti-fouling compositions.

METHODS OF TESTING

These are fully described in many of the books referred to at the end of this chapter. The retort test is the one most commonly used for all the naphthas and is very simple. Unfortunately it has not been properly standardised, so that different operators frequently obtain conflicting results. It is purely an empirical test, and consequently entirely dependent upon the method of operation and kind of apparatus employed. It is used, particularly in the case of crude benzole, as a basis of sale and purchase and should be thoroughly standardised.

The test which gives consistent results and is recommended by the author is as follows :—100 c.c. of light oil or crude benzole are placed in a 250 c.c. retort and attached to a 24-inch condenser, the whole apparatus having been previously rinsed with the oil to be tested. The thermometer is fixed so that the bulb is ⅜ of an inch from the bottom of the retort. The distillation is performed with a small naked flame shielded from draughts, the rate of distillation being 4 c.c. per minute. At 120° C. the flame is removed, the condenser allowed to drain, and the amount of distillate noted. The retort should be of the usual normal shape and the thermometer 15 inches long with a bulb ⅜ inch long, and graduated in one-fifth degrees from 70° to 130° C.

The corrections for barometric pressure and for inaccuracies of the thermometer may be made before each test by suspending the thermometer in the neck of a flask containing boiling water, taking care that the length of thread immersed in the steam is the same as the height of the retort. This is very important, as pointed out by Wheeler.[1] The same difference below or above 100° C. may be taken when stopping the distillation at 120° C., as it may usually be assumed that the error is the same at these two points.

It is most unfortunate that the retort test has grown to be the usual one, and it is very desirable that a test based on fractional distillation or even a distillation from an Engler flask (as used by some makers) be instituted and adopted as a standard.

Much useful work has been done recently on the analysis of the various naphthas but is too comprehensive to be described here. For those interested in this subject reference may be made to the under-mentioned papers :—

1. H. G. COLMAN. "Determination of Toluene in Commercial Toluol," *J.S.C.I.*, 1915, **34**, 168.

[1] *J. Soc. Chem. Ind.*, 1916, **35**, 1198.

2. H. M. JAMES. "Determination of Toluene, with a Note on Application of the Method to Benzene and Toluene," *J.S.C.I.*, 1916, **35**, 236.

3. P. E. SPIELMANN, E. G. WHEELER. "The Analyses of Commercial Benzoles," *J.S.C.I.*, 1916, **35**, 396.

4. A. EDWARDS. "The Estimation of Benzene and Toluene in Commercial Mixtures," *J.S.C.I.*, 1916, **35**, 587.

5. D. WILSON, L. ROBERTS (Gas Record, Chicago). "Determination of Benzene, Toluene and Solvent Naphtha in Light Oils, etc." *J. Gas Lighting*, 1916, No. 134, pp. 225-227; *Abst. J.S.C.I.*, 1916, **35**, 684.

6. P. E. SPIELMANN, F. BUTLER-JONES. "The Analyses of Benzole First Runnings," *J.S.C.I.*, 1916, **35**, 911.

7. E. G. WHEELER. "The Stem Correction of Thermometer," *J.S.C.I.*, 1916, **35**, 1198.

8. G. HARKER. "Estimation of Toluene and Benzene in Coal Tar Oils," *J. Royal Soc. N.S.W.*, 1916, No. 50, pp. 99-105; *Abst. J.S.C.I.*, 1917, **36**, 950.

9. P. E. SPIELMANN, G. CAMPBELL-PETRIE. "Observations on Crude Benzoles," *J.S.C.I.*, 1917, **36**, 831.

10. P. E. SPIELMANN, F. B. JONES. "Estimation of Xylene in Solvent Naphtha," *J.S.C.I.*, 1917, **36**, 480.

11. F. BUTLER-JONES. "Analyses of Commercial 'Pure' Benzole," *J.S.C.I.*, 1918, **37**, 324 T.

12. F. W. SPERR. "Method for Boiling Point Test of Benzole," *Met. and Chem. Eng.*, 1917, No. 017, pp. 586-588; *Abst. J.S.C.I.*, 1918, **37**, 26a and 50a.

13. W. G. ADAM. "Analyses of Crude Benzole," *Gas. Journal*, 1918, No. 14, p. 65; *Abst. J.S.C.I.*, 1918, **37**, 50.

14. J. M. WEISS. "Methods of Analysis used in Coal Tar Industry," *J. Ind. and Eng. Chem.*, 1918, No. 10, pp. 1006-1012; *Abst. J.S.C.I.*, 1919, **38**, 68a.

15. W. J. JONES. "Determination of Degree of Purity of Samples of Benzene," *J. Soc. Dyers and Col.*, 1919, No. 35, pp. 45-47; *Abst. J.S.C.I.*, 1919, **38**, 216a.

16. H. G. COLMAN, G. W. YEOMAN. "Determination of Benzene and Toluene in Coal Tar and Similar Products," *J.S.C.I.*, 1919, **38**, 57 T, 136 T, and 152 T.

17. W. J. JONES. "Determination of Benzene in Crude Benzole," *J.S.C.I.*, 1919, **38**, 128 T.

18. P. E. SPIELMANN, F. BUTLER-JONES. "Estimation of Carbon Di-sulphide," *J.S.C.I.*, 1919, **38**, 185-188 T.

For further information on this subject the following books may be consulted :—

LUNGE. *Coal Tar and Ammonia.*

WARNES. *Coal Tar Distillation.*

COOPER. *By-Product Coking.*

CHRISTOPHER AND BYROM. *Modern Coking Practice.*

MALATESTA. *Coal Tars and their Derivatives.*

E. HAUSBRAND. *Die Wirkungsweise der Rektificir- und Destillirapparate*, also *Verdampfen, Kondensieren, und Kühlen.*

C. MARILLER. *La Distillation fractionnée.*

S. E. WHITEHEAD. *Benzole : its Recovery, Rectification and Uses.*

THE DISTILLATION OF GLYCERINE

Lieut.-Col. E. BRIGGS, D.S.O., B.Sc.

TECHNICAL DIRECTOR, BROAD PLAIN SOAP WORKS, BRISTOL

CHAPTER XXXIX

THE DISTILLATION OF GLYCERINE

THE chemical substance glycerol, CH_2OH, $CHOH$, CH_2OH or glycerine as it is commonly known in commerce, is a colourless, odourless, viscid liquid having a sweet taste and a neutral reaction.

It is soluble in water in all proportions with evolution of heat, it is strongly hygroscopic and on exposure to the air as much as 50 per cent of its own weight of water is absorbed. It has powerful solvent properties, combining in this respect those of water and alcohol.

Its sp. gr. at $\dfrac{15^\circ}{15^\circ}$ C. $= 1 \cdot 26468$.

It boils under 760 mm. pressure at 290° C. with slight decomposition, but under reduced pressure it distils unchanged, its b.p. at 50 mm. pressure being 210° C. and at 12·5 mm. 179·5° C. It is not volatile at ordinary temperatures, but on concentration of a solution of it in water at ordinary pressures glycerol volatilises with the water vapours at a temperature of about 160° C., corresponding to a concentration of about 70 per cent.

When strongly heated it rapidly loses water with the formation of acrolein, leaving a residue of polyglycerols, chiefly diglycerol $C_6H_{14}O_5$.

Commercially pure glycerine has a sp. gr. of 1·260 and, apart from about 2 per cent of water, it contains only the minutest traces of impurities.

This practically pure substance is produced by the distillation of crude glycerine, a by-product of the soapmaking and candlemaking industries.

Fats and oils, on saponification with caustic soda, yield a dilute impure solution containing about 5 per cent of glycerine, and this, on treatment and purification, yields a product containing 80 per cent of glycerine and 10 per cent of salts, known as soap lyes crude glycerine.

In the candle industry fats are hydrolysed by the autoclave, Twitchell, or similar process, yielding fatty acids and a solution of glycerine, and this solution on treatment and concentration yields saponification glycerine containing about 90 per cent of glycerine and a small percentage of salts and other impurities.

Approximate analyses of these two crude products are given below, so that the problem of distilling them for the production of glycerine in its purer forms may be understood.

	Crude glycerine.	Saponification glycerine.
Glycerine T.A.V.	81·28	86·47
Specific gravity at 20°/20° C.	1·3010	1·2393
Total residue at 160° C.	12·45	0·80
Ash	9·73	0·42
Organic residue by difference	2·72	0·38
Arsenic parts per million	2	1

The organic residue consists of polyglycerols, fatty acids, resinous and colouring matters not removed in the treatment of the lyes.

The ash in crude glycerine consists mainly of sodium chloride with traces of sodium carbonate and oxides of iron, aluminium, and silicon derived from the materials used in the preliminary treatment of the lyes.

The ash in saponification glycerine is usually calcium or barium sulphate, also derived from the treatment of the " sweet water," as the solution of glycerine obtained on splitting fats is usually termed.

Crude glycerine is a clear, bright, rather viscous liquid, with a colour varying from orange-red to dark brown, while saponification glycerine is less viscous and lighter in colour.

The apparatus used for distilling each of these two products is identical, but owing to the larger amount of solids in crude glycerine greater precautions have to be taken to prevent overheating and the distillation is somewhat slower.

As by far the greatest bulk of glycerine produced comes from the soap lyes crude glycerine, the plant described will be that particularly applicable to the distillation of crude glycerine, though it is equally suitable for saponification glycerine.

Glycerine can only be distilled without loss under reduced pressure, and all the plants here described are worked under vacuum and free steam is invariably used to assist the distillation.

The earliest patents taken out for the distillation of crude glycerine in this country date from about 1881, as it was not until 1879 that the successful production of crude glycerine from soap lyes was effected on the manufacturing scale, and from this date onwards the efforts of the designers of glycerine distilling plants have been directed to the three points—

(1) Prevention of overheating the substance, and consequent loss of glycerine ;

(2) Condensation of the greatest amount of strong glycerine ;

(3) Economy in the use of fuel ;

and the various plants described below indicate the successive improvements that have been made in these directions, and latterly chiefly in the economy of fuel. The prevention of overheating and consequent loss due to the formation of polyglycerols and other products of the destructive distillation of crude glycerine were at once overcome when stills heated by means of steam at a definite and constant temperature took the place of fire-heated stills.

Fire-heated stills are now quite obsolete, but it may be of interest to describe one example so that the progress that has been made in glycerine distillation plant may be emphasised subsequently.

The illustration below (Fig. 181) shows the general arrangement of the plant. A is a cylindrical still with a flat bottom mounted on the brick-work setting B, which is fitted with a grate for the coal fire, the flames from which play on to the bottom of the still and around the sides below the level of the glycerine in the still, a current of free steam superheated in the same fire and controlled by the valve C bubbles through the liquid in the still in the form of fine jets.

The glycerine and water vapours pass through the still-head D and any entrainment is caught in the catchall E, and returns to the still by

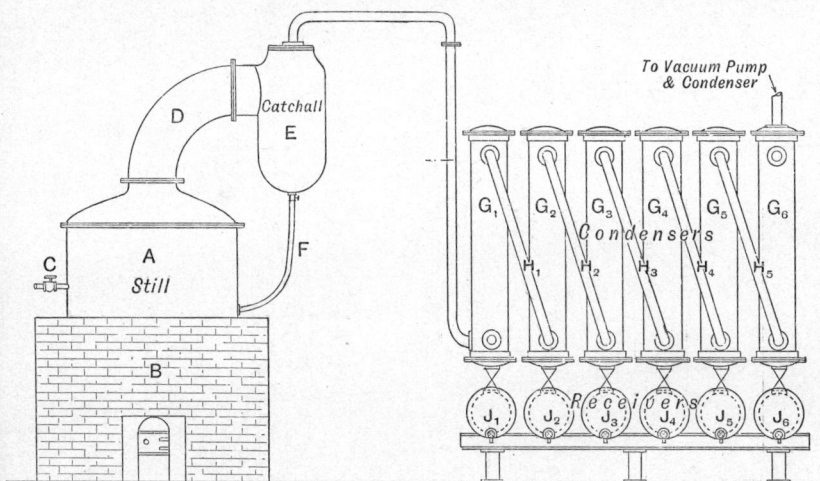

FIG. 181.—Glycerine distillation plant. Fire-heated still.

means of the pipe F, the vapours passing on to the series of air-cooled condensers G_1, G_2 to G_6 ; the vapours enter each condenser at the bottom and, emerging from the top, are carried to the bottom of the next condenser by means of the pipes H_1, H_2 to H_5.

Under each condenser is a receiver J_1, J_2 to J_6, and in these the condensed glycerine is collected ; from the first two the best quality is obtained, the quality gradually deteriorating and the glycerine becoming weaker towards the end condenser where the last of the glycerine should be condensed with much of the volatile impurities.

A water condenser and vacuum pump is connected to the top of the last condenser.

The loss of glycerine in such a plant is very heavy when crude glycerine is distilled ; the salt soon separating out and settling on the bottom of the still causes local overheating and decomposition of the glycerine with the formation of acrolein and other products

With saponification glycerine a fairly good yield can be obtained and the distilled glycerine is of good quality, though very great care is necessary in controlling the fire and the free steam, any excess of the latter causing the glycerine to froth over the still-head and contaminate the distillate already collected.

A much more efficient type of distillation plant that is largely used in this country is that of Van Ruymbeke, the distinctive characteristic of which is the heating of the glycerine in the still by means of high pressure steam circulating through a coil and the superheating of the expanded free steam by means of the steam from the close coil after it has passed through the still.

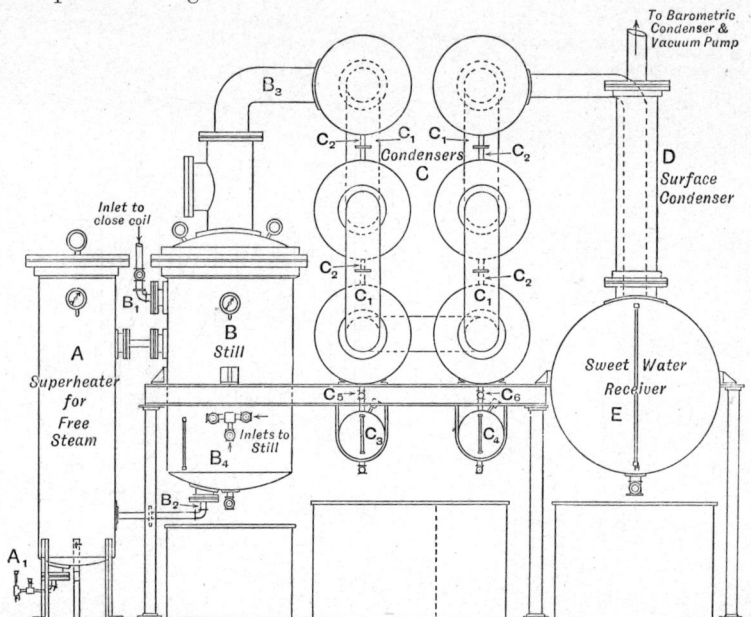

FIG. 182.—Glycerine distillation plant.   Van Ruymbeke system.

The plant (Fig. 182) consists of a steam superheater A, a still B, a set of condensers C, with suitable receiving vessels for collecting the condensed strong glycerine, a water-cooled surface condenser D connected to a large receiving vessel E in which the weak glycerine or " sweet water " collects, and which in turn is connected with a condenser and vacuum pump capable of maintaining a high vacuum.

The still B is heated by means of steam passing through a spiral coil of many turns which extends from the inlet B_1 to the outlet B_2, and then through a well insulated pipe to the superheater A, and the current of steam is so regulated that the pressure of the steam, and therefore also the temperature, is kept the same in the superheater as that of the steam entering the close coil; this pressure varies between 140 and 200 lb. per square inch, and the condensing surface of the plant is designed for a particular pressure of steam which should not be varied greatly. Under vacuum glycerine is not decomposed at a temperature of 200° C., and the greatest economy in working is obtained by distilling at the highest possible temperature. In practice, steam of 180 lb. pressure is found very suitable as it will give a still temperature of 195° C., and references in the description of this particular type of glycerine distilling plant will be made to one designed for use with steam of a pressure of 180 lb. per square inch. The free steam superheater is fitted with a coil spirally wound inside the vessel, and steam from the same main as that providing the close steam is admitted into this coil through a valve A_1. Half a turn of a $\frac{1}{2}$ in. valve is sufficient to admit the requisite quantity of steam into the 2 in. coil in the superheater. The loss of heat on expansion is made up by the high pressure steam in the superheater surrounding the coil, and, finally, when the steam reaches the end of the coil at the top of the vessel, although it is at a low pressure, it has the same temperature as the high pressure steam heating the contents of the still.

The function of the close coil in the still is to heat the contents of the still to the distilling temperature, and to maintain it at this temperature when distillation starts by supplying the heat required to make up for radiation losses from the still and that absorbed in vaporising the glycerine, and the function of the open steam is to keep the mass agitated, to supply the difference in pressure between the pressure at which the still is working and the vapour pressure of the glycerine at the temperature of the still, and to increase the velocity of the vapours so that they will be carried through the still-head and into the condensers before they can condense and drop back into the still.

In all cases of distillation carried out with the aid of currents of steam it is advisable that the temperature of the steam should not be less than that of the liquid undergoing distillation, otherwise priming occurs due to the rapid expansion of the lower temperature steam entering the hot liquid in the still ; and on the other hand, if the steam is hotter than the liquid there is a loss of efficiency and also danger of decomposing the liquid. In the ingenious manner just described both these dangers are overcome, and the distillation is controlled by the extremely simple method of noting that the steam pressure in the superheater is the same as that of the steam entering the close coil.

The condensers C which are connected to the still by the pipe B_3 are cylindrical drums connected to one another by the vapour pipes C_1 and by the smaller diameter pipes C_2, through which the condensed glycerine flows downwards, collecting in the receiving vessels C_3 and C_4.

These condensing drums can be arranged, as shown in the section of the plant illustrated, in two tiers, or they can be arranged in a horizontal row, in which case it is necessary either to have a receiving vessel connected beneath each drum, or, as is more common, to have no receiving vessels at all, but to allow the condensed glycerine to collect in the condensing drums until the distillation is complete, when all the drums can be emptied into tanks placed on the ground beneath them.

There are advantages in both methods, and it is largely a question of available space as to which arrangement is adopted. The vertical arrangement has the advantage of utilising the full condensing surface of the drums for the whole period of the distillation ; it is the most compact, and is to be recommended where floor space is limited and head room is available.

The horizontal arrangement has the disadvantage of a gradually reduced condensing surface due to the drums partially becoming filled with condensed glycerine, but there is the advantage of fractionating the glycerine into six or more fractions, depending on the number of condensing drums employed, instead of obtaining two fractions only in the case of the vertical arrangement.

On the whole where the head room is available the advantage lies with the vertical arrangement, as it is seldom necessary to fractionate the strong glycerine ; in practice all the fractions are mixed together for concentration to produce dynamite glycerine, or for a second distillation to produce chemically pure glycerine. After passing through the series of air-cooled condensers where 85 per cent to 90 per cent of the distillate should be condensed, the vapours pass through the tubular condenser D, which is cooled by a stream of water passing through it on the outside of the tubes, the glycerine vapours passing through the tubes where they condense, together with most of the steam employed in assisting the distillation, and collect in the sweet water receiver E. These sweet waters are therefore a weak solution of glycerine containing all the volatile condensable impurities of the crude glycerine.

The uncondensed vapours then pass to a barometrical condenser where the last traces of water vapour are condensed and the incondensable vapours, air, etc., are removed by means of a vacuum pump. Other types of condensers and vacuum pumps may, of course, be employed.

Having described the plant, we can now pass to an account of the operation of distillation. The first operation is to get the whole plant under vacuum, and the pump is started and the air evacuated. A good dry air pump will maintain an absolute pressure, while the still is working, of one and a half inches to two inches of mercury, and for efficient work it is necessary to have this vacuum.

Crude glycerine is then drawn into the still by means of a pipe connected to the valve B_4, and steam let into the close coil by opening the valve B_1. When the glycerine has attained the temperature of the steam, free steam is admitted by opening the valve A_1.

The amount of crude glycerine fed into the still is sufficient to cover the perforated pipe from which the free steam issues, and as the distillation proceeds glycerine is fed in more or less continuously at a rate equal to the rate of distillation until the solids and non-volatile constituents of the crude glycerine have accumulated to a volume approximately equal to the original charge fed into the still. In practice it amounts to the total charge being about seven times the weight of the original amount fed into the still.

On admitting the free steam, distillation starts at once and proceeds rapidly and uniformly ; the amount of free steam used varies considerably, but in an efficient plant it should be about equal to the weight of crude glycerine in the charge.

The distilled glycerine collects in the receivers c_3 and c_4, about twice as much in the former as in the latter, and as they become full they are emptied by closing the valves c_5 and c_6, which connect them to the condensers, admitting air to the receivers through a small cock, and opening the valves at the bottom. When emptied, the valves at the bottom and the small air cocks are closed, and the valves c_5 and c_6 are carefully and very gradually opened until the pressure in the receivers and the rest of the plant is equalised.

The specific gravity of the glycerine collected in the first receiver c_3 is 1·260 or higher, and that in the second receiver c_4 about 1·240, giving an average specific gravity for the whole distillate of about 1·253 = 95 per cent glycerol. This is afterwards concentrated in a vacuum pan to 1·260 specific gravity or 98 per cent glycerol, which is the standard concentration of commercial glycerine.

The sweet water collected in the receiver E is about equal in weight to the weight of the whole charge of crude glycerine, and contains approximately 10 per cent of glycerine and most of the volatile impurities in the crude. This is concentrated in a separate vacuum pan, and then undergoes further distillation and fractionation.

When no more glycerine will come over the distillation is stopped, all steam turned off, the pump stopped and air is slowly admitted to the plant until it is under atmospheric pressure. The residue in the still, or " glycerine foots," is run out as a viscous tarry mass ; the still is washed out with boiling water, and this water is added to the foots, which undergo subsequent treatment, concentration, and further distillation for the recovery of the glycerine they contain.

Table 118 gives the results of an actual distillation.

TABLE 118

Weight of crude glycerine charged to still, 149·26 cwt. at 82·8 per cent glycerol = 123·5 cwt. glycerol.

Yield.	Cwt.	Glycerine content.	Yield.
		Per cent.	Per cent.
Strongs	108·5	97·5	85·7
Weaks	138	9·2	10·3
Foots	23	14·2	2·6
		Total yield	98·6
		Loss . .	1·4

Duration of test 42 hours.
Vacuum   .   . 1·6 in. abs. pressure.
Steam pressure 180 lb. per square inch.

Another type of glycerine distillation plant is the " Garrigue."
This plant aims at the production of the maximum amount of strong

glycerine, and effects economy in the use of steam by employing the vapour from the evaporation of the sweet water as the free steam for the distillation ; the still, condensers, strong glycerine concentrator, and sweet-water evaporator being one self-contained unit.

The following illustration (Fig. 183) shows the general arrangement of the plant, and it is operated as now described.

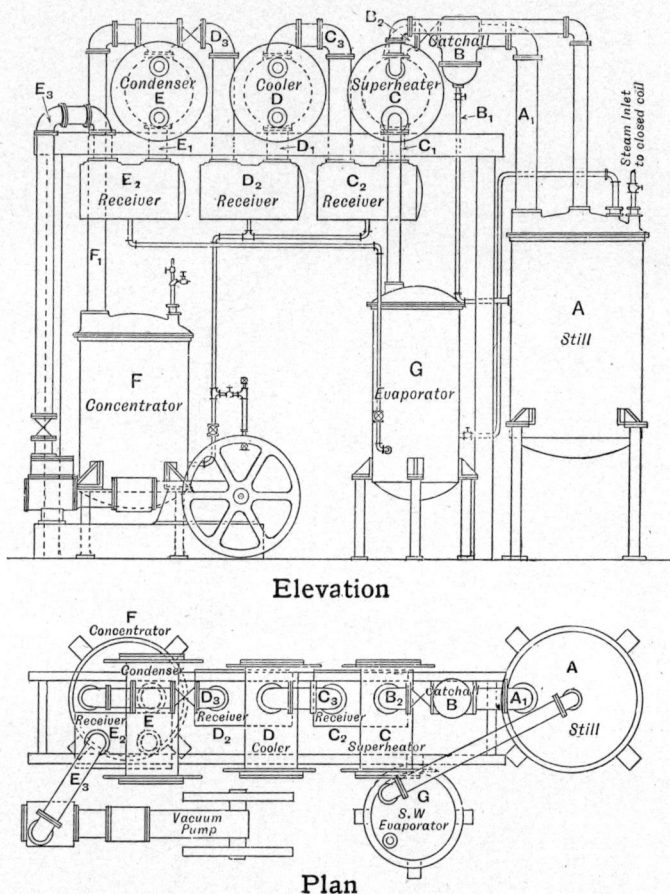

FIG. 183.—Glycerine distillation plant.   Garrigue system.

Crude glycerine is fed into the still A continuously, keeping the level a little above the free steam jets, a close coil supplying heat to the glycerine undergoing distillation.

The evaporator G contains sweet water from a previous distillation, or water in the case of starting the plant for the first time, and this is heated by the steam exhausting from the close coil in the still.   When the sweet water boils its vapours pass through the tubes in the superheater C, and issue from the jets below the surface of the glycerine in the still.

The glycerine and water vapours issuing from the still pass along the pipe A_1 and through the catchall B, where any entrainment from the still is deposited and returned to the still through the pipe B_1, then along the pipe B_2 through the superheater C on the outside of the tubes. The glycerine condensing on these tubes superheats the steam in the tubes from the evaporator, and the condensed glycerine passes down the pipe C_1 to the receiver C_2 and thence to the concentrator F.

The vapours uncondensed in the superheater pass along the pipe C_3 to the cooler D, through the tubes of which a very little water is flowing so as to maintain a temperature at which most of the glycerine and very little of the water vapour will condense, this condensate flows down the pipe D_1 and collects in the receiver D_2 and also flows to the concentrator F.

Steam is admitted to the close coil in the concentrator at such a rate as to produce dynamite glycerine of specific gravity 1·262 at the end of the distillation, the control being by means of a thermometer with its bulb immersed in the liquid, the water vapour passing along the pipe F_1 to the condenser E.

The vapour passing through the cooler travels along the pipe D_3 to the condenser E, through the tubes of which cold water is flowing in sufficient volume to condense all the vapour. The condensed sweet water runs down the pipe E_1 and collects in the sweet-water receiver E_2, and runs from there to the sweet-water evaporator G, in which it is being continuously concentrated as already described.

The vacuum pump draws from the top of the sweet-water receiver through the pipe E_3.

It is advisable to regulate the water in the cooler D so that some glycerine passes on to the condenser, and so to the evaporator. This glycerine will contain volatile impurities from the crude glycerine, such as the lower fatty acids, caproic and caprylic, and it is to keep these out of the strong glycerine that some glycerine is allowed to pass to the cooler.

Usually at the end of a week's run the sweet water from the evaporator is concentrated to about 80 per cent glycerine. The drain pipe from the sweet-water receiver E_2 is closed, and the water collected in the receiver is run to waste, as it should contain no glycerine but is highly contaminated with volatile impurities. The concentrated sweet water is fed to the still at the beginning of the next charge.

The residue or "foots" from the still is run out periodically, dissolved in water, or in partly concentrated sweet water from the evaporator, treated to remove impurities, concentrated, and redistilled.

With crude glycerine of ordinary good quality 90 per cent of the glycerine contained in the crude should be recovered as dynamite glycerine, specific gravity 1·260 to 1·262, while approximately 3 per cent of the glycerine will be contained in the sweet water, and 7 per cent in the foots.

The dynamite glycerine is subjected to a further distillation in an exactly similar plant to produce chemically pure glycerine. The

2 F

Garrigue plant gives good results with crude glycerine of good quality, but with poor crudes with a high residue at 160° C. the results obtained are undoubtedly poorer as regards the amount of glycerine left in the foots than those obtained with the Van Ruymbeke plant.

This result might be expected from a consideration of the working of the plant, as the free steam supply obviously tends to become less and to be at a lower temperature towards the end of the distillation, with the effect that the distillation slackens just at the point when a large volume of free steam at a high temperature is required to drive off the last traces of glycerine from the viscous residue in the still.

This, however, is not a source of loss of glycerine, as the foots can be treated and the glycerine subsequently recovered, but it means a lower yield of strong glycerine on the first distillation, and the subsequent treatment and redistillation involves expenditure on chemicals, labour, and steam.

Great care has to be taken in regulating the condensation of the weaks and strong glycerine, and careful control of the temperature of the cooler and of the cold water supply to the condenser is necessary. Good results are usually obtained when the specific gravity of the sweet water in the sweet-water receiver is between 1·002 and 1·003.

The fuel consumption of the Garrigue plant is estimated by competent observers to be about 0·5 lb. of coal per 1 lb. of crude glycerine distilled, and this includes the concentration of the weaks, while that of the Van Ruymbeke plant—not including the concentration of the weaks—is also about 0·5 lb. of coal per 1 lb. of crude, and if it is assumed that the weight of the weaks is equal to the weight of glycerine fed to the still, and that 0·4 lb. of steam, equal to 0·05 lb. of coal, is required in a double effect evaporator to concentrate 1 lb. of this sweet water to 80 per cent glycerine, the total fuel consumption of the Van Ruymbeke plant is 0·55 lb. of coal per 1 lb. of crude glycerine distilled.

The above quoted figures include the whole of the steam required for the distillation, including the steam for the vacuum pumps and the free and close steam for the still and dynamite glycerine concentrator, and are based on long periods of working and not on short test runs.

Another type of glycerine distilling apparatus is that of Frank J. Wood, U.S. Pat. 1098543, June 2, 1914, Eng. Pat. 24920, 1913, in which a novel method of bringing the free steam into intimate contact with the crude glycerine is employed, and it is claimed that economy in the use of steam is effected by using a single current of steam to distil the glycerine in a number of stills arranged in series. Each unit of the whole plant consists of a tubular feed heater for the crude glycerine, from which it is pumped into the upper portion of the still, where it meets a current of superheated steam, the vapours passing to a condenser designed in such a way that the glycerine is condensed but the water vapour passes on and supplies the free steam for the next still. A double effect evaporator for concentrating the sweet water condensed in the last still of the series forms an integral part of the whole plant, and supplies the free steam for the distillation of the glycerine in the first still.

If an objection is raised to the use of steam from the sweet water on the grounds that it is contaminated with volatile impurities from the glycerine, an additional evaporator fed with pure water may be provided.

Fig. 184 shows a still in perspective with the upper part broken

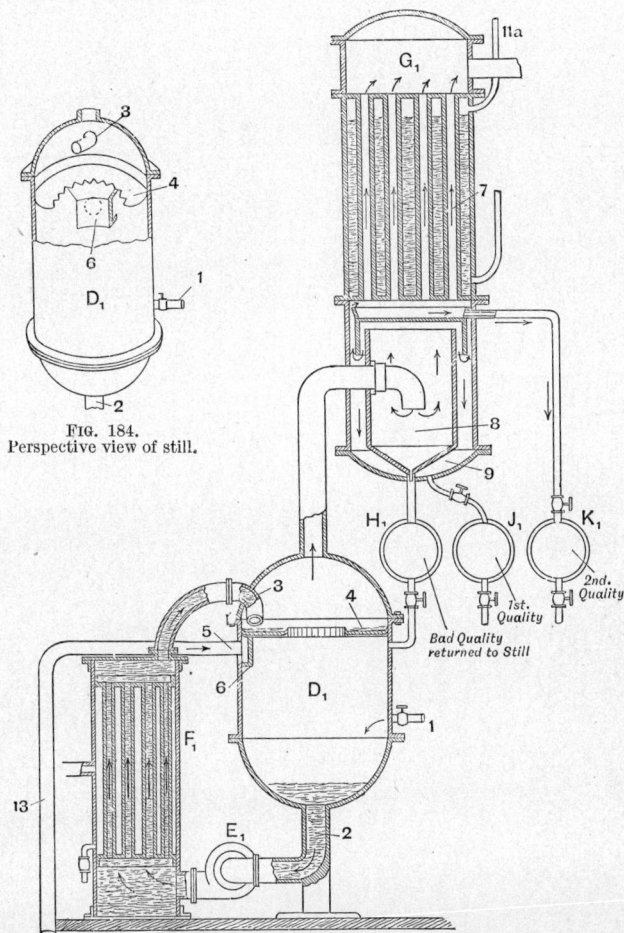

FIG. 184.
Perspective view of still.

FIG. 185.—Sectional elevation of one of the units.

away; Fig. 185 is a sectional elevation of one unit; Fig. 186 represents a complete plant of three stills in series.

The boiler " A " supplies high pressure steam to the preheaters F_1, F_2, F_3. Crude glycerine is fed to the still D_1 at 1, and the centrifugal pump E_1 connected to the outlet pipe 2 forces the glycerine through the preheater F_1 and into the upper portion of the still D_1, through the curved pipe 3, so that the glycerine is directed tangentially against the cylindrical wall of the still and thus caused to circulate within the still above the flat ring 4, the inner edge of which is serrated. Steam from the evaporator B_2 passes into the still at 5 below this ring and is caused to circulate round the still by means of the deflector plate 6. The glycerine falling over the edge of the ring forms a curtain through which the steam has to pass, and mingling with it carries away with it glycerine vapour, the combined vapours passing on to the condenser

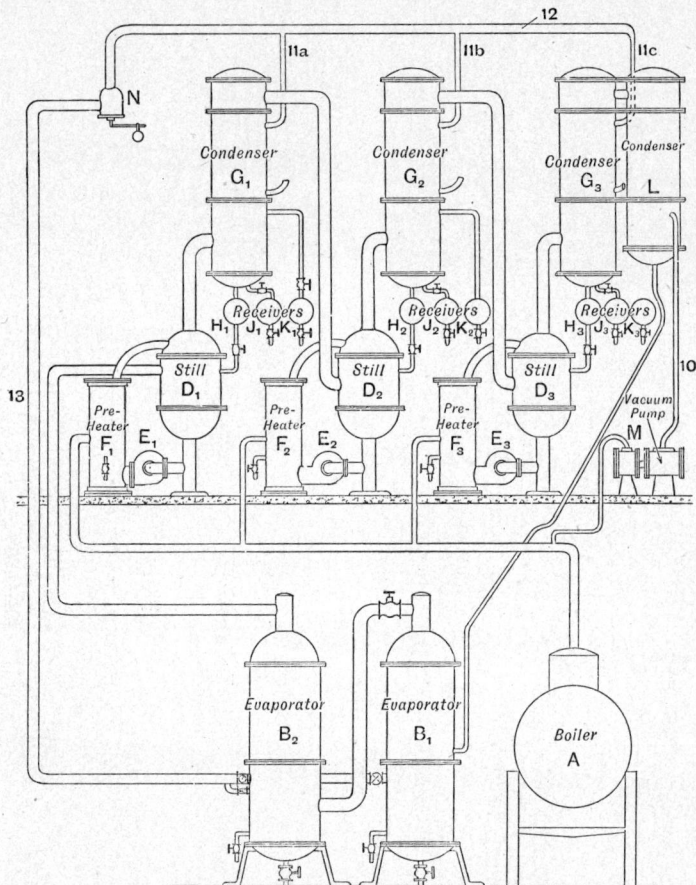

FIG. 186.—Glycerine distillation plant. Wood system.

G_1, which consists of two portions, the upper being a tubular hot water system 7 and the lower a non-condensing separator or catchall 8, arranged within an air-cooled condenser 9.

The arrows show the direction of the vapours through the condenser, and the direction of the condensed glycerine to the receivers J_1 and K_1.

The catchall 8, being surrounded by glycerine vapours, is kept hot, so that very little if any glycerine condenses therein, but it separates any entrained glycerine which collects in the receiver H_1 and is returned to the still.

Strong glycerine of first quality is condensed in the air-cooled condenser 9, and collects in the receiver J_1.

The water in the tubular condenser is kept at such a temperature that most of the glycerine in the vapours is condensed and collected in the receiver K_1. This glycerine is not of such good quality as that collected in the receiver J_1, being contaminated with volatile impurities—mostly fatty acids of low molecular weight. Very little water vapour is condensed, the greater part passing on and supplying the free steam to the next still D_2, which operates in exactly the same manner as the first. The vapours from the last condenser G_3 pass to a cold-water tubular condenser L from which the condensed sweet water flows to the evaporator B_1, the uncondensed vapour being carried away through the pipe 10 connected with a vacuum pump M.

From points above the water level in the condensers G_1, G_2, and G_3 pipes $11a$, $11b$, and $11c$ lead to the pipe 12, connected through a pressure regulator N with the pipe 13 leading to both evaporators, B_2 and B_1, with valves at the junction of the pipe with the evaporators to control the admission of steam to their heating systems.

The vacuum pump M is operated to maintain a pressure of about 2 in. of mercury in the evaporator B_2, the stills D_1, D_2, and D_3, the condensers G_1, G_2, and G_3, and the cold-water condenser L. The pressure regulator N is adjusted so that the pressure in the water spaces of the condensers is kept at 14 in. of mercury absolute, at which pressure the water will boil at about 80° C. ; this temperature is sufficient to condense the glycerine while maintaining the water in a state of vapour.

The method of mixing the steam and glycerine in the stills by spraying is very effective, and overcomes the difficulty often experienced with certain crude glycerines of foaming and priming when steam is used in the form of jets below the surface of the glycerine.

The usual method of feeding the stills is to supply the first still with fresh crude glycerine, the second still is fed from the residue from the first, and so on until the last still of the series is reached where the foots are accumulated and pumped away for treatment periodically. The illustration (Fig. 186) shows three stills in series, but as many as six have been operated successfully.

Table 119 gives the result of the distillation of 100 tons of crude glycerine.

TABLE 119

100 tons of crude glycerine at 80 per cent glycerol = 80 tons glycerol yield on one distillation.

	Tons.	Glycerine content.	Yield.
		Per cent.	Per cent.
Strongs . . .	69·5	99	86
Weaks . . .	70·0	5·71	5
Foots . . .	18·0	40	9

The foots on treatment and subsequent distillation will yield two-thirds of their content of glycerine, giving a final yield of strongs 92 per cent, weaks 5 per cent, and a loss of 3 per cent.

In considering the heat efficiency of the various types of glycerine stills it is not sufficient to take into account only the heat in the steam used for heating the glycerine up to distilling temperature and the heat in the free steam used in the distillation, but also account must be taken of the heat required to drive the various pumps and engines included in the plant. The yield of strong glycerine, steam required for concentrating the weaks, for the treatment and recovery of glycerine from the foots also must be considered.

There is an obvious saving of heat in the Garrigue and Wood processes over the Van Ruymbeke, as the two former utilise a portion of the latent heat of the glycerine vapour in condensing for subsequent operations and they do not condense the whole of the steam arising from the stills. The saving is approximately equal to the heat required for the evaporation of the weaks or, in other words, to the heat in the free steam required for the distillation. The claims for economy of heat in the Wood system as compared with the Garrigue, by reason of the utilisation of the same flow of steam in a series of stills, must be discounted by the fact that there must be a loss of heat after passing through each condenser, as the temperature of the steam cannot be higher than the temperature of the boiling water in the condenser, and therefore it must absorb heat from the crude glycerine pumped into the next still, which is fresh heat taken up in the glycerine heaters from the steam supplied direct from the boiler. There is also an additional consumption of heat by the pumps required to circulate the crude glycerine through the heaters and stills. The yield of strong glycerine on the first distillation is higher in the Garrigue and Wood systems than in the Van Ruymbeke by reason of the more efficient condensers, but the Van Ruymbeke system has the advantage of leaving at the end of the distillation considerably less glycerine as foots than either of the other two.

Comparisons of fuel consumption must be taken over long periods of working, where all factors are taken into account, and not on short test runs, and should be based on the amount of fuel required to produce a given weight of the final product and not on the amount required to distil a given weight of the crude material.

As all the above systems cannot be found working together under the same conditions in any one factory, comparisons are very difficult to make, but from figures supplied from different sources, and after making due allowances for the varying conditions, a careful computation gives the following result.

With steam of 180 lb. per sq. inch pressure at the plant, and assuming an evaporation of 1 to 8, for the production of 1 ton of once distilled glycerine, the Van Ruymbeke system requires 16 cwt. of coal, the Garrigue 14 cwt., and the Wood 13 cwt. These figures are capable of considerable reduction, and it ought to be possible to produce one ton of once distilled glycerine with 10 cwt. of coal.

A patent has recently been taken out by the Société Française des

Glycerines (English Patent Specification 125574, June 10, 1919) for a system of distilling glycerine without the use of free steam.

The process consists in atomising glycerine, previously heated to about 180° C., by passing it through suitable jets from one vessel to another at a lower pressure, the difference in pressure between the two vessels being relied upon to effect the atomisation. The atomised glycerine is vaporised at the reduced pressure of the second vessel and is collected in condensers, the non-volatile residue being drawn out through a cock at the bottom of the vessel.

The apparatus (Figs. 187 and 188) consists of a cylindrical vessel A

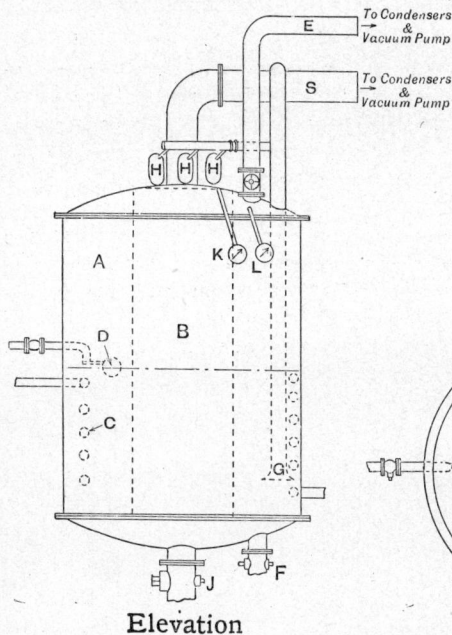
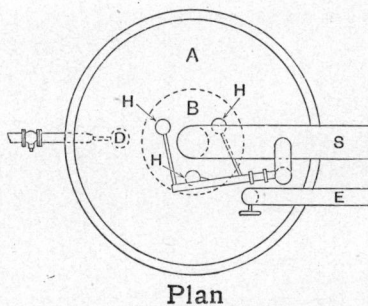

Elevation   Plan

FIG. 187.   FIG. 188.

Glycerine distillation plant.   Société Française des Glycerines.

containing the glycerine, in the interior of which is mounted a second concentric cylinder B, in which the atomising takes place.

Steam coils C provide sufficient heat to the vessels A and B to maintain a temperature of 180° C.

The glycerine is kept at a constant level in vessel A by means of a float valve D. The pressure in the vessel is reduced to 26 cm. of mercury, and any impurities in the glycerine volatile below 180° C. escape through the vapour pipe E, which is connected to a water condenser, and this in turn to the vacuum pump.

A drain cock F and a pressure gauge L are fitted, and a pipe G terminating in a rose supplies the jets or atomisers H with glycerine.

The cylinder B or atomising vessel has a drain cock J connected to a drum for collecting the residue, which is produced as the distillation proceeds and which must be discharged without having to stop the plant.

A vapour pipe s leads to a series of condensers where the glycerine vapours condense, and finally to the vacuum pump. The pressure in this second vessel B is indicated by the gauge K and is kept at about 4 cm. of mercury, so that there is a pressure difference of 22 cm. between the two vessels, which is said to be adequate to effect the atomisation. The atomisers H, three in number, are arranged symmetrically around the top of the vessel B with their jets directed downwards. The vessels A and B may be independent of each other instead of one being contained in the other.

It is claimed that the process is applicable to glycerine of all kinds, but no information is available concerning its application on a commercial scale, and it can be imagined that difficulties might arise due to the choking of the jets when crude glycerine is the product undergoing distillation ; it would also be a difficult matter to remove the practically solid residue from the second vessel B during the course of the distillation.

The method, however, is most ingenious, and provided that a sufficiently high vacuum and temperature can be maintained the principle of distilling without the aid of free steam and thus getting very little sweet water, and the economy effected in the use of steam and power by utilising the difference in pressure between the two vessels, is one that must commend itself to every one.

THE DISTILLATION OF ESSENTIAL OILS

By THOS. H. DURRANS, M.Sc. (Lond.), F.I.C.

OF MESSRS. A. BOAKE ROBERTS & CO., LTD., LONDON

CHAPTER XL

THE practice of steam distillation for the isolation and the purification of essential oils is world-wide and dates from the remotest antiquity. There is no fundamental difference between the distillation of essential oils and that of other liquids, but the nature of their origin and their somewhat delicate character render it necessary to employ special methods.

In general the scientific principles underlying these methods do not appear to be understood, and it will be well therefore briefly to state the more important theoretical considerations before dealing with the technical aspect. A few notes of a technical character have been introduced in the theoretical portion at suitable places in order better to bring out certain points.

Little or no reference is made to the theoretical side of ordinary or " dry " distillation, since this has been fully treated in another section of this book.

The mutual solubility of essential oils and water in one another is, in general, so slight that it can be neglected without serious error. The " steam distillation " of essential oils may in consequence be treated merely as an example of the distillation of completely immiscible liquids.

Dalton's Law.—The saturated vapours of such completely immiscible liquids follow Dalton's law, which states that when two or more gases or vapours which do not react chemically with one another are mixed, each gas exerts the same pressure as if it alone were present and that the sum of these partial pressures is equal to the total pressure exerted by the system. The law may be symbolised thus :—

$$P = p_1 + p_2 + p_3 + \ldots + p_n,$$

P being the total pressure of the system, and p_1, p_2, etc., the partial pressures of the components.

An important point to be noticed is that the total pressure is not influenced by the relative or the absolute amounts of the constituents.

Methods of Steam Distillation.—Steam distillation may be divided into two classes, the first comprising distillation at atmospheric pressure —as, for instance, when a mixture of an essential oil and water is boiled ;

the second, distillation with steam generated in a separate vessel. In this latter case the pressure of the steam has to be taken into account. These two classes are virtually one and the same when the steam employed in the second case is only under a low pressure (*e.g.* one or two pounds per square inch), the still is worked at atmospheric pressure and free water is also present.

Distillation of Immiscible Liquids.—In the first case, if a mixture of immiscible liquids be distilled, the boiling point is that temperature at which the sum of the vapour pressures is equal to that of the atmosphere ; this temperature is consequently lower than the boiling point of the most volatile constituent considered separately. When a mixture of immiscible liquids is distilled, the boiling point of the mixture remains constant until one of the constituents has been almost completely removed ; the boiling point then rises to that of the liquid remaining in the still. The vapour coming from such a mixture contains *all* the constituents in proportion (by volume) to the relative vapour pressure of each, and the distillate contains all the ingredients of the original mixture. It is impossible, therefore, by means of steam distillation completely to separate the constituents of an essential oil one from another, although a fair degree of separation may be attained in some instances.

If P_A and P_B are the vapour pressures of two completely immiscible liquids A and B at $t°$, the two substances will distil over in the proportion $P_A : P_B$ by volume of vapour ; if D_A and D_B are the respective vapour densities of the two vapours at the boiling point of the mixture, the relative quantities by weight, m'_A and m'_B, will be

$$\frac{m'_A}{m'_B} = \frac{P_A D_A}{P_B D_B}.$$

Inasmuch as the vapour density of a substance is a function of its molecular weight, viz.

$$M = 2D,$$

we can calculate the weights of the two components that will distil over from a knowledge of their molecular weights and vapour pressures at the boiling point of the mixture. From the above it follows that

$$\frac{m'_A}{m'_B} = \frac{P_A M_A}{P_B M_B},$$

and that the mixed vapour will consist of

$$\frac{M_A P_A}{P_A + P_B} \text{ of A and } \frac{M_B P_B}{P_A + P_B} \text{ of B.}$$

Or if we consider 1 litre of the mixed vapours, this will contain, according to Dalton's law, 1 litre of *each* component, and the weights of these will be respectively

$$\frac{0 \cdot 0896 \times M \times 273 \times P}{2 \times 760 \,(273 + t)}.$$

Given, then, the molecular weight of a substance and its vapour pressure at 100° C., we can calculate, approximately, the ratio of the amount of substance that will distil over with steam to the quantity of water required, and from this, as will be shown later, the amount of heat necessary.   Consider, for example, benzaldehyde, the chief constituent of oil of bitter almonds.   Its molecular weight is 106 and its vapour pressure P_B at 100° C. is 61 mm.   The partial pressures of water and benzaldehyde at 760 mm. and 100° C.[1] are respectively given by

$$P_A = \frac{P'_A \times 760}{P'_A + P'_B} = \frac{760 \times 760}{760 + 61} = 703 \cdot 5 \text{ for water}$$

and $\quad P_B = \frac{P'_B \times 760}{P'_A + P'_B} = \frac{61 \times 760}{760 + 61} = 56 \cdot 5 \text{ for benzaldehyde.}$

(Note $P_A + P_B = 703 \cdot 5 + 56 \cdot 5 = 760$.)

Substituting in the equation, we get

$$\frac{m'_A}{m'_B} = \frac{M_A P_A}{M_B P_B} = \frac{12670}{5988} = \frac{68}{32},$$

that is, the distillate contains 68 per cent of water and 32 per cent of benzaldehyde.

If an organic substance be not affected by water and have a vapour pressure of even so little as 1 mm. at 100° C., steam distillation is a feasible technical operation, especially so if the molecular weight of the substance be high.   It is this latter property that permits many essential oil ingredients of very low vapour pressure to be steam-distilled economically.   The following table is instructive :—

TABLE 120

Substance.	Boiling point.	P at 100° C.	M.	Per cent in distillate.
Carvone . . .	230° C.	9 mm.	150	9·7
Geraniol . . .	230	5	154	5·6
Anethole . . .	235	8	148	7·1
Eugenol . . .	250	2	164	1·7
α Santalol . . .	301	<1	220	0·5

The relation $m'_A : m'_B = M_A P_A : M_B P_B$ indicates the great value of steam distillation, since the smaller the product of $M_A P_A$ the larger the value of m'_B.   Water has an exceptionally low molecular weight and has only a relatively moderate vapour pressure, so that its value for MP is low ; all other substances have either a high molecular weight or a high vapour pressure and are generally miscible with essential oils.   One great advantage of steam distillation is that it allows the substance of high boiling point to be distilled at a temperature not exceeding 100° C.; this is of considerable importance when dealing with essential

---

[1] The mixture actually boils at a slightly lower temperature, since the boiling point of such a mixture is lower than that of the component of lower boiling point.

oils and other delicate or unstable substances. The boiling point of a system of immiscible liquids is lower than that of its most volatile component and it may be very much lower than that of its least volatile. Benzaldehyde under 760 mm. pressure boils at 178·3° C., but a mixture of water and benzaldehyde has a boiling point of 97·9° C.

Influence of Pressure on Composition of Distillate.—In a technical steam distillation it is generally desired that the amount of the substance of higher boiling point than water should be a maximum, or in other words, the conditions should be so chosen that the vapour pressure of the substance of high boiling point should be a maximum relatively to that of water.

The change of the vapour pressure of a substance with change of temperature is of a logarithmic nature and is expressed by the equation

$$\log P_1 - \log P_2 = \frac{ML}{R}\left(\frac{1}{T_2} - \frac{1}{T_1}\right),$$

where L is the latent heat, R the gas constant in calories, and T the absolute temperature. This equation may be rewritten

$$\log P_1 = \left(\frac{ML}{RT_2} + \log P_2\right) - \frac{ML}{R}\left(\frac{1}{T_1}\right),$$

and has been simplified by Sidgwick [1] to

$$\log P = A - \frac{B}{T},$$

A and B being constants determined empirically for each substance. Sidgwick shows that if $\log P$ be plotted against $\frac{1}{T}$ for any number of substances, a series of nearly straight lines is obtained radiating from points on the $\log P$ axis (where $\frac{1}{T} = O$ and hence $\log P = A$); that is to say, the lines approach one another as the temperature rises. The distance between two lines is the logarithm of the ratio of their vapour pressures. Hence as the temperature rises the value of $\log \dfrac{P_1}{P_2}$ approaches zero and the ratio $P_1 : P_2$ approaches unity. In other words, by increasing the temperature at which the distillation is carried out a larger proportion of the substance of high boiling point is obtained. The temperature of distillation can be increased by increasing the total pressure of the system. This is effected technically by working the still under pressure, a pressure valve being interposed between the vapour outlet of the still and the condenser.

A mixture of benzaldehyde and water at normal pressure boils at 97·9° C. and the distillate contains 31·4 per cent of benzaldehyde, but if the pressure be increased to four atmospheres the temperature rises to 140·7° C. and the percentage of benzaldehyde to 38·2. Conversely, by reducing the pressure the temperature and the relative amount of

[1] *Trans. Chem. Soc.*, 1920, **117**, 396.

benzaldehyde are lowered ; thus at 76 mm. the temperature is $45\cdot3°$ C. and the distillate contains $22\cdot5$ per cent of benzaldehyde.

Distillation with High Pressure Steam.—As an alternative to working the still under increased pressure, high pressure steam may be employed in a still worked at any pressure lower than that of the steam itself. When high pressure steam is injected into a still under a lower pressure the temperature of the steam diminishes, due to the work done by the steam in expanding against the pressure in the still. The following table indicates the drop in temperature over a range of one to ten atmospheres initial pressure :—

TABLE 121

Initial pressure, lbs. per sq. inch.	Initial temperature.	Temperature after expansion	
		to 760 mm.	to 76 mm.
14	$100\cdot00°$ C.	$100\cdot0°$ C.	$83\cdot3°$ C.
21	$111\cdot74$	$108\cdot1$	$91\cdot4$
28	$120\cdot60$	$113\cdot4$	$96\cdot7$
42	$133\cdot91$	$121\cdot9$	$105\cdot2$
56	$144\cdot00$	$128\cdot2$	$111\cdot5$
70	$152\cdot22$	$133\cdot3$	$116\cdot6$
84	$159\cdot22$	$137\cdot7$	$121\cdot0$
98	$165\cdot34$	$141\cdot4$	$124\cdot7$
112	$170\cdot81$	$144\cdot8$	$128\cdot1$
126	$175\cdot77$	$147\cdot9$	$131\cdot2$
140	$180\cdot31$	$150\cdot7$	$133\cdot9$

This drop in temperature may be calculated with the aid of Zeuner's equation,[1]

$$pv = BT - Cp^n,$$

where B, C, and n are constants, and

$$\frac{C}{B} = 38\cdot11, \; n = 0\cdot25,$$

p being the pressure in atmospheres, v the volume in cubic metres per kilogram, and T the temperature in degrees centigrade.

If the steam have the initial temperature of T_1, at a pressure p_1, and a volume v_1, and it be released into a still at T_2, p_2, and v_2 respectively, then

$$p_1 v_1 = p_2 v_2 = BT_1 - Cp_1{}^n = BT_2 - Cp_2{}^n,$$

whence

$$T_1 - T_2 = \frac{C}{B}(p_1{}^n - p_2{}^n).$$

For example, if we consider steam at 70 lb. per square inch (5 atmospheres) introduced into a still at 1 atmosphere we have, substituting figures taken from the table,

$$152\cdot22 - T_2 = 38\cdot11(5^{0\cdot25} - 1^{0\cdot25}),$$

whence

$$T_2 = 133\cdot3°\ \text{C.}$$

The use of superheated steam in a still worked at ordinary pressure

[1] Zeuner, *Technische Thermodynamik*, 1900, **2**, p. 221.

has the advantage that less condensation, due to the radiation of heat, takes place, the " superheat " of the steam causing the condensed water to evaporate, thus obviating the use of supplementary heat ; but it follows that beyond this no gain over steam used under ordinary pressure will result so long as condensed water is present in the still. If all condensation of the steam is prevented, then the higher temperature of the superheated steam will cause an increase of the proportion of the oil in the distillate.

Thus, if we employ steam at 5 atmospheres (56 lb. on the boiler gauge) to distil benzaldehyde under atmospheric pressure and allow no condensation to take place in the still, the temperature, as calculated above, will be $133 \cdot 3°$ C. At this temperature the vapour pressure of benzaldehyde is 220 mm.[1] The total pressure in the still is 760 mm., hence, using previous formulae,

$$P_A + P_B = 760,$$

$$P_A = 540 \text{ (water)}, \ P_B = 220 \text{ (benzaldehyde)} ;$$

whence
$$\frac{m'_A}{m'_B} = \frac{P_A M_A}{P_B M_B} = \frac{540 \times 18}{220 \times 106} = \frac{243}{583}.$$

The total distillate is $583 + 243$ parts and contains 583 parts of benzaldehyde, i.e. $70 \cdot 6$ per cent, whereas with steam at 100° C. and one atmosphere the distillate contained only $31 \cdot 4$ per cent of benzaldehyde.

If the benzaldehyde be heated above the temperature of the ingoing steam, the tendency is similarly to increase the proportion of benzaldehyde in the distillate.

Amount of Heat required.—In order to calculate the amount of heat required to steam-distil a given quantity of an essential oil it is necessary to consider each component separately. If we have to deal with an oil of comparatively simple composition such as that of almonds, cloves, caraway, or wintergreen, the calculation is within the region of practical application. We have to know, not only the proportion of each constituent, but also the amount of steam that is required to distil each of them separately. This latter can be calculated from a knowledge of the molecular weight of the substance and its vapour pressure at 100° C., as has already been indicated. These figures having been obtained we can calculate the amount of heat required to vaporise at constant pressure each component of the system—e.g. water, carvacrol, and cymene in the case of origanum oil—the total heat required being the sum of the heats necessary for the constituents. The heat of vaporisation may be calculated from the change of vapour pressure which accompanies a change of temperature.

If H be the heat of vaporisation in calories per molecule, T be the absolute temperature at pressure P, and dP the small rise in pressure which results from a small rise in temperature dT, then

$$H = \frac{dP}{dT} \cdot \frac{1 \cdot 985 T^2}{P}.$$

---

[1] This figure may be obtained by the interpolation of a boiling point-pressure curve or by actual experiment.

As an example, consider cymene. This boils at 174° C. under a pressure of 760 mm. For every 1 mm. increase of pressure (dP) the boiling point increases by $0.057°$ (dT), thus at normal pressure the heat of vaporisation is given by

$$H = \frac{1}{0.057} \cdot \frac{1.985(174+273)^2}{760}$$

$$= 9156 \text{ calories per molecule.}$$

By experiment H is found to be about 2 per cent lower, viz. 8978.

If we divide the value for H by the molecular weight of cymene, 134, we find the heat of vaporisation of 1 kilogram to be 68 (calculated).

The value of $dP : dT$ can be estimated from a boiling point-pressure curve taken over a small range in the neighbourhood of the pressure employed, either by evaluating the tangent drawn to the curve at that pressure or by averaging the difference between two points one on each side and equidistant from the point of which the value $dP : dT$ is required.   Alternatively, H may be estimated experimentally by well-known methods.   We may also employ the formula previously quoted,

$$\log p_1 - \log p_2 = \frac{ML}{R}\left(\frac{1}{T_2} - \frac{1}{T_1}\right),$$

since $ML = H$ and $R = 1.985$.

Very few experimental determinations have been published of the heats of vaporisation of the many bodies which occur in essential oils ; the following are available :—

TABLE 122

Substance.	Temperature ° C.	Heat of vaporisation.	
		1 kilogram.	1 kg. molecule.
Acetophenone	203·7	77·2	9.3×10^3
Anethole	...	71·5	10·6 ,,
Benzaldehyde	...	86·6	9·2 ,,
Cymene	175	67·0	9·0 ,,
Carvacrol.	...	68·1	10·2 ,,
Turpentine	159	69·0	9·4 ,,
Water	100	536	9·6 ,,

CHAPTER XLI

In this chapter the endeavour has been made to describe the general methods and apparatus employed in " winning " essential oils from their raw materials. The description of botanical and chemical characteristics is avoided, as these data are better obtained from books which treat this aspect specially.[1]

Where the method of treatment is unique or worthy of particular attention by reason of its wide use or technical interest, that method has been described in some detail ; these considerations also apply to the apparatus employed in the industry, but, in general, engineering details have been avoided and merely the type of machine or apparatus indicated.

So far we have only dealt with the steam distillation of the essential oil which has already been freed from the non-volatile organic matter with which it occurs in the natural state. The presence of this non-volatile matter introduces technical problems of some importance, and it is necessary so to prepare the raw material that the influence of the non-volatile matter is diminished as far as possible.

In any given distillation the rate will depend not only on the vapour pressure of the essential oil and the quantity of heat supplied, but also on the rate at which the volatile oil is liberated from the accompanying non-volatile matter. This last factor produces a lag which under unfavourable conditions may be of serious moment, and the object of this preliminary treatment of the raw material is to reduce this lag to a minimum.

It is possible roughly to estimate the value of the lag and thus to determine the best condition of the raw material for distilling. If the lag be *nil*, then the rate of distillation of the oil will be the same as that in the absence of the non-volatile matter, other conditions being equal ; any slower rate will be a measure of the lag.

In order to reduce the lag to a minimum the raw material must be reduced to such a state that the steam, introduced for the purpose of volatilising the essential oil, can readily penetrate the mass and yet can come into contact with every particle of essential oil in as short a time as possible.

[1] Cf. Parry, *The Chemistry of Essential Oils*, Scott Greenwood ; Schimmel's *Die Aetherische Oele*, Leipzig.

At first sight the obvious treatment would be to reduce the raw material to an exceedingly fine powder ; this, however, is seldom feasible and introduces subsequent difficulties through lack of porosity.

Frequently the raw material is best distilled without any previous treatment, as for instance in the case of leaves, twigs, petals, flowers, buds, grasses, and similar material, but occasionally the problem is one of considerable difficulty and a compromise between the fineness of division and the porosity of the mass is the best that can be attained. If the material were of a coarse texture, such as blocks of wood, the time necessary in order to distil out the oil would be excessive and a heavy consumption of fuel would result ; but if the other extreme had to be dealt with, such as a fine powder, the mass would become more or less impervious to the steam, which, in consequence, would find its way through the mass by means of fissures, leaving the greater part un-touched.   This latter difficulty can generally be overcome by diluting the charge with water in a quantity sufficient to render the resulting paste mobile.

Various methods for the treatment of large pieces of wood and branches present themselves. It is generally sufficient to chop branches up into short lengths, relying on expansion and capillary action to cause the oil to extrude during the distillation. The American practice in the case of the various pine oils is to chop the branches into pieces of one-half to one inch long and of less than half an inch in diameter ; pieces of this size yield their oil readily and the lag is comparatively small. Wood in the form of large blocks is one of the most difficult to deal with, and a large amount of power may have to be utilised to reduce it to a state suitable for distillation. The procedure generally is to saw it into suitable lengths and then to rasp it to a coarse powder ; alternative methods are to split it or tear it apart along the grain, or to shave it into chips, or to splinter or crush it by means of heavy crushing jaws similar to those used for crushing rocks. Rosewood and sandal-wood are generally reduced to small chips by means of a " raboteuse " chipping or planing machine.

Seeds and berries can frequently be distilled without preliminary treatment in spite of the enclosing epidermis, the heat of the steam and the increase of internal pressure due to the expansion of the contents and to osmosis being sufficient to rupture the epidermis and to set free the contents, but if the epidermis be too tough it is necessary to rupture it mechanically before distillation.

Schimmel found that uncrushed Ajowan seeds yielded 20 per cent less oil than crushed seeds, although the former were distilled twice as long as the latter. It is desirable that seeds similar to Ajowan, such as aniseed, caraway, and thyme, should be crushed.

Essential oils being volatile, care must be taken to ensure that the materials which bear them are not unduly exposed to conditions that would cause loss on this account.[1] Schimmel found that caraway seeds which under ideal conditions yielded 6·7 per cent of volatile oil gave only

---

[1] Cf. *Perfumery and Essential Oil Record,* 1921, p. 290.

5·45 per cent when they had been crushed and exposed to the air to dry, although the quality of the seeds was the same in both cases; this difference represents a loss of 18·7 per cent of the oil.   There is one compensating fact, namely, that the most volatile constituents of essential oils are usually the comparatively worthless terpenes, and the loss is, in consequence, one of quantity rather than of quality and is therefore not so serious as would appear at first sight.   Exposure to air may, however, occasion loss other than by evaporation, since the conditions may be such that oxidation or resinification can ensue, and the loss may therefore not be confined to the terpene constituents of the oil but may extend to the constituents on account of which the oil is chiefly of value.   The possibility of destructive bacterial fermentation must also not be overlooked, although the powerful antiseptic properties of most essential oils render loss on this account unlikely.

It is obvious therefore that during the reduction of the raw material to a state requisite for the subsequent distillation, care must be exercised in order to prevent undue exposure of the material to the air or to heat. This excludes the use of machines that rely on rubbing surfaces for their crushing action or on very swiftly revolving knives or centrifugal disintegrators, since the first generate heat and the latter pass large currents of air over the material.   Centrifugal disintegrators can, however, be arranged so that the same air circulates again and again through the machine and the loss of volatile oil is thereby limited.

The best type of mill for seeds of the caraway type is that employing enclosed crushing rolls which run slowly at the same or at very nearly the same speed.

In the case of peach and apricot kernels and bitter almonds these conditions do not apply, the essential oil not being present as such at the time of grinding, being subsequently developed by spontaneous fermentation; excessive heat should, however, be avoided, as this tends to kill the ferment.

Roots such as vertivert, ginger, and orris do not contain any serious quantity of very volatile substance and are best treated by drying and grinding in an ordinary drug mill.   Schimmel found that whereas a certain batch of cut vertivert root yielded 1·09 per cent of essential oil on distillation, a similar batch of uncut root yielded only 0·3 per cent.

In the case of orris and patchouli it is necessary to dry the material in order to develop the odour.

It is the practice in some countries to dry the raw material before distilling, especially if it is necessary to delay the distillation, since if the material be kept too long in a moist condition bacterial action sometimes tends to take place with consequent deterioration.   As has been shown above, drying results in a distinct loss of oil in the case of crushed caraway seeds, but it is stated that the loss occasioned by drying peppermint in the United States is negligible; the drying in this case is necessary since violent fermentation may ensue.   With lavender, however, the loss is quite serious, as the following results show:[1]—

---

[1] Messrs. Schimmel & Co.

TABLE 123

	Fresh.		Dried.	
	1.	2.	1.	2.
Time of drying in hours . . .	...	...	36	120
Total loss of material per cent . .	...	...	35·0	47·0
Oil yield per cent	0·84	0·80	0·79	0·72
Loss of oil per cent	...	...	5·7	10·0
Ester content of oil per cent . .	50·3	47·1	51·8	51·1
Loss of ester per cent on total oil loss .	...	...	26·0	11·0

In the case of roses and other flower petals the loss and deterioration are so serious that the effort is always made to distil them as soon as possible after gathering.

When drying is resorted to it must not be carried to excess, since

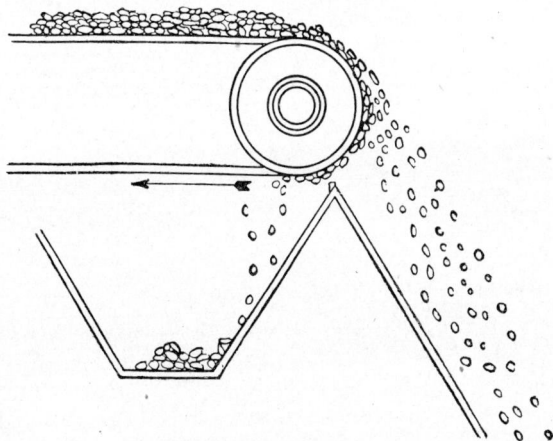

FIG. 189.

Igranic Electric Co., Ltd., Bedford.

This diagram illustrates the operation of a magnetic separator pulley. Non-magnetic material is projected beyond the pulley while magnetic material is held in contact with the conveyor belt until the latter passes under the pulley.

this tends to make the material friable and causes consequent loss in handling, this being particularly the case with peppermint.

When grinding, cutting, or otherwise reducing raw material to a condition suitable for distillation, considerable danger exists on account of the possible presence of stray iron, usually in the shape of nails, screws, or bolts. Two dangers are present, the first through the possible and generally costly damage to the machine, and the second through explosive ignition of the material under treatment, especially if this be dry or if it have a high content of essential oil. The use of a magnetic device for removing the iron is highly desirable and is now being widely practised. The most suitable form is that of a magnetic pulley, the

action of which is shown in Fig. 189.[1]   The magnetising of the pulley is
accomplished by the passage of a direct electric current through the
windings in the interior of the pulley.   Any iron or steel is attracted
by the pulley and is held in contact with the belt until the latter leaves
the pulley underneath ; it then drops clear of the other material, which is
projected beyond the pulley.

[1] By kind permission of Messrs. The Igranic Electric Co., Ltd., Bedford.

CHAPTER XLII

DISTILLATION

THE raw material having been reduced to the requisite state, it may be distilled by one of two methods or a combination of them, viz. :—

(1) Distillation by boiling with water.

(2) Distillation by means of " live " steam generated in a separate vessel or boiler.

The first is the more simple method if fire-heating be used, and is therefore largely adopted by itinerant distillers who take their stills to the fields and forests in search of their raw material. This method has some advantages over the method of distilling with steam, the apparatus is more portable and less costly, and the labour required to run it is less skilled, but it is open to the serious objection that there is danger of overheating the still in parts not in actual contact with the water; any raw material touching such a hot part is destructively distilled and gives rise to bodies having objectionable odours. The yield of oil obtained by this method is not usually so good as that obtained by means of live steam, the highest boiling fractions—which are sometimes the most valuable—tending to remain behind.

Experiments by Schimmel with lavender illustrate this point :—

Method.	Yield per cent.		Ester content per cent.	
	1.	2.	1.	2.
Water distillation . . .	0·71	0·75	44·0	43·6
Steam distillation . . .	0·81	0·82	50·9	53·7

It is possible also that the much slower water distillation tends to cause hydrolysis of the ester due to the prolonged heating with water, but the experiments do not definitely show this.

DISTILLATION WITH WATER

Typical French Itinerant Distillery.—Fig. 190[1] shows a typical itinerant distillery of South France. The distillation is being carried out by boiling the lavender with water, heat being supplied by means of a wood fire.

[1] Kindly lent by Messrs. Roure Bertrand Fils, Grasse.

The ready mobility of these stills is noticeable, and the chief difficulty that has to be overcome is that of finding a sufficient supply of cooling water for the condensers.

Permanent Spanish Field Installation.—For this reason it is not unusual to erect the still, more or less permanently, at a point where water is available and to transport the raw material over the comparatively short distances involved. This allows a somewhat more "scientific" arrangement to be adopted although at a loss of portability. This is well illustrated in Fig. 191,[1] which shows an installation used in Spain for the distillation of spike lavender, thyme, rosemary, sage, marjoram, etc.

FIG. 190.—Distillation of lavender in the environs of Castellane.   Roure Bertrand Fils.

This still is of considerable interest. It is built in the earth on the hill-side, a flue and chimney being arranged in the soil; the cooling water runs on to the dome-shaped top and effects a partial condensation; it then collects in the gutter and runs through the small pipe to the coil condenser, the vapour and condensed vapour passing through the larger pipe to the interior of the coil, finally being collected in the liquid state at the bottom of the condenser. The supernatant oil layer is then separated automatically from the condensed water by means of the well-known "Florentine flask."

The dome condenser has the advantage that a large quantity of heat can be dissipated by the aerial evaporation of the thin film of water flowing over it, but the subsequent utilisation in the coil condenser of the hot water resulting from this preliminary cooling is of doubtful economy. The reverse sequence would be preferable, but

[1] Kindly lent by Mr. Felix Gutkind, Malaga.

involves mechanical difficulties.   The most logical method is to apply
the coldest cooling water to the coldest distillate and the hottest cooling
water to the vapour on the counter current principle.

FIG. 191.—Felix Gutkind, Malaga.

Bulgarian Distillery for Roses.—The native stills of Bulgaria are of a
more permanent character, this being possible since the roses, for which
these stills are used, are cultivated year after year in the same fields.

The accompanying photograph (Fig. 192) [1] of an installation in the
Balkan mountains shows the stills employed.   These are quite small,
being of about 25 gallons capacity, and are constructed of copper tinned
on the inside ; partial condensation is effected by means of the large
air-cooled still-head and is afterwards completed in the usual " worm "

[1] Messrs. Shipkoff & Co., Rahmanlari.

condenser, which can be seen in the second photograph (Fig. 193), showing the method of collecting the distillate. No separating device is
employed, as little or no free oil is obtained in the first distillation ;
the reason for this will be explained later. It will also be noted that
the incoming cooling water is supplied to the top of the condenser, and
there mixing with the hot water loses much of its effectiveness
in consequence.

American Still for Oil of Wintergreen.—Stills very similar to these
are in use at Unionville, Pennsylvania, for the distillation of oil of
wintergreen.[1] The body of the still is surrounded by a roughly built

FIG. 192.—Shipkoff & Co., Rahmanlari.

shelter of stones, and is usually capable of holding 200 to 500 pounds of
leaves ; the condenser is the usual coil immersed in a barrel of running
water.

Stills in Tunis and Ceylon.—The small size of these stills is in great
contrast to those used in other parts of the world. In Southern Tunis
stills ranging up to 150 gallons capacity are employed. Two stills are
generally employed with one condenser, so that one still is in operation
while the other is being cleansed and recharged with fresh herbs. This
convenient method is also employed in Ceylon for the distillation of
lemon-grass and citronella oils. The photograph (Fig. 194) shows the
arrangement in use in Tunis. The two stills have a common " swan-
neck," by means of which the vapour is conducted to the condensing
coil immersed in a concrete tank of water ; the operation of changing

[1] From information kindly supplied by John T. Stolz, Broadheadsville, Pa.

the still-head from one still to the other is facilitated by means of the overhead gantry.

Crude Javanese Plant.—In some instances itinerant distillations are carried out in the crudest imaginable manner with makeshift plant of the most unsuitable character. It is not unusual for the still to consist of an old locomotive boiler set on end, a pipe, running

FIG. 193.—Shipkoff & Co., Rahmanlari.

for a considerable distance along the bed of a stream, serving for the condenser. A very usual type of native still is shown in the illustration (Fig. 195), the one in particular being used in Java for distilling Cananga oil.

Portable Fire-heated French Still.—Very adaptable and convenient stills are readily to be obtained in Europe; a portable fire-heated type is depicted in Fig. 196.[1]

---

[1] Still constructed by Deroy Fils Aîné, Paris.

A highly interesting "portable" steam-distilling plant is shown in Fig. 197. This plant is in use for distilling rosemary and consists

FIG. 194.—*Perfumery and Essential Oil Record*, 1916.

of a fire-heated steam generator, which supplies steam under low pressure to the bottom of a cylinder containing the herb supported

FIG. 195.—Schimmel & Co., Leipzig.

on a false bottom; two such cylinders are employed, one being recharged while the other is in use, and each being provided with a detachable head, which can be removed in order to introduce or remove the herb. These detachable heads can be connected as required to a common condenser.

DISTILLATION WITH LIVE STEAM

Distillation with " live " steam generated in a separate boiler is more suitable for the established distillery, since, in general, better results are to be obtained than with water distillation.

If " live " steam be used its pressure must not be excessive, as there exists a strong tendency deleteriously to alter the oil if it contain delicate constituents such as aldehydes or esters.

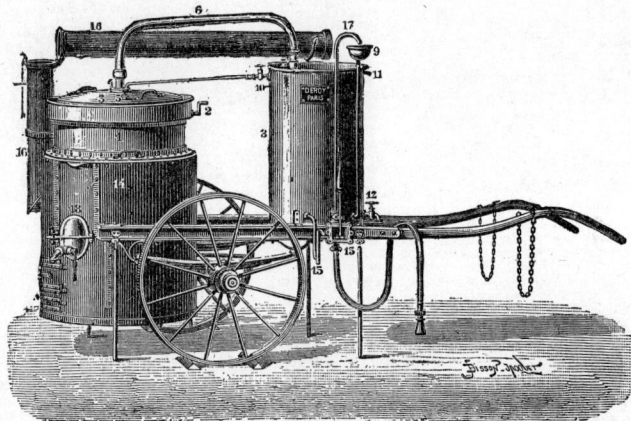

FIG. 196.—Deroy Fils Aîné, Paris.

The following experiment [1] demonstrates the hydrolysing action of " 50 lb. steam " on lavender oils :—

Oil.	Ester content per cent.	
	Before distillation.	After distillation.
Lavender M.B. .	40	35
Spike lavender .	7	2

In general a boiler-steam pressure of as little as 2 lb. per square inch is sufficient, but if it is necessary to use a higher pressure the hydrolysing effect can be mitigated by coiling the steam inlet pipe round the inside on the bottom of the still after the manner of a volute or " watch spring " spiral, and ensuring that sufficient free water to cover the spiral be in the still.

The danger of destructively distilling the raw material is avoided with steam distillation, and is not likely to arise if a closed steam coil or a steam jacket is employed for boiling the water in the still, since the highest local temperature possible is that of the steam itself.

[1] A. Boake Roberts & Co., Ltd., London.

The best distillery practice is to fit the stills for both water and steam distillation, steam being used as the heating medium in both cases.

Material for Still.—Copper is the most usual material of which these stills are constructed, but it is not necessary to employ such expensive

FIG. 197.—Schimmel & Co., Leipzig.

material ; iron is generally quite as satisfactory, and aluminium is now being used to a large extent ; even wooden vats can be successfully employed, and are in regular use in Australia for the production of eucalyptus oil, and in America and England for peppermint. Some of these vats are capable of holding two tons of material.

It is customary in America to coat the inside of wooden stills with a waterproofing compound, which not only protects the wood, but eliminates, to a large extent, the danger of contaminating one oil with another if the still is used for more than one preparation. Such stills have been found to be quite suitable for distilling sassafras, cedar, spruce, etc., the heating medium being steam generated in a separate boiler. Wooden vats having copper or galvanised iron bottoms have been used successfully for distilling birch with water by direct heat ; but lead linings are unsatisfactory, since the lead tends rapidly to become distorted.

Shape of Still.—The shape of the vats is usually cylindrical or nearly so, of slightly less diameter at the top than at the bottom ; they are built up of staves held in place by bands of iron after the manner of a

FIG. 198.—Tombarel Frères, Grasse.

barrel. The taper produced by the varying diameter assists in keeping the iron bands in position and preventing leaks.

In America, for distilling a brushwood such as birch, cedar, or spruce, the wooden still is of an oblong, rectangular shape, this shape permitting the branches to be laid in without cutting them into short lengths.[1]

The exact shape of a still is usually not of very great importance and it depends largely on the mechanical facilities that may be available. A spherical still is the most economical as regards loss of heat by radiation, but the constructional difficulties are excessive.

French Distillery for Peppermint, etc.—A useful type of still, largely used in South France, is shown in Fig. 198.[2] The stills are in use for

---

[1] Information kindly supplied by Chas. V. Sparhawk, N.Y.
[2] By kind permission of Messrs. Tombarel Frères, Grasse.

peppermint, a heap of which can be seen in the foreground.   The herb
is introduced into the still by way of the manhole at the top, and the
exhausted charge is removed through a similar manhole in the side and
near the bottom of the still ; this latter manhole protrudes through the

FIG. 199.—Stafford Allen & Sons, Ltd., Long Melford.

wall behind the still, so that the exhausted charge is discharged into a
yard outside the stillroom, thus avoiding dirt and disorder.   These
stills are of general utility and are regularly used for roses, lavender,
geranium, peppermint, rosemary, etc.

The relative height and diameter of the still will depend on the porosity of the material with which it has to deal ; if the charge be of an " open " character its depth may be considerable, but if of a close texture too great a depth should be avoided.

Shallow Still for Powders.—For powders or similar material a shallow type of still is the best, being operated with boiling water instead of live steam.

Fig. 199 shows a well designed still of this type used by Messrs. Stafford Allen & Sons, Ltd. The following points are worthy of notice :

1. The large diameter of the still compared with its height.

2. The careful lagging of the still so as to avoid loss of heat by radiation.

3. The overhead hopper by means of which the charge is introduced into the still.

Stills for Orris and Almonds.—Stills of radically different designs to the above are used in the south of France for distilling orris and almonds, both of which are reduced to a fine powder before distillation.

The methods involved in the production of these two oils are unique and of considerable interest. It will be well to deal with them at this point.

Oil of Bitter Almonds.—In the case of oil of bitter almonds, so called in order to distinguish it from the non-volatile sweet oil of almonds, the almonds have first to be crushed and submitted to an enzymic fermentation for the purpose of generating the volatile oil. The method adopted in the south of France is to crush them by means of large stone edge-runners ; these edge-runners develop no sensible heat and deal quite efficiently with the material. After crushing and before distilling it is necessary to extract the fixed oil, which consists mainly of the glycerine esters of oleic and linoleic acids, and is practically non-volatile.

Hydraulic Press.—For this purpose hydraulic presses of the open type are used, the meal being placed in a number of square bags made of camel hair ; these bags withstand very well the strain due to the great pressure employed. The yield of fixed oil varies from 38 to 50 per cent. A

FIG. 200.
Rose, Downs & Thompson, Hull.

more modern and improved press is the Anglo-American oil press, a photo of which (Fig. 200) is shown. This press does away with the

2 H

necessity of using the costly camel-hair bags, and in consequence allows a larger quantity of material to be pressed at one time. The press illustrated has a 16-inch ram and can treat nearly three hundredweight of material at one charge ; the pressure employed is 2 tons per square inch.

Fermentation.—After " drawing " the fixed oil in this manner, the press cakes—which are odourless—are broken up and placed in a still with cold or lukewarm water and there allowed to ferment for about twenty-four hours at a temperature not exceeding 40° C. During the process benzaldehyde is liberated by the action of the enzyme ferment, emulsin, on the glucoside amygdalin, glucose and hydrocyanic acid also being formed.

Distillation.—The stills used for this purpose are conical in shape, the apex of the cone being at the bottom and being provided with a tap. Live steam is the only heating medium employed, the distillation being of comparatively short duration on account of the relatively high vapour pressure of the benzaldehyde. A diagrammatic sketch of the type of still in use at Grasse is shown in Fig. 201.

In this process one of the products of the fermentation is the highly poisonous prussic acid which comes over mainly at the start of the distillation ; this can be rendered innocuous by employing a closed receiver provided with a pipe to lead away the gas to some place where

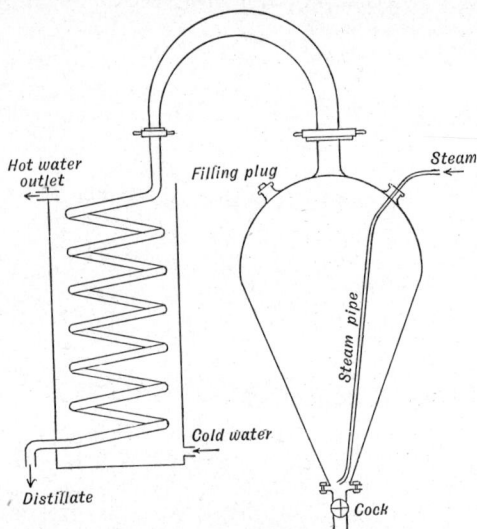

FIG. 201.

it can be safely discharged into the atmosphere. The prussic acid may alternatively be absorbed by placing a mixture of slaked lime and ferrous sulphate in the still. For every 100 lb. of almond meal there is required at least 6 oz. of slaked lime and 8 oz. of ferrous sulphate crystals. In order to remove the last traces of prussic acid, the oil from the distillate should be thoroughly agitated with a similar mixture and water.

Orris Root.—In the case of orris powder, stills similar to those used for herbs generally, as shown on p. 463, are employed. The distillate is a paste known as concrete orris, and consists largely of myristic acid (b.p. 318° C.) with a small quantity of the valuable ketone irone (b.p. about 260° C.), both of which on account of their low vapour

Fig. 202.—Usine de Grasse. Appareils de Distillation. Antoine Chiris.

pressures distil over only very slowly with steam, thus involving the distillation of a large proportion of water.

Cohobation.—In spite of the low degree of solubility of irone in water the loss of this valuable body, if the aqueous portion of the distillate were discarded, would be considerable, but by returning this water to the still and using it to distil over a further quantity of oil the loss is minimised. This method is known as "cohobation," and is frequently employed when the oils concerned are difficult to separate from the aqueous portion of the distillate by reason of their solubility or tendency to emulsify in water.

In order to effect the return of this water several devices are employed. A usual method is to elevate the condenser and receiver at a level higher than that of the liquid in the still and to return the aqueous portion of the distillate to the still by means of a gravity feed. This method can only be used if the pressure inside the still is quite small, and in any case a liquid seal must be provided in order to prevent the egress of vapour by this inlet. The sealing may be effected by continuing the inlet pipe to well below the level of the liquid in the still or, better, by passing the incoming liquid through a U-tube of considerable height, each 2 feet of height being equivalent to 1 lb. per square inch pressure inside the still.

The Antoine Chiris Plant. — This is the method employed by Messrs. Antoine Chiris at Grasse, and can be seen in the accompanying photograph (Fig. 202), which is of general interest. The essential details of the plant are shown in the diagram, Fig. 202A.

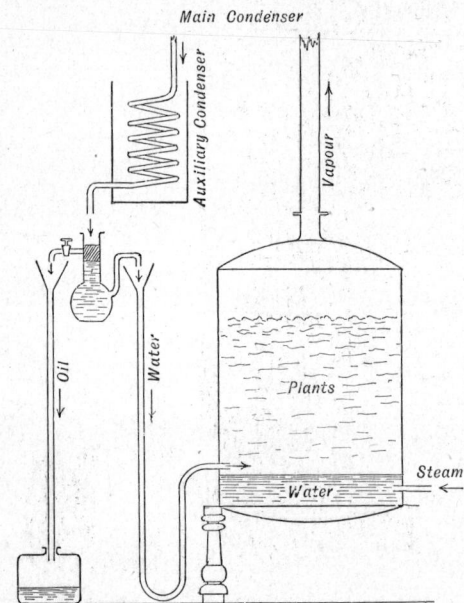

FIG. 202A.

The vapour from each still ascends by the wide pipe to a condenser on the roof, and the partly cooled distillate returns by the small pipe to a small coil condenser and thence it flows into a glass separating flask placed on a level with the top of the still. In this flask the essential oil separates from the water by flotation and is run at suitable intervals into the small storage "coppers" which can be seen standing on the floor. The aqueous layer passes from the bottom of the separating flask by means of an S-shaped side tube into a funnel, which is connected to a U-shaped pipe reaching to the floor and rising again to a short distance above the bottom of the still, finally entering at the side.

John Dore & Co.'s Still.—The accompanying design (Fig. 203)[1] shows a still which is very suitable for distilling substances such as orris or cloves where cohobation is desirable. The still is provided with a stirring gear, which keeps the mass in motion and ensures that it is completely treated. The condenser is designed to present a thin film of large area of vapour to the cooling water, thus ensuring rapid conden-

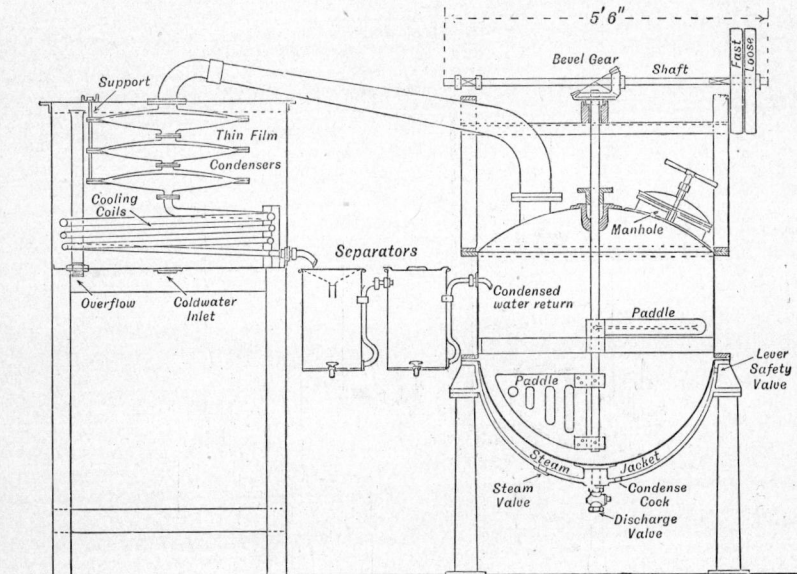

FIG. 203.—John Dore & Co.'s Still.

sation. Two Florentine separating receivers collect the distillate, the second one serving as a liquid seal by means of which the aqueous portion of the distillate is returned to the still for cohobation.

Economy of Heat and Water.—The distillation of a large quantity of water involves the consumption of a large quantity of heat which is ultimately dissipated by the condenser water, and the process of distillation finally resolves itself into a transference of heat from the fuel employed to the condenser water which is run to waste. The process is obviously very extravagant and many schemes have been devised in order to minimise this waste of energy; frequently, also, economy in condenser water is of considerable moment, especially in districts where at times droughts prevail. In certain industries which involve the evaporation of large quantities of liquids the multiple effect apparatus

[1] Designed by John Dore & Co., London.

has been employed with considerable advantage, but no attempt seems to have been made to use this device in the essential oil industry, although the difficulties involved are by no means insuperable. An economy which is, however, widely practised is that of feeding the still with water heated beforehand by the hot vapour coming from the still. Such a device is depicted in the accompanying design (Fig. 204),[1] and consists essentially of a preliminary condenser (18), the cooling water for which is obtained from the receiver (25) after separation of the oil, obtained during the distillation, by means of the separators (24). A pump (26) is used to circulate this water, which becomes heated in the preliminary condenser (18) and returns to the still by the valve (27). The water

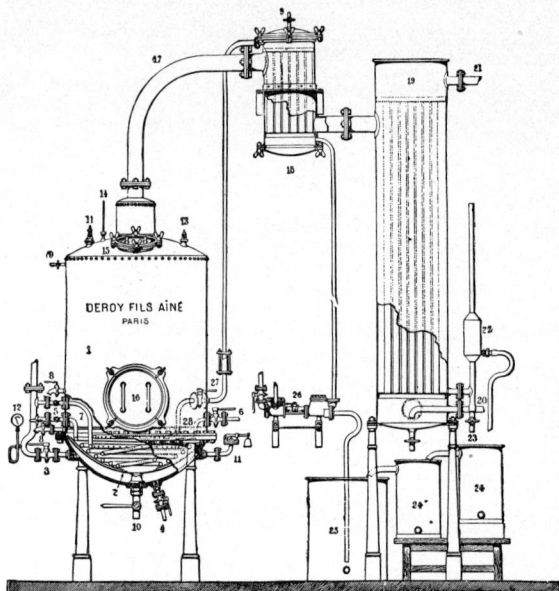

FIG. 204.

is thus circulated cohobatively and on entering the still has a temperature neighbouring on 100° C., so that the quantity of heat saved is equal to that required to heat the water from the temperature of the distillate in the receiver to 100° C. The whole of the latent heat of the steam is dissipated in the main condenser (19) and is lost, but about 13 per cent of the total heat, neglecting radiation losses, is saved and a corresponding economy of cooling water is effected.

A similar, well-known method is to feed the still with part of the hot water coming from the condenser, but this does not effect any economy of cooling water and only a very little of the heat. Occasionally small steam injectors are used to return the condensed aqueous layer to the still for cohobation, but this effects little or no economy of heat and none at all of cooling water.

[1] Designed by Deroy Fils, Aîné, Paris.

An economy of cooling water can readily be made by resorting to aerial radiation for an initial condensation. In the distillation of orris the condensation is almost entirely by aerial radiation, many yards of wide bore copper pipe being erected in a suitably elevated position. A better product is said to be obtained in this instance than with water-cooled condensers.

Support of Raw Material in Still.—A point to be considered in the design of a still is the manner in which the raw material is to be supported in the still. In the cases of orris and almonds, mentioned above, no support of the nature of a false bottom is necessary, but a perforated false bottom greatly aids the even distribution of the steam, and is highly desirable if fire heating be used, since it prevents contact between the raw material and the hot bottom of the still. In the case of coarse material, the false bottom presents no difficulty, consisting merely of a perforated plate, but substances such as small seeds may readily obstruct circular orifices or not be retained by the plate. These difficulties may be overcome by using very narrow slits, an inch or two long, instead of circular perforations. These slits need not be more than one or two millimetres wide and will retain the smallest seeds without unduly obstructing the passage of the steam. If desired a mat of coconut fibre may be placed on the top of the tray, but this is seldom necessary and not an altogether satisfactory procedure.

During distillation the material may alter considerably in volume, either increasing through swelling or diminishing on account of settlement, although frequently, on the other hand, no appreciable change in volume takes place. If the material has the property of settling into an impervious mass it is necessary to introduce additional perforated plates at fixed distances, one above the other, so as to relieve the low portions of the material from the whole weight of that above it. Such an arrangement necessitates that the top of the still be removable. The perforated plates are best kept apart by means of suitably arranged legs, so that the arrangement is similar to a series of tables standing piled one on another. For a still of general utility a design such as the above is probably the best.

As an alternative to the false bottom for supporting the raw material a gauge basket is frequently employed; this also necessitates a still having a removable top. Care, however, must be taken so to construct the basket that steam cannot readily escape past the sides, but penetrates through the material in the basket.

The use of a basket facilitates the speedy removal of the spent charge and its replacement by a fresh charge in a similar basket. Sometimes the spent charge is removed by lifting the false bottom. A widely used still of this type is shown in the sketch [1] (Fig. 205) with a specification appended.

Otto of Roses.—One branch of the essential oil industry that deserves special consideration is the production of otto of roses. Until

---

[1] Designed by Bennett Sons & Shears, Ltd., London.

recent years it has been practically a Bulgarian monopoly, the mountain regions of Bulgaria being especially suitable for the cultivation of roses. The industry is now also being established in France, but the otto produced there appears to be of a slightly different nature. It has been estimated that the output of otto in Bulgaria reached as much as 5000 kilos in 1907 and involved the distillation of roughly ten thousand million roses.

The distillation of roses takes place in two stages. First, the roses,

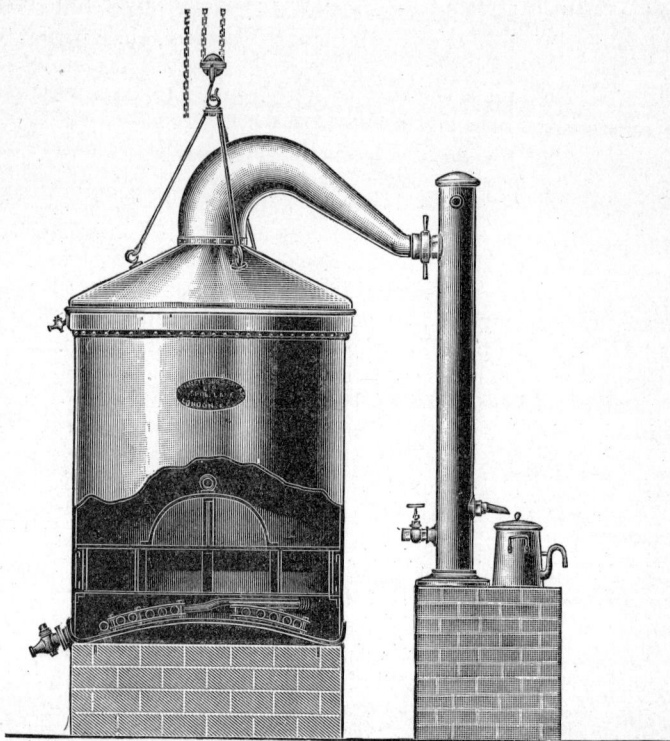

FIG. 205.—Bennett Sons & Shears, Ltd., London, E.C.

A copper still, 1200 gallons capacity, to take a charge of 1 ton of plants and 400 gallons of water, 6 feet diameter by 7 feet deep, with removable copper dome and waterseal joint with suitable fastenings, fitted with discharge pipe and cock, powerful copper steam heating coil with copper stays and steam valve, galvanized iron false bottom and frame resting on gunmetal lugs, with lifting rods and ring to enable the false bottom, etc., to be gradually lifted out of the still and the spent leaves discharged. A copper still-head and lower end connected by a union to an "Ideal" patent condenser with water valve and oil separator.

while in a perfectly fresh state, are distilled by boiling with water, and there results a saturated solution of the essential oil of the rose, little or no free oil separating out gravitationally. This phenomenon is probably due to the solubility ratio of the oil in the water being of the same order as the rate of the liberation of the oil from the petals or what might be termed the lag-ratio. To obtain the oil as such, advantage is taken of the phenomena pertaining to the distillation of mixed liquids. On distilling the solution the essential oil comes over in the first portions of the aqueous distillate, and being now present in a larger proportion separates as an oily layer on top of the water. The more

FIG. 206.—Shipkoff & Co., Kazanlik, Bulgaria.

soluble constituents of the oil remain dissolved in the water—the so-called "rose water"—and hence this water contains a relatively large proportion of phenylethyl alcohol.

The type of still used in the peasant distilleries has been shown on

p. 458 ; its manner of operation is as follows : [1]   Into each still is placed one part by weight of fresh roses with five to six parts of water ; about 25 to 30 per cent of the water is distilled off in the course of about 45 minutes.   This water is then redistilled and a fraction of about 30 per cent collected ; this fraction is very strong in odour and quite turbid in appearance, due to tiny globules of pale yellow oil, which in the course of time unite on the top of the water.   The oil is removed by means of a small conical spoon having a hole in the bottom, large enough to allow the passage of the water, yet sufficiently small to retain the oil.

Shipkoff & Co.'s Plant.— The modern installation of Messrs. Shipkoff & Co., constructed by Egrot of Paris, is shown in the accompanying photograph (Fig. 206).   Two types of still are employed, one for the first distillation of the roses and the second for the redistillation of the rose water resulting from the first.   The former stills are 1·35 metres in diameter and are heated by means of a steam jacket at the bottom ; the latter are 0·85 metres in diameter and are heated by means of steam coils.   Vertical tubular type condensers are used to condense the vapours, which are led from each still by pipes of ample dimensions.   At the bottom of the respective condensers are storage cylinders and a series of separating receivers.   The rose water from the first distillation is transferred to the rectifying stills by means of air pressure or steam injectors, and the hot water from the condensers is used to feed the stills wherein the distillation of the roses takes place in the first instance.

[1] From information kindly supplied by Messrs. Shipkoff & Co., Bulgaria.

CHAPTER XLIII

AN essential oil, having been "won" from the plant, is frequently in need of purification. Oils which have been produced by itinerant or native distillers can generally be vastly improved by redistillation. The colour may be bad, or the oil may have an earthy or a "burnt" odour, or, if it be old oil, it may be resinous. The usual practice is to steam-distil the oil in much the same manner that was employed to win it, but "dry" distillation may also be employed. This redistillation is generally conducted in a properly equipped distillery, where refined methods may be applied, depending on the nature of the oil.

A most important point to be observed in distilling essential oils is that, with few exceptions, they are deleteriously affected by heat. It is probably for this reason that an essential oil almost invariably has a distinctly different odour to that of the plant from which it was produced, the "cooking" to which the oil has been subjected having in some way chemically altered its constituents.

The constituents of essential oils, with a few exceptions, such as camphor, safrol, menthol, thymol, are relatively unstable bodies and tend to decompose or polymerise under the action of heat; thus geraniol isomerises to cyclogeraniol, anethole forms various polymers, citral decomposes to cymene and methyl heptenone, citronellal isomerises to isopulegol, cinnamic aldehyde resinifies, and so on. Obviously therefore, it is desirable as far as possible to limit both the temperature and the duration of a distillation. The temperature can be reduced by conducting the distillation under diminished pressure. Theoretically the diminution of temperature may be so great that the boiling point of the substance could approach the absolute zero. Practically, however, we are limited by the efficiency of our vacuum pumps, the air-tightness of our stills and the efficiency of our condensers. It is a task of very great difficulty on the technical scale to obtain a diminished pressure or "vacuum" of one-fifth of a millimetre, and more often the vacuum available is of the order of 10 or 20 mm. Under a pressure of 20 mm. the majority of the constituents of essential oils boil at temperatures considerably in excess of 100° C., hence unless very efficient and refined apparatus be available it is better to resort to steam-distillation, since in this case the temperature cannot exceed 100° C.

In order still further to reduce the temperature, steam distillation

under reduced pressure may be employed. This method is not frequently practised on account of two drawbacks. In the first place the
amount of oil coming over with a given quantity of water is less than if
a higher pressure be used,[1] provided the boiling point of the oil when
distilled alone is not much below that of water at the same pressure.
This is shown in the following table :

TABLE 124

Substance.	Pressure. Mm.	Temperature. ° C.	Per cent substance in distillate.
Benzaldehyde	76	45·3	22·5
Boiling point 178° C. . . .	760	97·9	31·4
	3040	140·7	38·2
Acetophenone	76	45·9	10·4
Boiling point 201° C. . . .	760	99·05	18·5
	3040	142·32	23·1
Camphor	76	46·0	9·82
Boiling point 206° C. . . .	760	99·08	21·5
	3040	142·6	24·2

If, however, the boiling point at normal pressure is much below
100° C., the reverse holds.

Benzene	76 mm.	16° C.	95·3 per ct.
Boiling point 80° C. . . .	760	69·2	91·0
	3040	112·6	86·9

Table 126 in the Appendix gives other examples.

The second drawback is the difficulty of condensing the vapour.
The boiling point of water under pressure much below 50 mm. approaches
the normal atmospheric temperature and there is then a strong tendency
for its vapour to pass through the vacuum pumps carrying with it
a proportionate quantity of the essential oil vapour. This is both
wasteful and very bad for the pumps and must be carefully avoided.
The boiling points of water under diminished pressures may be seen at
the end of the boiling point table in the Appendix.

If the oil with which we have to deal be stable and not deleteriously
affected by heat, considerable technical advantage is to be obtained by
conducting the steam distillation under pressures higher than atmospheric. The theoretical aspect has already been considered.[2]

The use of increased pressure is of considerable value with oils
having valuable constituents of very low vapour pressure such as oils
of ginger, vertivert, sandalwood, cedar, and camphor, which at normal
pressure only yield a low proportion of oil in a steam distillate, neces-
sitating excessive quantities of steam for their complete treatment.

---

[1] Cf. Theoretical section, p. 446 *et seq.*   [2] *Ibid.*

As an alternative or an addition to the use of increased pressure, superheated steam may be employed with considerable advantage,[1] but, as has already been pointed out, superheated or high pressure steam has a tendency to cause the hydrolysis of esters and must be used with caution if the oil contain any chemically delicate body.

" Dry " distillation has the advantage over steam or water distillation that, other things being equal, it can be conducted more rapidly, owing to the much smaller quantity of vapour to be dealt with, also, if it be desired to isolate any particular portion of the oil, this can be more accurately done by dry fractional distillation. The disadvantages are, firstly, the usually much higher temperature required ; secondly, the loss of material occasioned by vapour and residue remaining in the still, and lastly, the more costly nature of the apparatus. The first defect may largely be overcome by conducting the distillation under very low pressure ; the second, by employing steam at the end of the dry distillation to steam-distil the residue and to drive out the vapour remaining in the still.

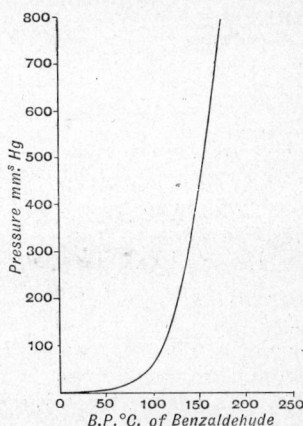

FIG. 207.
Boiling point curve of benzaldehyde.

Excepting a few instances, essential oils cannot be dry distilled under ordinary pressure without spoiling them, since the temperature required and their low conductivity for heat results in high local temperatures causing decomposition to occur, the bodies foreign to the oil produced in consequence generally possessing disagreeable odours. Such oils are said technically to be " burnt." Excessive temperature also tends to cause polymerisation or resinification, or, if the temperature be sufficiently excessive, the oil may be " cracked " and gaseous bodies produced.

There is no doubt but that the subjection of an essential oil to the action of heat acts deleteriously on its odour, and it follows that the lower the temperature at which an oil can be distilled the less is its odour damaged.

The relation between the boiling point of a substance and the pressure of its vapour is typically shown in Fig. 207, which illustrates the curve for benzaldehyde.

The relatively large reduction in boiling temperature that can be obtained by the use of very low pressure does not seem generally to be appreciated. In the instance illustrated a diminution in pressure from the atmospheric to say 20 mm. — in all a diminution of 740 mm.— lowers the boiling point by 103° C., but a further diminution of only 19 mm. causes an additional drop in the boiling point of 46°. To take another example, carvone under normal pressure boils at 230°, and under 20 mm. at 115°, whereas at the low pressure of 0·069 mm. the boiling

[1] *Vide* p. 447 *et seq.*

point is 22·6°. Such an extremely low pressure as this last is not practicable on a large scale, but a pressure of 1 mm. and even lower is quite within the range of modern appliances.

The duration of the distillation also has an important bearing on the odour of the distillate, and should be as short as possible. If it be desired to isolate any particular constituent of an essential oil, it is necessary to employ a fractionating column. One of the effects of a column is continually to return some of the condensed vapour to the still, where it again undergoes conversion into vapour. It follows that certain portions of the oil are subjected many times to a deleterious reflux boiling. The greater the efficiency of the column the less does this action occur, and it is a minimum with a column of absolute efficiency; that is to say, a column which allows the passage of only the lowest boiling constituent in the column and does not return any of this constituent to the still.

Columns for the separation of fractions of relatively wide-boiling range, such as terpeneless and sesquiterpeneless oils, need not be of very great efficiency, if certain oils be excepted, since the terpene and sesquiterpene constituents have boiling points differing considerably from those of the desired fraction. When, however, the terpene is present to a preponderating extent, as in the cases of lemon and orange oils, a column of high efficiency is desirable if a good yield of the terpeneless fraction is to be obtained.

Essential oils, especially those obtained by expression, frequently contain valueless bodies of very low vapour pressure, termed stearoptenes. These stearoptenes, if present to any considerable extent, render it difficult to obtain the last portions of the terpeneless fractions without resorting to excessive temperatures. The separation is generally best effected by distilling with steam.

Another method occasionally used to separate the terpenes from an essential oil depends on the relative insolubility of terpenes in dilute spirit compared with that of the terpeneless portions. The oil is distilled with spirit of about 40 per cent strength, using a plain stillhead; the distillate consists of a non-homogeneous mixture of dilute alcohol and essential oil, and the alcohol layer is cohobatively returned to the still, carrying with it in solution the more soluble portions of the distilled oil, while the insoluble terpene remains in the receiver. This operation is continued until the distillate coming over is homogenous, after which the alcohol in the still is distilled off and the remaining oil steam-distilled and collected separately.

A very suitable still for the preparation of terpeneless and sesquiterpeneless oils is shown in Fig. 208. The fractionating column consists of a number of tubes, through which the vapour passes, surrounded by a common water jacket F. The flow of the water through this jacket, and consequently the temperature of the column, can be regulated by means of the valve v at the bottom. A chain is suspended in each vapour tube in order to increase the surface exposed to the vapour. The still is worked under diminished pressure, and the water jacket F kept in operation until all the terpene has been distilled,

after which the water is removed and the terpeneless fraction
collected separately. Two receivers J and I are connected in series
with the condenser G in order to facilitate the removal of the fractions without " breaking " the vacuum in the whole plant. The still
is also provided with filling and emptying valves H and D and with

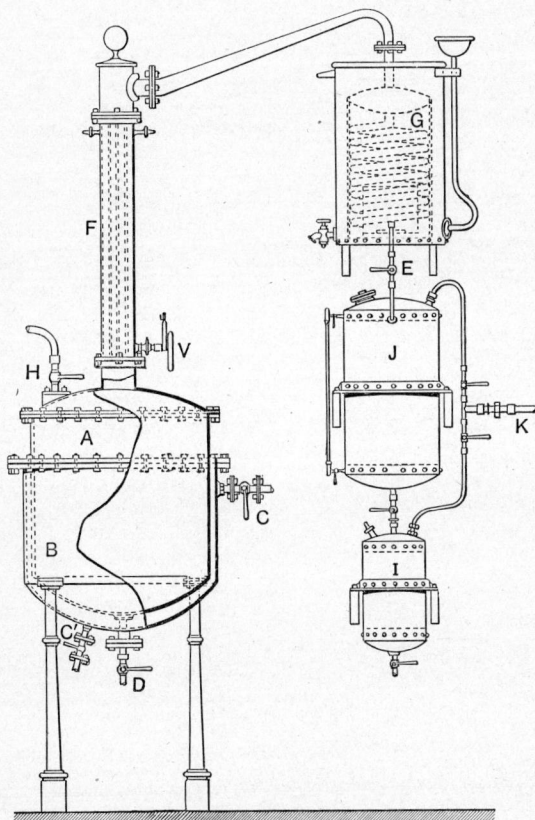

FIG. 208.—Deroy Fils Aîné, Paris.

steam valves c and c′, while the evacuation is effected through the
valve K.

Another type of column largely used in the industry is a modification
of the Hempel column, and consists merely of a wide tube filled with
broken glass or similar material. It is open to three serious objections,
channeling, priming, and bad draining ; but it can be easily and cheaply
constructed and gives quite good results. The use of Raschig's or of
Lessing's rings in the place of the broken glass tends considerably to
reduce the defects.

The Condenser

The next part of the distilling plant to be considered is the condenser; the function of the condenser is to abstract heat from the vapour, and, in consequence, to cause it to condense to a liquid state. This is not difficult to do, but a good condenser should also cool the condensed liquid so as to reduce its vapour pressure in order that loss may not occur through aerial evaporation, or if a reduced pressure is employed, in order that any important quantity of the distillate may not be lost in the vapour form through the pumps. This latter loss can occur either through inefficient condensation or on account of air leaks in the apparatus. If, as not infrequently happens with bad plant, both of these defects exist, the loss of material may be very serious.

The first demand on a condenser is, therefore, that it should be capable of extracting the heat of the vapour and of the condensed liquid very completely; the second is that it should do so with the employment of as small a quantity of cooling water as possible.

The problem of calculating the size of a condenser is not at all an easy one as many factors are involved, some of which are of uncertain magnitude.

The only safe and most economical plan is to err on the large size. This means a bigger initial outlay, but results in economy and ease of working.   We can calculate the minimum quantity of water required :

If L = Latent heat of vaporisation of the substance,
    S = Its specific heat in the liquid state,
    W = The weight of substance to be distilled,
    M = The weight of water required,
    T = The boiling point of the substance,
    t = The final temperature of the distillate, $i.e.$, the temperature of the cooling water entering the condenser,
    T_1 = The temperature of the water leaving the condenser,

we have

    Heat given up on condensation = WL,
    Heat lost on cooling from T to t = $W(T - t)S$,
    Heat taken up by water = $M(T_1 - t)$.

Then
$$M(T_1 - t) = WL + WS(T - t),$$
whence
$$M = \frac{WL + WS(T - t)}{T_1 - t}; \quad . \quad . \quad . \quad . \quad (1)$$

or for condensing the vapour only
$$M = \frac{WL}{T_1 - t}. \quad . \quad . \quad . \quad . \quad (2)$$

Formula (1) necessitates the supposition that the condenser is sufficiently long to allow the condensed liquid to come to the same temperature as the cooling water entering the condenser at the bottom; if this temperature be fixed, then T_1 will depend on the velocity of the water, and will be equal to T as a limit.

To take an example, suppose 500 kilos of oil of turpentine to be distilled at 20 mm., when its boiling point is 53° C., we have

$$W = 500 \; ; \; L = 69 \; ; \; S = 0 \cdot 5 \; ;$$
$$T = 53° \text{ C.} \; ; \; t = \text{(say) } 20° \text{ C.} \; ; \; T_1 = \text{(say) } 53° \text{ C.} = T.$$

Then

$$M = \frac{(500 \times 69) + (500 \times 0 \cdot 5 \times 33)}{33} \text{ kilos}$$

$= 280$ gallons approximately as the minimum quantity necessary.

This is supposing the coil to be infinitely long. The rate of cooling will depend on the difference in temperature between the condensed liquid and the cooling water, neglecting the heat lag through the metal of the condenser.

A rough guide is to allow 1 square foot of cooling surface for every 1 gallon of liquid to be condensed and cooled to normal temperature per hour.

The cooling surface of a coil can be calculated from the formula $\frac{22}{7} dln$, where d is the diameter of the pipe, l the length of each turn, and n the number of turns.

In the case of a steam distillation, more water is required on account of the high latent heat of vaporisation of water, and the fact that, with essential oils at any rate, a large quantity of water is required to distil a relatively small quantity of oil. Thus, to steam-distil 500 kilos of limonene, which is one of the most volatile of essential oil ingredients, we require 750 kilos of water (see Appendix). To condense and cool to 20° C. these 750 gallons of water we require cooling water to the amount given by

$$M = \frac{(750 \times 536) + (750 \times 80)}{80 - 20} \text{ kilos,}$$

the water leaving the condenser at 80°, whence

$$M = 1690 \text{ gallons.}$$

This is required for the purpose of condensing the steam only, and a further quantity is required to condense the limonene, which can be calculated in a similar manner.

In the average worm condenser there is a dead space in the centre. This is bad, because convection currents are set up, and this tends to reduce the efficiency of the condenser on account of the cold water at the bottom mixing with the hot water from the top, resulting in increased consumption of cooling water and incomplete cooling of the condensed liquid.

This dead space should be filled up; for instance, by a cylinder exposed on the inside to the air, or by a substance of low conductivity.

For the same reason a cylindrical containing tank is better than one of rectangular plan.

Cooling water should not be injected into the bottom of the condenser in a rapid stream, but should be broken up or baffled so as to avoid mixing with the hot water above it. Economy of water may be exercised by using aerial radiation for the first few turns of the coil before entering into the water.

In places where water is scarce, aerial radiation has to be relied on entirely ; this is quite feasible, if a sufficient surface be exposed.

In a coil condenser the top coils should be of ample internal diameter in order to allow for the large volume of vapour that has to be dealt with there ; the subsequent coils can diminish progressively in internal diameter. Since the ratio between the surface exposed to the cooling water and the volume of the coil is given by $\dfrac{2}{r}$, where r is the internal radius of the coil, it follows that the smaller the radius of the coil the greater the *relative* surface exposed.

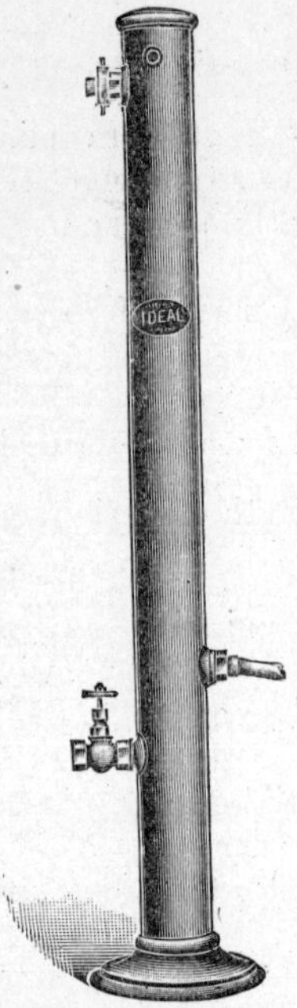

A tall condenser is better than a broad one, since the heat gradient is not so steep, that is, the difference in temperature between adjacent coils is not so great.

Tubular condensers deal efficiently with the vapour, but the liquid resulting is apt to flow down too quickly and be inadequately cooled in consequence. This defect is sometimes overcome by raising the outlet to above the bottom of the tubes, so that they become sealed by the condensed liquid, which thus remains in contact with the cooling surface for a longer time. Where accurate fractionation is required this is bad practice.

The " Ideal " tubular condenser,[1] shown in Fig. 209, overcomes this defect by retarding the flow of the hot liquid.

FIG. 209.—Bennett Sons & Shears, Ltd.

This condenser effects the rapid condensation and the cooling of the resulting liquid between narrow annular spaces, formed by fixing one tube inside another of slightly larger diameter, a wire being wound spirally in the space between the two tubes. The vapour to be condensed enters the annular space between the two tubes and is forced to take a spiral path by the wire ; a long path for the vapour and condensed liquid is thus provided and the cooling is effected from both sides of the annular space.

Economy of cooling water may be obtained by taking advantage of the latent heat of evaporation of water by causing the water to flow in a thin film over a chamber or coil containing the hottest vapour so that aerial evaporation of the water occurs.

---

[1] Bennett Sons & Shears, Ltd., London.

The Receiver

The last item of the distilling plant needing consideration is the receiver. In the case of steam distillation the receiver should be so constructed that it will automatically separate the oil from the aqueous portion of the distillate.

Two types are necessary, according to whether the oil has a greater or a less specific gravity than water. With some oils, such as thyme, the problem is complicated by the fact that the first portions of the distillate, which contain the terpenes, are lighter than water, whereas the later portions, containing the thymol, are heavier. Thymol also has a tendency to solidify, and would choke the various pipes of the receiver; this can be overcome by allowing the condenser to run hot. Thymol melts at 51° C.

Florentine Flask.—The usual " Florentine flask " is too well known

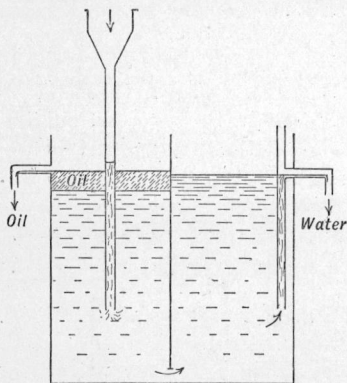

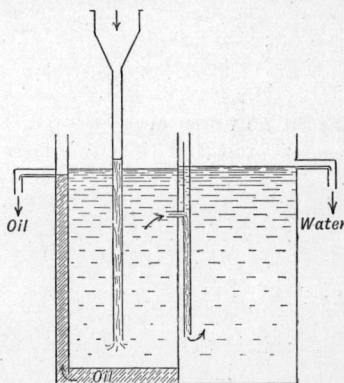

FIG. 210A.   FIG. 210B.

to need description, it can be seen in operation in the illustrations on pages 468, 469, and 473. Improved designs are shown in Figs. 210, A and B. These receivers can be adjusted to suit the varying specific gravities of the various oil layers by slightly tilting them one way or the other. They afford a double safeguard against loss of oil, having two settling chambers, and have proved to be very satisfactory in operation.

Receiver for Vacuum Work.—For vacuum work the receiver is of an entirely different character, and consists essentially of a closed vessel connected to the condenser on the one hand and to the exhausting pump on the other. Receivers do not call for much comment, but of course must be sufficiently strong to withstand the external atmospheric pressure, which may rise as high as 15 lb. per square inch or thereabouts. The construction of these vessels is an engineer's problem. Provision should be made for gauging the amount of the contents, and for this purpose circular disc sight-glasses fixed in the lid are the most satisfactory; the contents can quite readily be estimated by means of graduations marked on the inside of the receiver and viewed through the sight-glasses. As a refinement, one of the sight-glasses may be in the shape of a " well " ; this will allow an electric light to be lowered so as to be completely within the receiver ; this arrangement will illuminate the interior very thoroughly.

Tube gauge-glasses set after the manner of boiler gauges are frequently unsatisfactory, being easily broken, difficult to clean, and apt to become choked up. When two layers of liquid are present the readings registered are quite erroneous, and it must always be remembered that it is the heavier liquid at the bottom of the receiver which is pushed up into the gauge-glass.

Pressure in Still and Receiver.—The pressure or "vacuum," as it is generally termed, in the still and in the receiver should be measured separately, for, if the distillation is very rapid or air leaks exist in the apparatus, the pressure as indicated by the two gauges will differ : the cause of any big difference should at once be sought ; with distillates like oil of thyme, which are prone to solidify, the condenser may become choked ; the pressure in the still would increase in consequence, upsetting the boiling temperature; there is also a possibility of dangerous " plus " pressures arising in the still. It is the pressure in the still that should be taken account of rather than that in the receiver, but the correct practice is to measure the pressure at the same point that the temperature is measured, *e.g.* at the top of a fractionating column. Dial gauges are not suitable for accurate measurements, and change their zeros with use. The most suitable gauge consists of a glass tube 1 metre high, connected at the top to the apparatus, and dipping into a trough of mercury at the bottom, the difference in level between the mercury in the trough and in the tube being measured with a metre scale ; it is well not to have the tube too narrow, as in this case the capillarity introduces an error—this error is inappreciable with tubes of $\frac{1}{4}$-in. bore or more. The pressure indicated should be compared with the barometer, the difference between the readings being the pressure in the apparatus.

I have to express my indebtedness to A. L. Bloomfield, Esq., B.A. of New College, Oxford, for kindly correcting my manuscript, and to Messrs. A. Boake Roberts & Co., Ltd., Stratford, for permission to publish it ; also to the following gentlemen and firms for much original information and assistance, which I gratefully acknowledge :

Felix Gutkind, Esq., Malaga, Spain.
John T. Stolz, Esq., Monroe Co., U.S.A.
Messrs. Stafford Allen and Sons, Ltd., London, E.C.
     ,,   Tombarel Frères, Grasse, France.
     ,,   Roure Bertrand Fils, Grasse, France.
     ,,   Antoine Chiris, Grasse, France.
     ,,   Lautier Fils, Grasse, France.
     ,,   Shipkoff & Co., Kazanlik, Bulgaria.
     ,,   Schimmel & Co., Leipzig, Germany.
     ,,   Chas. V. Sparhawk, Inc., N.Y., U.S.A.

Messrs. Ungerer & Co., N.Y., U.S.A.
    ,,   John Dore & Co., London, E.
    ,,   Igranic Electric Co., Ltd., Bedford.
    ,,   Rose, Downs & Thompson, Hull.
    ,,   Bennett Sons & Shears, Ltd., London, E.C.
    ,,   J. Harrison Carter, Ltd., Dunstable.
    ,,   Deroy Fils, Paris XV.

And to the following publications :

The Perfumery and Essential Oil Record, London,
The American Perfumer and Essential Oil Record, N.Y., U.S.A.,
The Chemistry of Essential Oils, E. J. Parry (Scott Greenwood),
Rechenberg's *Gewinnung und Trennung der ätherischen Öle* (Schimmel
   & Co.).

APPENDIX

TABLE 125

YIELDS OF ESSENTIAL OILS

Material.	Yield per cent (approximate).	Material.	Yield per cent (approximate).
Ajowan . . .	3·0—4·0	Lavender, French . .	0·5—1·0
Aniseed, Russian . .	1·5—6·0	,, English . .	0·8—1·7
Backhousia citriodoria leaves	0·7	,, Spike . .	0·5—1·1
Bay leaves . . .	1·1—1·4	Linaloe wood, Mexican .	7—9
Bitter almonds . . .	0·5—1·0	Lemongrass, E. Indian .	0·2—0·26
Black pepper . . .	1·0—2·3	,, W. Indian .	0·24—0·4
Bois de rose . . .	0·6—1·6	Mustard, black seeds .	0·5—1
Buchu leaves . . .	0·8—2·5	Myrrh	2—10
Calamus root (dried) . .	1·5—3·5	Neroli flowers . . .	0·086—0·15
Camomile flowers, English .	0·2—0·35	Nutmegs	4—15
,, ,, German .	0·35—0·51	Origanum, Italian . .	2—3
Cananga	1·5—2·0	,, Smyrna .	1·4—2·4
Caraway seeds . . .	3·1—7·0	Orris root . . .	0·1—0·2
Camphor leaves . .	0·23—3·0	Parsley	2—7
Cardamom seeds, Malabar .	3·5—8·0	Patchouli, Singapore, dried leaves	2·6—4
Cedarwood . . .	2·5—5·0	Petitgrain, leaves and twigs	0·2—0·4
Celery seeds . . .	2·5—3·0	Pimento berries . .	3—4·5
Cinnamon wood . . .	0·5—1·0	Palmarosa grass . .	0·3—1·74
,, leaves . .	1·8	Pennyroyal leaves, N. American	0·6—1·0
Citronella grass, Ceylonese .	0·25—1·0	Peppermint, Japanese .	1—1·6
,, ,, Javan .	0·5—0·7	,, American .	0·1—1·0
Cloves	15—23·1	,, English .	0·4—1·0
,, stems . .	5—6	,, Italian .	0·02—0·025
Coriander fruit, Russian .	0·8—1·0	Rosemary leaves . .	1—2
,, ,, E. Indian .	0·15—0·2	Rue	0·06
Cubebs fruit, Javan .	10·0—18·0	Rose, Otto, Bulgarian .	0·02
Cumin fruit . . .	2·5—4·5	Sage leaves, Dalmatian .	1·3—2·5
Dill seeds . . .	2·5—4·0	Sandalwood, W. Indian .	1·3—3·5
Estragon . . .	0·1—0·4	,, Indian . .	2·5—6·0
Eucalyptus (various) . .	0·18—4·2	Spearmint, N. American .	0·3
,, Amygdalina .	4·2	Savin leaves . . .	3—5
,, Globulus .	0·92	Snake root . . .	1—3
Frankincense . . .	5—9	Star aniseed, Chinese .	3—3·5
Flebane . . .	0·26—0·66	Sassafras rind . .	3—9
Fennel seeds . . .	4—6	,, whole root . .	1
Geranium rose . . .	0·1—0·2	Storax	1
Ginger root, Jamaican .	2—3	Tansy, N. American . .	0·1—0·2
,, ,, Cochin .	1·5	Thyme, French . .	0·2 0·5
Gingergrass . . .	0·1	Thuja leaves . . .	0·4—0·65
Hops	0·3—1	Valerian root . . .	0·5—1
Hyssop . . .	0·07—0·29	Vetivert root . . .	0·4—2
Juniper berries, Italian .	1—1·5	Wintergreen, gaultheria .	0·55—0·8
,, ,, French .	2	,, birch . .	0·5—0·7
,, ,, Hungarian .	0·8—1	Wormseed . . .	0·35—2
Laurel leaves . .	1—3		

TABLE 126

AMOUNT OF SUBSTANCE IN A STEAM DISTILLATE *

Substance.	Per cent by weight of substance.		
	76 mm.	760 mm.	3040 mm.
Styrene .	...	57·0	...
Pinene .	..	55·6	...
Benzaldehyde	22·5	31·4	38·2
Cymene	...	45·7	...
Limonene	...	40·0	...
Linalool	...	18·2	...
Acetophenone	10·4	18·5	23·1
Menthol	...	12·0	...
Benzyl alcohol	...	10·2	17·1
Carvone	...	9·7	...
Camphor	9·8	21·5	24·2
Anethole	...	7·1	...
Cinnamic aldehyde	...	3·0	...
Eugenol	...	1·7	...
Santalol	...	0·5	...
Geraniol	...	5·6	...

* Rechenberg, pp. 349, 360, 363.

TABLE 127

AVERAGE OIL-CONTENT OF STEAM DISTILLATES FROM UNDRIED ESSENTIAL
OIL-BEARING RAW MATERIALS *

Material.	Per cent oil.	Material.	Per cent oil.
Ajowan seeds	0·77	Cloves	0·6—0·9
Angelica seeds	0·2	Patchouli leaves .	0·12
Aniseeds	0·8—1·2	Peppermint	0·1
Bay leaves .	0·76	Pimento	0·18
Cedarwood .	1—1·4	Sandalwood, W.I.	0·2—0·34
Cubebs	1·2	Celery seeds	0·17
Ginger root	0·3	Vertivert root	0·02
Coriander seed	0·56	Cinnamon .	0·32
Caraway seeds	2·2—3		

* Cf. Rechenberg, p. 362.

TABLE 128
489

BOILING POINTS

Substance.	Mm. pressure.							
	760	100	50	30	20	10	5	
	°	°	°	°	°	°	°	
α-Pinene	156	92	79	63	53	38	22	
Camphene	160·5				62	46		
p-Me-cresyl ether	171			76				
Me. heptenone	173							
p-Cymene	174·3					57		
Sylvestrene	175							
α-Phellandrene	175					59	47	
β-Phellandrene	175					55		
Limonene	175					54		
Me. heptenol	175-6						65	
Eucalyptole	176				66	54		° °
Benzaldehyde	178·3	112·5	95·3	83·5	75·2	62·0	50·1	42·5(3)29·3(1)
Terpinolene	186				87	69		
Fenchone	191				83	69		
Me-heptyl ketone	195				86	74		
Androl	198							
Linalool	198			101	93	81	68	
Thujone	201				93	80		
Acetophenone	201·5	133·2	114·7	102·4	93·4	79·0	65·0	
iso-Pulegone						87		
iso-Pulegol						87		
Citronellal	205			106	96	84		
Camphor	206	138·7	118·8	104·0	95·8	82·6	71·1	
Menthone	207							
β-Terpineol	209·5					90		
Borneol	212							
Menthol	213·6					96	90	
Me. chavicole	216			113	93	81		
Phenyl ethyl alcohol	219		136	124	116	101		
α-Terpineol	219				110	98		
Linalyl acetate	220			120	112	96		
Pulegone	221		126	114	106	94		
Pulegol					111	98		
Dihydrocitronellol					115	103		
Dihydrocarveol	222				118	105	94	
Dihydrocarvone	223				106			
Bornyl acetate	223				113	98		
Me. salicylate	223					95		
Citronellol	226				122	111	99	
Nerol	226·5			128·7	120·3			
Menthyl acetate	227				114	102		
Neryl formate				124	115			
Citral	228				118	106		°
Citronellyl formate						ca 99		
Carvone	230	137	125	115	102	90		34·5(0·2)
Geraniol	230				124	110	101	
Et. salicylate	231·5	139·1	127·1	119·0	104·8			
Thymol	231·8					109		
Cumic aldehyde	232	158·0	138·1	125·5	117·9	103·5	91·0	
Safrole	233				123	107	93	
Anethole	235					106	89	
Ph-Pr-alcohol	235					116		
Geranyl formate					118·5	102·5		
Neryl acetate				137·5	129·5			
Carvacrol	237					113		
Chavicole	237							
Citronellyl acetate					126	113		
Geranyl acetate	245				133	119		
Cinnamic aldehyde	247				138	121		
Me. eugenole	248					127		
Anisic aldehyde	248						94	
Ionone	250					123		
Eugenol	250·5			146		123		
α-Santalene	253							
iso-Safrol	254						107·5	
Betelphenol	254					127		
Cinnamyl alcohol	257·5				°	ca138		
Irone					149	133		
Caryophyllene	260					120		
iso-Eugenol	261	193			150	132	117	
Heliotropine	263				149	135		
Me-iso-Eugenole	263							
Anise ketone	263							
β-Santalene	263							
Me. anthranilate					141	124		
Cadinene	274						131	
Vanillin	285					155		
Myristicin					158	142		
Cumarin	290·5					152		
α-Santalol	301					160		
β-Santalol	309					168		
Water	100	51·7	38·3	29·9	22·3	11·4	1·3	

NAME INDEX

SUBJECT INDEX

THE END

Printed in Great Britain by R. & R. CLARK, LIMITED, *Edinburgh.*